Modelling Ocean Climate Variability

Modelling Ocean Climate Variability

by

Artem S. Sarkisyan
Institute of Numerical Mathematics, Moscow, Russia

and

Jürgen E. Sündermann
University of Hamburg, Germany

Springer

Prof. Dr. Artem S. Sarkisyan
Russian Academy of Sciences
Institute of Numerical Mathematics
Gubkina 8
Moscow
Russia 119991
sarkis@inm.ras.ru

Prof. Dr. Jürgen E. Sündermann
Institute of Oceanography
University of Hamburg
Bundesstr 53
20146 Hamburg
Germany
juergen.suendermann@zmaw.de

ISBN 978-1-4020-9207-7 e-ISBN 978-1-4020-9208-4

DOI 10.1007/978-1-4020-9208-4

Library of Congress Control Number: 2008942129

Printed on acid-free paper

9 8 7 6 5 4 3 2 1

springer.com

Preface

This book presents our understanding of the main achievements and problems in numerical modelling of variations in ocean/sea climatic characteristics. Trying to do it one meets several difficulties:

1. There is a variety of simple and very complicated models and their number is growing rapidly, so how to choose the best one for analysis?
2. The results of calculations by one and the same model depend on methodology applied by users.
3. Choice of even the best modelling results would bring to a very thick volume of a reference book.

The latter idea is possibly useful but we have chosen another way: out of hundreds of different results we pay attention mainly to those which perform nonlinear thermodynamical model data synthesis rather than direct modelling. Besides, we have chosen papers the authors of which make a maximally possible use of the observed data and simulate basins' geometry as accurately as a high resolution numerical model enables one to do.

Finally, we pay more attention to publications in Russian and to papers that have appeared in journals less known for the English-speaking oceanographers.

The content of the book is constructed in the following way.

Chapter 1 is devoted to the general theory and mathematical problems of ocean/sea dynamics modelling, that is solvability of a general system of equations, possibility of total energy conservation, etc. We present different versions of the background system of equations (classical, σ-coordinate, etc.) and numerical methods (finite difference, finite element), forward and adjoint models of the data assimilation problem, etc. In short, Chapter 1 serves as a theoretical background for the subsequent five chapters.

Chapter 2 presents several simple models for diagnostic calculations. Their qualitative analysis has explained many aspects of ocean/sea dynamics. Namely, they show organic interconnections between strong horizontal density gradients and intense jet-like currents. Moreover, both of them strongly depend on the basins' bottom topography.

It is shown that the JEBAR (Joint Effect of Baroclinicity and Bottom Relief) is an absolutely necessary factor and should accurately be taken into account, directly or indirectly, in any model. Finally, they show that any further simplification of the models brings unrealistic and even contradictory results. Vivid examples are the Sverdrup relation and the reference level model. It is shown that despite their popularity they are contradictory.

In case the sea surface height is used explicitly instead of the mass transport stream function another very essential factor appears, that is BARBE (baroclinic β-effect). This factor exists also in flat-bottom ocean models.

In geophysical literature the vorticity equation and the wind-stress vorticity as one of main driving forces of ocean currents are very common. In Chapter 2, we present an other version too, that is the divergence equation for the sea surface height (SSH). Based on it, we show that wind-stress divergence may be considered as wind part of the driving force as well. This version of SSH equation is valid for the equatorial area too. It helps to explain why the equatorial undercurrents exist.

Chapter 3 turns exclusively to a fully nonlinear non-stationary system of ocean thermo-hydrodynamics because we argue that the linear system of equations is valid only for qualitative analysis. That kind of analysis was already made, so it is useful only for students.

Several versions of primitive equation models (PEM) are presented and analyzed in this chapter.

It is shown that wind stress participation in the full versions is threefold: divergence, vorticity and simple, even constant, wind stress.

Chapter 3 shows the adjustment stage of calculations, the first stage or short model time of prognostic calculation, analysis as a necessary step to be undertaken after the diagnostic calculations before passing to a very long model time. The adjustment, adaptation, calculation stage have proven all the results of diagnostic calculations (fields of SSH, mass transport stream function, horizontal and vertical flow fields, energetics, the role of JEBAR and BARBE, etc.).

By analysis of the model calibration results we show that the duration of prognostic calculations depends on the model quality, especially the grid mesh and parametrization of turbulent processes. Inadequately long model-time of integration means the ocean water baroclinicity "washes up", underestimates of the jet-like currents, overestimates of the effect of wind-stress vorticity bring the baroclinic ocean integral mass transport to the one controlled by the Sverdrup transport.

Let us list part of the main results presented in Chapter 4. Usually any modeller starts prognostic calculations with the observed, i.e. time and space averaged T and S fields. We put a question: what does the model do with these data? Chapter 3 presents the over-filtering effect of longer model-time integration. Detailed analysis of the initial stage energetics (Paragraph 4.1 of Chapter 4) shows that the dramatic fall of kinetic energy happens during the first six hours of integration. Here and in many other parts of the book we emphasize the necessity to monitor the calculation energetics from the very initial stage. The other results contained in this chapter are as follows. A method used by WOCE and other programs is presented of T, S sections data processing which results in generating the three velocity components and enriching the observed T, S data; then based on Kalman filtering, a model of a four-dimensional analysis is presented, its accuracy was tested and the method was applied for the analysis of the observed data from several regions of the North Atlantic.

Chapter 5 deals with specific ocean variability modelling. Two papers which consider the whole World Ocean are mentioned in brief, but the chapter refers mainly to selected ocean basins, namely: (1) Arctic-North Atlantic system, (2) North Atlantic, (3) Pacific, and (4) Indian Ocean. The reason for this is that for the time being there has been no paper on high resolution modelling for the whole World Ocean while each of the basins considered has its own peculiarities which should be discussed separately.

For the first of the basins, consideration is being given to the problems of the sea-ice and snow thermodynamics, ice thickness, vertical multilayer distribution and the dynamics of ice-water phase transition modelling. Besides, the Arctic is strongly connected with the North Atlantic and depends on the surrounding river discharge. The effect of rivers is two-fold: first, it should be taken into account for the ocean water volume variation and, second, rivers bring fresh water and possible contamination which propagates up to the North Atlantic.

The results of calculations by models which take into account most of the enumerated factors are presented with the corresponding charts. For the time being the resolution performed varies from 35 km in the Arctic to one degree in the North Atlantic Ocean. This resolution is crude for both basins. A paragraph devoted to critical analysis of advantages and drawbacks of the North Atlantic high resolution modelling is presented. The calculation analysis shows that for the Gulf Stream even 0.3° resolution is crude. The DYNAMO group (1997) calculations shifted the Gulf Stream separation point northwards after four-year model-time integration, while Smith et al. (2000) managed to continue integration by 0.1° resolution for 15 years to obtain realistic results. The Gulf Stream and the North Atlantic current area is one of the best regions for comparing calculation with the observed data.

The next paragraph of Chapter 5 is devoted to the North Pacific circulation and contaminated water propagation. A short model-time of integration with high resolution (0.12°) enabled to keep the results of calculations at the realistic stage (Antipov et al., 2006) but more interesting are their results on contaminated water propagation.

We present part of their calculations on the process of propagation of contaminated water after possible wreckage of URF (Used Radioactive Fuel) carrying ships.

An analysis of successful modelling of the Kuroshio extension and the Kuroshio itself by quite high resolution and adequate model-time of integration by Japanese scientists is presented in this paragraph too.

The last paragraph of Chapter 5 contains the Indian Ocean descriptive hydrometeorology and thermo-hydrodynamics modelling. We have analyzed tens of papers devoted to the Indian Ocean Dipole (IOD) "positive" and "negative" phases, Indo-Pacific interaction and their behavior during ENSO, the process of Indonesian throughflow water propagation, the oceans' thermo-hydrodynamics strong seasonality, etc. Some of the papers were taken from the *Journal of Climate* (2006), especially from the one published in 2007 and devoted exclusively to the Indian Ocean.

Chapter 6 describes some modelling results on variability of the dynamics of the North Sea, the White Sea, the Caspian Sea and the Black Sea. Each of these four seas has its own peculiarities.

The decisive factors of the White Sea dynamics are strong tidal processes. The main tide comes from the Barents Sea with an amplitude of almost 6 meters; it is the semi-diurnal wave M_2. There are other kinds of tides (K_1, M_4, MS_4) too; the vertical displacement of isotherms excited by internal waves of tidal period reaches up to 20 m. The Barents Sea saline water and the Dvina river fresh water form frontal zones and eddies. Therefore, one should be careful in averaging the observed data for longer than one day time, it would filter out the oscillating processes.

Quite a big paragraph of this chapter is devoted to modelling seasonal variability of the Caspian Sea, the largest enclosed water body on Earth with meridionally elongated geometry (1000 km long and 200–300 km wide) and complicated bottom topography, the maximum depth of the northern shallow part is only 20 m while the southern deeper parts reach 1 km and more. The sea surface heat flux components undergo very strong seasonal variations following the surrounding continental climate. Even the water budget is sharply variable; its main component is evaporation which is being balanced by river runoff and precipitation. These and other peculiarities were taken into account in PEM and seasonal variations of all thermo-hydrodynamic characteristics were calculated. Due to high resolution an undercurrent was revealed along the eastern coast sea. Some results of the calculations are analyzed in this paragraph and presented in tens of figures.

Seasonal variations of the Black Sea thermo-hydrodynamical characteristics are described in the last and the largest paragraph. Quite sufficient volume of the paragraph is devoted to descriptive oceanography of the Black Sea. Besides, modern models and numerical methods describe the seasonal variability of the Black Sea internal physical characteristics as Cold Intermediate Layer (CIL), transformation of the effect of river fresh water discharge, as well as water exchange with the Mediterranean Sea. But the main result is the new four-dimensional analysis-double-correction method of data assimilation. It is a quite sophisticated multivariant model which combines the assimilation of internal characteristics with satellite data.

The privilege and accuracy of the method are discussed and relevant figures are presented. It is shown, on the whole, that models are very successful in making short-term forecasts and describing the ocean/sea seasonal variability, while there are problems in describing long-term (years to ten years) variability. We consider the model-data synthesis to be one of the most perspective ways for understanding the ocean/sea long-term variability.

Moscow, Russia A.S. Sarkisyan
Hamburg, Germany J.E. Sündermann

Acknowledgements

This book was supported in part by the Russian Foundation for Basic Research, project no. 08-05-00949a. The authors are grateful to N. Diansky, R. Ibrayev, V. Knysh, E. Ozsoy, T. Pohlmann, E. Semenov, and V. Zalesny for providing us with their unpublished materials that was helpful for keeping some sections of the book in state-of-the-art. Many interesting discussions in the institutions of both authors were very fruitful. We thank also to V. Arutyunyan and A. Schneehorst for their assistance in the preparation of the manuscript for publication, and to Nina Bennink and Rajasekar Subramaniam for their comments and corrections of the manuscript; yet we ask a pardon from a careful reader who may still find some mistakes in the text.

Contents

Chapter 1
Mathematical Background and Methods of Ocean Modelling

Abstract The oceans and marginal seas play an important role in the global climate system. Their surface and deep currents redistribute heat, salt, and chemical compounds all over the world. Ocean and shelf sea thermohaline circulations have a complicated vertical and horizontal structure which is determined by atmospheric forcing, sea ice, distribution of continents, and bottom relief.

Modelling of the ocean general circulation includes a number of aspects of geophysical, mathematical, and algorithmic nature, each of them being of scientific interest and thus deserving individual consideration. As mentioned by K. Bryan (Manabe and Bryan, 1969), the history of baroclinic ocean numerical modelling began with the work of A. S. Sarkisyan (1961, 1962).

Two basic areas of physical and mathematical nature can be distinguished in the theoretical description of sea and oceanic currents.

The first area is associated with the study of physical processes controlling the large-scale marine and oceanic circulations and their variability. A great number of theoretical and experimental works are dedicated to this field.

The second area involves the aspects associated with mathematical and numerical modelling of the ocean and sea dynamics. It is the subject which we discuss here (Marchuk et al., 2001a).

Essential items are the discovery and study of the Joint Effect of Baroclinicity and Bottom Relief (JEBAR), the introduction of sea surface height (instead of stream integral function) calculations into numerical models, the development of the method of dynamical adaptation of hydrological fields (currents, temperature, and salinity), the calculation of dynamically correlated hydrological fields in the world ocean (Sarkisyan, 1969a,b; Sarkisyan and Ivanov, 1971; Sarkisyan and Demin, 1983).

Keywords Theory · Energy conservation · Parameterization · Solvability · Approximation

A.S. Sarkisyan, J.E. Sündermann, *Modelling Ocean Climate Variability*,
DOI 10.1007/978-1-4020-9208-4_1, © Springer Science+Business Media B.V. 2009

1.1 Classical Problems of Marine Dynamics: Equations, Initial and Boundary Conditions

The ocean general circulation (OGC) models are mainly based on the nonlinear equations of large-scale fluid dynamics. The model equations follow from the Reynolds equations that describe the baroclinic motion of a rotating fluid with certain approximations, which are traditional in oceanology.

The Reynolds equations are obtained by averaging the general equations governing the motion of an ideal fluid. The motion in oceanic and sea water is always turbulent. Qualitative analysis of this motion by means of the Eulerian equations would consider the whole spectrum of turbulent pulsations, chaotic flows, regular mesoscale structures, and large-scale circulations. This is practically impossible and also not of interest for a view of the global ocean circulation. Therefore the approach proposed by Reynolds in 1895 is employed together with a parameterization of turbulent fluxes when constructing mathematical models.

The construction of these averaged equations is performed as follows. A sought vector function φ with such components as velocity, pressure, density, etc. is represented as a sum of rapidly and slowly varying components φ', $\hat{\varphi}$

$$\varphi = \varphi' + \widehat{\varphi}, \widehat{\varphi} = \frac{1}{\tau}\int_0^\tau \varphi dt.$$

The time axis is divided into regular intervals of length τ, and the original equations describing the nonlinear dynamics of an ideal fluid are integrated with respect to time on each interval. For the new sought vector function we choose the slowly varying component of the old vector function $\widehat{\varphi}$ which is equal to the corresponding average values on each time interval τ.

The integral of the rapidly changing components or that of the pulsations of all quantities within the intervals τ is zero ($\hat{\varphi}' = 0$). The integrals of their nonlinear interactions are expressed by combinations of averaged functions on the basis of certain physical hypotheses of turbulent closure.

1.1.1 Equations, Initial and Boundary Conditions of Ocean and Sea General Circulation

Equations of the dynamics of seas and oceans, which describe the averaged large-scale evolution of turbulent thermohaline fields on a rotating sphere, have the form

$$\frac{du}{dt} + muv\cos\theta - fv = -\frac{m}{\rho_0}\frac{\partial p}{\partial \lambda} + \frac{\partial}{\partial z}K_M\frac{\partial u}{\partial z} + A_M\Delta u \tag{1.1.1}$$

$$\frac{dv}{dt} + mu^2\cos\theta + fu = -\frac{n}{\rho_0}\frac{\partial p}{\partial \theta} + \frac{\partial}{\partial z}K_M\frac{\partial v}{\partial z} + A_M\Delta v \tag{1.1.2}$$

$$\frac{\partial p}{\partial z} + g\rho = 0 \tag{1.1.3}$$

$$m\left[\frac{\partial}{\partial \theta}\sin\theta v + \frac{\partial u}{\partial \lambda}\right] + \frac{\partial w}{\partial z} = 0 \tag{1.1.4}$$

$$\frac{dT}{dt} = \frac{\partial}{\partial z}\frac{K_T}{\delta}\frac{\partial T}{\partial z} + A_T \Delta T \tag{1.1.5}$$

$$\frac{dS}{dt} = \frac{\partial}{\partial z}\frac{K_S}{\delta}\frac{\partial S}{\partial z} + A_S \Delta S \tag{1.1.6}$$

$$\rho = \rho\,(T, S, p) \text{ in } D\,(x, y, z), \tag{1.1.7}$$

where

$$\frac{d}{dt} = \frac{\partial}{\partial t} + mu\frac{\partial}{\partial x} + nv\frac{\partial}{\partial y} + w\frac{\partial}{\partial z}$$

$$\Delta = m^2\frac{\partial^2}{\partial \theta^2} + mn\frac{\partial}{\partial \theta}\frac{n}{m}\frac{\partial}{\partial \theta}$$

$$f = 2\omega \sin\theta.$$

Equations (1.1.1)–(1.1.7) are written in the right-handed coordinate system λ, θ, z. The coordinate λ is directed along the latitude (eastwards), the coordinate θ is directed along the longitude (northwards), and the coordinate z is directed upwards from the ocean bottom to undisturbed sea surface. System (1.1.1)–(1.1.7) is considered on the time interval $(0, t]$ in a three-dimensional domain $D(\lambda, \theta, z)$. The domain D is bounded by the boundary ∂D that consists of the undisturbed sea surface $z = 0$, the lateral (coastal) surface $\sum$, and the bottom relief $H\,(\lambda, \theta)$ with a normal n_H.

Here u, v, w are the components of the velocity vector; T is the potential temperature; S is the salinity; p is the pressure; ρ is the density; K_M, K_T, K_S are the coefficients of vertical turbulent viscosity and diffusion; A_M, A_T, A_S are the coefficients of horizontal turbulent viscosity and diffusion; f is the Coriolis parameter. The values m and n are metric functions; in the spherical coordinate system $m = 1/R\sin\theta$ and $n = 1/R$, where R is the Earth radius; ω is the angular velocity of the Earth rotation. The parameter $\delta = 1$ when the vertical stratification is stable; when in process of calculation an instability occurs, δ should be specified as a big quantity, say 10^3. That kind of technique converses the instable stratification into neutral one.

The boundary conditions for equations (1.1.1)–(1.1.7) can be formulated as follows:
in the vertical direction at the surface ($z = \zeta$):

$$K_M \frac{\partial u}{\partial z} = \frac{\tau_\lambda}{\rho_0}, \; K_M \frac{\partial v}{\partial z} = \frac{\tau_\theta}{\rho_0}, \; w = 0, \; \left(\text{or } w = \frac{\partial \zeta}{\partial t}\right) \tag{1.1.8}$$

$$K_T \frac{\partial T}{\partial z} = \gamma_T (T - T_S) + Q_T, \; K_S \frac{\partial S}{\partial z} = \gamma_S (S - S_S) + Q_S \tag{1.1.9}$$

where $z = \zeta$ denotes the water surface elevation.

- in the vertical direction at the bottom $H(z = -H(\lambda, \theta))$:

$$w = m \frac{\partial H}{\partial \lambda} u + n \frac{\partial H}{\partial \theta} v, \; (Du, n_H) = 0, \; (Dv, n_H) = 0 \tag{1.1.10}$$

$$(DT, n_H) = 0, \; (DS, n_H) = 0 \tag{1.1.11}$$

- at the lateral surface $\sum$:

$$u = 0, \; v = 0 \tag{1.1.12}$$

$$K_T \frac{\partial T}{\partial n} = 0, \; K_S \frac{\partial S}{\partial n} = 0, \tag{1.1.13}$$

where n is a normal to $\sum \cdot \zeta$ is the deviation of the actual sea level from an undisturbed one. The boundary condition (1.1.10) for w will be commented below. Equations (1.1.1)–(1.1.7) are supplemented by the initial conditions for $t = 0$:

$$u = u^0, \; v = v^0, \; T = T^0, \; S = S^0 \tag{1.1.14}$$

The function $\rho(T, S, p)$ that defines the equation of state for sea water is chosen on the basis of empirical relations. In some cases, for example in shallow seas, the density may be assumed to depend only on temperature and salinity.

The above equations are considered in the domain D with a fixed boundary. The domain D can be multiply connected because of the presence of continents and islands.

In oceanology the system of equations (1.1.1)–(1.1.7) is traditionally called the system of "primitive" equations of the general circulation. It is obtained from the classic Reynolds equations of hydrodynamics for a rotating fluid with the use of traditional approximations: the Boussinesq approximation, hydrostatics, incompressibility, and linear turbulence closure with respect to momentum, heat, and salt.

The setting of a "rigid lid" boundary condition $w = 0$ at the undisturbed sea surface (1.1.9) is also an approximation which filters out external gravity waves.

A sea medium is considered as a two-component balanced thermodynamical system within the chosen diffusion approximation. We assume that seawater stratification is always stable by density. This requires an appropriate parameterization of convection processes, which is not explicitly described in equations (1.1.1)–(1.1.7).

One possible formulation is the following. The solution of the initial-boundary value problem (1.1.1)–(1.1.4) is considered in the subspace of possible solutions such that $\frac{\partial \rho}{\partial z} < 0$. Additionally, we assume that K_T, K_S are nonlinear functions of the potential density gradient, for example,

$$K_T = K_S \equiv K_{\min} + \frac{(K_{\max} + K_{\min})}{2}\left(1 + sign\left(\frac{\partial \rho}{\partial z}\right)\right) \tag{1.1.15}$$

or (see Segschneider and Sündermann, 1997)

$$K_T = K_S \equiv K_{\min} \exp\left(\alpha_1\left[1 + \tanh\left(\alpha_2 \frac{\partial \rho}{\partial z}\right)\right]\right) \tag{1.1.16}$$

where $K_{\max}, K_{\min}, \alpha_1, \alpha_2$ are positive constants, $K_{\max},\ K_{\min},\ \alpha_1 = \frac{1}{2}\ln\left[K_{\max}/K_{\min}\right]$.

1.1.2 Some Comments on the Boundary Conditions

The general circulation problem for shelf seas and oceans is formulated as an initial boundary value problem for the system of equations (1.1.1)–(1.1.14). When setting the boundary conditions the boundary of the domain D is divided into several parts: the upper interface surface (atmosphere-sea), the lower bottom surface, and the lateral coastal boundary. In most of the general circulation problems it is assumed that all the parts of the boundary are fixed.

It is also assumed that the upper boundary coincides with the undisturbed sea surface $z = 0$ and that the constraint $H(\lambda, \theta) \geq H_0 > 0$ holds for the bottom surface, so that the coastal contour is also fixed. In the case of shelf sea circulation mostly the upper boundary is defined as a free surface $z = \zeta(\lambda, \theta, t)$ (Sündermann, 1994).

Sometimes moving lateral boundaries are also considered $H + \zeta < 0$. These models are used when examining the mesoscale dynamics of coastal zones with shallow basins (Burchard, 2002).

The kinematic boundary condition $w = 0$ for the vertical velocity at the upper undisturbed boundary $z = 0$ is well known as the rigid lid approximation. The milder, linearized dynamic condition $w = \frac{\partial \zeta}{\partial t}$ at $z = 0$ allows for including long external gravity waves of tidal type into the spectrum of the model solution. In the general case one may use the exact kinematic condition with an additional condition for pressure:

$$w = \frac{\partial \zeta}{\partial t} \text{ at } z = \zeta(\lambda, \theta, t) \tag{1.1.17}$$

$$p = p_a, \tag{1.1.18}$$

where p_a is the atmospheric pressure.

The boundary conditions for temperature and salinity at the upper boundary (1.1.9) are written in the general form. Their concrete representation is chosen as the situation requires. Starting with the Dirichlet condition, when the asymptotics $\gamma_T, \gamma_S \to \infty$ assert, they may change up to the so-called "mixed conditions", the Newton conditions for temperature

$$K_T \frac{\partial T}{\partial z} = \gamma_T (T - T_S)$$

and the Neumann conditions for salinity ($\gamma_S = 0$)

$$K_S \frac{\partial S}{\partial z} = Q_S.$$

It seems that the mixed boundary conditions reflect a real situation more adequately. The heat flux into the ocean is defined by local conditions at the surface, and the third kind condition (the Newton condition) holds for it. The salt flux is related with T_0 precipitation formation which may occur far from the sea surface. S_0 is possible to turn the value γ_S to zero and use the Neumann condition for salinity. In this case salinity, as well as pressure, is determined up to a constant, which must be prescribed.

1.1.3 Some Comments on the Initial Conditions

The lack of data on the 3D-fields of horizontal velocity components u^0, v^0, temperature T^0 and salinity S^0 is the main difficulty in formulating the initial conditions (1.1.14).

Observational data on temperature and salinity for oceans and shelf seas as a rule describe their mean climatic values: annual average (sometimes averaged over a season) and monthly averaged data. The latter allow it for reconstructing their seasonal cycle.

Observational data on the velocity field are scantier: they are practically unavailable for many water basins. In this connection the problem of constructing appropriate current fields from the measured data of temperature and salinity arises.

The main requirement to the restored (by various methods) fields of temperature, salinity, and currents is that the so-called initialization shock must be eliminated in the model solutions. The disturbance implies a sharp change of the model solution during its initial integration interval. This initialization shock is caused by inconsistent not dynamically correlated initial conditions in the fields of currents, temperature, and salinity.

The initialization problem can be solved on the basis of forward modelling (Ivanov et al., 1997), of adjoint equations or optimal control theory (Marchuk et al., 2001b).

1.1.4 Total Energy Conservation

Suppose, energy sources and energy sinks are absent in the model, i.e. neglect the terms describing the turbulent exchanges in (1.1.1)–(1.1.13). In addition, let the state function $\rho(T, S, p)$ for sea water be such that from the temperature and salinity equations follows

$$\frac{\partial \rho}{\partial t} = 0. \tag{1.1.19}$$

Then the conservation law for the total energy (Zalesny, 1997)

$$\frac{\partial}{\partial t} \int\limits_D \left[\rho_0 \frac{u^2 + v^2}{2} + g\rho z \right] dD = 0 \tag{1.1.20}$$

holds. It is still valid when the terms describing the horizontal turbulent exchange ρ are included into the equation for the density of sea water:

$$\frac{\partial \rho}{\partial t} = F^{\rho}. \tag{1.1.21}$$

1.1.5 Parameterization of Subscale Physical Processes

The processes of pulsation, which are small as compared to the phenomena described by the model, are called the "subgridscale" processes.

In the Reynolds equations (1.1.1)–(1.1.7), the terms describing the processes of turbulent exchange were chosen in the simplest way. It was assumed that the macro-scale diffusion of momentum, heat, and salt is similar to molecular diffusion with large coefficients $K_M, A_M, K_T, K_S, A_T, A_S$.

However, the examination of natural processes, sea and ocean motions shows that the simple analogy with chaotic molecular motions is not sufficient in many cases. In seas and oceans, alongside irregular chaotic pulsations, there are dynamically active highly energetic wave processes, micro-scale and meso-scale structures possessing certain regularity and their own features. Their dynamics, the interaction between them and the large-scale circulation are of a complex character. Wherever possible, all that must be parameterized in general circulation models (Thomas et al., 2001).

Turbulent processes in the ocean are divided into vertical and horizontal exchanges due to their characteristic scales ($H << L,\ H$ is vertical and L is horizontal scale) and due to the presence of density stratification ($\rho = \rho(z)$). Sometimes turbulent exchanges along and across isopycnal surfaces are taken into consideration.

The smallness of vertical scales in shelf seas and oceans basins (H) as compared to horizontal ones (L) allows to radically simplify the third equation of motion and reduce it to the hydrostatics equation (1.1.3). This simplification, together with the introduction of the Earth rotation effect (the Coriolis force), distinguishes the primitives of equations as compared to the Navier-Stokes equations widely used in fluid dynamics. This is reflected in the qualitative behavior of model solutions, in the modifications of physical parameterizations, and in analytical and numerical methods of solution.

When neglecting the term that describes the vertical acceleration (dw/dt), certain small scale processes are excluded. These are short surface and internal waves, convection and double-diffusion processes, Langmuir meso-scale circulations, etc. These processes with a typical length scale comparable to the depth of ocean layers regulate the vertical exchange of momentum, heat, and salt. For example, the process of breaking wind waves redistributes heat and momentum in the surface layer of the ocean. The destruction of internal waves leads to intensification of mixing in the layers of thermo- and haloclines. The Langmuir circulation generates an additional heat and salt exchange in the upper layer.

The parameterization of subgridscale processes is performed in the models of large-scale ocean circulation in several ways. Most frequently used is the closure on the zero level, i.e. an approach relating the coefficients of vertical turbulent exchange with the macroscale parameters momentum, heat, and salt K_M, K_T, K_S, for example, the gradients of currents and density in the vertical. In this case the form of the terms describing the vertical turbulent exchange in (1.1.1)–(1.1.7) is the same but the exchange coefficient is a function of the solution, that is, the linear diffusion term is replaced by a nonlinear one. The exchange coefficient can be expressed with the help of algebraic or integral relationships depending on whether the process to be parameterized is of a local or nonlocal character.

Another approach for parameterizing the subscale turbulent processes is based on higher order closure procedures. In this case the model is supplemented by new equations such as the equations for turbulent energy, dissipation, etc. which describe the dynamics of the first, second and higher moments of pulsation components of the solution.

1.2 Solvability of Ocean/Sea Dynamics Problems

Intensive investigations on modelling the marine general circulation were started in the middle of the 20th century by Schtockman (1946), Sverdrup (1947), Stommel (1948), Munk (1950), Sarkisyan (1954), Hansen (1959), Bryan (1969), Friedrich (1970) and others. They proposed various mathematical models ranging from simple two-dimensional and linear ones to nonlinear models of the three-dimensional thermohaline circulation.

First studies on mathematical aspects of geophysical fluid dynamics problems were contributed by Marchuk and Demidov (1966), Arakawa (1966), Lilly (1965), Kreiss (1959), Ovsyannikov (1966) and continued by Lions et al. (1992), Kazantsev et al. (1998), and other mathematicians. They proved of existence and uniqueness of solutions for various formulations of hydrodynamic problems. Let us consider some of them.

1.2.1 Linear Model of Small Oscillations in the Ocean with Stable Stratification

Consider a simple linear model, which describes the large-scale motion in a stratified rotating fluid in a Cartesian coordinate system (x,y,z):

$$\frac{\partial u}{\partial t} - fv + \frac{1}{\rho(z)}\frac{\partial p}{\partial x} = 0 \tag{1.2.1}$$

$$\frac{\partial v}{\partial t} + fu + \frac{1}{\rho(z)}\frac{\partial p}{\partial y} = 0 \tag{1.2.2}$$

$$\frac{\partial p}{\partial z} + g\rho = 0 \tag{1.2.3}$$

$$\frac{\partial u}{\partial x} + \frac{\partial v}{\partial y} + \frac{\partial w}{\partial z} = 0 \tag{1.2.4}$$

$$\frac{\partial \rho}{\partial t} + \Gamma w = 0 \text{ in } D\,(x, y, z) \tag{1.2.5}$$

where D is a cylindrical domain, with surface area D_S, $D = D_s \times [0, H\,(x, y)]$, $\Gamma = \frac{\partial \rho\,(z)}{\partial z}$, and the mean density $\rho(z)$ is a given function.

The boundary and initial conditions for (1.2.1)–(1.2.5) are chosen as follows:

$$\frac{\partial \rho}{\partial t} - g\rho(z)w = 0 \quad \text{for} \quad z = 0$$

$$w = 0 \quad \text{for} \quad z = -H$$

$$(u, n) = 0 \text{ at the external boundary } \sum \tag{1.2.6}$$

$$u = u^0, \; v = v^0, \; p = p^0 \quad \text{for} \quad t = 0. \tag{1.2.7}$$

A solution to (1.2.1)–(1.2.5) exists and is unique in appropriate functional spaces (Bubnov and Kazhikhov, 1971).

1.2.2 Linear Stationary Model of Ocean Dynamics With Turbulent Exchange Parameterization

Consider the linear steady-state ocean model in a Cartesian frame:

$$-fv + \frac{1}{\rho_0}\frac{\partial p}{\partial x} = A_M \Delta u + K_M \frac{\partial^2 u}{\partial z^2} \tag{1.2.8}$$

$$fu + \frac{1}{\rho_0}\frac{\partial p}{\partial y} = A_M \Delta v + K_M \frac{\partial^2 v}{\partial z^2} \tag{1.2.9}$$

$$\frac{\partial p}{\partial z} + g\rho = 0 \tag{1.2.10}$$

$$\frac{\partial u}{\partial x} + \frac{\partial v}{\partial y} + \frac{\partial w}{\partial z} = 0 \tag{1.2.11}$$

$$\Gamma w = A_T \Delta\rho + K_T \frac{\partial^2 \rho}{\partial z^2} \quad \text{in } D(x, y, z) \tag{1.2.12}$$

$$K_M \frac{\partial u}{\partial z} = \frac{\tau_x}{\rho_0}, \; K_M \frac{\partial v}{\partial z} = \frac{\tau_y}{\rho_0}$$

$$w = 0, \; K_T \frac{\partial \rho}{\partial z} = Q_\rho \quad \text{for } z = 0$$

$$u = v = w = 0, \; \rho = \rho_H \text{ for } z = -H$$

$$u = v = 0, \; \frac{\partial \rho}{\partial n} = 0 \quad \text{on } \sum, \tag{1.2.13}$$

where the wind stress components τ_x, τ_y are the given functions of $C\ (D_S) \cap W_0^{1,2}(D_S)$ and ρ_H, Q_ρ are specified too. C means a space of continuous functions given in a two-dimensional domain D_S, $w^{1,2}$ – Sobolev space of functions possessing a generalized (weak), square integrable first order derivatives.

One can eliminate two unknown functions from (1.2.8)–(1.2.13): the pressure p and the vertical velocity w.

Then the theorem on the existence and uniqueness of the solution to (1.2.8)–(1.2.13) is valid with $(u, v, w) \in W^{1,2}(D)$ (D is three-dimensional domain), $p_S = p(x, y, 0, t)$ i.e. ρ_0 is unique up to an additive constant.

If one considers a nonstationary problem with additional terms $\partial u/\partial t$, $\partial v/\partial t$, $\partial \rho/\partial t$ which corresponds to (1.2.8)–(1.2.12), then existence and uniqueness of the solution also are given and the solution converges to the solution of the stationary problem as $t \to \infty$.

1.2.3 Prognostic Nonlinear Problem of Ocean General Circulation

Consider the nonstationary nonlinear problem of ocean general circulation in a Cartesian frame:

$$\frac{du}{dt} - fv + \frac{1}{\rho_0}\frac{\partial p}{\partial x} = A_M \Delta u + \frac{\partial}{\partial z} K_M \frac{\partial u}{\partial z} \tag{1.2.14}$$

$$\frac{dv}{dt} + fu + \frac{1}{\rho_0}\frac{\partial p}{\partial y} = A_M \Delta v + \frac{\partial}{\partial z} K_M \frac{\partial v}{\partial z} \tag{1.2.15}$$

$$\frac{\partial p}{\partial z} + g\rho = 0 \tag{1.2.16}$$

$$\frac{\partial u}{\partial x} + \frac{\partial v}{\partial y} + \frac{\partial w}{\partial z} = 0 \tag{1.2.17}$$

$$\frac{dT}{dt} + \Gamma_T w = A_T \Delta T + \frac{\partial}{\partial z} K_T \frac{\partial T}{\partial z}, \; \Gamma_T = -\frac{d\hat{T}(z)}{dz} \tag{1.2.18}$$

$$\frac{dS}{dt} + \Gamma_S w = A_S \Delta S + \frac{\partial}{\partial z} K_S \frac{\partial S}{\partial z}, \; \Gamma_S = -\frac{d\hat{S}(z)}{dz} \tag{1.2.19}$$

$$\rho = \alpha_T T + \alpha_S S \text{ in } D(x, y, z) \tag{1.2.20}$$

(where α_T, α_S are given constants, $\hat{T}$ and $\hat{S}$ are vertical profiles of T and S) with boundary and initial conditions,

$$\frac{\partial u}{\partial z} = \frac{\partial v}{\partial z} = 0;\ w = 0;\ \frac{\partial T}{\partial z} = 0;\ \frac{\partial S}{\partial z} = 0 \quad \text{for} \quad z = 0$$

$$u = v = w = 0,\ T = 0,\ S = 0 \quad \text{for} \quad z = -H$$

$$u = v = 0;\ \frac{\partial T}{\partial n} = 0;\ \frac{\partial S}{\partial n} = 0 \quad \text{on} \quad \sum \tag{1.2.21}$$

$$u = u^0,\ v = v^0,\ T = T^0,\ S = S^0 \quad \text{for} \quad t = 0 \tag{1.2.22}$$

For this model the theorem on the existence of a strong solution is proven (Kordzadze, 1982) with

$$(u, v, T, S) \in C^2(D) \cap C^1\left(\bar{D}\right), \qquad \rho, w \in C^1(D),$$

where $C^1\left(\bar{D}\right)$ is a set of functions with continuous first order derivatives; $C^2(D)$ is a set of functions with continuous second order derivatives; $\overline{D}$ is a bounded domain (domain together with boundaries); D is a three-dimensional domain.

1.2.4 Prognostic Nonlinear Problem of Ocean General Circulation With Parameterization of Deep Convection

Finally, we consider the equations (1.1.1)–(1.1.6) and add the linear state equation

$$\rho = \rho_0 + \alpha_T T + \alpha_S S$$

with boundary and initial conditions (1.1.8)–(1.1.14) and nonlinear parameterization of convective adjustment (1.1.16). Then the following theorem holds (Kazantsev et al., 1998):

Let $\gamma_t = \gamma_s$, $A_T = A_S$ and $K_T = K_S \equiv K(\partial\rho/\partial z)$ be such that

$$0 < K_{\min} \leq K\left(\frac{\partial \rho}{\partial z}\right) \leq K_{\max} < +\infty$$

and the function $r \rightarrow v(r)r$ is an increasing function.

Suppose that $u^0, v^0, T^0, S^0 \in L_2(D)$ and $\tau_x, \tau_y \in L_2(D_0)$. Then for all positive moments of time $\hat{t}$ the problem under consideration admits the solution

$$u, v \in C\left(0, \hat{t}, L_2(D)\right) \cap L_2\left(0, \hat{t}, H^1(D)\right),$$
$$T, S \in C\left(0, \hat{t}, L_2(D)\right) \cap L_2\left(0, \hat{t}, H^1(D)\right),$$

where H^1 is Hilbert space.

The problem admits no more than one solution

$$u, v, T, S \in L_2\left(0, \hat{t}, W^{1,\infty}(D)\right)$$

The study of solvability for a nonlinear equation of state is not completed yet.

1.3 Alternative and Generalized Models of the General Circulation in Oceans and Shelf Seas

There are other model formulations which are adapted to specific applications in ocean dynamics. By convention these models can be divided into three groups: conceptual or simplified models, alternative, and generalized ones. The conceptual models follow from the original model under additional simplifying assumptions. The models considered in the preceding section can serve as an example. The alternative models are close to the original model but are written for other independent variables and/or for other sought functions. The alternative models can be equivalent in a certain sense to the original model but may also differ from it. The difference is often due to the changes in the boundary and initial conditions and also due to the presence of the terms describing the turbulent exchange. The generalized models are certain extensions of the original model and describe a wider spectrum of admissible solutions.

1.3.1 Model Based on the Nonlinear Shallow-Water Equations

A mathematical model of a shallow sea with a pronounced barotropic component of currents can be based on the nonlinear shallow-water equations. In a Cartesian coordinate system these equations have the form

$$\frac{\partial u}{\partial t} + u\frac{\partial u}{\partial x} + v\frac{\partial u}{\partial y} - fv = -g\frac{\partial \zeta}{\partial x} - Ru + Q_x \tag{1.3.1}$$

$$\frac{\partial v}{\partial t} + u\frac{\partial v}{\partial x} + v\frac{\partial v}{\partial y} + fu = -g\frac{\partial \zeta}{\partial y} - Rv + Q_y \tag{1.3.2}$$

$$\frac{\partial \zeta}{\partial t} + \left[\frac{\partial (H+\zeta)u}{\partial x} + \frac{\partial (H+\zeta)v}{\partial y}\right] \quad \text{in} \quad D(x, y) \tag{1.3.3}$$

$$\left(\vec{u}, \vec{n}\right) = 0 \quad \text{on} \quad \partial D(x, y) \tag{1.3.4}$$

$\vec{u} = (u, v)$; u, v are vertically averaged velocity components.

The system of equations (1.3.1)–(1.3.4) can be obtained by integrating the equations of an ideal rotating fluid over the vertical direction from the surface to the bottom under certain assumptions (Hansen, 1956; Marchuk et al., 1987). The main assumptions are: the density of sea water is constant ($\rho = \rho_0 \equiv \text{const}$), the motion is almost plane. In system (1.3.1)–(1.3.3) the bottom friction described by the terms Ru and Rv is taken into account. The terms Q_x, Q_y describe the external forcing, for example, the wind stress τ_x, τ_y:

$$Q_x = \frac{\tau_x}{\rho_0 (H + \zeta)}, \quad Q_y = \frac{\tau_y}{\rho_0 (H + \zeta)} \tag{1.3.5}$$

In the absence of energy sources and sinks ($Q_x = Q_y = 0,\ R = 0$), the conservation law

$$\frac{\partial}{\partial t} \int\limits_D \left[(H + \zeta) \left(u^2 + v^2 \right) / 2 + g\zeta^2 \right] dD = 0 \tag{1.3.6}$$

holds for the system of equations (1.3.1)–(1.3.4).

The model based on the shallow-water equations is of interest both from the practical and theoretical standpoints. Using this model, one can calculate wind-driven currents, tidal waves, and storm pileups in shallow-water areas. The shallow-water formulation has become a classic one for developing and testing new numerical algorithms. The main difficulties associated with its numerical solution are the following: the realization of the boundary conditions at a moving boundary, complexity of developing efficient methods for calculating the sea surface height, which is due to the presence of supersonic regimes at small thickness of a liquid layer $(H + \zeta)$ (Marchuk et al., 1987). Note that due to the Coriolis effect of rotation (for the linearized shallow-water equations) the problem can be transformed into a boundary sea level value problem for the sea with a Poincare condition at the solid boundary (Marchuk, 1969).

1.3.2 Ocean General Circulation Model in the σ – Coordinate System

An alternative formulation of ocean general circulation models is written in depth-following σ – coordinates with $\sigma = z/H(\lambda, \theta)$.

In many cases this system is most convenient from the physical viewpoint. Its main advantage is that it adequately reproduces the dynamic processes over a complicated bottom relief. The system allows to model the regimes of a low-viscosity flow over relief, sloping bottom convection and adequately resolves the bottom layer structure.

In a new system of coordinates $t_1 = t,\ \lambda_1 = \lambda,\ \theta_1 = \theta,\ z_1 \equiv \sigma = z/H(\lambda, \theta)$ equations (1.1.1)–(1.1.7) have the form (Sarkisyan, 1969a; Zalesny, 1997)

$$H\frac{\partial u}{\partial t_1} - H[f - m\cos\theta u]v + H\frac{m}{\rho_0}\frac{\partial p}{\partial \lambda_1} + \frac{gmH}{\rho_0}z_1\frac{\partial H}{\partial \lambda_1}p = \frac{\partial}{\partial z_1}\frac{K_M}{H}\frac{\partial u}{\partial z_1} + F_1{}^u \tag{1.3.7}$$

$$H\frac{\partial v}{\partial t_1} + H\left[f - m\cos\theta u\right]u + H\frac{n}{\rho_0}\frac{\partial p}{\partial \theta_1} + \frac{gnH}{\rho_0}z_1\frac{\partial H}{\partial \theta_1}p = \frac{\partial}{\partial z_1}\frac{K_M}{H}\frac{\partial v}{\partial z_1} + F_1{}^v \tag{1.3.8}$$

$$\frac{\partial p}{\partial z_1} + gH\rho = 0 \tag{1.3.9}$$

$$m\left[\frac{\partial Hu}{\partial \lambda_1} + n\frac{\partial}{\partial \theta_1}\left(\frac{Hv}{m}\right)\right] + \frac{\partial w_1}{\partial z_1} = 0 \tag{1.3.10}$$

$$H\frac{dT}{dt_1} = \frac{\partial}{\partial z_1}\frac{K_T}{H}\frac{\partial T}{\partial z_1} + F_1{}^T \tag{1.3.11}$$

$$H\frac{dS}{dt_1} = \frac{\partial}{\partial z_1}\frac{K_S}{H}\frac{\partial S}{\partial z_1} + F_1^S \tag{1.3.12}$$

$$\rho = \rho(T, S, p) \text{ in } D_1(\lambda_1, \theta_1, z_1) \tag{1.3.13}$$

Here

$$\frac{d}{dt_1} = \frac{\partial}{\partial t_1} + mu\frac{\partial}{\partial \lambda_1} + nv\frac{\partial}{\partial \theta_1} + \frac{w_1}{H}\frac{\partial}{\partial z_1}$$

and the domain D_1 is a cylinder with a plane base $z_1 = 1$ and w_1 is a new vertical velocity:

$$w_1 = w + mz_1\left(\frac{\partial H}{\partial \lambda_1}u + \frac{n}{m}\frac{\partial H}{\partial \theta_1}v\right) \tag{1.3.14}$$

The kinematic boundary conditions with respect to the new vertical coordinate z_1, which follow from (1.1.8), (1.1.10), have the simple form

$$w_1 = 0 \quad \text{for} \quad z_1 = 0 \tag{1.3.15}$$

$$w_1 = 0 \quad \text{for} \quad z_1 = -1. \tag{1.3.16}$$

The equations written in bottom-following coordinates are used in modern models of ocean general circulation, in models of shelf sea and coastal dynamics. From the computational viewpoint the advantage of this system [in comparison with the z-system (1.1.1)–(1.1.7)] is a simple form of the computational domain over the vertical direction. This simplifies the construction of projection schemes for solving (1.3.7)–(1.3.13). For example, one can use two-dimensional finite elements over the horizontal direction and approximate the equations by the finite difference method over the vertical direction.

A disadvantage of the σ-system is a more complex form of the operator of the problem: the presence of additional constraints over (λ, z) and (θ, z)-coordinates and of mixed derivatives with respect to these pairs of coordinates in the terms describing the turbulent transport. Due to the complicated form of governing equations a more careful choice of their discrete analogs and methods of integration with respect to time is necessary. The authors' experience suggests to adjust the form of the last terms on the left-hand side of (1.3.7), (1.3.8) to the form of the terms describing the density (temperature and salinity) transport (Sarkisyan and Zalesny, 2000). Here the important condition is that the discrete analog of the conservation law (1.1.20) is valid. For getting a more efficient computing algorithm it is also necessary to make the method of integration with respect to time more implicit.

1.3.3 Generalized Model of Sea Dynamics With a Free Surface in the σ-Coordinate System

Suppose that the solution domain D with the bottom relief $H(x, y)$ includes shallow subdomains, which is typical for shelf dynamics problems. In this case one cannot disregard the deviations of the sea level from the undisturbed surface. The exact kinematic condition as well as an additional condition for pressure must be used at the upper free surface:

$$w = \frac{\partial \zeta}{\partial t} \tag{1.3.17}$$

$$p = p_{\text{atm}} \text{ at } z = \zeta\,(\lambda, \theta, t)\,. \tag{1.3.18}$$

In this case it σ-coordinates are introduced by the relation

$$z_1 = \sigma \equiv \frac{z + \zeta}{H + \zeta}. \tag{1.3.19}$$

In the new coordinate system the equations of the model are close to equations (1.3.7)–(1.3.13) in which, instead of $H(\lambda, \theta)$, we have a new unknown function $G = (H + \zeta)$ equal to the thickness of the water layer. The function G depends on spatial coordinates λ_1, θ_1 and time.

In the new coordinate system the continuity equation has the form

$$\frac{\partial G}{\partial t_1} + m\left[\frac{\partial G u}{\partial \lambda_1} + \frac{\partial}{\partial \theta_1}\left(\frac{G v}{m}\right)\right] + \frac{\partial w_1}{\partial z_1} = 0 \tag{1.3.20}$$

When using the variable substitution (1.3.19) the solution domain represents a cylinder of unit height as in the previous case. The obtained equations are more complicated in comparison with the z-system, the calculated domain and kinematic conditions in the vertical are simpler. The following conditions instead of (1.3.17), (1.3.18) are satisfied:

$$w_1 = 0 \tag{1.3.21}$$

$$p = p_{atm} \text{ at } z_1 = 0 \tag{1.3.22}$$

$$w_1 + w + \left[\frac{\partial \zeta}{\partial \lambda} + z_1 \frac{\partial (H+\zeta)}{\partial \lambda}\right] u + \left[\frac{\partial \zeta}{\partial \theta} + z_1 \frac{\partial (H+\zeta)}{\partial \theta}\right] v + (1 - z_1)\frac{\partial \zeta}{\partial t} = 0$$

where w_1 is a new vertical velocity.

1.4 Numerical Methods

Numerical modelling is the main method for examining the processes of circulation variability in oceans and shelf seas. Simulation of the general ocean and shelf sea dynamics has several specific features.

The problem is described by a complicated system of nonlinear partial differential equations whose solution is sought in multiply connected domains of complex form. Characteristic features are:

- This is an initial boundary value problem in which the Earth rotation effect plays a crucial role. The solution has narrow boundary layers near the walls and within domains. The rotation effect generates dispersion waves of small size, which slowly move in the horizontal plane. Applied problems, for example, propagation of contaminants from different sources, often require fine resolution (Rolinski et al., 2001).
- The characteristic size of energy disturbances (for the ocean, in midlatitudes) is about 100km, while the size of the domain under study is about 100 times larger.
- There is a diversity of time scales for individual characteristic phenomena: from 2 to 3 hours (inertial fluctuations in high latitudes) to 1000 years (deep-sea thermohaline variability) (Segschneider and Sündermann, 1998).

The works on numerical modelling of shelf sea and ocean dynamics show that the performance of models are significantly improved by an increase in resolution. Small scale processes (which nevertheless control the global dynamics) as friction, convection or inertial motion are adequately represented. This implies, for example, to resolve motions whose sizes are of the order of the baroclinic Rossby radius which changes from 300 km at the equator to 1 km for sea basins in high latitudes.

From the mathematical viewpoint it is necessary to construct and use efficient numerical models, which describe a totality of physical processes with large spread in space scales and time frequencies. Such models should approximate well the governing equations in space, have low computational costs, and be stable as $t \to \infty$.

Considering numerical models of oceans and shelf seas there are out three important components: the choice of a differential formulation of the problem; the approximation of the differential problem with respect to spatial coordinates; the choice of a method for integrating the problem with respect to time.

Free surface models for shallow waters require a stable behavior of the model solution in the subsonic regime at small values of $G = (H + \zeta)$. In this case the use of the splitting method and the method of ε-regularization of equations at some stages of splitting may be highly efficient.

1.4.1 The Choice of a Differential Formulation of the Problem

The choice of a differential statement of the problem can consist of a number of steps. Among them are the choice of a coordinate system convenient for solving the formulated problem in a given geometry; the choice of a special form of equations convenient for the subsequent approximation in space; and finally the ε-regularization of the problem for improving the properties of its solutions.

The choice of a special form, for example, the symmetrization of the equations, can be associated with the retention of the properties of the differential problem in its discrete analog. For instance, a symmetrized form of the equations for describing the transport of a passive substance φ in a divergence-free flow is

$$\frac{\partial \varphi}{\partial t} + \frac{1}{2}\left[u\frac{\partial \varphi}{\partial x} + \frac{\partial u\varphi}{\partial x} + v\frac{\partial \varphi}{\partial y} + \frac{\partial v\varphi}{\partial y} + w\frac{\partial \varphi}{\partial z} + \frac{\partial w\varphi}{\partial z}\right] = 0$$

This form is called semi-divergent. It satisfies the following requirements, which are efficient, both for forward and adjoint equations:

- The operator of the problem is represented as a sum of three simple non-negative one-dimensional transport operators with respect to each of x-, y-, and z-coordinates.
- This form permits a simple finite difference approximation that retains the skew-symmetry properties for each original differential transport operator in the x-, y-, and z-coordinates.
- The adjoint equation operator coincides with the original operator.

The second example illustrates the symmetrization technique. It is known that the numerical solution of ocean equations written in the σ-coordinates has some difficulties associated with the calculation of the pressure gradients terms. These difficulties can be overcome by formulating the pressure gradient terms (in the momentum equations) in accordance with the semi-divergent form of the density (temperature and salinity) transport operator. For example, one can use the following equivalent form (Sarkisyan and Zalesny, 2000)

$$\frac{\partial}{\partial \alpha}\left(p + \frac{g}{2}Hz\rho\right) - \frac{g}{2}\left[(zH)\rho_\alpha - (zH)_\alpha \rho\right] \equiv \frac{\partial p}{\partial \alpha} + g(Hz)_\alpha \rho \qquad (1.4.1)$$

where α is one of the coordinates x_1, y_1 or z_1.

The procedure of ε-regularization is often useful for increasing the stability of the numerical solution and/or for decreasing the sensitivity of the numerical solution to the external forcing variations. The method of ε-regularization involves the supplement of certain terms with a small coefficient $\varepsilon << 1$ to the original equations. It is similar to the method of artificial compressibility, which was proposed by Yanenko (1967) for solving the equations of a viscous incompressible fluid. The idea of this method is to replace the continuity equation by a nonstationary equation in the form (for the two-dimensional case)

$$\frac{\varepsilon}{\rho_0 g H}\frac{\partial p}{\partial t} + \frac{\partial u}{\partial x} + \frac{\partial v}{\partial y} = 0, \quad 0 < \varepsilon << 1 \tag{1.4.2}$$

Note that the transition from the rigid lid model to the model with the boundary condition $w = \frac{\partial \zeta}{\partial t}$ can be considered as regularization of the two-dimensional problem for an incompressible fluid with $\varepsilon = 1$. Besides, the efficiency of the calculation of depth-averaged velocities increases.

1.4.2 Methods of Spatial Approximation

In most of numerical models of ocean dynamics spatial derivatives are approximated using the finite difference method (FDM) or the finite element method (FEM) (Mesinger and Arakawa, 1982; Marchuk, 1980; Marchuk et al., 1987).

Finite difference method: Suppose we will approximate the ocean general circulation system of differential equations (1.1.1)–(1.1.7), which is represented in an abstract form by the Cauchy problem:

$$B\frac{\partial \varphi}{\partial t} + L\varphi = f, \quad B\varphi(0) = \varphi^0. \tag{1.4.3}$$

Here L is the operator of differential equations (1.1.1)–(1.1.7) [or (1.3.7)–(1.3.13)] with corresponding boundary conditions, which acts in a Hilbert space H. The finite difference method implies the replacement of each term of the operator L by its finite difference analog, after which the problem (1.4.3) is transformed into the discrete problem

$$B_h\frac{\partial \varphi_h}{\partial t} + L_h\varphi_h = f_h, \quad B_h\varphi_h(0) = {\varphi_h}^0. \tag{1.4.4}$$

Index h denotes a value at a grid point h.

The discrete analog of the conservation law must hold if it holds for the original differential problem (1.4.3); this is the most important requirement to the problem (1.4.4).

The FDM is simple to construct, and the discrete models are easily realized; this is their most attractive advantage. There are some limitations in this method: the polygonal approximation of oceanic basins with complex boundaries is not always efficient; in some cases there are certain difficulties with discrete analogs of integral invariants.

Finite element method: In the last decades of the 20th century the finite element method has become an alternative method for approximating the equations of ocean dynamics (Grotkop, 1982; Kuzin, 1985). Considerable progress has been made, for example, in calculating tidal motions and the circulations in marginal seas and shelf zones of oceans. The FEM has attracted widespread attention due to some of its advantages as compared to other methods.

The finite element method belongs to the family of Galerkin methods. The Galerkin approach to the solution of the differential problem (1.4.3) is to find a weak solution $\hat{\varphi}(x,t)$ which satisfies the equations

$$\begin{aligned} &\left(B\frac{\partial\hat{\varphi}}{\partial t},\psi\right)+(L\hat{\varphi},\psi)=(f,\psi) \quad \text{for} \quad t>0, \\ &(B\hat{\varphi},\psi)=\left(\varphi^0,\psi\right) \quad \text{for} \quad t=0, \end{aligned} \tag{1.4.5}$$

where $\psi(x)$ is an arbitrary function that belongs to the space H.

The representation of the approximate solution $\varphi^{(n)}$ to (1.4.5) as a series of basic functions $\psi_i(x)$

$$\varphi^{(n)}(t,x)=\sum_{i=1}^{n}a_i(t)\psi_i(x), \tag{1.4.6}$$

where $\psi_i(x)$ are functions with a local support or shape functions (nonzero functions only in a small part of the whole spatial domain), leads to the FEM.

If in this case the space of the trial functions (ψ) coincides with the space of the basic functions ($\psi_i(x)$), we arrive at the classic Galerkin method; if the above spaces differ, we have the Petrov-Galerkin method.

The FEM is constructed in the same way for regular and irregular grids; many well-known schemes are extended to the case of variable-resolution grids thereby allowing to maximize the resolution when necessary. In addition, the FEM that belongs to projective methods is based on a generalized (or weak) statement of the problem (1.4.5). The problem is formulated here on the basis of integral statements instead a system of partial differential equations. Thus, the principal conservation laws intrinsic in the original physical problem hold, and it is easier to realize the natural boundary conditions for the schemes of the FEM than for those of the FDM (Kuzin, 1985; Marchuk et al., 1987).

Weak formulations of problems are more natural for complicated ocean dynamics problems because their solvability is usually proven in a weak sense (see Sections 1.2.4 and 1.2.5).

The algorithmic complexity of calculations and the tendency to lose local properties of the solution to the approximated differential problem are, however, principal disadvantages of the FEM.

The use of the FEM requires the elaboration of special algorithms intended for the efficient solution of difference equations in time, including implicit and semi-implicit schemes (Kuzin, 1985), as well as the elaboration of algorithms for solving difference equations (algebraic systems of equations) at each time step. The lumping procedure for the time derivative terms should be used in order to apply explicit schemes.

When considering ocean dynamics using spatial discriminations FDM or FEM, a number of questions arises. These aspects are closely associated with the specific character of particular solutions and external conditions.

One of these questions is connected with the balance approximation of the model equations with respect to space.

The smallness of spatial scales of ocean variability and the order of typical time scales lead to the construction of numerical models on the basis of nondissipative (or weakly dissipative) spatial approximations. With this aim in view, symmetric balance schemes (satisfying finite difference analogs of the conservation laws) are generally employed. Their construction starts from the approximation of the primitive equations with the use of staggered grids. In this case it is easy to ensure the energy balance, which provides a possibility to integrate the model over a long time period.

However, when approximating the original system we try to preserve the integral balance; this may sometimes result in a distortion of local properties of the solution. In particular, this may manifest itself in the loss of the solution monotonicity or in the generation of imaginary, artificial boundary layers by the model (Delecluse and Zalesny, 1996). In this case the numerical noise effects are dramatic, especially when spatial resolution is not fine enough.

One can suggest several ways to overcome this difficulty, for example, to use quasi-spectral spatial approximations for theoretical studies and the solution of problems in the domains of simple forms, to use a modified model instead of the original one with special procedures of filtering or lumping discrete governing equations, to apply the high-order accuracy numerical schemes (Delecluse and Zalesny, 1996), to increase the grid resolution together with implicit time stepping.

1.4.3 Methods for Discrete Time Integration

In most numerical models explicit methods for integrating with respect to time are employed. The fulfillment of the stability conditions leads to a substantial reduction in the numerical efficiency. It should be emphasized that the variance ratio for the energetic ocean disturbances (for Rossby waves) has the form of an inversely proportional relation between time and space scales. Therefore when simulating similar processes one may not decrease the time step when employing implicit methods. The problem in this case is reduced to a search for the efficient realization of implicit schemes.

The method of weak approximation which splits the problem at the differential level, the method of domain decomposition which solves the problem separately in different subdomains, the method of splitting by physical processes, and the method of component by component splitting which integrates the equations with respect to time can be assigned to such methods. These methods can be efficiently realized at the grid level.

1.4.4 The Method of Weak Approximation (Splitting of the Differential Problem)

The method of weak approximation is the method of splitting the equations in differential form. Consider the abstract Cauchy problem

$$B\frac{\partial \varphi}{\partial t} + L\varphi = f \qquad (1.4.7)$$

$$B\varphi = \varphi^0, \quad t = 0 \qquad (1.4.8)$$

in the domain $Q = D \times [0, \bar{t}]$, where

$$L(t) = \sum_{\alpha=1}^{K} L_\alpha(t). \qquad (1.4.9)$$

L and L_α are non-negative operators with

$$\left\| L_\alpha L_\beta \right\| \leq M = \text{const} < \infty \quad \text{and} \quad f = \sum_{\alpha=1}^{K} f_\alpha \qquad (1.4.10)$$

Introduce an auxiliary function φ_α, $\alpha = 1, 2, \ldots, K$, on the interval $\left[t_n, t_{n+1}\right]$ and reduce the problem to the sequence of problems (Marchuk, 1988)

$$\frac{\partial \varphi_\alpha}{\partial t} + L_\alpha \varphi_\alpha = f_\alpha \qquad (1.4.11)$$

$$\varphi_\alpha = \varphi_{\alpha-1}, \alpha = 1, 2, \ldots, K, \qquad (1.4.12)$$

The solutions of these problems approximate the solution of the original problem. Some main properties of the splitting method are:

- The process of the solution of the original problem reduces to the solution of a number of separate subproblems of a simpler structure. The module principle is used for the implementation of the splitting method: a separate module for a separate subproblem. Each module can have its adjoint analog.
- The splitting schemes based on an implicit approximation are unconditionally stable.
- Each module may have an approximation of a different type and different accuracy. It is possible to combine FDM and FEM schemes and approximate separate problems of higher order of accuracy.
- The splitting method is used to solve systems with non-negative operators. This property must be established *a priori* for the problem formulated.

Of major importance in the formulation of a splitting algorithm is the splitting of the original problem into a sum of simple subproblems with non-negative operators. In some cases the choice of this splitting is not trivial and not unique.

1.5 Forward and Adjoint Models

There is a specific feature of oceanological field investigations: the information relevant to the processes in sea basins is expensive to obtain. This is true for both ship and satellite observations. The *in situ* shipboard measurements are localized in space, while the satellite measurements are global but they do not describe the vertical structure of the processes. The data arrays of ocean observations are scanty; real physical processes are complicated and characterized by a wide range of spatial and temporal scales, and they depend on intricate coastal lines and bottom relief.

One of the urgent issues in oceanology is the observational data assimilation in numerical models. The problem of finding the correspondence between observational data and simulation results is important and interesting both from a theoretical and practical standpoint (Marchuk et al., 1987).

The following aspects should be mentioned.

- The updating of observational data arrays in space and time and the four-dimensional data analysis.
- The construction of initial fields with the use of historical archive information and short-term observational data over the preceding period.
- The assimilation of observed data in numerical models and the compilation of unified observed-calculated data arrays.

Mathematical methods based on the theory of adjoint equations and the optimal control theory became very popular as an effective tool for assimilating and analyzing the observational data (Marchuk, 1995). When solving complicated nonlinear data assimilation problems on the basis of these theories a number of problems arise:

- Their physical aspects are often connected with the formulation of the physical problem and with the choice of informative optimization functionals.
- Their mathematical aspects are associated with the investigation of the equivalence of variational problems and differential problems obtained after the transformations, as well as with the examination of their solvability.
- Their computational aspects involve the choice of highly accurate discrete approximations and the development of efficient algorithms for solving discrete problems.

1.5.1 Data Assimilation Problem

One of the most frequently used formulations of data assimilation is as follows.

Consider the equation written in the operator form

$$A\psi = f, \tag{1.5.1}$$

where A is a linear operator acting in the Hilbert space H with the domain $D(A)$; $\psi, f \in H$. For the sake of simplicity we assume that A is either a matrix or an operator corresponding to a certain system of ordinary differential equations (ODE).

Equation (1.5.1) appears, for example, in approximating with respect to space variables within linear boundary value problem (in this case A is a matrix) or within evolutionary systems of partial differential equations (then A is an ODE operator). Assume that there is some arbitrariness in prescribing the right-hand side in (1.5.1). For instance, some components of the vector f are unknown or they are prescribed within a certain error range. This is the so-called problem with incomplete information; it is quite often encountered in practice, including marine dynamics problems.

Assume, however, that there are some measurement data, i.e. some components of the solution vector ψ are known. Let $\overline{\psi}$ denote a somehow redefined solution vector, which has already incorporated the known components.

Consider the problem: find a solution to (1.5.1) that minimizes the functional

$$J(\psi) = (B(\psi - \bar{\psi}), (\psi - \bar{\psi})), \tag{1.5.2}$$

where B is a certain linear positive definite operator acting in H and $(\bullet, \bullet)$ is the inner product in H.

It should be noted that when posing the problem (1.5.1)–(1.5.2) the arbitrariness in prescribing the right-hand side f is used, namely the right-hand side is selected so that the corresponding solution should minimize the functional (1.5.2).

This is the problem of finding the conditional minimum of the functional $J(\psi)$ with the constraints (1.5.1). There are many methods for solving this problem. For example, the problem can be reduced to that of finding the absolute minimum of an extended functional $L(\psi, \lambda_0, \lambda)$

$$L(\psi, \lambda_0, \lambda) = \lambda_0(B(\psi - \bar{\psi}), (\psi - \bar{\psi})) + (A\psi - f, \lambda) \tag{1.5.3}$$

by using the method of Lagrange multipliers. Then one of the numerous methods for finding the absolute minimum of $L(\psi, \lambda_0, \lambda)$ is employed.

Sometimes it is convenient to pose the data assimilation problem in a different manner. Suppose that in (1.5.1) A is a matrix and ψ and f are vectors. Assume that there is no complete information about the right-hand side f but K values of the functional $Jp_k(\psi)$ are known:

$$Jp_k(\psi) = (\psi, p_k), \qquad k = 1, \ldots, K, \tag{1.5.4}$$

where $p_k(\psi)$ are the prescribed "weight" vectors characterizing the subdomains with data measurements. With the use of (1.5.1) the vector f should be restored and then ψ should be found.

To solve this problem, consider K equations adjoint to (1.5.1):

$$A^*\psi^* = p_k, \quad k = 1, \ldots, K, \tag{1.5.5}$$

where A^* is the transpose of the matrix A. From (1.5.1), (1.5.5) follows the identity

$$(f, \psi_k^*) = (p_k, \psi), \quad k = 1, \ldots, K. \tag{1.5.6}$$

The right-hand side in (1.5.6) is known [see (1.5.4)], and ψ_k^* can be found by solving (1.5.5). Thus, f can be found from (1.5.6).

System (1.5.6) for the unknown vector f can be rewritten in the matrix form

$$Lf = F, F = (p_k, \psi), \quad k = 1, \ldots, K, \tag{1.5.7}$$

where L is an $N \times K$ rectangular matrix, N is the dimension of the vector f, and K is the dimension of the vector F. As a rule, in practice holds $N > K$. The solution of (1.5.7) is understood as the pseudo-solution f_1 satisfying the equation

$$L^* L f_1 = L_* F. \tag{1.5.8}$$

1.5.2 Initialization Problem

One of the important problems in the dynamics of shelf seas and oceans is the construction of initial conditions. Consider the equation

$$\frac{\partial \psi}{\partial t} + A\psi = f \tag{1.5.9}$$

$$\psi = \eta \text{ for } t = 0 \tag{1.5.10}$$

where $\psi = \psi(t)$ is an unknown function, A is an operator (generally nonlinear) acting in the Hilbert space H with the definition domain $D(A) \in H, \eta \in H$, and $f = f(t)$ is a known right-hand side. Assume that the initial condition η for (1.5.9) is unknown or is known with a certain error: for example, it is in the neighborhood of $\bar{\psi}^0$. We need to solve (1.5.9) in the absence of information or with incomplete information about the initial condition η.

The absence of information about the initial condition can be compensated by certain additional data on the solution. This is possible, in particular, if there is apriori information about the behavior of the solution on the time interval $0 < t \leq t_1$. Suppose we know the average value of ψ on this interval:

$$\bar{\psi} = \int_0^{t_1} \psi dt.$$

Introduce the functional J:

$$J(\psi) = \frac{1}{2}\int_0^{t_1} (C_1(\psi - \bar{\psi}), (\psi - \bar{\psi}))_H dt$$

$$+\frac{1}{2}\int_0^{t_1} (C_2(\psi(0) - \bar{\psi}^0), (\psi(0) - \bar{\psi}^0))_H dt \tag{1.5.11}$$

where C_1, C_2 are bounded self-adjoint positive semi-definite operators and $(\bullet, \bullet)_H$ is an inner product in H.

The initialization problem can be formulated as follows: find ψ and η, such that they satisfy (1.5.9)–(1.5.10), and the functional (1.5.11) takes the minimum value on the set of solutions to equation (1.5.9). Write this problem as

$$\frac{\partial\psi}{\partial t} + A\psi = f, \text{ for } t \in (0, t_1] \tag{1.5.12}$$

$$\psi = \eta \text{ for } t = 0 \tag{1.5.13}$$

$$J(\psi) = \inf J(\bar{\psi})_{\bar{\eta}\in H}, \tag{1.5.14}$$

where η is the solution of (1.5.9) when $\bar{\psi}(0) = \bar{\eta}$.

Problems in the form (1.5.12)–(1.5.14) were analyzed by Pontryagin et al. (1962), Marchuk et al. (2001b), etc.

The necessary optimality condition reduces (1.5.12)–(1.5.14) to the following boundary value problem of the Pontryagin maximum principle (for functions ψ, ψ^*, η):

$$\frac{\partial\psi}{\partial t} + A\psi = f, \quad t \in (0, t_1], \quad \psi(0) = \eta \tag{1.5.15}$$

$$-\frac{\partial\psi^*}{\partial t} + (A')^*\psi^* + C_1(\psi - \bar{\psi}) = 0, \quad t \in (0, t_1], \quad \psi^*(t_1) = 0 \tag{1.5.16}$$

$$C_2(\eta - \bar{\psi}^0) - \psi^*(0) = 0, \tag{1.5.17}$$

where $\left(A'\right)^*$ is the operator adjoint to the Frechet derivative of A.

One of the approaches to the solution of (1.5.15)–(1.5.17) involves its reduction to a boundary value problem for ψ^*. Therefore, assuming sufficient smoothness of the functions under consideration and eliminating η, ψ by using (1.5.15) and (1.5.17), we arrive at the boundary value problem for the adjoint function ψ^*:

$$\left(\frac{d}{dt}+A\right)C_1^{-1}\left(-\frac{d}{dt}+A'^*\right)\psi^* = \left(\frac{d}{dt}+A\right)\bar{\psi} - f, \quad t \in (0, t_1) \quad (1.5.18)$$

$$C_2C_1^{-1}\left(-\frac{d}{dt}+A'^*\right)\psi^*(0) + \psi^*(0) = 0, \quad \psi^*(t_1) = 0. \quad (1.5.19)$$

The problem (1.5.18)–(1.5.19) is the second-order functional-differential equation with respect to t with two boundary conditions at $t = 0$ and $t = t_1$. Note that when $C_2 = 0$, this is the Dirichlet problem with respect to time, and the introduction of C_2 plays the role of a "penalty" and leads to a problem with the boundary conditions of the third kind. The introduction of C_2 can be considered as a regularization of the variational initialization problem, which leads to a more efficient numerical algorithm for its solution.

It is possible to formulate a number of numerical algorithms on the basis of (1.5.18)–(1.5.19) for solving the initialization problem.

The problem (1.5.18)–(1.5.19) illustrates the physical meaning of the initialization problem as a boundary value problem with respect to time and space. Therefore such a problem is often called the problem of four-dimensional data analysis.

1.6 Advances and Trends in Mathematical Modelling of Marine and Oceanic Currents

Mathematical modelling of oceanic currents is a rather new direction in oceanology. It emerged in the latter half of the 20th century.

This direction was formed and is developing at the interface between several classical sciences: hydrodynamics, the theory of differential equations, and computational mathematics. In recent years two new trends have been evident: the use of the adjoint-equation method and the optimal control method; qualitative studies of attractors of model solutions.

The current state of the art and the trends in the development of mathematical modelling of marine and oceanic currents can be briefly outlined as follows.

- **The formulation of initial boundary value problems and the study of their solvability**: The proof of the global solvability (as $t \to \infty$) of nonlinear differential problems of general circulation, including parameterization of deep convection, may be considered as a major achievement.
- **Numerical methods**: Studies of differential problem solvability show that the projection methods (the Galerkin-type methods) based on the generalized (weak) formulation of problems (including the FEM) are very promising. When using these methods one should overcome the difficulties associated with a distortion of local properties of the solution (monotonicity) and the complexity of the FEM scheme realization. Due attention should be given to a search for efficient and stable time stepping schemes. Here symmetrized implicit splitting schemes and explicit methods with variable time steps (Lebedev, 1997) hold much promise.

- **Variational problems of observational data assimilation and of data processing**: The development of new methods of modelling and analysis of shelf sea and ocean dynamics is stimulated by the fact that input information is scanty. The above problems belong to the family of problems with incomplete informations, which is associated with the absence of initial fields and inexact assignment of the external forcing. Unlike the prediction problem (the Cauchy problem) these problems can be represented as two-point (with respect to time) boundary value problems, their numerical solution is much more complicated. From the mathematical standpoint they are close to ill-posed problems and require a search for efficient regularizations. Here particular attention should be given to further development of the adjoint-equation method and the optimal control method. What is wanted are appropriate functionals with good regularizing properties and efficient numerical methods with small sensitivity to the input parameters.

References

Arakawa A., 1966. Computational design for long term numerical integration of the equations of field motion: Two-dimensional incompressible flow, Part I. *J. Comp. Phys. Res.* 1: 119–163.

Bryan K., 1969. A numerical method for the study of the circulation of the world ocean. *J. Comp. Phys.* 4: 347–376.

Bubnov V. F. and F. V. Kazhikhov, 1971. Existence of a unique solution of the fundamental boundary value problem in the linear theory of oceanic circulation. *Sov. Phys. Dokl.* 16: 429–431.

Burchard H., 2002. *GETM – A General Estuarine Transport Model. Scientific Documentation.* Technical Report EUR 20253 EM, European Commission, 157pp.

Delecluse P. and V. B. Zalesny, 1996. Problems of numerical modelling of equatorial dynamics. *Oceanology* 36(1): 26–42.

Friedrich H. J., 1970. Preliminary results from a numerical multilayer model for the circulation in the North Atlantic. *D. Hydr. Zeit.* 23(4): 145–164.

Grotkop G., 1982. Finite element analysis of long-period water waves. *Comp. Math. Appl. Med. Eng.* 2: 89–112.

Hansen W., 1956. Strömungen in Randmeeren nebst Anwendungen. *Tellus* 3: 283–300.

Hansen W., 1959. Wind und Massenverteilung als Ursache der Meeresströmungen. The Rossby Memorial Volume, Oxford University Press, 102–106.

Ivanov Yu. A., K. V. Lebedev, and A. S. Sarkisyan, 1997. Generalized hydrodynamic adjustment method (GHDAM). *Izv Ross. Akad Nauk, Fit. Atmos. Okeana* 33(6): 812–818 (in Russian).

Kazantsev Ch., S. N. Moshonkin, and V. B. Zalesny, 1998. Mathematical modelling of the global ocean dynamics: solvability, numerical algorithm, calculations. In: *Proc. of 4-th Conf. 'Variability and Predictability of Atmospheric and Oceanic Circulations'*. Moscow, Russia, 81–95.

Kordzadze A. A., 1982. *Mathematical Questions for Solving the Problem of Ocean Dynamics.* Novosibirsk, Sib. Branch, USSR Academy of Science (in Russian), 148 pp.

Kreiss H.-O., 1959. Über die Lösung des Cauchy Problems für lineare partielle Differentialgleichungen mit Hilfe von Differenzen gleichungen. *Acta Math.* 101: 179–199.

Kuzin V. I., 1985. *The Finite Element Method in Modelling of Oceanic Processes.* Novosibirsk, Sib. Branch, USSR Academy of Science (in Russian), 189 pp.

Lebedev V. I., 1997. Explicit difference schemes with variable time steps for solving stiff systems of equations. In: *Numerical Analysis and its Applications.* Lecture Notes in Computer Science 1196. Springer, 274–283.

Lilly D. K., 1965. On the computational stability of numerical solutions of time-dependent nonlinear geophysical fluid dynamics problems. *Monthly Weather Rev*. 93: 11–26.
Lions J. L., R. Temam, and S. Wang, 1992. On the equations of the large scale ocean. *Nonlinearitv* 5: 1007–1053.
Manabe S. and K. Bryan, 1969. Climate and the ocean circulation. *Monthly Weather Rev*. 97(11): 739–774.
Marchuk G. I., 1969. On the numerical solution of the Poincaré problem for oceanic circulations *Dokl. Akad. Nauk SSSR* 185(5): 1041–1044 (in Russian).
Marchuk G. I., 1980. *Methods of Computational Mathematics*. Nauka, Moscow, 536 pp.
Marchuk G. I., 1988. *Splitting-Up Methods*. Nauka, Moscow, 263 pp.
Marchuk G. I., 1995. *Adjoint Equations and Analysis of Complex Systems*. Kluwer Academic Publishers, Dordrecht, 466 pp.
Marchuk G. I. and G. V. Demidov, 1966. Theorem for existence of a solution for the short-range weather prediction problem. *Dokl. Akad. Nauk SSSR* 7(5): 1006–1008 (in Russian).
Marchuk G. I., J. Sündermann, and V. B. Zalesny, 2001a. Mathematical modelling of marine and ocean currents. *Russ. J. Numer. Anal. Math. Modelling* 16(4): 331–362.
Marchuk G. I., V. P. Dymnikov, and V. B. Zalesny, 1987. *Mathematical Models in Geophysical Fluid Dynamics and Numerical Methods of Their Realization*. Leningrad, Gidrometeoizdat (in Russian), 296 pp.
Marchuk G. I., V. P. Shutyaev, and V. B. Zalesny, 2001b. Approaches to the solution of data assimilation problems. In: Menaldi J. L. et al. (eds.), *Optimal Control and Partial Differential Equations*. IOS Press, Amsterdam, 489–497.
Mesinger F. and A. Arakawa, 1982. *Numerical Methods Used in Atmospheric Models*. GARP Publ. Ser. 17, I, WMO, 64pp.
Munk W. H., 1950. On the wind-driven ocean circulation. *J. Meteorol*. 7: 79–93.
Ovsyannikov L. V., 1966. *Theorem on a Unique Solution for Linearized System of Ocean Dynamics Equations*. Preprint of the Novosibirsk State University (in Russian), 12 pp.
Pontryagin L. S., V. G. Boltyanskii, R. V. Gamkrelidze, and E. F. Mishenko, 1962. *The Mathematical Theory of Optimal Processes*. John Wiley, New York, 360 pp.
Rolinski S., J. Segschneider, and J. Sündermann, 2001. Long-term propagation of tailings from deep-sea mining under variable conditions by means of numerical simulations. *Deep Sea Res*. II: 3465–3485.
Sarkisyan A. S., 1954. Calculation of steady-state wind currents in the ocean. *Izv. Akad. Nauk SSSR, Ser. Geofiz*. 6: 554–561 (in Russian).
Sarkisyan A. S., 1961. On the role of the density advection by wind in dynamics of baroclinic ocean. *Izv. Ross. Acad. Nauk SSSR* 9: 1396–1407.
Sarkisyan A. S., 1962. On dynamics of wind-driven currents in a baroclinic ocean. *Oceanologia*. II(3): 393–409.
Sarkisyan A. S., 1969a, *Theory and computation of ocean currents*, U.S. Dept. of Commerce and the NSF, Washington, DC, 90 pp.
Sarkisyan A. S., 1969b. Deficiencies of barotropic models of ocean circulation. *Izv. Acad. Nauk SSSR, Ser. Fiz. Atmos. Okeana* 5(8): 818–835 (AGU English translation).
Sarkisyan A. S. and Ju. L. Demin, 1983. A semidiagnostic method of sea currents calculation. *Large-Scale Oceanographic Experiments in the WCRP* 2(1): 210–214.
Sarkisyan A. S. and V. B. Zalesny, 2000. Splitting-up method and adjoint equation method in the ocean dynamics problem. *Russ. J. Numer. Anal. Math. Modelling* 15(3–4): 333–347.
Sarkisyan A. S. and V. F. Ivanov, 1971. Joint effect of baroclinity and bottom topography as an important factor in sea dynamics. *Izv. Acad. Nauk SSSR, Ser. Fiz. Atmos. Okeana* 2(6): 818–835 (AGU English translation).
Schtockman V. B., 1946. Equations of full flow fields induced by wind in nonhomogeneous sea. *Dokl Akad. Nauk SSSR* 54(5): 403–406 (in Russian).
Segschneider J. and J. Sündermann, 1997. Response of a global ocean circulation model to real time forcing and implications to Earth's rotation. *J. Phys. Ocean*. 27: 2370–2380.

Segschneider J. and J. Sündermann, 1998. Simulating large scale transport of suspended matter. *J. Mar. Syst.* 14: 81–97.

Stommel H., 1948. The westward intensification of wind-driven ocean currents. *Trans. Amer. Geophys. Union* 29: 202–206.

Sündermann J., 1994. *Circulation and Contaminant Fluxes in the North Sea.* Springer, 654pp.

Sverdrup H. U., 1947. Wind-driven currents in a baroclinic ocean; with application to the equatorial currents of the Eastern Pacific. *Proc. Nat. Acad. Sci. Wash.* 33(11): 318–326.

Thomas M., J. Sündermann, and E. Maier-Reimer, 2001. Consideration of ocean tides in an OGCM and implications for polar motion. *Geophys. Res. Lett.* 28(12): 2457–2460.

Yanenko N. N., 1967. *Method of Fractional Steps for Solution of Multidimentional Problems of Mathematical Physics.* Nauka, Novosibirsk (in Russian), 195 pp.

Zalesny V. B., 1997. Variability and equilibrium states of the world ocean circulation. *Russ. J. Numer. Anal. Math. Modelling* 12(6): 547–567.

Chapter 2
Simple Linear Models for Diagnostic Calculation of Ocean Climate Characteristics

Abstract We present several simple models for diagnostic calculations of ocean climate characteristics. Their qualitative analysis has explained many aspects of ocean/sea dynamics. Namely, they show intrinsic interconnections between strong horizontal density gradients and intense jet-like currents. Moreover, both of them strongly depend on the basins' bottom topography. It is shown that the JEBAR (Joint Effect of Baroclinicity and Bottom Relief) is an absolutely necessary factor and should accurately be taken into account, directly or indirectly, in any model. Finally, they show that any further simplification of the models brings unrealistic and even contradictory results.

Additionally, we discuss the divergence equation for the sea surface height (SSH). Based on it, we show that wind-stress divergence may be considered as driving force as well. This version of SSH equation is valid for the equatorial area too. It helps to explain why the equatorial undercurrents exist.

Keywords Simple models · Equatorial undercurrent · JEBAR · Model alternatives

2.1 Basic Equations

We start with general formulations of World Ocean climatic models followed by some simplifications.

Two versions of system of equations in spherical coordinates and several simplified ones in Cartesian coordinates are presented in Chapter 1. We start with one comparatively general system of equations in Cartesian coordinates.

The momentum equations take the following form:

$$\frac{\partial u}{\partial t} + L(u) - fv + \frac{1}{\rho_0}\frac{\partial p}{\partial x} = \frac{\partial}{\partial z}K_M\frac{\partial u}{\partial z} + A_M\Delta u \tag{2.1.1}$$

$$\frac{\partial v}{\partial t} + L(v) + fu + \frac{1}{\rho_0}\frac{\partial p}{\partial y} = \frac{\partial}{\partial z}K_M\frac{\partial v}{\partial z} + A_M\Delta v \tag{2.1.2}$$

A.S. Sarkisyan, J.E. Sündermann, *Modelling Ocean Climate Variability*,
DOI 10.1007/978-1-4020-9208-4_2, © Springer Science+Business Media B.V. 2009

The equations for temperature and salinity look as follows:

$$\frac{\partial T}{\partial t} + L(T) = \frac{\partial}{\partial z}\frac{K_T}{\delta}\frac{\partial T}{\partial z} + A_T \Delta T \tag{2.1.3}$$

$$\frac{\partial S}{\partial t} + L(S) = \frac{\partial}{\partial z}\frac{K_S}{\delta}\frac{\partial S}{\partial z} + A_S \Delta S \tag{2.1.4}$$

where the advection operator of some variable μ is

$$L(\mu) = \frac{\partial u\mu}{\partial x} + \frac{\partial v\mu}{\partial y} + \frac{\partial w\mu}{\partial z} \tag{2.1.5}$$

The operator Δ is the two dimensional Laplacian, i.e. $\Delta = \frac{\partial^2}{\partial x^2} + \frac{\partial^2}{\partial y^2}$, and the Coriolis parameter $f = 2\omega \sin\varphi$.

The equation of continuity (on the assumption of sea water incompressibility) commonly used to determine w has the simple form

$$L(1) = 0 \tag{2.1.6}$$

Finally, simple equation of statics and state are added

$$\frac{\partial p}{\partial z} = -\rho_1 g \tag{2.1.7}$$

$$\rho_1 = \rho(T, S, p) + \rho_0 \tag{2.1.8}$$

where p is the pressure, ρ – the density anomaly.

The latter equation is usually represented by a simple algebraic expression relating water pressure and salinity anomalies to the anomalies of potential temperature and potential density. Such are, for example, the UNESCO formula (Fofonoff and Millard, 1983) or the formula of Bryden et al. (1999).

The system of equations (2.1.1)–(2.1.4), (2.1.6)–(2.1.8) is not closed. To close it different analytical formulations of subgrid-scale processes and boundary conditions will be made in this and the following chapters.

2.2 Simple Diagnostic Models for Ocean Characteristics Calculation

2.2.1 Geostrophic Flow

Assuming characteristic values it is easy to show that for large-scale flows the main terms of equations (2.1.1) and (2.1.2) are the accelerations caused by horizontal pressure gradients and the Coriolis effect. So the first order approximation is

$$u = -\frac{1}{\rho_0 f}\frac{\partial p}{\partial y}, \; v = \frac{1}{\rho_0 f}\frac{\partial p}{\partial x}, \tag{2.2.1}$$

which are the formulae for geostrophic velocity calculations.

First order validity holds also for (2.1.7). Denoting the sea surface atmospheric pressure anomaly by p_α and the sea surface height (SSH) by ζ_1, we get by integration

$$p = p_\alpha + \rho_0 g \zeta_1 + \int_z^0 g\rho dz \tag{2.2.2}$$

or

$$p = \rho_0 g \zeta + \int_z^0 g\rho dz \tag{2.2.3}$$

where

$$\zeta = \zeta_1 + \frac{p_\alpha}{\rho_0 g} \tag{2.2.4}$$

For practical purposes we can often assume: $\zeta \approx \zeta_1$.

Assuming that there is a depth H_1 in the ocean where the horizontal pressure gradients are negligible (the so-called "zero surface" or reference level), one gets from formula (2.2.3)

$$\zeta^d = -\frac{1}{\rho_0}\int_{-H_1}^0 \rho dz \text{ and, consequently } p^d = -g\int_{-H_1}^z \rho dz \tag{2.2.5}$$

Using (2.2.1) one comes to formulae

$$u = \frac{1}{\rho_0 f}\frac{\partial}{\partial y}\int_{-H_1}^z g\rho dz, \; v = -\frac{1}{\rho_0 f}\frac{\partial}{\partial x}\int_{-H_1}^z g\rho dz \tag{2.2.6}$$

By diagnostic calculations we mean calculations of flow velocity components from specified density anomalies. Formulae (2.2.6) represent the simplest possible model for direct calculations in extraequatorial ocean areas. This model type is called dynamic or reference level method. The model was presented more than a century ago (Sandström and Helland-Hansen, 1903).

Actually, for the calculations of this simplest type of geostrophic currents the horizontal pressure gradients and the "reference level" $z = -H_1$ have to be specified.

2.2.2 *Simple Linear Model for Diagnostic Calculation in the Ocean Equatorial Region (The Model* $\mathbf{D}_1$*)*

2.2.2.1 Equatorial Undercurrents

Consider the following simplified version of the momentum balance equations (2.1.1) and (2.1.2)

$$K_M \frac{\partial^2 u}{\partial z^2} + f v = \frac{1}{\rho_0} \frac{\partial p}{\partial x} \tag{2.2.7}$$

$$K_M \frac{\partial^2 v}{\partial z^2} - f u = \frac{1}{\rho_0} \frac{\partial p}{\partial y} \tag{2.2.8}$$

or

$$K_M \frac{\partial^2 M}{\partial z^2} - i f M = F_1 \tag{2.2.9}$$

where

$$M = u + iv \text{ and } F_1 = \frac{1}{\rho_0} \left[\frac{\partial p}{\partial x} + i \frac{\partial p}{\partial y} \right] \tag{2.2.10}$$

The boundary conditions for M are the following:

$$\text{at } z = 0 : \rho_0 K_M \frac{\partial M}{\partial z} = -\left(\tau_x + i\tau_y\right) \tag{2.2.11}$$

$$\text{at } z = -H : \frac{\partial M}{\partial z} \equiv 0 \tag{2.2.12}$$

$$\text{or } \rho_0 K_M \left[\frac{\partial u}{\partial z}, \frac{\partial v}{\partial z} \right] = \left(\tau_x^{-H}, \tau_y^{-H}\right), \tag{2.2.12'}$$

where $\tau_x{}^{-H}$, $\tau_y{}^{-H}$ are the bottom stress components.

In the case F_1 depends also on z, the solution of equation (2.2.9) with boundary conditions (2.2.11), (2.2.12) (2.2.12′) is of the form (Sarkisyan, 1977):

$$M = \frac{sh\,[(1+i)\alpha(H+z)]}{(1+i)\alpha k_m \rho_0 ch(1+i)\alpha H}(\tau_x + i\tau_y) - \\ - \frac{k_m}{2(1+i)\alpha} \left[\int_z^0 F_1 e^{(1+i)\alpha(\xi+z)} d\xi + \int_{-H}^0 F_1 e^{-(1+i)\alpha(\xi+z)} d\xi \right]$$

$$- \frac{e^{(1+i)\alpha H)}}{2(1+i)\alpha ch\,[(1+i)\alpha H]} \int\limits_{z}^{0} F_1 ch\,[(1+i)\alpha(\xi + z)] \cdot d\xi$$

$$- \frac{k_m}{2(1+i)\alpha ch\,[(1+i)\alpha H]} \int\limits_{z}^{0} F_1 sh\,[(1+i)\alpha(H - \xi + z)] \cdot d\xi \qquad (2.2.13)$$

where

$$\alpha = \sqrt{\pm \frac{f}{2K_M}} \qquad (2.2.14)$$

In formula (2.2.14) the upper sign corresponds to the northern hemisphere and the lower one to the southern one.

Let us make qualitative analysis of formula (2.2.13) for a narrow equatorial zone (the equator +1–2°). In the equator environs, all expressions like αH, $\alpha(\xi + z)$ etc., are infinitesimal quantities tending to zero when approaching the equator. We can therefore decompose all the exponentials on the right-hand side of formula (2.2.13) into degrees of the corresponding arguments and retain only the small quantities of the first order, that is, assume $\exp[\pm(1+i)\alpha H] \approx 1 \pm (1+i)\alpha H$. After this substitution and some simple manipulations part of expression (2.2.13) takes the final form, but some terms of it at the equator become singular. Expanding according to l'Hospital's rule and considering $f \to 0$, we finally come to the following simple formula:

$$M = \frac{H+z}{K_M \rho_0}\left(\tau_x + i\tau_y\right) - \frac{1}{K_M \rho_0}\left[\int\limits_{z}^{0} (z-\xi)\, F_1 d\xi + \int\limits_{-H}^{0} (H+\xi)\, F_1 d\xi\right] \qquad (2.2.15)$$

or

$$u = \frac{H+z}{K_M \rho_0}\tau_x - \frac{1}{K_M \rho_0}\left[\int\limits_{z}^{0} (z-\xi)\frac{\partial p}{\partial x} d\xi + \int\limits_{-H}^{o} (H+\xi)\frac{\partial p}{\partial x} d\xi\right] \qquad (2.2.16)$$

$$v = \frac{H+z}{K_M \rho_0}\tau_y - \frac{1}{K_M \rho_0}\left[\int\limits_{z}^{0} (z-\xi)\frac{\partial p}{\partial y} d\xi + \int\limits_{-H}^{o} (H+\xi)\frac{\partial p}{\partial y} d\xi\right]$$

Formulae (2.2.16) can serve for the approximate calculations of u and v at the equator and in its narrow environs. At the equator, they satisfy equation (2.2.9) with $f = 0$ and boundary conditions (2.2.11) and (2.2.12) (2.2.12′). Poleward from the equator these formulae become inaccurate and at some critical latitude they are unsuitable.

It is seen from formulae (2.2.16) that there are neither boundary layers nor an Ekman spiral at the equator (on Ekman theory see below in Section 2.2.3). Here pure drift currents coincide with wind direction and the effect of bottom friction encompass the whole ocean column.

Let us investigate the direction of gradient currents at the equator. Setting

$$\frac{\partial p}{\partial y} = 0, \frac{\partial p}{\partial x} = \left(\frac{\partial p}{\partial x}\right)_0 e^{\gamma z}, \gamma = const > 0 \tag{2.2.17}$$

for the sake of simplicity, and using formulae (2.2.16) and (2.2.17) we obtain for the components of the gradient part of the flow velocity

$$u_g^{(0)} = -\frac{1}{K_M \rho_0 \gamma^2}\left[\gamma (H + z) + e^{\gamma z} + e^{-\gamma H}\right]\left(\frac{\partial p}{\partial x}\right)_0 < 0, v_g^{(0)} = 0 \tag{2.2.18}$$

We assume that formula (2.2.17) is valid not only at the equator but also in a narrow equatorial zone. Then from geostrophic relations and the assumed simplification we obtain the following.

For the Northern hemisphere:

$$u_g{}^N = 0, v_g{}^N = -\frac{e^{\gamma z}}{2\omega \sin\varphi \rho_0}\left(\frac{\partial p}{\partial x}\right)_0 < 0 \tag{2.2.19}$$

For the Southern hemisphere:

$$u_g{}^S = 0, v_g{}^S = -\frac{e^{\gamma z}}{2\omega \sin(-\varphi)\rho_0}\left(\frac{\partial p}{\partial x}\right)_0 = -\frac{e^{\gamma z}}{2\omega \sin\varphi\rho_0}\left(\frac{\partial p}{\partial x}\right)_0 > 0 \tag{2.2.20}$$

In formulae (2.2.17–2.2.20) the lower index “0” means the sea surface and the upper one the equator.

Thus, in the Northern hemisphere a gradient current deviates to the right from the pressure gradient, i.e. is directed to the South, and in the Southern hemisphere, to the left, i.e. to the north. At the equator a gradient current takes the direction of the pressure gradient vector. This can be expressed schematically as in Fig. 2.1.

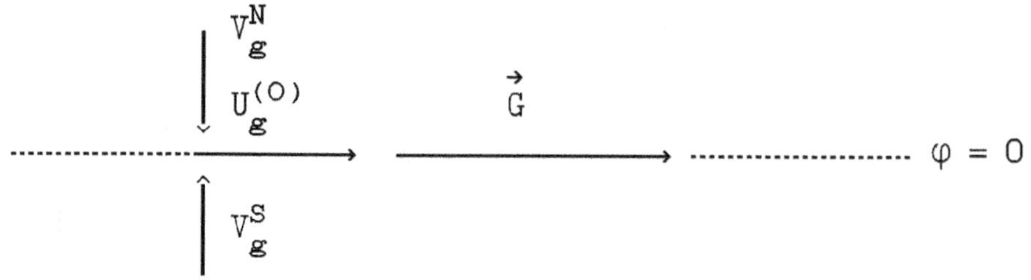

Fig. 2.1 Schematic picture of the equatorial countercurrent (Sarkisyan, 1977)

Due to the Easterly winds, in equatorial regions the sea level is higher at the West coast than at the East one, therefore a gradient current is directed from West to East. At the sea surface and the very surface layers the wind-driven current dominates the gradient currents, so, the current coincides with the wind direction, that is, from East to West. With increasing depth the drift components of the current fade more rapidly than do gradient components. At some depth the gradient current overcomes the wind drift and the total current is directed to the East. That is why the equatorial undercurrents exist.

Thus, the qualitative analysis of momentum balance equations gives explanation of the Equatorial Undercurrents.

2.2.2.2 The Model D_1

Let us now integrate equations (2.2.7) and (2.2.8) with respect to z from $-H$ to 0, and take into account boundary conditions (2.2.11), (2.2.12"). We obtain

$$\frac{1}{\rho_0}\left(\tau_x - \tau_x^{-H}\right) + \int_{-H}^{0}\left(fv - \frac{1}{\rho_0}\frac{\partial p}{\partial x}\right)dz = 0 \tag{2.2.21}$$

$$\frac{1}{\rho_0}\left(\tau_y - \tau_y^{-H}\right) - \int_{-H}^{0}\left(fu + \frac{1}{\rho_0}\frac{\partial p}{\partial y}\right)dz = 0 \tag{2.2.22}$$

By applying the divergence operator together with condition (2.2.12) (2.2.12′), we come to the following equation

$$\frac{1}{\rho_0}\left(\frac{\partial p_{-H}}{\partial x}\frac{\partial H}{\partial x} + \frac{\partial p_{-H}}{\partial y}\frac{\partial H}{\partial y}\right) = \int_{-H}^{0}\frac{\partial}{\partial x}\left(fv - \frac{1}{\rho_0}\frac{\partial p}{\partial x}\right)dz -$$
$$-\int_{-H}^{0}\frac{\partial}{\partial y}\left(fu + \frac{1}{\rho_0}\frac{\partial p}{\partial y}\right)dz - \frac{1}{\rho_0}\left[div\left(\vec{\tau} - \vec{\tau}^{(-H)}\right)\right] \tag{2.2.23}$$

Suppose the right-hand side of (2.2.23) is somehow specified, then it is an equation for determining the bottom pressure anomaly p_{-H}.

Keeping in mind that the flow velocity is close to geostrophy and supposing that $\vec{\tau}$ and $\vec{\tau}^{(-H)}$ are small with respect to the baroclinic pressure gradients. So whole right-hand side of (2.2.23) is small as a result of small differences of big values. We conclude that the bottom pressure torque (BPT) i.e. the left-hand side of (2.2.23) is a small quantity in comparison to $\frac{1}{\rho_0}\frac{\partial p}{\partial x}$ or $\frac{1}{\rho_0}\frac{\partial p}{\partial y}$.

Let us now split the BPT into two terms by using the equation of statics (2.2.3).

$$p_{-H} = \rho_0 g\zeta + g\int_{-H}^{0} \rho dz \tag{2.2.24}$$

$$\frac{\partial\zeta}{\partial x}\frac{\partial H}{\partial x} + \frac{\partial\zeta}{\partial y}\frac{\partial H}{\partial y} = \underbrace{-\frac{1}{\rho_0}\int_{-H}^{0}\left(\frac{\partial\rho}{\partial x}\frac{\partial H}{\partial x} + \frac{\partial\rho}{\partial y}\frac{\partial H}{\partial y}\right)dz}_{JEBAR-2^*} +$$
$$+\frac{1}{g}\int_{-H}^{0}\frac{\partial}{\partial x}\left(fv - \frac{1}{\rho_0}\frac{\partial p}{\partial x}\right)dz - \frac{1}{g}\int_{-H}^{0}\frac{\partial}{\partial y}\left(fu + \frac{1}{\rho_0}\frac{\partial p}{\partial y}\right)dz - \tag{2.2.25}$$
$$+\frac{1}{\rho_0 g}div\left[\vec{\tau} - \vec{\tau}^{(-H)}\right]$$

On the right-hand side of (2.2.25) appears now JEBAR-2, which, in contrast to right-hand terms of (2.2.23), is not small in comparison to $\frac{1}{\rho_0}\frac{\partial p}{\partial x}$ or $\frac{1}{\rho_0}\frac{\partial p}{\partial y}$. Due to baroclinic compensation p_H is small in comparison to any of the two terms of the right-hand side of (2.2.24). The splitting reveals the available potential energy, there is a big source for kinetic energy in the right-hand side of (2.2.25) – that is JEBAR-2.

The other two integrals of the right-hand side of (2.2.25) are smaller than JEBAR-2 because of "geostrophic compensation".

As mentioned above the wind and bottom stresses are smaller than the pressure gradients in a baroclinic ocean.

We apply the splitting procedure to the right-hand side of equation (2.2.25): express the pressure anomaly by the formula (2.2.3), transform double integrals by formula $\int_{-H}^{0}\int_{z}^{0}\rho d\zeta dz = \int_{-H}^{0}(H+z)\rho dz$, etc. Finally we come to the following equation for ζ

$$\underbrace{\Delta\zeta}_{I} + \underbrace{\frac{1}{H}\left(\frac{\partial H}{\partial x}\frac{\partial\zeta}{\partial x} + \frac{\partial H}{\partial y}\frac{\partial\zeta}{\partial y}\right)}_{II} = \underbrace{-\frac{1}{\rho_0 H}\int_{-H}^{0}(H-z)\Delta\rho dz}_{I} -$$
$$\underbrace{-\frac{1}{\rho_0 H}\left[\frac{\partial H}{\partial x}\int_{-H}^{0}\frac{\partial\rho}{\partial x}dz + \frac{\partial H}{\partial y}\int_{-H}^{0}\frac{\partial\rho}{\partial y}dz\right]}_{\substack{II \\ JEBAR-2}} - \frac{\beta}{gH}S_x - \tag{2.2.26}$$

---

*JEBAR-2 another version of the Joint Effect of Baroclinicity and Bottom Relief (JEBAR), see also below the vorticity equation (2.2.46)

$$-\frac{f}{gH}\left(\frac{\partial S_y}{\partial x}-\frac{\partial S_x}{\partial y}\right)+\frac{1}{gH\rho_0}\left[div\vec{\tau}-div\vec{\tau}^{(-H)}\right]$$

where the operator Δ is the Laplacian $\Delta=\frac{\partial^2}{\partial x^2}+\frac{\partial^2}{\partial y^2}$, $S_x=\int\limits_{-H}^{0}udz$, $S_y=\int\limits_{-H}^{0}vdz$.

In the right-hand side of (2.2.26) the β-effect is small compared with terms, containing the gradient of ρ, because this is the divergence equation; the role of β-effect in different versions of vorticity equation will be discussed in following sections.

Near the equator f is small so the terms with S_x and S_y may be taken by successive approximations. The main terms are the groups I and II in the right-hand side of (2.2.26), they are sources of available potential energy accumulated by ocean during centuries. This energy never diminishes because the ocean regenerates it due to the sun.

The terms with ζ are generated by the first term of the right-hand side of (2.2.24), the ones with ρ – by the second terms. The terms caused by the same physical mechanism are denoted by the same roman number I, II on both sides of (2.2.26).

We need boundary condition to solve the equation (2.2.26). We assume that the flow velocity components are specified at the side boundaries of a narrow equatorial belt. Function ζ does not depend on z, so we have to specify only $S_x\equiv S_x^b$ and $S_y\equiv S_y^b$ at the side of boundaries.

We use the relations (2.2.21) and (2.2.22) for this purpose. After splitting the pressure p by formula (2.2.3) and some simple transformations we get:

$$\frac{\partial\zeta}{\partial x}=-\frac{1}{H\rho_0}\int\limits_{-H}^{0}(z+H)\frac{\partial\rho}{\partial x}dz+\frac{f}{gH}S_y^b+\frac{\tau_x}{\rho_0 gH}-\frac{\tau_x^{-H}}{\rho_0 gH} \tag{2.2.27}$$

$$\frac{\partial\zeta}{\partial y}=-\frac{1}{H\rho_0}\int\limits_{-H}^{0}(z+H)\frac{\partial\rho}{\partial y}dz-\frac{f}{gH}S_x^b+\frac{\tau_y}{\rho_0 gH}-\frac{\tau_y^{-H}}{\rho_0 gH}$$

By using again (2.2.3) equation (2.2.16) can be transformed into the following form:

$$u=\frac{H+z}{K_M\rho_0}\tau_x+\frac{g}{2K_M}\left(z^2-H^2\right)\frac{\partial\zeta}{\partial x}+$$

$$+\frac{g}{K_M\rho_0}\left[\int\limits_{z}^{0}(z-\xi)\int\limits_{\xi}^{0}\frac{\partial\rho}{\partial x}dzd\xi-\int\limits_{-H}^{0}(H-\xi)\int\limits_{\xi}^{0}\frac{\partial\rho}{\partial x}dzd\xi\right] \tag{2.2.28}$$

$$v = \frac{H+z}{K_M\rho_0}\tau_y + \frac{g}{2K_M}\left(z^2 - H^2\right)\frac{\partial\zeta}{\partial y} +$$

$$+\frac{g}{K_M\rho_0}\left[\int_z^0 (z-\xi)\int_\xi^0 \frac{\partial\rho}{\partial y}dzd\xi - \int_{-H}^0 (H-\xi)\int_\xi^0 \frac{\partial\rho}{\partial y}dzd\xi\right]$$

Thus, the model is formulated. The equations can be solved by the following method of successive approximations. As a first approximation, equations (2.2.26), (2.2.27), are solved numerically omitting all terms with flow velocity components.

The first approximation of flow velocity and bottom stress components can be achieved from (2.2.28). Inserting it into right-hand side of equation (2.2.26) we obtain the second approximation of ζ, etc. In Section 2.2.5 below we shall dwell on peculiarities of using equations (2.2.27) to define ζ at side boundaries. Note that equation (2.2.26) and boundaries relations for defining ζ at side (2.2.27) are valid for extraequatorial regions as well, while formulae (2.2.28) are valid only for a very narrow equatorial belt. The vertical component of flow velocity usually is obtained from equation of continuity.

An essential drawback of formulae (2.2.28) is the dependence of u and v on the coefficient of turbulence K_M the value of which is indefinite. Still, these formulae are useful for qualitative conclusions and for approximate quantitative calculations. One of such conclusions has been made above when explaining the cause of an equatorial undercurrent.

2.2.3 *Simple Linear Model for Diagnostic Calculations in the Ocean Extraequatorial Regions (The Model* $\mathbf{D}_1{}'$*)*

To construct the model needed we return to equation (2.2.9). Assume that the pressure gradients on the right-hand side of (2.2.9) do not depend on z. In this case the equation can easily be resolved relative to the complex velocity (Sarkisyan, 1977):

$$M = \frac{1}{f\rho_0}\left(i\frac{\partial p}{\partial x} - \frac{\partial p}{\partial y}\right) + c_1 e^{-\alpha(1+i)z} + c_2 e^{\alpha(1+i)z} \qquad (2.2.29)$$

Expression (2.2.29) can be used for the solution of a few auxiliary problems.

Consider the wind vertical distribution in the nearwater boundary layer of the atmosphere. Place the origin of coordinates at the sea surface, direct the z-axis vertically upward, substitute the sea level atmospheric pressure p_a for p denote the corresponding atmospheric characteristics by primes and, assume the following boundary conditions: at the ocean surface the wind velocity is equal to zero, and it should be limited when $z \to \infty$, i.e. $c_2 = 0$.

As a result from (2.2.29) it is easily obtained

$$u' + iv' = \frac{1}{f\rho_0'} \left(i\frac{\partial p_a}{\partial x} - \frac{\partial p_a}{\partial y} \right) \left(1 - e^{-(1+i)\alpha' z} \right) \quad (2.2.30)$$

where $\alpha' = \sqrt{f/2K'_M}$, K'_M is the coefficient of the vertical turbulent viscosity of the air, and ρ_0 now got the meaning of the mean air density at the sea level. Formula (2.2.30) means that the velocity grows exponentially from 0 to geostrophic wind velocity at the top of the nearwater atmospheric boundary layer. We are only interested in the tangential wind stress at the sea level. By differentiating (2.2.30) with respect to z and using boundary condition (2.2.11) we obtain:

$$\tau_x + i\tau_y = \rho_0 K'_M \left. \frac{\partial M'}{\partial z} \right|_{z=0} = -\frac{K'_M}{f} \alpha' (1+i) \left(i\frac{\partial p_a}{\partial x} - \frac{\partial p_a}{\partial y} \right) \quad (2.2.31)$$

or

$$\tau_x = \frac{1}{2\alpha'} \left(\frac{\partial p_a}{\partial x} + \frac{\partial p_a}{\partial y} \right), \tau_y = \frac{1}{2\alpha'} \left(-\frac{\partial p_a}{\partial x} + \frac{\partial p_a}{\partial y} \right) \quad (2.2.32)$$

The formulae (2.2.32) may be used to calculate wind stress components by sea level atmospheric pressure.

Now we apply (2.2.29) to describe the vertical distribution of currents in the ocean near surface boundary layer. We take again the coordinate origin at the sea surface, and direct the z-axis vertically upward, apply condition (2.2.11) at the sea surface and the condition of limited flow velocity at the lower boundary of the friction layer.

Thus, we obtain:

$$M = \frac{1}{f\rho_0} \left(i\frac{\partial p}{\partial x} - \frac{\partial p}{\partial y} \right)_{z=0} + \frac{(1-i)}{2\rho_0 K_M \alpha} \left(\tau_x + i\tau_y \right) e^{(1+i)\alpha z} \quad (2.2.33)$$

The ocean pressure gradients in this expression are the values at the sea surface. The formula itself is meaningful only in a thin upper ocean boundary layer about 20–30 m thick; the second term of (2.2.33) describes the Ekman spiral. Particularly, we use this formula when determining the vertical turbulent viscosity in the ocean nearsurface friction or Ekman layer:

$$K_M \left. \frac{\partial^2 M}{\partial z^2} \right|_E = \frac{\alpha}{(1+i)} \left(\tau_x + i\tau_y \right) e^{(1+i)\alpha z} \quad (2.2.34)$$

Let us dwell in brief on the Ekman theory. In the case of $\rho \equiv 0$ pressure anomaly p does not depend on z and the system (2.2.7), (2.2.8) becomes equations of the Ekman theory. Indeed, any model of ocean circulation transforms into the Ekman model if one puts $\rho \equiv 0$. For example, equation (2.2.48) presented below is the equation for sea surface topography of the Ekman model in the case $\rho \equiv 0$. In scientific literature the Ekman spiral of pure wind-driven currents M_d, i.e. the second group of the terms in formula (2.2.33) is more popular. After simple transformation it takes the form

$$M_d = \left[\left(\tau_x + \tau_y\right)\cos\alpha z - \left(\tau_y - \tau_x\right)\sin\alpha z\right] + \\ + i\left[\left(\tau_x + \tau_y\right)\sin\alpha z + \left(\tau_y - \tau_x\right)\cos\alpha z\right]e^{\alpha z} \tag{2.2.35}$$

At the surface $M_d = \left(\tau_x + \tau_y\right) + i\left(\tau_y - \tau_x\right)$.

It means that in the Northern hemisphere the wind drift declines by 45° to the right from wind direction whereas in the Southern hemisphere the declination is to the left by the same angle. According to (2.2.35) the flow velocity vector revolves many times with depth and forms the well-known Ekman spirals. Simultaneously due to $e^{\alpha z}$ the module of the flow velocity decays with depth and at depth $H_e = -\frac{\pi}{\alpha}$ it practically vanishes. The value $H_e = -\pi\sqrt{\frac{K_M}{\omega \sin\varphi}}$ is called the depth of Ekman friction layer.

Returning gain to formula (2.2.24), finally, we apply the same procedure to the bottom boundary layer, using two types of boundary conditions for the ocean bottom: (2.2.12) and (2.2.12′) and, consequently, obtaining two versions of solution.

In this case we present the general solution (2.2.29) in the form

$$M = \frac{1}{f\rho_0}\left(i\frac{\partial p}{\partial x} - \frac{\partial p}{\partial y}\right) + c_1 e^{-\alpha(1+i)(H+z)} + c_2 e^{\alpha(1+i)(H+z)}$$

where c_2 should again be equal to zero to keep flow velocity limited for $z \to \infty$.

The final form of solution is as follows

$$M = \frac{1}{f\rho_0}\left(i\frac{\partial p}{\partial x} - \frac{\partial p}{\partial y}\right)_{z=-H}\left[1 - \delta e^{-\alpha(1+i)(H+z)}\right] \tag{2.2.36}$$

where $\delta = 1$ stands for non-slip and $\delta = 0$ for free slip condition at the sea bottom. From this relation we can easily obtain an expression for the bottom friction:

$$\tau_x^{-H} + i\tau_y^{-H} = \rho_0 K_M \left.\frac{\partial M}{\partial z}\right|_{z=-H} = -\frac{(1+i)\alpha\delta}{f\rho_0}\left(i\frac{\partial p}{\partial x} - \frac{\partial p}{\partial y}\right)_{z=-H} \tag{2.2.37}$$

$$\tau_x^{-H} = \frac{\delta}{2\alpha}\left(\frac{\partial p}{\partial x} + \frac{\partial p}{\partial y}\right)_{z=-H}, \quad \tau_y^{-H} = \frac{\delta}{2\alpha}\left(\frac{\partial p}{\partial x} + \frac{\partial p}{\partial y}\right)_{z=-H} \tag{2.2.38}$$

From (2.2.36) we can also obtain an expression for the vertical turbulent viscosity in the bottom friction layer:

$$\left(K_M\frac{\partial^2 M}{\partial z^2}\right)_{-H} = \frac{\delta}{\rho_0}\left(\frac{\partial p}{\partial x} + i\frac{\partial p}{\partial y}\right)e^{(1+i)\alpha(z+H)} \tag{2.2.39}$$

and simplify it. We take into account the vertical turbulent viscosity in the thin nearsurface and near bottom friction layers by the approximate formula (2.2.34) and neglect it in the main ocean body. As a result the equation (2.2.9) transforms into formula (2.2.38):

$$M = \frac{1}{f\rho_0}\left(i\frac{\partial p}{\partial x} - \frac{\partial p}{\partial y}\right) + \frac{(1-i)}{2\rho_0 K_M \alpha}\left(\tau_x + i\tau_y\right) e^{-(1+i)\alpha z} -$$

$$- \frac{\delta}{f\rho_0}\left(i\frac{\partial p}{\partial x} - \frac{\partial p}{\partial y}\right)_{z=-H} e^{-(1+i)\alpha(z+H)} \qquad (2.2.40)$$

Formula (2.2.40) gives the current velocity in terms of the pressure gradient and specified tangential wind stress field. It has a very clear and simple structure. The first term represents the geostrophic currents in the whole ocean body because the pressure gradients are functions of both, horizontal and vertical coordinates. The second term represents the direct effect of wind stress, it decreases very rapidly in depth and vanishes at the nearsurface Ekman layer's lower boundary. Depending on the boundary condition at the ocean bottom, the sum of the first and the last term may vanish (when $\delta = 1$) or not (when $\delta = 0$). In case of a non-slip condition ($\delta = 1$) the last group exists only in the thin Ekman bottom boundary layer, has its maximum value at the ocean bottom and vanishes at a distance of tens of meters above the bottom.

It is easy to show that (2.2.40) approximately satisfies equation (2.2.9) and the boundary conditions. For example, to fulfill boundary condition (2.2.11), one should assume that the pressure gradients do not depend on z (only in the upper Ekman layer) and that the last group does not exist in the upper Ekman layer. The same is true for the Ekman bottom layer: in case of a free-slip $\frac{\partial M}{\partial z}\big|_{z=-H} \approx 0$. In other words, the nearsurface and nearbottom Ekman's layers do not overlap (when the basin's depth is more than some 50 m).

The above-mentioned formula (2.2.40) gives only the horizontal components of the flow velocity. To get the vertical component we use equation (2.1.6) and the "rigid lid" condition $w|_{z=0} = 0$; then we have

$$w = \int_z^0 \left(\frac{\partial u}{\partial x} + \frac{\partial v}{\partial y}\right) dz. \qquad (2.2.41)$$

The easiest way for deriving an analytical expression for w is integration of formula (2.2.40) with respect to z before separating it into real and imaginary components. Considering that the coefficient K_M is constant the vertical profiles of u, v and especially w may not be realistic in the boundary layers. So, we present a formula for w, which is valid only beyond the upper and the lower boundary layers.

$$w = \frac{1}{\rho_0 f} rot_z \vec{\tau} + \frac{\beta}{f^2}\tau_x - \frac{\beta}{\rho_0 f^2}\int\limits_z^0 \frac{\partial \rho}{\partial x} dz \tag{2.2.42}$$

$$\text{where } rot_z \vec{\tau} = \frac{\partial \tau_y}{\partial x} - \frac{\partial \tau_x}{\partial y} \tag{2.2.43}$$

The first two terms of formula (2.2.42) represent the approximate value of w at the lower end of the surface Ekman boundary layer.

Formulae (2.2.40) and (2.2.42) represent all three components of flow velocity due to pressure anomaly, and wind stress $\vec{\tau}$. For diagnostic calculations the only thing left is the construction of an equation for the sea surface height ζ by means of formula (2.2.3) for p. One version of such an equation, which is based on divergence of the momentum balance equations was already presented as (2.2.26). As mentioned above, (2.2.26) is applicable to extraequatorial regions too. Here we construct an equation for ζ based on the vorticity equation as it is common for oceanographic and meteorologic literature.

First we use another boundary condition for w at $z = -H(x, y)$:

$$w = -\left(u_H \frac{\partial H}{\partial x} + v_H \frac{\partial H}{\partial y}\right) \tag{2.2.44}$$

For the large-scale circulation, the non-slip boundary condition (when $\delta = 1$ and u, v do not vanish at the ocean's bottom) and calculation of w by formula (2.2.44) are the most realistic approach. At the other depths the w may be calculated by (2.2.44), ignoring the effect of the near-bottom Ekman layer.

Use of two boundary condition for the first order equation (2.1.6) is mathematically allowable because actually $z = 0$ is replaced by $z = \zeta$ with ζ being an additional unknown function. Integrating (2.1.6) from $-H$ to 0 and using boundary condition (2.2.44) and $w|_{z=0} = 0$ we come to the well-known simple equation

$$\frac{\partial S_x}{\partial x} + \frac{\partial S_y}{\partial y} = 0. \tag{2.2.45}$$

Now let us construct the equation for sea surface height based on the integrated vorticity equation. By cross differentiation together with (2.2.45) we first obtain the equation for the bottom pressure torque

$$J(H, P_{-H}) = -\rho_0 \beta S_y - rot_z(\vec{\tau} - \vec{\tau}^{-H}) \tag{2.2.46}$$

where $J(a, b) = \frac{\partial a}{\partial x}\frac{\partial b}{\partial y} - \frac{\partial b}{\partial x}\frac{\partial a}{\partial y}$ is the Jacobian operator.

Scale analysis yields characteristic balances in the ocean. If we balance $-\rho_0 \beta S_y$ with $rot_z \vec{\tau}$ we come to Sverdrup dynamics, which underestimates the meridional mass transport. Generally, (2.2.46) is a delicate balance of equally small terms that makes the essential difference against the momentum balance equations (2.1.1), (2.1.2) where two major terms form the geostrophic balance.

The question arises how to get flow velocity and mass transport of realistic magnitude? Both, the Sverdrup and the geostrophic balance have their limitations.

To overcome this difficulty we turn to the equation of statics (2.2.3). At the sea surface ($z = \zeta \approx 0$) the pressure anomaly is maximal. The integral $g \int\limits_{z}^{0} \rho dz$ grows with depth and mainly is of opposite sign to $\rho_0 g \zeta$. As a result $p_{-H} = \rho_0 g \zeta + g \int\limits_{-H}^{0} \rho dz$ becomes so small that the assumption of the reference level method about decaying p_{-H} brings surface gradient currents to a realistic order of magnitude.

In other words, p_{-H} is small because it is the difference of two big almost equal quantities. In late 1960s the idea of Sarkisyan was to split p_{-H} and, consequently the bottom pressure torque into two terms and place them on different sides of the equation. As a result (2.2.46) transforms into

$$J(H, \zeta) = \underbrace{-\frac{1}{\rho_0} J\left(H, \int\limits_{-H}^{0} \rho dz\right)}_{JEBAR} - \frac{\beta}{g} S_y - rot_z \left(\vec{\tau} - \vec{\tau}^{-H}\right) \qquad (2.2.47)$$

and one out of the two almost mutually compensating terms – the JEBAR – becomes a source of available potential energy for the other – the $J(H, \zeta)$, which is a source of kinetic energy.

So the JEBAR appears after splitting the bottom pressure torque as the atomic energy appears after splitting the atom. This explanation is necessary because in oceanographic literature many authors consider the JEBAR as analog of the bottom pressure torque. Moreover, we saw above in point 2.2.3 that there is another JEBAR, the JEBAR-2, which has nothing in common with $J(H, p_{-H})$ but again is result of splitting the p_{-H}.

To finish constructing the equation for SSH we replace in (2.2.47) S_y by usual geostrophic formulae, apply splitting of p and other transformations like in deriving equation (2.2.26).

After these transformations we obtain the final form of the vorticity equation for ζ.

$$\frac{f}{2\alpha H}\Delta\zeta + \frac{f}{H}J(H, \zeta) + \beta\frac{\partial \zeta}{\partial x} = \qquad (2.2.48)$$

$$= -\frac{f}{2\alpha\rho_0 H}\int\limits_{-H}^{0} \Delta\rho dz \underbrace{- \frac{f}{\rho_0 H}\int\limits_{-H}^{0} J(H, \rho) dz}_{JEBAR} \underbrace{- \frac{\beta}{\rho_0 H}\int\limits_{-H}^{0} (H - z)\frac{\partial \rho}{\partial x} dz}_{BARBE^{\dagger}} + \frac{f}{\rho_0 g H} rot_z \vec{\tau}$$

† BARBE- Baroclinic Beta Effect

Equation (2.2.48) is actually valid for calculation of SSH not only in the extraequatorial but also in the equatorial regions of the World Ocean. Though in the latter case it is necessary to consider α to be constant within the equatorial strip (say, the equator $\pm 2^{\circ}$). The equations (2.2.27) are valid for calculating ζ at any basin side boundaries for further solving both equations (2.2.26) and (2.2.48). Let us show it briefly.

Usually, a region's boundary is approximated by a line traced piece-wise along one of the coordinate axes. Therefore it is enough to specify at side boundaries only the mass transport component normal to the corresponding axis. At the sections parallel to meridians, S_x is specified and we use the second of equations (2.2.27).

Consequently, at the zonal sections we use the first of equations (2.2.27). Because at every section we use only tangential differentials of ζ, they may be written in a general simple form, which is valid for any shape of a horizontal boundary

$$\frac{\partial \zeta}{\partial s} = f\,(x,\,y) \tag{2.2.49}$$

where the right-hand side is specified and S is the basin's coastline.

In principle, ζ may be found at all points of a closed boundary, but there is one obstacle to overcome.

If one writes a finite difference version of (2.2.49) one will get n equations with n unknown quantities the determinant of which is equal to zero. It means that this problem is not correctly set because the integral $\int f'\,(x,\,y)\,dx$ along a closed boundary line is not necessarily equal to zero. To proceed let us define a function $\zeta'\,(x,\,y)$, minimizing an operator $\left(\left(\frac{\partial \zeta}{\partial s}\right) - f'\right)$ and call it a quasi-solution of the problem. It has been proven that, if

$$\left\| \left(\frac{\partial \zeta}{\partial s}\right) - f' \right\| < \varepsilon,$$

then a maximum of the absolute value of the difference between ζ and the solution of a correctly formulated problem is less than ε_1.

There is no problem either in correcting the specified observed data, so that the sum $\sum f_{i,j} = 0$ holds because we deal with a rather simplified model anyway. A numerical method of calculation of ζ at the grid points of a closed boundary has been proposed by Sarkisyan (1977). As SSH is being found with an accuracy of an additive constant we can put $\zeta = 0$ at one of the boundary grid points, say, $\zeta_{0,0}$ and delete one of the equations containing $\zeta_{0,0}$; as a result we shall have a system of n − 1 equations with n − 1 unknowns which has a single solution. Then, beginning with one of the neighboring points ($\zeta_{0,1}$ or $\zeta_{1,0}$), we proceed clockwise or counterclockwise around the basin and find step by step $\zeta_{i,j}$ at all the boundary grid points.

2.2.4 Final Formulation, Comparison and Qualitative Analysis of Models $\mathbf{D}_1$ and $\mathbf{D}'_1$

The anomaly of density (temperature, salinity) may be considered as the main internal physical characteristics, while the pressure anomaly is the main characteristics of ocean dynamics. The density anomaly is specified in diagnostic calculations, that is why ζ becomes the ocean dynamics main characteristics. So we turn our attention to equations (2.2.26) and (2.2.46).

The divergency equation (2.2.26) is valid for both, equatorial and the extraequatorial regions. Moreover, the use of (2.2.26) is preferable because the main dynamic characteristics the pressure field was kept during cross-differentiation and has generated the main terms of equation (2.2.26), i.e. Laplacians from ζ and ρ. As for the joint effect of baroclinicity and bottom relief – JEBAR-2, it will be discussed later.

The vorticity equation (2.2.48) was constructed for the extraequatorial ocean region. As mentioned above to use it for the equatorial zone one should consider α to be constant in the equatorial belt. In contrast to (2.2.26), the terms containing Laplacians in (2.2.48) are secondary. The main terms on the right-hand side of (2.2.48) are JEBAR and BARBE. Thus the main task in diagnostic calculations by the above two models is, first of all, to solve one of the two equations: either (2.2.48) or (2.2.26). In principal, both are valid, for any region of the World Ocean. In case equation (2.2.26) is being solved, the terms, containing flow velocity components, should be omitted as a first approximation. As for equation (2.2.48), the above-mentioned restriction for α in the equatorial belt applies.

The determination of ζ at closed ocean basin's side boundaries (or at an island coast) via equation (2.2.27) is identical for both equations (2.2.26), (2.2.48).

Concluding the qualitative analysis of equations (2.2.26) and (2.2.48) let us show that the determination of ζ by the dynamic method is a particular case of these two equations. Supposing that we are dealing with a basin of minimum depth 0.5–1 km we can neglect the density anomaly at the bottom of the basin; that is we assume that

$$\int_{-H}^{0} \frac{\partial \rho}{\partial x} dz \approx \frac{\partial}{\partial x} \int_{-H}^{0} \rho dz, \quad \int_{-H}^{0} \frac{\partial \rho}{\partial y} dz \approx \frac{\partial}{\partial y} \int_{-H}^{0} \rho dz \tag{2.2.50}$$

Further, remembering that the anomalies of density decrease with depth, we can assume

$$\left| \int_{-H}^{0} \frac{\partial \rho}{\partial x} dz \right| > \frac{1}{H} \left| \int_{-H}^{0} z \frac{\partial \rho}{\partial x} dz \right| \tag{2.2.51}$$

$$\int_{-H}^{0} (H - z) \frac{\partial \rho}{\partial x} dz \approx H \int_{-H}^{0} \frac{\partial \rho}{\partial x} dz$$

Finally, let us drop the last term of the right-side of equation (2.2.48). After these simplifications (2.2.48) takes the form:

$$L(\zeta) = \frac{1}{2\alpha}\Delta\zeta - J(H,\zeta) - \frac{\beta H}{f}\frac{\partial \zeta}{\partial x} = -L\left(\int_{-H}^{0} \rho dz\right) \qquad (2.2.52)$$

The solution of equation (2.2.52) is

$$\zeta_d = -\frac{1}{\rho_0}\int_{-H}^{0} \rho dz \qquad (2.2.53)$$

which represents the formula for calculation of sea surface topography by the dynamic method ($-H$ is accepted as a reference level (Peredery and Sarkisyan, 1972)). This solution can also be applied to the boundary conditions. Indeed, suppose that water flux across the lateral boundaries is given by the dynamic method; i.e. u and v are determined at the contour by formulae (2.2.8), where $-H_1$ is removed by $-H$. Then integrating (2.2.6) for z from $-H$ to 0 we get

$$S_y^d = \frac{g}{\rho_0 f}\int_{-H}^{0} \frac{\partial \rho}{\partial x} z dz, \qquad S_x^d = -\frac{g}{\rho_0 f}\int_{-H}^{0} \frac{\partial \rho}{\partial y} z dz \qquad (2.2.54)$$

Then, remembering simplification (2.2.50) and omitting terms with $\vec{\tau}$ and $\vec{\tau}^{-H}$ we can readily see that formula (2.2.53) satisfies boundary conditions (2.2.21) and (2.2.22) (Peredery and Sarkisyan, 1972). In the same way it can be shown that (2.2.53) is valid for a simplified version of equation (2.2.26) as well.

Inasmuch as wind stress and density anomalies are specified and ζ is calculated by equation (2.2.26) or (2.2.48), the correlations (2.2.28) may be considered as simple formulae for calculating flow velocity horizontal components at the very equator and within a narrow equatorial strip (equator $\pm 1°$). Having calculated u and v one can calculate w by formula (2.2.41). Thus, we have finished formulation of a simple equatorial diagnostic model.

Let us estimate the order of magnitude of the horizontal flow based on formulae (2.2.28).

The easiest way is the evaluation of the first two terms for the sea surface ($z = 0$), for example

$$\frac{H}{K_M \rho_0}\tau_x \text{ and } \frac{gH^2}{2K_M}\frac{\partial \zeta}{\partial x}.$$

We have to choose $-H$ and K_M appropriately, the flow velocity has a realistic value, say, 30–50 cm/s.

A suitable choice is $K_M = 10^2\,\text{cm}^2/\text{s}$ and $H = 50\,\text{m}$.

It means that either formulae (2.2.28) are valid only for the upper ocean layer, or K_M should be much larger for deeper layers. The strong dependence of the flow velocity on ocean depth and on the uncertain value of K_M means that the simple formulae (2.2.28) are suited for a qualitative assessment rather than for reliable quantitative calculations.

Let us finish with the formulation of a simple extraequatorial model. For calculation of SSH, as mentioned above, both (2.2.26) and (2.2.48) are valid, but which one is optimal to get the flow velocity components? The formula (2.2.40) presents the flow velocity horizontal components as a sum of gradient currents and both surface and bottom Ekman spirals. In the real ocean there are no pure Ekman spirals, besides, we have considered a simple model without nonlinear terms. Thus, the models are valid only for numerical calculation of large-scale currents when vertical distances between two neighbouring layers are larger than thickness of the Ekman layer. So, using formulae (2.2.3) and (2.2.40) we come to the following simple expressions for u and v. At the sea surface:

$$\begin{aligned} u &= \frac{1}{2\rho_0 K_M \alpha}\left(\tau_x + \tau_y\right) - \frac{g}{f}\frac{\partial \zeta}{\partial y} \\ v &= \frac{1}{2\rho_0 K_M \alpha}\left(\tau_y - \tau_x\right) + \frac{g}{f}\frac{\partial \zeta}{\partial x} \end{aligned} \tag{2.2.55}$$

At the intermediate levels

$$\begin{aligned} u &= -\frac{g}{f}\frac{\partial \zeta}{\partial y} - \frac{g}{\rho_0 f}\int_z^0 \frac{\partial \rho}{\partial y} dz \\ v &= \frac{g}{f}\frac{\partial \zeta}{\partial x} + \frac{g}{\rho_0 f}\int_z^0 \frac{\partial \rho}{\partial x} dz \end{aligned} \tag{2.2.56}$$

In case of a free slip condition u and v at the bottom are calculated by the same formulae (2.2.56) after replacing the integrals lower limit by $-H$, instead of z.

In case of a non-slip condition, the depth of the nearest to the bottom calculation layer should be $\leq -H + D$, where $D = \frac{\pi}{\alpha}$ is the thickness of the Ekman layer. In that near-bottom layer u and v are given by:

$$\begin{aligned} u &= -\frac{g}{f}\frac{\partial \zeta}{\partial y} - \frac{g}{\rho_0 f}\int_{-H+D}^0 \frac{\partial \rho}{\partial y} dz \\ v &= \frac{g}{f}\frac{\partial \zeta}{\partial x} + \frac{g}{\rho_0 f}\int_{-H+D}^0 \frac{\partial \rho}{\partial x} dz. \end{aligned} \tag{2.2.57}$$

A short note about the vertical component. At the sea surface, due to a “rigid lid” condition, w is equal to zero. In all intermediate layers w is calculated by formula (2.2.42) and at the bottom – by formula (2.2.44).

At the lower end of the surface Ekman layer w is given by:

$$w|_{z=-D} = -\frac{1}{\rho_0 f} rot_z \vec{\tau} \tag{2.2.58}$$

Let us compare the sea surface wind drift currents obtained by the equatorial model and the extraequatorial one. According to formulae (2.2.28), wind drift currents are directed exactly as the wind while according to extraequatorial formulae (2.2.55) they deviate by 45° to the right from the wind direction (on the Northern hemisphere). It is interesting to note that in reality the surface flow, as a rule, chooses “the golden mean”. The multipliers of wind stress components in (2.2.55) are of the form $\left(\sqrt{2fK_M}\right)^{-1}$, which means that surface currents obtained by the extraequatorial model depend only slightly on K_M, while, according to (2.2.28), surface wind drift is very strongly dependent on K_M and H. This fact is an essential drawback of the equatorial model and therefore flow velocity calculated by the equatorial model is less reliable.

The results of calculations reviewed below in the Chapter 3 show that the optimum combination is calculation of ζ by the equatorial equation of divergence (2.2.26) and of the flow velocity – by the extraequatorial model. For the sake of short reference we called this version the Model D_2.

Actually, equation (2.2.48) is quite useful too, and many calculations of SSH have been made and will be made by using it (we shall call it D_2').

From the formal mathematical point of view, both divergence and vorticity operations are equally correct but when deriving (2.2.48) from momentum balance equations the main part of the pressure anomalies is canceled, so in the resulting vorticity equation the balance is more “delicate”, less stable. Consequently, it makes higher demand on the observed density anomalies and on numerical models. These drawbacks are essential especially in the equatorial area. Still the comparison of calculations made by equations (2.2.48) and (2.2.26) is worthwhile.

The smallest step forward from the dynamic method is just adding vertical turbulent viscosity with a constant coefficient enabling to exclude all the above-mentioned and other drawbacks of the dynamic method. As was shown above, it is enough to solve numerically only the one two-dimensional quasilinear equation for ζ in D_1 or D_2 – type models and simple formulae enable to calculate other characteristics of a closed oceanic or sea basin with arbitrary geometry and boundary conditions. The dynamic method is a simplified case of the D_1 model, it may be used only as a first approximation of ζ for assessment of the current's gradients in the upper ($\approx$ 100, 200 m) part of the ocean, but it rapidly loses its accuracy with depth.

2.2.5 Simple Linear Model for Diagnostic Calculations by Using Mass Transport Stream Function (The Model $\mathbf{D}_1''$)

After the pioneering papers of Shtockman (1946, Chapter 1), Sverdrup (1947, Chapter 1), Munk (1950, Chapter 1) and many others the use of the mass transport stream function Ψ for modelling ocean circulation has become very popular, especially in the Western countries. So, it makes sense to pass from ζ to Ψ in some of the above equations and formulae and to perform a qualitative analysis, first of all. Based on equation (2.2.45), we can use the well known formulae

$$S_x = -\frac{\partial \Psi}{\partial y}, \quad S_y = \frac{\partial \Psi}{\partial x} \tag{2.2.59}$$

to transform equations (2.2.27) into the form

$$\frac{\partial \zeta}{\partial x} = \frac{f}{gH}\frac{\partial \Psi}{\partial x} - \frac{1}{\rho_0 H}\int\limits_{-H}^{0}(H+z)\frac{\partial \rho}{\partial x}dz + \frac{\tau_x}{\rho_0 g H} \tag{2.2.60}$$

$$\frac{\partial \zeta}{\partial y} = \frac{f}{gH}\frac{\partial \Psi}{\partial y} - \frac{1}{\rho_0 H}\int\limits_{-H}^{0}(H+z)\frac{\partial \rho}{\partial y}dz + \frac{\tau_y}{\rho_0 g H} \tag{2.2.61}$$

The bottom stress $\left(\tau_x^{-H}, \tau_y^{-H}\right)$ has been neglected. It does not mean completely neglecting the bottom stress effect. To derive an equation for Ψ we put the expressions for $\frac{\partial \zeta}{\partial x}$ and $\frac{\partial \zeta}{\partial y}$ from (2.2.60), (2.2.61) into (2.2.48), make simple transformations, keep the main parts of the effects of bottom topography, bottom stress, baroclinicity, etc., but neglect secondary parts. As a result we obtain:

$$\begin{aligned} \frac{f}{2\alpha H}\Delta\Psi + \frac{f}{H}J(H,\Psi) + \beta\frac{\partial \Psi}{\partial x} = \\ = -\frac{f}{2\alpha\rho_0 H}\int\limits_{-H}^{0}\Delta\rho dz - \frac{f}{\rho_0 H}\int\limits_{-H}^{0} zJ(H,\rho)dz + \\ + \underbrace{\frac{rot\vec{\tau}}{\rho_0} - \frac{1}{\rho_0 H}\left(\frac{\partial H}{\partial x}\tau_y - \frac{\partial H}{\partial y}\tau_x\right)}_{JEWAR^{\dagger\dagger}} \end{aligned} \tag{2.2.62}$$

The first term on the left – and right side of this equation (i.e. the Laplacian of Ψ) is generated due to bottom stress.

††JEWAR – Joint Effect of Wind And Relief

The side boundary condition for solving (2.2.62) is very simple $-\Psi$ is specified.

It is very easy to make equation (2.2.62) quite close to the equation of other models that use integral mass transport function. For example, taking into account horizontal turbulent viscosity and accepting the free-slip condition at the ocean bottom one would replace the first term of the left-hand side by $A_M \Delta\Delta\Psi$ and would drop the first term of the right-hand side of this equation. But this is not our aim here.

In Chapter 3 we shall repeat the main results of the original paper where both horizontal mixing and nonlinear terms were taken into account (Sarkisyan, 1977). But one should keep in mind, that neither horizontal, nor vertical viscosity describes the real turbulent processes anyway, and none of them is the main term in equations of the (2.2.62) type. Here we make a qualitative analysis and comparison of different simple linear models, reveal the drawbacks of the models paying attention to the main terms rather than to horizontal mixing or bottom stress.

All the necessary formulae for flow velocity components of Model D_1'' may easily be derived by using the respective equations of Model D_1'. Namely, we substitute the expressions for sea surface slopes (2.2.60), (2.2.61) into formulae (2.2.42), (2.2.55)–(2.2.57) and after simple transformations come to relations for the calculation of u, v, w via mass transport stream function Ψ and specified $\vec{\tau}$, ρ. We have: the flow velocity horizontal components at the sea surface,

$$
\begin{aligned}
u &= \frac{1}{2\alpha\rho_0 K_M}\left(\tau_x + \tau_y\right) - \frac{1}{H}\frac{\partial\Psi}{\partial y} + \frac{g}{\rho_0 f H}\int\limits_{-H}^{0}(H+z)\frac{\partial\rho}{\partial y}dz \\
v &= \frac{1}{2\alpha\rho_0 K_M}\left(\tau_x - \tau_x\right) + \frac{1}{H}\frac{\partial\Psi}{\partial x} + \frac{g}{\rho_0 f H}\int\limits_{-H}^{0}(H+z)\frac{\partial\rho}{\partial x}dz
\end{aligned}
\tag{2.2.63}
$$

and at the intermediate levels,

$$
\begin{aligned}
u &= -\frac{1}{H}\frac{\partial\Psi}{\partial y} + \frac{g}{\rho_0 f H}\int\limits_{-H}^{0}(H+z)\frac{\partial\rho}{\partial y}dz \\
v &= \frac{1}{H}\frac{\partial\Psi}{\partial x} + \frac{g}{\rho_0 f H}\int\limits_{-H}^{0}(H+z)\frac{\partial\rho}{\partial x}dz.
\end{aligned}
\tag{2.2.64}
$$

In the near bottom layer, as in case of Model D_1', there are two versions of respective formulae

(a) in case of a non-slip condition

$$u = -\frac{1}{H}\frac{\partial \Psi}{\partial y} + \frac{g}{\rho_0 f H}\int\limits_{-H+D}^{0}(H+z)\frac{\partial \rho}{\partial y}dz$$
$$v = \frac{1}{H}\frac{\partial \Psi}{\partial x} + \frac{g}{\rho_0 f H}\int\limits_{-H+D}^{0}(H+z)\frac{\partial \rho}{\partial x}dz \tag{2.2.65}$$

(b) in case of a free-slip condition, flow velocity at the bottom equals to the value of the lowest calculation level.

Next, we consider the vertical component of flow velocity. At the sea surface is $w = 0$; at the lower boundary of the near-surface Ekman layer formula (2.2.58) for w is valid; at all the other calculation levels we get the formula for w by using the formulae (2.2.42), (2.2.3) and (2.2.60)

$$w = -\frac{1}{\rho_0 f} rot_z \vec{\tau} + \frac{\beta z}{fH}\frac{\partial \Psi}{\partial x} - \frac{g\beta z}{\rho_0 f^2 H}\int\limits_{-H}^{0}(H+z)\frac{\partial \rho}{\partial x}dz +$$
$$+ \frac{g\beta}{\rho_0 f^2}\int\limits_{z}^{0}(z+\zeta)\frac{\partial \rho}{\partial x}d\zeta \tag{2.2.66}$$

At the ocean bottom formula (2.2.44) may be used for calculation of w.

2.3 Are The Sverdrup Relation and the Dynamic (Reference-Level) Method Still Up-To-Date?

2.3.1 Westward, Equatorward, Coastward and Other Kinds of Intensifications in Baroclinic Basins with Variable Depth

Historically, the word "westward intensification" appeared first and this is natural because the most intense currents (Gulf Stream, Kuroshio, Somalian current) are located at the Western coasts of oceans and this fact called for explanation. The first papers explaining the westward intensification by the β-effect appeared in the late forties and the early fifties (Sverdrup, 1947, Stommel, 1948, Munk, 1950, in Chapter 1). Since the early sixties (Sarkisyan, 1960, 1974, see also references to Chapter 1 for papers of Sarkisyan 1961, 1962, 1969a, b) additional elements have been proposed for the word "intensification".

1) Almost all major currents are intense only within the ocean's upper layer of one to few hundred meters.
2) There are deep-water intense currents which usually are directed opposite to the upper ocean currents. It means that any integrated mass transport, including the

Sverdrup simplest relation $S_y = \left(rot_z\vec{\tau}\right)/(\rho_0 g)$, may veil the upper westward layer or any coastward intensification.
3) Intense currents are closely related to high values of cross-current density gradients while the Sverdrup relation does not include the baroclinicity at all.
4) When taking into account just bottom topography one can see that the Sverdrup transport becomes much less than in the idealized flat-bottom ocean which itself results in an underestimated value of Ψ (Sarkisyan, 1977).
5) Stommel (1948, Chapter 1) considered a homogeneous ocean model and obtained indeed a relative westward intensification, but, when taking into account the real depth of the ocean, the flow velocity becomes 1–2 orders of magnitude smaller than the observed values.

In summary, the β-effect (caused by the spherical shape of the Earth) is not the only factor of westward intensification. Let us describe briefly of only one of those contributions (Sarkisyan, 1960). In that paper the variables were presented as a sum of zonal parts (denoted by a bar) and nonzonal anomalies. The equation of density transport was linearized relative to zonal flows, then the nonzonal pressure anomaly $p(x, y, z)$ was presented in the form of exponential function

$$p = p_0(x, y)\, e^{-k(x,y)z}$$

The resulting equation for the surface layer pressure anomaly took the form

$$\frac{k^2}{g}\frac{\partial p_0}{\partial t} - \left[\frac{k}{g}\frac{\partial \bar{\rho}}{\partial y} - \bar{u}\frac{k^2}{g} - \frac{\beta}{f^2 \rho_0}\frac{\partial \bar{\rho}}{\partial z}\left(e^{kz} - 1\right)\right]\frac{\partial p_0}{\partial x} = F \tag{2.3.1}$$

A comparison showed that all the three terms in brackets are almost of the same order of magnitude with a reasonable value of k and limited ocean depth. Assuming the gradient of the zonal density $\frac{\partial \bar{\rho}}{\partial y}$ to be positive and $\frac{\partial \bar{\rho}}{\partial z}$ to be negative, a conclusion was made about the role of the zonal flow velocity component in westward intensification: in the Southern latitudes is $\bar{u} < 0$, so the advection of density ($\bar{u}\frac{\partial p_0}{\partial x}$ is generated by the advection of density) favors westward intensification; in the mid-latitudes – vice-versa, $\bar{u} > 0$ and the density advection favors separation of jet-like currents from the west coast.

In the models of Sverdup-Stommel-Munk, β-effect is dominating because a barotropic flat bottom was considered. In a baroclinic ocean, despite linearization of the equation for density anomaly, and despite neglect of the ocean bottom relief, there is competition of β-effect with other factors- at the minimum with "free" ocean zonal flow and density anomaly. The expression in brackets of (2.3.1) becomes negative and the Gulf Stream separates from the coast in the right place. We shall discuss this aspect below when considering other factors of intensification.

In the late 1960s and early seventies when the important role of JEBAR and BARBE in ocean dynamics was proven (Sarkisyan, 1969a, Sarkisyan, 1974,

Sarkisyan and Ivanov, 1971, in references to Chapter 1) it became obvious that, besides the advection of density anomaly, there are at least two other "rivals" to β -effect.

In the mid-1970s A. Sarkisyan put a question to several well-known oceanographers: what is the main cause of the westward intensification? The answers were the following.

K. Bryan: I believe it to be β-effect, but the direction of the prevailing density gradient may be essential too. For example, the Western boundary currents would be much weaker if the density in the subpolar regions were less than in the equatorial region (if the gradient of zonal density was of the opposite sign).

N. Fofonoff: The western intensification of currents means evidence of the law of conservation of potential vorticity. I think the inertial theories present the main mechanism of intensification.

J. Veronis: The western intensification may be generated by both β-effect and gradients of temperature. I think the β-effect is more essential, though joint effect of these two factors is possible as a result of their nonlinear interaction.

W. Holland: I think the western intensification is modeled well on the basis of the so-called inertial theories. The Gulf-Stream jet crossing the North Atlantic is baroclinically unstable, it deteriorates before reaching the Eastern boundary as a result of the separation of eddies. That is why there is Western intensification but there is no Eastern intensification.

P. Lineykin: The β-effect serves as the main cause of the Western intensification, but thermohaline factors also play a definite role. The variation of ocean depth with latitude is essential too.

V. Kozlov: β-effect, variation of depth near the coast and variation of density near the coast (intensification of JEBAR) are the factors that may cause coastal intensification. The main cause of intensification depends on the prevalence of one of these factors. If $H = const$ (JEBAR $= 0$) we have the classical β-effect. In case of $\rho = const$ the relation $^{1}/_{H}$ acts. Possibly the role of JEBAR in intensification has been studied much less than the role of other factors.

Many of these answers are not out of fashion even now and are worthy of comments. Many scientists (not only the named above) surely regard the β-effect as the main cause of westward intensification. Let us look at the problem from a pure formal mathematical point of view. To have an asymmetry in the solution of a differential equation there should be terms with an odd degree of differentials. In our case the first order differential of this kind is $\beta \frac{\partial \zeta}{\partial x}$ –, with $\beta > 0$, and it favors westward intensification. Then there should be a boundary, in our case this is a meridional boundary. This is also a cause favoring westward intensification. The major part of the continental coastlines is almost meridional. Now let us look at general equations for ζ, (3.3.6) and (3.3.8) of the next chapter, which show that besides β-effect there are two other terms generated by bottom topography, which depending on their sign and value, may be either favorable or unfavorable to westward intensification. They may separate intense currents from the Western boundary. The effect of zonal currents was discussed above based on some previous calculations (Sarkisyan, 1960, 1974). K. Bryan, G. Veronis and P. Lineykin mentioned it in a general form.

Furthermore, looking at equations for temperature and salinity one takes notice of the terms $u\frac{\partial T}{\partial x}$, $u\frac{\partial S}{\partial x}$ - as analogous to $\beta\frac{\partial \zeta}{\partial x}$. The role of JEBAR was considered above and mentioned by V. Kozlov.

Discussion of westward or eastward intensification may naturally lead to the question: what about northward or southward intensification? The question is surely reasonable and for the answer one could look at equation (3.3.6) where one finds the term $a_3\frac{\partial \zeta}{\partial y}$, and at expression (3.3.9) containing the effect of bottom topography, and the β-effect. Both of them may favor northward or southward intensification.

Lots of intense currents exist near coasts oriented neither meridionally nor zonally.

Summing up the above discussion, we may speak about coastward, instead of westward intensification. For any intensification there should, first of all, exist obstacles in the ocean as well as a flow component normal to an obstacle. Such obstacles are given by underwater mountains, continental coastlines with sharp slopes, islands, peninsulas, straits, bays. Only several well-known currents at the oceans Western boundary (Gulf Stream, Somalia Current, Kuroshio, etc.) are more intense than others due to β-effect, but flow velocity in them would be at least an order of magnitude smaller without baroclinicity, JEBAR, bottom topography and flows in the open ocean directed normal to Western meridional coast.

Whereas the most intense currents are at the Western boundaries, the most intense upwelling likewise intensified constantly by baroclinicity and JEBAR are at the Eastern boundaries.

Finally, we should not forget about equatorward intensification and upwelling which may be intense enough even without β-effect but may not be so intense without baroclinicity. That is why, we shall, give more attention to the baroclinicity of ocean water and introduce the term coastward intensification (of which a part is westward intensification) and mention other factors specified, which are able not only to excite intensification but also to separate intense currents from coasts or from the equator, islands, peninsulas, straits, bays.

2.3.2 Contradiction Between the Dynamic (Reference Level) Method and the Sverdrup Relation

To analyze the dynamic method we begin to compare it with the method of total mass transport. The integrated meridional transport in the baroclinic layer of the ocean are determined in the dynamic method exclusively by the density field. In the theory of total mass transport the same quantity is calculated only by the wind field. Let us present the simple formulae of these methods.

According to Sverdrup's relation we have

$$S_y^{(SV)} = \frac{1}{\rho_0 \beta} rot_z \vec{\tau} \qquad (2.3.2)$$

According to the dynamic method based on the formulae (2.2.6), we derive

$$S_y{}^d = \frac{g}{\rho_0 f_1} \int\limits_{-H_1}^{0} z \frac{\partial \rho}{\partial x} dz \tag{2.3.3}$$

where $f_1 = const$ is some average value of the Coriolis parameter, and H_1 is the depth of a reference level. These two formulae are fundamentally different. When calculating the meridional transport by equation (2.3.2) we must assume that the wind field and Coriolis parameter are variable, but the seawater may be homogeneous. On the contrary, for the calculation of $S_y{}^d$ by equation (2.3.3) the seawater must be inhomogeneous, but the wind field and Coriolis parameter may be constant. Of course, the wind field is indirectly present in formula (2.3.3), to some extent, taking part in the redistribution of the density field. However, we do not have an indirect effect of ρ in formula (2.3.2); neither do we have the effect of heat and salt exchange processes. On the contrary, in formula (2.3.3) we have only the indirect effect of the wind, the effect of heat and salt exchange, and other factors due to the density field anomaly ρ.

From the very beginning of last century calculations have been carried out by the dynamic method in an attempt to approximate the observed data.

Since 1947 the calculations of Ψ have been performed with the aid of Sverdrup's relation and the results also seemed to correspond to the observed data. Thus two contradictory relations repeatedly used in the theory of ocean currents on the one hand, and in the practice of oceanographic calculations on the other, have existed for more than half a century. In the presented above general equations for ζ and Ψ functions we have all the factors existing in both contradictory relations. We therefore have the opportunity to compare these relations and decide which of them better reflects the dynamics of stationary currents. We shall do it, first by a qualitative comparison. To do this we return to formula (2.2.46) neglecting the effect of bottom stress. As a result we come to the simple relation

$$\beta S_y = -\frac{1}{\rho_0} J(H, p)_{z=-H} - \frac{1}{\rho_0} rot_z \vec{\tau} \tag{2.3.4}$$

Let us now consider two limiting cases.

1. $H = const$. In this case the Jacobian vanishes in equation (2.3.4) and we come to the relation of Sverdrup (2.3.2).
2. $H \neq const,\ f = const\ (\beta = 0)$. Let us also neglect the second term on the right-hand side of equation (2.3.4), which is small in comparison with the first term. We thus obtain

$$J(H, p)_{z=-H} = 0 \tag{2.3.5}$$

Relation (2.3.5) is satisfied if we assume

$$\left(\frac{\partial p_H}{\partial x}\right)_{z=-H} = \left(\frac{\partial p_H}{\partial y}\right)_{z=-H} = 0 \tag{2.3.6}$$

We thus obtain the dynamic method in which the real bottom of the ocean serves as a reference level. In this case the sea-surface topography ζ^d and p^d are determined by formula (2.2.5), $S_y{}^d$ – by formula (2.3.3), and the total mass transport function by formula

$$\Psi^d = -\frac{g}{\rho_0 f}\int\limits_{-H}^{0} \rho z dz$$

$$\psi_d = -\frac{g}{\rho_0 f}\int\limits_{0}^{H} \rho z dz \tag{2.3.7}$$

The formula (2.3.7) may be also obtained from relations (2.2.60) and (2.2.61) if we take into account formula (2.2.5) and neglect the small terms generated by wind stress. Formula (2.3.3) easily follows from formula (2.3.7) if we introduce the definition of the total mass transport function. Thus formulae (2.3.2) and (2.3.3) are two extreme cases of relation (2.3.4). Because these two formulae do not agree physically, either both of them are wrong or only one of them is suitable. It is shown on the obasis of diagnostic calculations (see Sarkisyan, 1977) that $S_y{}^d$ is more or less close to the value of S_y, obtained from the general equation for ζ or Ψ, but $S_y{}^{(SV)}$ is smaller 2–3 times.

A comparison of orders of magnitude shows that the Jacobian of the right-hand side of equation (2.3.4) is of the same order of magnitude as $rot_z\vec{\tau}$ in regions apart from the continental slope and one order of magnitude greater in the vicinity of the slope. So the Jacobian should not be neglected. Furthermore, accepting (2.3.6) means making three equations from (2.3.4) which is mathematically nonsense.

To look at simplification (2.3.6) from the physical point of view let us return to formula (2.2.46) and use the formula (2.2.24). We have.

$$\beta S_y = -gJ(H,\zeta) - \frac{g}{\rho_0}J\left(H,\int\limits_{-H}^{0}\rho dz\right) - \frac{1}{\rho_0}rot_z\vec{\tau}. \tag{2.3.8}$$

Keeping only $\frac{1}{\rho_0}rot_z\vec{\tau}$ on the right-hand side of (2.3.8) means to assume that the meridional water flux does not depend on baroclinicity and on the sea surface slope that is obviously wrong. Finally, there is another drawback of the Sverdrup transport approach: wind stress acts only in case, when $rot_z\vec{\tau} \neq 0$. In Chapter 3 we show that in general models wind stress effect is three-fold: $div\vec{\tau}$, $rot_z\vec{\tau}$ or simply $\vec{\tau}$.

Deficiencies of the dynamic method have already been listed above. While in the ocean's upper layer this method gives a first approximation of gradient part of the flow velocity, it rapidly loses its accuracy with depth, that is why calculation of $S_y{}^d$ by formula (2.3.3) is erroneous. But (2.3.2) is not better than (2.3.3). Since 1969–1974 (Sarkisyan, 1969a) we know that in (2.2.62) the JEBAR is the main term and disregarding it was the main shortcoming of the original papers of Shtockman (1946), Sverdrup (1947) (both in references to Chapter 1), Munk (1950) and others.

Some of the authors neglected this effect, assuming the ocean to have flat bottom or the effect of baroclinicity and bottom topography to compensate each other in such a way that the vertically averaged baroclinic ocean behaves as a flat-bottom barotropic ocean (Welander, 1959).

It was shown (Sarkisyan 1977) that the above-mentioned and other simplifications just resulted in excluding all the terms with ρ from the right-hand side of the (2.2.62)-type equations. Actually, assumptions of this kind extremely simplify the problem mathematically; as a result one gets one equation with a single variable Ψ. Inversely, taking JEBAR into account is necessary not only because its weight is quite big but also because links Ψ with other variables. In case of the diagnostic calculation by keeping both effects – wind stress and baroclinicity one can see which of the two contradictory formulae (2.3.2), (2.3.3) is close to the complete solution.

The following can be concluded:

1) Equations (2.2.26), (2.2.48) and (2.2.62) are the simplest possible equations for diagnostic calculations close to reality.
2) For any sea or ocean basin and for a baroclinic lake, one out of these three differential equations should be solved in diagnostic calculations.
3) There are some factors and terms that should be taken into account with a possibly high accuracy in any diagnostic calculation. Those are: baroclinicity, wind stress and bottom topography. Mathematically it means the necessity of taking into account the first term and $div\vec{\tau}$ on the right-hand side of (2.2.26), JEBAR – in any of these equations, BARBE – in equation (2.2.46), $rot_z\vec{\tau}$ – in (2.2.48) and (2.2.62), JEWAR – in (2.2.62) along with other terms.
4) One should be very careful in considering idealization of any problem, unless it is geophysically justified. Finally, there is only one correct way: to write the complete equation and by scale analysis to drop really insignificant terms. Moreover, there is no problem to keep all terms to the very end when solving the problem numerically.

There was one argument in favor of the dynamic method -usually it served as an exclusive method of getting information on the flow velocity from a single hydrographic section. In Chapter 3 an alternative method is presented for the calculation of absolute flow velocity by using a single hydrographic section. The method makes it possible to avoid relations (2.3.6) and many other simplifications.

So, our answer to the question posed in the title of this section is negative. The simple Sverdrup relation served half a century and the dynamic method – the whole 20th century. We have proposed above three alternative simple models, and there are many others.

2.3.3 A Short Note on the Ocean Modelling First Steps

The calculation of the real ocean flow velocity from fields of observed density (temperature, salinity) begun by Sandström and Helland-Hansen (1903). The calculation of that kind results in a geostrophic flow velocity relative to some indefinite depth of baroclinic layer's lower boundary. Since Defant (1941) many attemptions have been made to define as reference the depth of the "Warmwassersphäre", then scientists begun just to specify the depth of the reference level ("level of no motion") quite differently and somewhat doubtful. Despite of the deficiencies of the dynamic method there is a main advantage: in areas with essential stratification the dynamic method gives a realistic order of magnitude of the surface layer gradient currents.

Ekman made an attempt to put the flow velocity calculation on real mathematical and hydrodynamical background with necessary boundary conditions, but in the homogeneous ocean the gradient currents are of one order of magnitude smaller then the dynamic method gives.

Scientists understood the necessity to take into account sea water baroclinicity in mathematical modelling in late forties at the previous century, but the analytical solution of the relevant equations were (and still are) very difficult. Besides the computer possibilities were very limited in that time.

Shtockman, Sverdrup and Munk considered the vertically integrated mass transport instead of vertical distribution of flow. They thought their idea enables to calculate the baroclinic ocean integral transport by avoiding the flow velocity and density vertical stratification.

As it was understood later, it was an unsuccessful attempt to calculate the BAROCLINIC ocean integral mass transport. Let us dwell on this theory in greater detail.

It is convenient to begin the discussion of the first works on the method of integral mass flux with the following equation by Munk:

$$\left(A_H \nabla^4 - \beta \frac{\partial \psi}{\partial x} \right) \psi = -rot_z \vec{\tau} \tag{2.3.9}$$

where ψ is the function of integral mass flux, $\vec{\tau}$ is the wind tangent stress, and A_H is the coefficient of the horizontal mixing; $\nabla^4 = \frac{\partial^4}{\partial x^4} + 2\frac{\partial^4}{\partial x^2 \partial y^2} + \frac{\partial^4}{\partial y^4}$ is the biharmonic operator, and equation (2.3.9) is a vertically integrated variant of the vorticity equation.

Being a special case of Munk's equation, the main equation of Shtockman is obtained by neglecting the beta-effect in the form:

$$A_H \nabla^4 \psi = -rot_z \vec{\tau} \tag{2.3.10}$$

The relation of Sverdrup is obtained from equation (2.3.9) by neglecting the effect of the horizontal mixing and is known to have the following simplest form:

$$\beta S_y = rot_z \vec{\tau}, \tag{2.3.11}$$

where $$S_y = \frac{\partial \psi}{\partial x} = \int_{-H}^{\zeta} \rho v dz \approx \rho_0 \int_{-H}^{\zeta} v dz. \tag{2.3.12}$$

It is interesting in itself that, as early as in the late 1940s, with the existing difficulties in data exchange at the time, the paper published by Shtockman in the USSR was immediately criticized by some foreign authors.

Indeed, Sverdrup's commentary was: "The theoretical results by Shtockman are inapplicable to oceanic conditions because of his assumption of water homogeneity." Let us notice immediately that Shtockman made no such assumption and his paper included the word "baroclinicity"; however, the reason why Sverdrup used such wording will become apparent in the further discussion. The attitude of Munk to Shtockman's paper can be expressed trough the following sentence: "The solution of equation (2.3.10) provided by Shtockman is of formal character, because he neglects such an important factor as the planetary vorticity." Notice that the latter term means the β-effect. However, Munk criticizes not only Shtockman, but also Ekman and Stommel. This criticism deserves special attention, because it includes the most complete substantiation of the works considered to be the theory of integral current of the BAROCLINIC ocean layer. The essence of Munk's reasoning is as follows: "In Ekman's and Stommel's model the ocean is assumed homogeneous, a case in which the currents extend to the very bottom". Not only is this in contrast with observations, according to which the bulk of the water transport in the main ocean currents takes place in the upper thousand meters, but it also leads to mathematical complications which rendered Ekman's analysis very difficult, and forced Stommel to resort to a rather arbitrary frictional force along the bottom. To avoid these difficulties, we retain Sverdrup's integrated mass transport as the dependent variable. The device makes it possible to examine the more general case of a BAROCLINIC ocean without having to specify the nature of the vertical distribution of density and current. In recognition of the evidence that currents essentially vanish at great depths, we shall depend on lateral friction for the dissipative forces. By this, Munk gives a clear and detailed explanation of the postulate shared also by Sverdrup and Shtockman. Let us notice that the title of Sverdrup's paper also contains the word "baroclinicity".

To facilitate the derivation of the equation for integral mass flux Shtockman, Sverdrup and Munk assumed that there is a surface at a great depth with virtually phenomenal properties. Namely, they believed that the flow velocity, pressure gradients, and vertical friction attenuate at this depth; in addition, this depth of the baroclinic layer is constant over the entire ocean, and isobaric surfaces intersect isosteric ones.

A correct argument made by the authors of the method of integral mass flux is that the density gradients decrease with depth like the vertical friction. However, these variables do decrease rather than attenuate, while the depth of the baroclinic layer is not constant. Moreover, all the three originators of the theory of integral mass flux were erroneous in their belief that they calculated the integral transport of the baroclinic ocean layer.

In fact, their model ocean is barotropic because the governing equation used by them has no density gradients and the only driving force for the function of integral mass flux is the curl of the wind tangent stress, which has no relation to water baroclinicity. It should be noted that a similar error was made in the paper of Sarkisyan (1956), a follower of the aforementioned originators. He believed that calculations were made for the baroclinic ocean layer; however, it turned out he had calculated the integral mass flux of wind-driven currents (the function ψ) of a homogeneous ocean.

Still what is the case of their failure? Formally the integral of the type $\int_0^H u\rho_1 dz$ which was used by the founders of the theory is an integral of mass transport of a baroclinic ocean because ρ_1 is variable. But $\rho_1 = \rho_0 + \rho$ and ρ is smaller than ρ_0 by three orders of magnitude. Therefore $\int_0^H u\rho_1 dz \approx \rho_0 \int_0^H u dz$, so this kind of integration does not help to pass from a homogeneous ocean to a baroclinic one. Actual baroclinicity is in gradients of ρ_1 or ρ.

Another idea of baroclinic ocean modelling simplification was put forward by Neumann (1955, 1958). He supposed that, at least in oceans mid-latitudes, a vertically mass compensation of baroclinic ocean may exist, therefore (as Welander thought too) the baroclinic ocean with variable depth behaves as the barotropic ocean with constant depth. Sarkisyan (1969a) showed that realistic modelling of the baroclinic ocean must be based on the general equations with due regard of water density horizontal gradients. The above situations show that the classics understood the drawbacks of the homogeneous ocean models and believed that they have passed to baroclinic models but in the era of analytical methods of solving the geophysical problems it was impossible. The Munk's model became very popular and many scientists tried to improve it. For example Neumann (1995, 1958) and Ivanov and Kamenkovich (1959) added the effect of bottom topography to the left hand side of equation (2.3.9), Carrier and Robinson (1962) added nonstationarity and nonlinearity, but each of them added only some terms to the left- hand side of equation (2.3.9) without passing to a baroclinic ocean model. That was impossible inasmuch as on the right- hand side of the equation $rot_z\vec{\tau}$ was the only driving force in their models. Within the model framework of a homogeneous ocean it is impossible to get the right point of the Gulf Stream separation, and its transport is several times smaller than the observation shows.

Based on progress in numerical mathematics and electronic computing it seemed prospective to join the Ekman's hydrodynamic model, advantages of the dynamic method and Lineykin's (1957) equation for the density evolution to make prognostic calculation of both the velocity and density fields of a baroclinic ocean.

As mentioned by Bryan (1969, in references to Chapter 1), the first calculation of that kind was done in early 1960s by Sarkisyan for a deep flat-bottom stratified ocean (see Sarkisyan, 1961, 1962 in references to Chapter 1).

The main results of those early calculations were the following:

1. The evolution of the density field is of several orders of magnitude slower that of the wind-driven circulation.
2. The flow velocity gradient currents are generated mainly via the density horizontal gradients, while the purely wind-driven gradient currents are of one order of magnitude smaller.
3. The oceans bottom topography should be taken into account.
4. Due to coarse resolution, the prognostically calculated gradient currents as well as the resulted density field gradients are several times smaller than the observed ones.

So one should carefully examine the terms in the equations for the sea surface topography ζ and the mass transport stream function ψ to keep the water baroclinicity. It appears that the main terms possessing the baroclinicity are the BARBE and JEBAR in equation for ζ and only JEBAR – in equation for ψ.

2.3.4 Brief List of the Main Results of Diagnostic Calculations

The results of diagnostic calculations made by the above three models have been analyzed in detail in the paper by Sarkisyan (1977), in the book of Marchuk and Sarkisyan (1988) and in many other papers. Here we shall deal with only the main conclusions.

1. The main indicator of climatic characteristics is ocean water baroclinicity, i.e. the horizontal variation of temperature and salinity. The resulting anomaly of ρ must be specified as accurate as possible.
2. Taking the observed and processed data on the ρ-field one has to avoid underestimation of the gradients of ρ. This requires the use of a high-resolution model and specification of H as accurate as possible.
3. When these two fields are specified wind stress should accurately be included in the calculation procedure as additional essential factor.
4. The calculations show that neither the dynamic method nor the Sverdrup relation may serve for accurate calculations of ocean climatic characteristics. Any out of the three equations (2.2.26), (2.2.48) and (2.2.62) is a further higher stage in comparison with the Sverdrup transport and/or the dynamic method.
5. The observed data on T and S are reliable only in the upper 0.5–1.5 km layer. The sea surface height (SSH) is formed mainly by the density anomaly of this upper layer and is less sensitive to ρ value of a deeper layer. The Ψ – function, vice-versa, is formed mainly by deeper layer data where the anomaly of ρ is small and indefinite, Fig. 2.2 shows that Ψ is much more sensitive to baroclinic ocean bottom topography than ζ. Figure 2.3 shows the degree of sensitivity of ζ and Ψ to different methods of extrapolation of ρ from 1.5 km depth to the ocean bottom. It is easy to see that Ψ is very sensitive, while ζ is much less sensitive, to the bottom layer baroclinicity. As a result of this sensitivity two visually

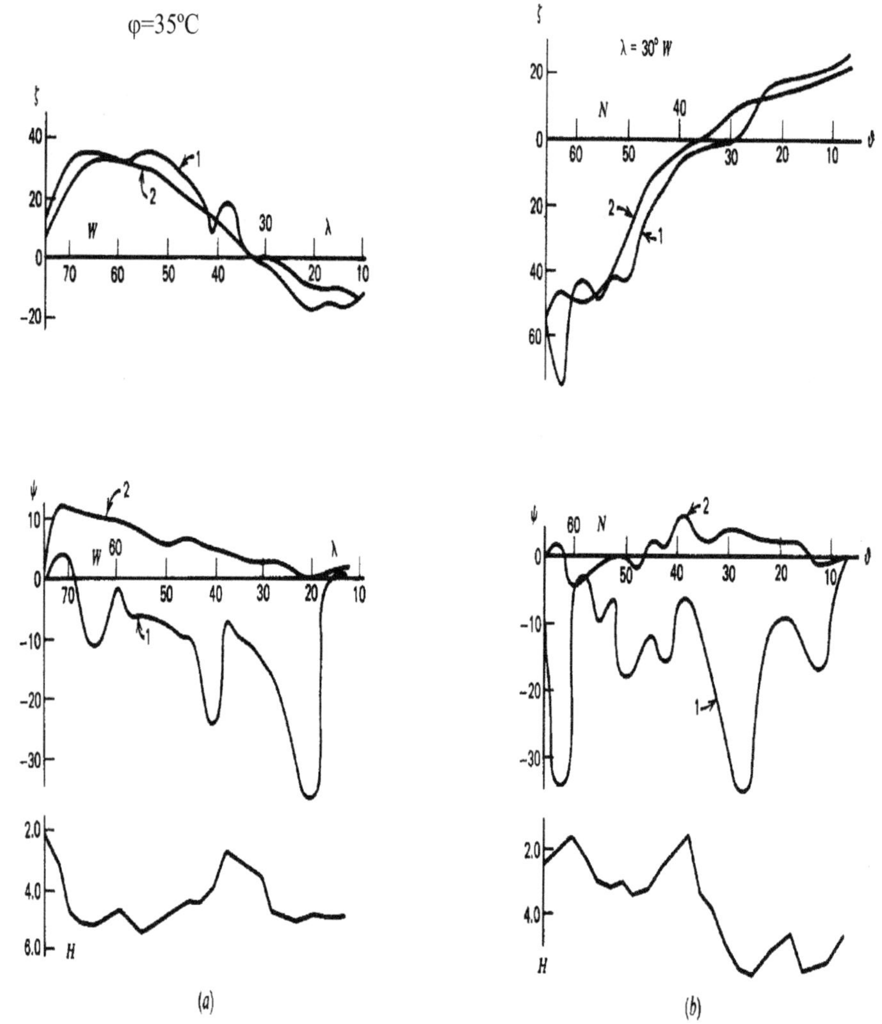

Fig. 2.2 Sea-surface topography ζ (cm) and mass transport stream function ψ ($10^6\,m^3/s$) at (**a**) sections along the 35 °N latitude and (**b**) 30 °W longitude of the North Atlantic. 1 – the case of real bottom topography; 2 – for $H = const = 1.5$ km. The bottom topography (km) along the sections is shown below (Sarkisyan, 1977)

different pictures of the Ψ – field may actually represent horizontal flow-fields which are almost similar in the upper but differ essentially from each other in the deeper layers. But any Ψ-field, even the one obtained without consideration of deeper layer baroclinicity, differs very much from the Sverdrup transport if bottom topography is taken into account because JEBAR plays the main role in the formation of Ψ.

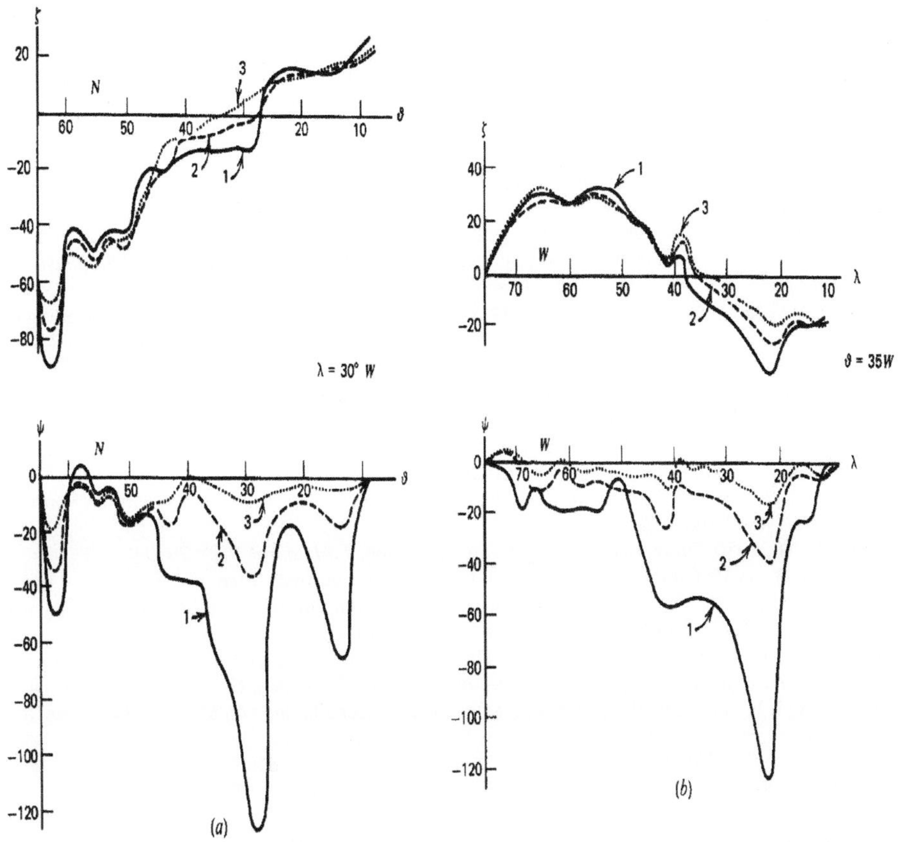

Fig. 2.3 Sea-surface topography and mass transport stream function along (**a**) 30 °W longitude and (**b**) 35 °N latitude of the North Atlantic by different methods of extrapolation of density anomalies at great depths. 1 – first method of extrapolation; 2 – second method of extrapolation; 3 – the case when the ocean is taken to be homogeneous below 1.5 km (Sarkisyan, 1977)

6. The vertical component of flow velocity is very sensitive to inaccuracies of $\vec{\tau}$, ρ, model resolution and model formulation. Two visually similar ζ or Ψ fields and consequently two similar surface gradient current fields may represent very different w-fields. It means that the main indicator of data plus model's accuracy is the vertical component of flow velocity, especially in the oceans' near-coastal areas.
7. With accurate observed data and high resolution simulation, all the three linear models D_1, D'_1 and D''_1, result in similar reliable circulation patterns, even in the coastal jet-like current areas. The only exception is the equatorial belt (equator $\pm 2°$) which needs nonlinear terms and is much more sensitive to model and data deficiencies.

References

Bryden D., S. San, and R. Bleck, 1999. A new approximation of the equation of state for seawater, suitable for numerical ocean models. *J. Geoph. Res.* 104(C1): 1537–1540.

Bryan K., 1969. A numerical method for the study of the circulation of the world ocean. *J. Comp. Phys.* 4: 347–376.

Carrier G. F. and A. R. Robinson, 1962. On the theory of the wind-driven ocean circulation. *J. Fluid Mech.* 12(1): 49–80.

Defant A., 1941. Die absolute Topographie des physikalischen Meeresniveaus und der Druckflächen sowie die Wasserbewegungen im Atlantischen Ozean. *Deutsche Atlantische Exped. "Meteor"* (1925–1927) Wiss. Ergeb. Bd. 6, Teil 2, IFR 5, 260 pp.

Ivanov Y. A. and V. M. Kamenkovich, 1959. The bottom topography as the main factor forming the Atlantic circumpolar current's non-sonality. *Doklady Akademii Nauk SSSR.* 128(6): 1167–1170 (in Russian).

Fofonoff N. P. and R. C. Millard, 1983. *Algorithms for Computation of Fundamental Properties of Seawater.* UNESCO, Tech. Pap. In Mar. Sci., 44, 53 pp.

Lineykin P. S., 1957. *The Main Problems of the Dynamic Theory of the Ocean Baroclinic Layer.* Gidrometeoizdat, Leningrad, 139 pp. (in Russian).

Marchuk G. I. and A. S. Sarkisyan, 1988. *Mathematical Modeling of Ocean Circulation.* Springer-Verlag, New York, 262 pp.

Munk W. H., 1950. On the wind-driven ocean circulation. *J. Meteorol.* 7(2): 79–93.

Neumann G., 1955. On the dynamics of wind-driven ocean currents. *Meteorol. Pap.* 2(4): 1–33.

Neumann G., 1958. On the mass transport of wind-driven currents in a baroclinic ocean with application to the North Atlantic. *Z. Meteorol.* 12(4–6): 138–147.

Peredery A. I. and A. S. Sarkisyan, 1972. The exact solution of several transformed equations of the sea currents' dynamics. *Izv. Acad. Nauk SSSR. Ser. Fiz. Atmos. Okeana* 8(10): 1073–1079.

Sandström I. W. and B. Helland-Hansen, 1903. Über die Berechnung von Meeresströmungen. *Res. On Norw. Fish. And Mar. Inst.*

Sarkisyan A. S., 1956. Calculation of stationary wind-driven currents in oceam baroclinic layer. *The USSR Acad. Sci., Proc. Of Geophysics Institute.* 37(164): 50–61.

Sarkisyan A. S., 1960. The density advection and westward intensification of the wind-driven currents. *Doclady Acad. Nauk SSSR.* 134(6): 1339–1342 (in Russian).

Sarkisyan A. S., 1961. On the role of the density advection by wind in dynamics of baroclinic ocean. *Izv. Ross. Acad. Nauk SSSR* 9: 1396–1407.

Sarkisyan A. S., 1962. On dynamics of wind-driven currents in a baroclinic ocean, *Oceanologia.* II(3): 393–409.

Sarkisyan A. S., 1969a, *Theory and computation of ocean currents*, U.S. Dept. of Commerce and the NSF, Washington, DC.

Sarkisyan A. S., 1969b. Deficiencies of barotropic models of ocean circulation. *Izv. Acad. Nauk SSSR, Ser. Fiz. Atmos. Okeana* 5(8): 818–835 (AGU English translation).

Sarkisyan A. S., 1974. Mechanism of the general oceanic circulation. *Izv. Aca. Nauk SSSR, Ser. Fiz. Atmos. Okeana.* 10(12): 1293–1308.

Sarkisyan A. S., 1977. The diagnostic calculation of a large scale oceanic circulation. *The Sea, Marine Modelling*, Vol 6, Wiley and Sons, New York, 363–458.

Sarkisyan A. S. and V. F. Ivanov, 1971. Joint effect of baroclinity and bottom topography as an important factor in sea dynamics. *Izv. Acad. Nauk SSSR, Ser. Fiz. Atmos. Okeana* 2(6): 818–835 (AGU English translation).

Shtockman V. B., 1946. Equations of full flow fields induced by wind in nonhomogeneous sea. *Dokl Akad. Nauk SSSR* 54(5): 403–406 (in Russian).

Stommel H., 1948. The westward intensification of wind-driven ocean currents. *Trans. Amer. Geophys. Union* 29: 202–206.

Sverdrup H. U., 1947. Wind-driven currents in a baroclinic ocean; with application to the equatorial currents of the Eastern Pacific. *Proc. Nat. Acad. Sci. Wash.* 33(11): 318–326.

Welander P., 1959. On the vertically integrated mass transport in the oceans. In: *The Atmosphere and the Sea in Motion.* B. Bolin (ed.), Rockefeller Institute Press, New York, 95–101.

Chapter 3
Nonlinear Models for Diagnostic, Prognostic and Adjustment Calculations of Ocean Climate Characteristics

Abstract In the following we will take into account nonlinear and nonstationary terms in the equations for momentum balance, temperature and salinity with. Moreover the simplified version (in terms of molecular mixing and diffusion) of horizontal and vertical turbulence will be considered. The models are different with respect to

- the method of calculation of the pressure anomaly (usually via sea surface height ζ or mass transport function Ψ),
- the numerical method of solution

Simplifications concern:

- the comparison with linear models of Chapter 2,
- the description of turbulent viscosity and/or diffusivity.

Keywords Nonlinear models · Prognostic calculations · Adjustment · Divergence model · Calibration

3.1 Nonlinear Models Based on Sea Surface Height Calculations

3.1.1 The Equation of Vorticity

In the following we will consider fully nonlinear models (3DPEM). For the sake of simplicity we will use Cartesian coordinates (Sarkisyan and Sündermann, 1995). We start from (2.1.1), (2.1.2) and (2.2.3) and proceed with the derivation of equations for ζ and ψ.

$$\frac{\partial}{\partial z} K_M \frac{\partial u}{\partial z} - \frac{\partial u}{\partial t} + fv - g\frac{\partial \zeta}{\partial x} - \frac{g}{\rho_0}\int_z^0 \frac{\partial \rho}{\partial x} dz + A_M \Delta u = A \tag{3.1.1}$$

DOI 10.1007/978-1-4020-9208-4_3,

$$\frac{\partial}{\partial z}K_M\frac{\partial v}{\partial z}-\frac{\partial v}{\partial t}-fu-g\frac{\partial \zeta}{\partial y}-\frac{g}{\rho_0}\int\limits_z^0\frac{\partial \rho}{\partial y}dz+A_M\Delta v=B \tag{3.1.2}$$

where

$$A=u\frac{\partial u}{\partial x}+v\frac{\partial u}{\partial y}+w\frac{\partial u}{\partial z} \tag{3.1.3}$$

$$B=u\frac{\partial v}{\partial x}+v\frac{\partial v}{\partial y}+w\frac{\partial v}{\partial z} \tag{3.1.4}$$

We now integrate equations (3.1.1) and (3.1.2) from $-H$ to 0 taking into account boundary conditions (2.2.11) and (2.2.12″)

$$-gH\frac{\partial \zeta}{\partial x}+fS_y+\frac{1}{\rho_0}\left(\tau_x-\tau_x^{-H}\right)-\frac{g}{\rho_0}\int\limits_{-H}^0(H+z)\frac{\partial \rho}{\partial x}dz-$$
$$-\frac{\partial S_x}{\partial t}+A_M\Delta S_x=\int\limits_{-H}^0 Adz \tag{3.1.5}$$

$$-gH\frac{\partial \zeta}{\partial y}-fS_x+\frac{1}{\rho_0}\left(\tau_y-\tau_y^{-H}\right)-\frac{g}{\rho_0}\int\limits_{-H}^0(H+z)\frac{\partial \rho}{\partial y}dz-$$
$$-\frac{\partial S_y}{\partial t}+A_M\Delta S_y=\int\limits_{-H}^0 Bdz \tag{3.1.6}$$

Cross differentiation of the equations results in the vorticity equation

$$-\beta S_y-J(H,\zeta)-\frac{\partial}{\partial t}rot\vec{S}+\frac{1}{\rho_0}\left(rot\vec{\tau}-rot\vec{\tau}^{-H}\right)$$
$$+\frac{g}{\rho_0}\int\limits_{-H}^0 J(\rho,H)dz-A_M\Delta rot\vec{S}+\frac{\partial}{\partial y}\int\limits_{-H}^0 Adz-\frac{\partial}{\partial x}\int\limits_{-H}^0 Bdz=0 \tag{3.1.7}$$

After that we substitute $\vec{\tau}^{-H}$ by ζ and ρ using formulae (2.2.37) and (2.2.3), use geostrophic correlations for an approximate value of S_y, S_x in expressions for the effect of horizontal mixing, ignoring in them the secondary part of both β-effect and the effect of bottom topography, and come to the following final equation for ζ (Sarkisyan, 1977).

$$\frac{\partial}{\partial t}\Delta\zeta + \frac{1}{H}J(H,\zeta) + \beta\frac{\partial\zeta}{\partial x} = \frac{1}{\rho_0 g H} rot_H \vec{\tau} - \underbrace{-\frac{f}{\rho_0 H}\int\limits_{-H}^{0} J(H,\rho)\,dz}_{JEBAR} \underbrace{-\frac{\beta}{\rho_0 H}\int\limits_{-H}^{0}(H-z)\frac{\partial\rho}{\partial x}dz}_{BARBE} + f_1 \tag{3.1.8}$$

where

$$f_1 = \frac{f}{gH}\left[\frac{\partial}{\partial y}\int\limits_{-H}^{0} A dz - \frac{\partial}{\partial x}\int\limits_{-H}^{0} B dz\right] - \frac{1}{\rho_0 H}\int\limits_{-H}^{0}(H-z)\frac{\partial\Delta\rho}{\partial t}dz + \\ + A_M\Delta\Delta\zeta - \frac{f}{2\alpha H}\Delta\zeta - \frac{1}{\rho_0 g H}\left[\beta\tau_x - \frac{\partial}{\partial t}div\vec{\tau} + \frac{\beta}{f}\frac{\partial\tau_y}{\partial t}\right] + \\ \frac{A_M}{\rho_0 H}\int\limits_{-H}^{0}(H-z)\Delta\Delta\rho dz - \frac{f}{2\alpha\rho_0 H}\int\limits_{-H}^{0}\Delta\rho dz \tag{3.1.9}$$

Equation (3.1.8) is written in the form, which is valid not only for diagnostic but also for prognostic and adjustment calculations. Moreover, the experience (which will be discussed in the following chapters) shows that even in diagnostic calculations all terms in momentum equations, especially in the vorticity equation for ζ (or for Ψ), should be taken into account without simplifications. An exception is made for the linear parts of the inertial acceleration $\left(\frac{\partial u}{\partial t}, \frac{\partial v}{\partial t}\right)$ and horizontal turbulent mixing, which we take into account in this model in a simplified form (quasigeostrophic approximation). The first term of the left-hand side is simple and generated due to these approximations.

Concerning the right-hand side of equation (3.1.8) JEBAR and BARBE (baroclinic β-effect), play a definitive role in climate modelling. Further important are the linear and nonlinear terms generated by the integral of A and B in (3.1.9) plus $\frac{\partial}{\partial t}\Delta\rho$. As for the other terms in (3.1.9), they are secondary; in many cases they may be ignored without generating an essential error.

Comparing equations (3.1.8) and (2.2.46), the common peculiarity is that the main terms are BARBE and JEBAR in the right-hand parts, as well as their "relatives" – β-effect and the effect of bottom topography in the left-hand parts. With reasonable values of the turbulence coefficients characteristic values of all other terms (linear or nonlinear) are one or several orders of magnitude smaller. In any case, once horizontal and vertical mixing effects are taken into account the value of the coefficients K_M, A_M should be taken as small as possible, because the terms

$A_M \Delta\Delta\zeta$, $\frac{1}{2\alpha H}\Delta\zeta$ can not reflect the real turbulent mixing and large values of K_M, A_M result in an overly smoothed solution.

Thus, the main difference between (3.1.8) and (2.2.46) is not just the appearance of the nonlinear terms on the right-hand side of (3.1.8) but also the existence of the first term on the left side. This term enables one to ignore both horizontal turbulent mixing and bottom stress. This is a principal advantage of (3.1.8) against the linear stationary vorticity equation (2.2.46). The term $\frac{\partial}{\partial t}\Delta\zeta$ allows to take into account side boundary conditions for ζ, that is why the equation is written in an unusual form-there is no turbulence effect on its left-hand side.

Actually, there is no need to ignore the effect of turbulence, despite its idealized form, besides, it is useful for a more accurate presentation of side boundary conditions for flow velocity. Still equation (3.1.8) allows one at least to operate with smaller values of A_M and K_M to avoid over filtering.

The principal scheme of performing calculations by this model is as follows.

For diagnostic calculations one should first solve equation (3.1.8) ignoring A and B as a first approximation. Then by the simple formula (2.2.3) p is calculated, and then, by using equations (2.1.1), (2.1.2), (2.1.6), the flow velocity components. Usually, within quite a short model time period (weeks to months) the variables reach a steady state with high accuracy. Although a variety of numerical methods for solving the equations considered for adjustment or prognostic calculations we advise first to make diagnostic calculations, then, using specified T and S and calculated u,v,w as initial data and adding the equations for T and S. The details of calculations, their analysis and a short review will be given in the following chapters.

3.1.2 The Divergence Equation

To derive another version of the nonlinear model which also uses the sea surface height we return to equations (3.1.5), (3.1.6) and make the divergence, instead of vorticity operation. It means differentiating (3.1.5) by x and adding (3.1.6) differentiated by y. The resulting equation is the following:

$$\Delta\zeta + \frac{1}{H}\left(\frac{\partial H}{\partial x}\frac{\partial \zeta}{\partial x} + \frac{\partial H}{\partial y}\frac{\partial \zeta}{\partial y}\right) = \frac{1}{\rho_0 g H} div\vec{\tau} - \frac{1}{\rho_0 H}\int_{-H}^{0}(H+z)\,\Delta\rho dz -$$

$$-\frac{\beta}{gH}S_x - \underbrace{\frac{1}{\rho_0 H}\int_{-H}^{0}\left[\frac{\partial H}{\partial x}\frac{\partial \rho}{\partial x} + \frac{\partial H}{\partial y}\frac{\partial \rho}{\partial y}\right]dz}_{JEBAR-2} + \frac{f}{gH}\left(\frac{\partial S_y}{\partial x} - \frac{\partial S_x}{\partial y}\right) -$$

$$-\frac{1}{gH}\left[\frac{\partial}{\partial x}\int_{-H}^{0}Adz + \frac{\partial}{\partial y}\int_{-H}^{0}Bdz\right] + \frac{1}{\rho_0 g H} div\vec{\tau}^{-H} \tag{3.1.10}$$

During these transformations we ignored the secondary effect of bottom topography in terms, generated by horizontal mixing. The result of transformation of the terms $\frac{\partial S_x}{\partial t}, \frac{\partial S_y}{\partial t}$ and the main effect of horizontal mixing took the form of total mass divergence and, in accordance with equation (2.2.44) vanished. In (3.1.10) the Laplacians on the left- and the right-hand sides are generated by pressure gradients; the β-effect, the effects of bottom topography and JEBAR have other forms than those in (3.1.8). Direct wind effect is represented by $div\vec{\tau}$, instead of $rot\vec{\tau}$; the last term may be kept or omitted depending on the boundary condition at the ocean bottom. The divergence equation was formulated for the equatorial belt and naturally it does not degenerate when f becomes zero (Demin and Sarkisyan, 1977). The nonlinear terms should necessarily be taken into account. In the following chapters we show that (3.1.10) is valid for the extraequatorial latitudes too.

The principal scheme of diagnostic, prognostic or adjustment calculations based on divergence equation (3.1.10) is the same as described above for the vorticity equation (3.1.8).

It should be mentioned that despite the absence of time derivative in (3.1.10), ζ depends on time parametrically via time variation of $\vec{\tau}$ and $\rho\,(T, S)$.

3.2 Nonlinear Extraequatorial Model Based on the Mass Transport Stream Function Calculation

In the previous section two equivalent models were presented for calculation of the physical characteristics of ocean climate (with specified nonstationary conditions at the ocean free surface). We have discussed one of them based on the ζ-equation.

In principle there is now an easy way to use Ψ, instead of ζ. For example, we can differentiate (2.2.60) by x, add to it (2.2.61) differentiated by y and make transformations to obtain a very simple equation for Ψ

$$\frac{\partial}{\partial x}\frac{f}{gH}\frac{\partial \Psi}{\partial x} + \frac{\partial}{\partial y}\frac{f}{gH}\frac{\partial \Psi}{\partial y} = \Delta\zeta + \frac{\partial A_1}{\partial x} + \frac{\partial B_1}{\partial y} \tag{3.2.1}$$

where

$$A_1 = \frac{1}{\rho_0 H}\int_{-H}^{0}(H+z)\frac{\partial \rho}{\partial x}dz - \frac{\tau_x}{\rho_0 g H}$$

$$B_1 = \frac{1}{\rho_0 H}\int_{-H}^{0}(H+z)\frac{\partial \rho}{\partial y}dz - \frac{\tau_y}{\rho_0 g H} \tag{3.2.2}$$

A_1, B_1 might contain nonlinear terms too but practically this is not necessary because ζ is calculated by taking into account all nonlinear terms. In principle the equation (3.2.1) allows to calculate Ψ, once ζ is calculated by one of the above presented models. Despite of it, there is a demand to construct an equation for Ψ which does not contain ζ explicitly because:

- it is common in oceanographic literature to use an integrated vorticity equation based on Ψ;
- quite interesting qualitative conclusions arise when comparing the equation for Ψ with the equation for ζ presented above.

To get the equation for Ψ we return to the simplified relations (2.2.60), (2.2.61), insert them into equation (3.1.8), make some transformations and further simplifications (ignore the secondary parts of β-effect and the effects of bottom topography in smaller terms, keep only the highest differentials).

The resulting vorticity equation has the following complicated form:

$$\begin{aligned}
&-A_M \Delta\Delta\Psi + \frac{\partial}{\partial t}\Delta\Psi + \frac{f}{H}J(H,\Psi) + \beta\frac{\partial\Psi}{\partial x} = \frac{1}{\rho_0} rot_z \vec{\tau} + \\
&+\frac{1}{\rho_0}\frac{\partial}{\partial t} div\vec{\tau} + \frac{\beta}{\rho_0 f}\left(\tau_x - \frac{\partial \tau_y}{\partial x}\right) + \underbrace{\frac{\beta}{\rho_0 f}\left(\frac{\partial H}{\partial y}\tau_x - \frac{\partial H}{\partial x}\tau_y\right)}_{JEWAR} - \\
&\underbrace{-\frac{g}{\rho_0 H}\int\limits_{-H}^{0} zJ(H,\rho)\,dz}_{JEBAR} + f_2
\end{aligned} \tag{3.2.3}$$

where

$$f_2 = \frac{\partial}{\partial y}\int\limits_{-H}^{0} A dz - \frac{\partial}{\partial x}\int\limits_{-H}^{0} A dz - \frac{f}{2\alpha H}\Delta\Psi - \frac{g}{2\rho_0 \alpha H}\int\limits_{-H}^{0} z\Delta\rho dz \tag{3.2.4}$$

There is a new joint effect on the right-hand side of equation (3.2.3) which consists in the multiplication of wind stress components with gradients of bottom relief. The JEWAR does exist implicitly in equations (3.1.8), (3.1.10) too, the explicit version of this factor in (3.2.3) shows, that the effect of bottom relief is essential even in a coastal shallow homogeneous water area.

Many other terms in (3.1.8) and (3.2.3) are alike. In (3.2.3) the effect of horizontal mixing is kept on the left-hand side just because that form is common in the papers where Ψ is used instead of ζ.

But there are two principal differences between (3.1.8) and (3.2.3):

(1) There is no BARBE in the equation for Ψ. That is why in (3.2.3) the presence of JEBAR is absolutely necessary. In (3.1.8) BARBE serves as another essential source of potential energy to favour the westward intensification. In Section 2.3 we already discussed the problem of coastward intensification and, in particular,

westward intensification and we shall be returning to it in the following chapters more than once because until now in scientific literature there exists a misunderstanding concerning this aspect of modelling.

(2) The JEBAR's shape itself in (3.2.3) differs essentially from its analog in (3.1.8) and, in general, it differs from any JEBAR in equatorial or extraequatorial, linear or nonlinear models based on the calculation of SSH.

The difference is generated by the multiplier z existing under the integral of equations (3.2.3) right-hand side.

In the thin upper ocean layers, density gradients are larger and are known to a higher accuracy but z is small. In the lower layers, which represent the major part of the oceans 'water body it is exactly vice-versa. So, Ψ is formed mainly by an erroneous deeper layer density field, while ζ, on the contrary, by more reliable information of the upper density field. This aspect should be kept in mind when comparing charts of these two functions.

Figures 2.2 and 2.3 presented above at the end of Chapter 2 prove this on the basis of diagnostic calculations. Same kind of analysis was done based on adjustment or prognostic calculations and brought to analogous conclusions (Sarkisyan, 1977; Marchuk and Sarkisyan, 1988).

Here are some further conclusions:

1. All its linear terms, apart from the effect of baroclinicity, are expressed in terms of Ψ; in addition, if the first term on the right-hand side, i.e., the joint effect of baroclinicity and relief (JEBAR) and the effect of bottom stress are neglected, then (3.2.3) is transformed into a single equation in one unknown, which is very convenient.
2. When solving problems for a simply connected closed domain there are the following simple boundary conditions along the horizontal:

$$\Psi \equiv 0, \qquad \frac{\partial \Psi}{\partial n} \equiv 0.$$

It is likely that these two circumstances resulted in the fact that general ocean circulation models relying on an equation of form (3.2.3) are in wide use. But there are a number of inconveniences associated with the employment of this equation.

1. Once it has been shown that the neglect of the JEBAR is erroneous, the first convenience has dropped out. Now, it is necessary to solve mutually the overall system of equations of hydrothermodynamics even in the linear setting.
2. As discussed above the equation form with the JEBAR on the right-hand side of (3.2.3) is correct from formal mathematical viewpoint but leads geophysically to inaccurate results, since the Jacobian is multiplied by z in the integrand expression.
3. When modelling the Word Ocean circulation at a more or less satisfactory resolution, a multiply connected domain is dealt with. The condition $\Psi \equiv 0$ may prescribed only along the contour of one island (continent); in this case the values

of Ψ (which is kept constant along the contour of each island) for all other islands must be found through cumbersome calculations (Bryan, 1969).

The next equation to be analyzed is (3.1.8). Like (3.2.3) it is a vorticity equation, and the operator with respect to ζ on its left-hand side is similar to the one with respect to Ψ in (3.2.3). It remains to compare the right-hand sides of these equations and ways of determining the functions along the contours of basins or islands. It is easy to preassign Ψ along the boundary provided a simply connected domain (a sea or a part of an ocean without islands) is dealt with, whereas from corresponding boundary conditions for velocity components, ζ should be determined at each point of a boundary calculation grid. However, when there are island(s) one should calculate as many additional functions Ψ as the number of islands are, where ζ is being calculated locally along each island separately, and the number of islands, does not impose additional requirements on the procedure of calculations or the computer memory (Marchuk, 1969; Sarkisyan, 1977); this is convenient. Another circumstance is more important. On the right-hand side of (3.1.8) the density gradients in integral expressions of the basic terms are not multiplied by z, i.e., ζ depends slightly on inaccurate values of the density gradients in deep layers. Furthermore, on the right-hand side of (3.1.8) the baroclinic β-effect (BARBE) is an important source of available potential energy in addition to the JEBAR. For these and other reasons mentioned above, as numerous diagnostic calculations have shown (Marchuk and Sarkisyan, 1988; Sarkisyan, 1977), ζ is formed in the main by the density gradients of basin upper layers and depends on the bottom relief to a lesser degree. For (3.2.3), the explicit influence of seawater baroclinicity on Ψ is lost, provided it is assumed that $H = \text{const}$. Finally, there is one more advantage of (3.1.8). If the small terms on the right-hand side of (3.1.8) are neglected (the terms with the multiplier z in the integrand expressions are omitted, and the action of wind and the nonlinear terms are neglected) as a first approximation, then it is easily seen that an exact solution of (3.1.8) with the simplified right-hand side is the formula:

$$\zeta = -\frac{1}{\rho_0} \int_{-H}^{0} \rho dz \tag{3.2.5}$$

As mentioned in Chapter 2 this formula presents a dynamical method for determination of the sea level in the case when the basin floor serves as a reference level. It may be shown that (3.2.5) is valid with lesser simplification of the right-hand side of (3.1.8) as well, provided the Ekman spiral is taken as velocity profile at the basin bottom and the nonlinear terms are replaced by their geostrophic values. So, (3.1.8) with the simplified right-hand side, unlike (3.2.3), has an important indicator, a useful test for its numerical computations.

The diagnostic calculations have shown that (3.2.5) comprises almost two thirds of a numerical solution of (3.1.8) without any simplifications. In other words, the dynamic method is a good first approximation of problems concerning dynamics of a baroclinic ocean with complete consideration for all the factors.

Equation (3.1.10) was radically different from (3.2.3) and (3.1.8). When deriving this equation the idea was to hold the pressure rather than to exclude it, since the pressure is a basic element in the dynamics of seas and oceans. Thus cross the differentiation in obtaining (3.1.10) is the divergence of the velocity but not its curl. The terms with the Laplace operator on the right- and left-hand sides of the equation are basic. The second group of terms on the right-hand side of (3.1.10) gives another form of the JEBAR, i.e., the integrand expression

$$\frac{\partial H}{\partial x}\frac{\partial \rho}{\partial x} + \frac{\partial H}{\partial y}\frac{\partial \rho}{\partial y}, \tag{3.2.6}$$

which is of the same order as the main terms. Unlike the JEBAR existing on the right-hand side of (3.2.3), the expression (3.2.6) is not the only source of the potential energy, i.e., even with $H = \text{const}$ one of the two main sources of baroclinicity on the right-hand side remains and, what is more, is qualitatively important. Equation (3.1.10) was constructed with consideration for all these advantages; it formed the basis of many diagnostic and adjustment investigations of dynamics of seas and oceans.

However, one disadvantage of (3.1.10) is worth noting. The integral effect of Earth's rotation in this equation is rearranged to the right-hand side and, as is the case with the nonlinear inertial terms, is taken into account by the method of successive approximations. This is natural in the case when the curl of the horizontal velocity is small. Such a condition is well-satisfied near the equator. Thus (3.1.10) is particularly efficient in the vicinity of the equator where $f \to 0$. It was precisely for this purpose when this equation was derived almost 40 years ago (Demin and Sarkisyan, 1977; Sarkisyan and Serebryakov, 1969). Subsequent repeated applications of (3.1.10) for diagnostic and adjustment calculations of the overall World Ocean and many seas have revealed its efficiency in off-equatorial regions as well. At the same time, this equation has not been employed for long-time prognostic calculations. It should be also mentioned that the time is taken into consideration in (3.1.10) in a parametric way, and the effect of horizontal turbulent viscosity is reduced by cross differentiation. Finally, it should be noted that solution (3.2.5) satisfies this equation as well, provided only the basic part of terms containing the anomaly of the density ρ is kept on the right-hand side of (3.1.10) as a first approximation; this fact proves the applicability of the dynamical method for sea level calculations also for the equator. Naturally this is not the case with calculations of the flow velocity proper. The mentioned convenience in solving problems for multiply connected domains is relevant not only to (3.1.8) but to (3.1.10) as well.

So, we have given the three variants of equation for integral functions. Any one of them is applicable for the solution of problems of the dynamics of the seas and oceans. Some simplifications (different for each equation) have been made in their derivation. This does not mean that these equations are of different degrees of accuracy. First, these simplifications are not necessary at all; they are justified only by the fact that the awkward expressions of the second order in accuracy, which are not needed in a qualitative analysis, are excluded. Second, these equations differ from each other not from the mathematical standpoint but geophysically, depending on the problem being solved.

A final remark. In a number of studies analytical expressions of all terms of equations like (3.2.3), (3.1.8), or (3.1.10) are not derived in an explicit form. Rather than differentiate the equations, researchers switch to finite differences without any simplification, which is needed only for a visual illustration. In these cases not all advantages and disadvantages are evident to readers (and, probably, to authors as well).

3.3 Quasianalytical Version of One Nonlinear Model

Two out of the three nonlinear models presented above become less effective at the equator: equation (3.2.3) for Ψ loses practically the direct effect of bottom topography, and equation (3.1.8) for ζ loses JEBAR. Furthermore, there are no time derivatives in equations (2.2.26), (3.1.10) constructed originally for the ocean's equatorial area. This means that they are valid for investigation of slow processes only. Thus they cannot be considered as quite general ones, despite their usefulness in extraequatorial areas as well. There is a need for a more general model and a respective numerical scheme.

To construct the dynamical part of a more general model we follow the paper of Sarkisyan and Serebryakov (1969).

Let us rewrite the momentum balance equations (3.1.1) and (3.1.2) with small redistributions:

$$\frac{\partial u}{\partial t} - fv = -g\frac{\partial \zeta}{\partial x} - \frac{g}{\rho_0}\int_z^0 \frac{\partial \rho}{\partial x}dz + \frac{\partial}{\partial z}K_M\frac{\partial u}{\partial z} + A_2 \tag{3.3.1}$$

$$\frac{\partial v}{\partial t} + fu = -g\frac{\partial \zeta}{\partial y} - \frac{g}{\rho_0}\int_z^0 \frac{\partial \rho}{\partial y}dz + \frac{\partial}{\partial z}K_M\frac{\partial v}{\partial z} + B_2$$

where

$$A_2 = A_M\Delta u - \left[u\frac{\partial u}{\partial x} + v\frac{\partial u}{\partial y} + w\frac{\partial u}{\partial z}\right]$$

$$B_2 = A_M\Delta v - \left[u\frac{\partial v}{\partial x} + v\frac{\partial v}{\partial y} + w\frac{\partial v}{\partial z}\right]$$

During the subsequent transformations we shall use the following boundary conditions for u and v in a vertical direction: at the sea surface the conditions (2.2.11) and at $z = -H$ – the conditions (2.2.12′).

Altering the time-differentials in (3.3.1) by directed differences (just for simplicity) and denoting the time step by δt we have:

$$u^{t+1} - f\delta t v^{t+1} = u^t -$$

$$-\delta t \left(g\frac{\partial \zeta}{\partial x} - \frac{g}{\rho_0}\int\limits_z^0 \frac{\partial \rho}{\partial x} dz - \frac{\partial}{\partial z} K_M \frac{\partial u}{\partial z} - A_2 K \right)^{t+1} = A_3$$

$$v^{t+1} - f\delta t v^{t+1} = v^t -$$

$$-\delta t \left(g\frac{\partial \zeta}{\partial y} - \frac{g}{\rho_0}\int\limits_z^0 \frac{\partial \rho}{\partial y} dz - \frac{\partial}{\partial z} K_M \frac{\partial v}{\partial z} - B_2 K \right)^{t+1} = B_3. \quad (3.3.2)$$

For simplicity we drop the upper index $t + 1$, temporarily assume A_3 and B_3 to be specified functionals and consider (3.3) as two algebraic equations for u and v. Then we get:

$$u = \frac{A_3 + f\delta t B_3}{1 + (f\delta t)^2}, \; v = \frac{B_3 - f\delta t A_3}{1 + (f\delta t)^2} \quad (3.3.3)$$

or approximately

$$u \approx \frac{1}{1 + (f\delta t)^2} \Bigg\{ u^t - \delta t \Bigg[g\frac{\partial \zeta}{\partial x} + \frac{g}{\rho_0}\int\limits_z^0 \frac{\partial \rho}{\partial x} dz - \frac{\partial}{\partial z} K_M \frac{\partial u}{\partial z} -$$

$$-A_1 - f v^t + f g \delta t \frac{\partial \zeta}{\partial y} + \frac{f g \delta t}{\rho_0}\int\limits_z^0 \frac{\partial \rho}{\partial y} dz - f g \delta t \frac{\partial}{\partial z} K_M \frac{\partial v}{\partial z} - f\delta t B_1 \Bigg] \Bigg\} \quad (3.3.4)$$

$$v \approx \frac{1}{1 + (f\delta t)^2} \Bigg\{ v^t - \delta t \Bigg[g\frac{\partial \zeta}{\partial y} + \frac{g}{\rho_0}\int\limits_z^0 \frac{\partial \rho}{\partial y} dz - \frac{\partial}{\partial z} K_M \frac{\partial v}{\partial z} -$$

$$-B_1 - f u^t + f g \delta t \frac{\partial \zeta}{\partial x} + \frac{f g \delta t}{\rho_0}\int\limits_z^0 \frac{\partial \rho}{\partial x} dz - f g \delta t \frac{\partial}{\partial z} K_M \frac{\partial u}{\partial z} - f\delta t A_1 \Bigg] \Bigg\}$$

The terms with a higher order of δt were neglected when passing from (3.3.3) to (3.3.4). Actually it is not difficult to keep them, but there is no need in them, because presentation of an accurate equation for ζ is not our principal goal. The principal aim is to keep the main parts of all kinds of different factors just for a qualitative analysis, but not all the terms of all the factors.

Now we insert the expressions of u, v from (3.3.4) into the complete equation of continuity (without using simplified conditions of "rigid lid")

$$\frac{\partial \zeta}{\partial t} + u^{(0)}\frac{\partial \zeta}{\partial x} + v^{(0)}\frac{\partial \zeta}{\partial y} + \frac{\partial}{\partial x}\int\limits_{-H}^0 u dz + \frac{\partial}{\partial y}\int\limits_{-H}^0 v dz = 0 \quad (3.3.5)$$

make some transformations and simplifications to obtain an equation for SSH, ζ:

$$a_1 \Delta\zeta + a_2 \frac{\partial \zeta}{\partial x} + a_3 \frac{\partial \zeta}{\partial y} + \zeta = \zeta^t + F_1 \tag{3.3.6}$$

where

$$a_1 = \frac{gH(\delta t)^2}{1+(f\delta t)^2} \tag{3.3.7}$$

$$a_2 = \frac{g(\delta t)^2}{1+(f\delta t)^2}\left[f\delta t \frac{\partial H}{\partial y} - \frac{\partial H}{\partial x} + \beta H \delta t\right] + u^{(0)}\delta t \tag{3.3.8}$$

$$a_3 = \frac{-g(\delta t)^2}{1+(f\delta t)^2}\left[f\delta t \frac{\partial H}{\partial x} + \frac{\partial H}{\partial y} - \beta f H (\delta t)^2\right] + v^{(0)}\delta t \tag{3.3.9}$$

$u^{(0)}, v^{(0)}$ – are the components of the sea surface currents and

$$\begin{aligned} F_1 = \frac{\delta t}{1+(f\delta t)^2} \Bigg\{ & -\frac{fg(\delta t)^2}{\rho_0} \underbrace{\int_{-H}^{0} J(H,\rho)\,dz}_{JEBAR} - \\ & -\frac{g(\delta t)^2}{\rho_0} \underbrace{\int_{-H}^{0} \left(\frac{\partial H}{\partial x}\frac{\partial \rho}{\partial x} + \frac{\partial H}{\partial y}\frac{\partial \rho}{\partial y}\right) dz}_{JEBAR-2} + \frac{\beta g(\delta t)^2}{\rho_0} \underbrace{\int_{-H}^{0} (H+z)\left(\frac{\partial \rho}{\partial x} - \frac{\partial \rho}{\partial y}\right) dz}_{BARBE} - \\ & -\frac{g\delta t}{\rho_0}\int_{-H}^{0} (H+z)\,\Delta\rho\, dz + \frac{f(\delta t)^2}{\rho_0} rot_z \vec{\tau} - \frac{\beta(\delta t)^2}{\rho_0}\tau_x + \\ & + \frac{\delta t}{\rho_0} div \vec{\tau} - \frac{\partial}{\partial x}\int_{-H}^{0} u^t dz + \frac{\partial}{\partial y}\int_{-H}^{0} v^t dz \Bigg\} + F_2 \end{aligned} \tag{3.3.10}$$

where F_2 is a complicated functional containing the effects of nonlinear terms, horizontal mixing, the terms of a higher order of smallness and other kinds of combinations of factors, the main parts of which exist already in (3.3.6). The presentation of their full analytical expression is not needed for a qualitative analysis. Moreover, the neglected terms are actually small, and both adjustment and prognostic calculation may be made without them. Note the shape of equation (3.3.6) at the equator. Formulae (3.3.7)–(3.3.9) show that at the equator ($f = 0$), a_1 does not change essentially, a_2 and a_3, lose only part of their items, F_1 loses one of the two JEBAR's and one more term, and that is all. Thus the main terms of equation (3.3.6) and its right-hand side are not subject to essential variations at the equator. This was the main goal of constructing equation (3.3.6). Thus the system of equations of this general nonlinear model is the following. The equation for ζ is (3.3.6), the equations for calculation of u and v are presented by (3.1.1), the equations for T and S are naturally (2.1.3) and (2.1.4). The choice of the remaining equations is trivial: those are (2.1.8), (2.1.12) and (2.2.3).

We are sure that all the nonlinear terms of the equations for $u,\ v,\ w,\ \zeta,\ T,\ S$ should necessarily be taken into account without any exception, while bottom stress may be neglected (for deep ocean or sea) and the coefficients of horizontal mixing should be taken as small as possible.

The model enables us to use free-slip conditions at the coasts of the continents and islands, which is preferable in comparison with non-slip conditions, because large-scale processes (with the horizontal scale of more than 5–10 km) are considered.

One aspect of this model should be discussed. When constructing the previous three nonlinear models, the choice first of all is made of the main dynamical equation for the integral function: either equation of vorticity (chosen by most modelers) or the equation of divergence. The choice of one of these two versions by modelers is often arbitrarily. This is not the case in this model. As the reader has seen, simply from two momentum equations the components of flow velocity were defined via other functionals, and the expressions for u and v were inserted naturally into the equation of continuity. The resulting equation appeared to be an equation for ζ containing all the factors that existed in the previous three nonlinear models (two kinds of JEBAR, BARBE, β-effects and the effects of bottom relief) and many other terms. Thus, it is obvious that this is the most general of the four models and, besides, the structure of (3.3.6) shows that this is a divergence equation. Here is one more argument in favour of choosing the divergence rather than the vorticity equation and using ζ instead of Ψ. Actually, we have not chosen it ourselves, it naturally has become a divergence equation because the continuity equation (3.3.5) is a divergence equation.

Inasmuch as many modelers use Ψ it is worthwhile to construct an equation for the mass transport stream function, which, at least partly, may serve as alternative to (3.3.6). For this goal we transform the left-hand side of equation (3.3.6) by using the correlations (2.2.60), (2.2.61); we have

$$
\begin{aligned}
&-\frac{a_1}{g}\left(\frac{\partial}{\partial x}\frac{f}{H}\frac{\partial\Psi}{\partial x}+\frac{\partial}{\partial y}\frac{f}{H}\frac{\partial\Psi}{\partial y}\right)+\frac{f}{gH}\left(a_2\frac{\partial\Psi}{\partial x}+a_3\frac{\partial\Psi}{\partial y}\right)=\\
&=\frac{a_1}{\rho_0 gH}\left(\frac{\partial H}{\partial x}\tau_x+\frac{\partial H}{\partial y}\tau_y\right)-\frac{a_1}{\rho_0 gH}div\vec{\tau}-\frac{1}{\rho_0 gH}\left(a_2\tau_x+a_3\tau_y\right)+ \qquad (3.3.11)\\
&+\frac{a_1}{\rho_0 H}\int\limits_{-H}^{0}(H+z)\,\Delta\rho dz+\frac{a_1}{\rho_0 H}\int\limits_{-H}^{0}\left(\frac{\partial\rho}{\partial x}\frac{\partial H}{\partial x}+\frac{\partial\rho}{\partial y}\frac{\partial H}{\partial y}\right)dz+\\
&+\frac{1}{\rho_0 H}\int\limits_{-H}^{0}(H+z)\left(a_2\frac{\partial\rho}{\partial x}+a_3\frac{\partial\rho}{\partial y}\right)dz+\zeta^t-\zeta+F_1.
\end{aligned}
$$

Expressions (3.3.7)–(3.3.10) and the right-hand side of equation (3.3.11) show that both kinds of JEBAR, the Joint effect of wind and topography (JEWAR), $div\vec{\tau}$, $rot\vec{\tau}$ and even the constant $\vec{\tau}$ explicitly take part in formation of water mass transport.

It is easy to show by scaling that the effect of $div\vec{\tau}$ is not any smaller than that of $rot\vec{\tau}$. Moreover, in the next chapters it will be shown by concrete calculations that the effect of wind stress constant for the whole basin is of the same order of magnitude as the effect of $rot\vec{\tau}$. All these suggest that there is poor need for any of the two contradictory and overly simplified models - the half century old Sverdrup correlation and the century old dynamic method.

3.4 Comments on Different Approaches and Different Integral Equations Used in Ocean Modelling

Modelling of the ocean may be conventionally represented as two main "avenues" originating almost simultaneously. The first "avenue" concerns calculations of ocean characteristics on the basis of a system of equations with consideration for initial and boundary conditions but without using data of deep-ocean observations. Studies in line with this avenue are called the theory of oceanic flows or prognostic calculations, since a system of corresponding equations of thermodynamics and boundary conditions at the ocean surface are most significant with this approach, while initial conditions are of secondary importance or do not play any considerable role at all. The studies of Marchuk's school (Marchuk, 1969, 1980; Sarkisyan, 1969a,b), on the one hand, and those of Bryan's school (Bryan, 1969; Cox, 1970), on the other hand, have shown that there are yet insuperable difficulties along this avenue of investigations. Scientifically, a principal difficulty is the parametrization of turbulent processes; technically, limited potentialities of supercomputers present a barrier to researchers.

The second "avenue" considered below means a synthesis of the complete system of equations of thermodynamics (prognostic models) and observational data obtained both at the surface and inside the ocean. The method has a claim on the construction of initial data as well as the diagnosis and reconstruction of climatic characteristics of the ocean. The first stage of the method consists in diagnostic calculations of characteristics of ocean dynamics, i.e., in calculations of the three components of the flow velocity from values of the temperature (T), salinity (S) and the tangential wind friction ($\vec{\tau}$) specified on grid points with allowance for realistic geometry of a basin. Despite the assumption on stationarity of the fields of T and S, the equations of motion may be nonstationary; in the case they are solved until the flow velocity is completely (not only from the geophysical standpoint but also mathematically) determined with an *a priori* prescribed high accuracy. We will not concentrate on this problem and refer readers to the corresponding literature (see, for example, Sarkisyan, 1969a; Sarkisyan, 1977; Demin et al., 1990, etc.).

Despite of a number of interesting results, the method of diagnostic calculation is of limited scientific interest. The next step is mutual adjustment of all physical characteristics. This stage of modelling was given several different names, namely "semidiagnostic method" (Sarkisyan and Demin, 1983), adjustment (initialization) (Marchuk and Sarkisyan, 1988; Sarkisyan and Sidorova, 1998), diagnostic-prognostic calculation (Ezer and Mellor, 1994; Mellor, 1996), etc.

The common practice of calculations presented in this series of papers is as follows.

Taking specified *T, S* and the corresponding results of diagnostic calculation as an initial state, short model-time prognostic calculations are made using the whole system of relevant equations and boundary conditions. The procedure of short model-time of integration makes mutual adjustment of diagnostically obtained dynamic characteristics with *T* and *S* and does not affect essentially the specified temperature and salinity fields. The results of such calculations pretend to serve as observed data rather than the data of theoretical-numerical experiments. The time evolution of kinetic energy shows that the process of mutual adjustment of thermodynamic characteristics via current numerical models is connected with a drastic decrease in kinetic energy. The typical behavior of kinetic energy versus time both in diagnostic and prognostic (adjustment) calculations is presented in Fig. 3.1 (Sarkisyan, 2001).

Special numerical experiments showed that the main part of available potential energy loss happens at the very beginning of the adjustment process (Bulushev and Sarkisyan, 1996). This loss happens sooner or later, may be very big or not too big, depending on the model resolution and/or quality, but usually the loss is big and the decrease is drastic. Actually the moment of transfer of kinetic energy decay from rapid to slow is used as a stopping criterion for the adjustment process. The adjustment process filters out the subgrid-scale features (eddies, loops, waves, fronts), make smoother the isolines of *T* and *S*, but still keeps the main large-scale features of specified hydrological elements. An example of the sea surface transformation from the diagnostic stage to the adjusted version is shown in Fig. 3.2. All the above conclusions made as a result of diagnostic calculations still stand after adjustments. Because of short model-time of integration today's computers allow to make adjustment calculation with sufficiently fine resolution.

The adjustment calculation allows to extrapolate the statistically consistent observed data to the neighboring area. One example of such nonlinear thermohydrodynamic extrapolation is the method of isolated hydrologic sections data processing (Sarkisyan and Stepanov, 1999; Levitus and Sarkisyan, 2001). It has a great advantage because the major part of ocean deep-layer hydrographic measure ments including those made by the WOCE programme consists of single *T, S* sections.

The method of thermohydrodynamic extrapolation enables to step forward from the century-old "dynamic" (reference-level) method of data processing. The first experiments showed the undeniable success of the method.

Here, the emphasis is on the second stage of the method, namely, on adjustment calculations. Studies are based on the overall available hydrologic information

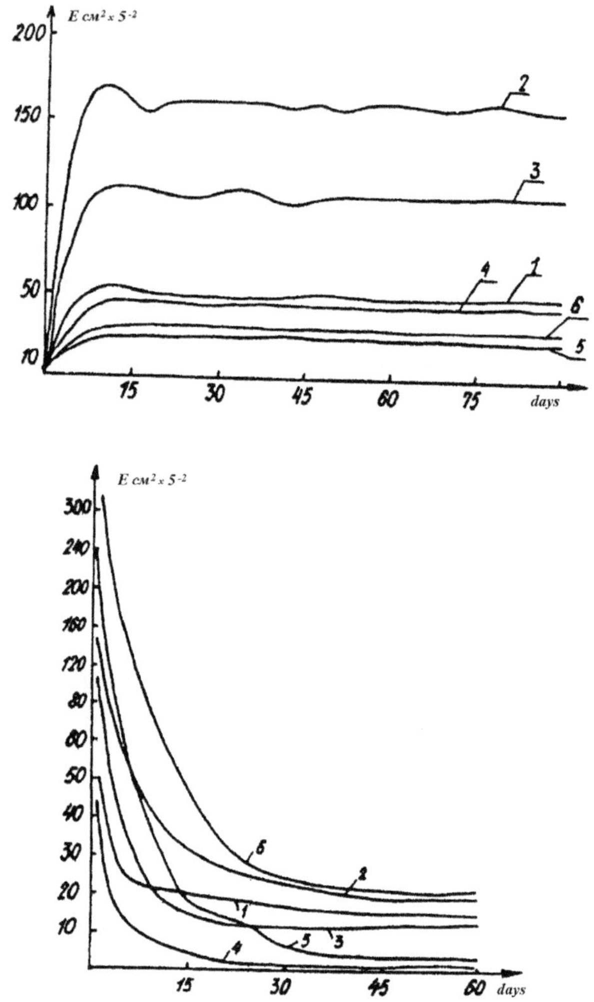

Fig. 3.1 Typical behavior of normalized kinetic energy (KE) versus time in (*upper*) diagnostic and (*lower*) prognostic (adjustment) calculations. The numbers of curves denote: 1 – at 5 m ($\times 10^2$) depth; 2 – at 50 m ($\times 10$); 3 – at 100 m ($\times 10$); 4 – at 500 m ($\times 10$); 5 – at 4000 m ($\times 10$); 6 – vertically averaged ($\times 20$). The multipliers in brackets should be used to obtain the actual values of KE from the corresponding curves (Sarkisyan, 2001)

and the same complete system of equations as that for prognostic investigations; however, unlike these latter, integration is performed over a short interval of model time. Furthermore, as opposed to diagnostic calculations, in adjustment calculations the equations for T and S are solved; in so doing preassigned values of T and S play the role of initial data and may be incomplete, i.e., values of T and S should not necessarily be prescribed at all grid points at the initial time moment. In other words, adjustment calculations are a nonlinear thermohydrodynamic interpolation in space and extrapolation in time aimed at the reconstruction of necessary fields (of flows and partially of T and S), their adjustment to the equations, boundary conditions, and geometry, and their mutual adjustment. It should be mentioned that this adjustment is incomplete, and mathematically inaccurate, and the result strongly depends on initial data, the purpose at hand, and even on the level of geophysical

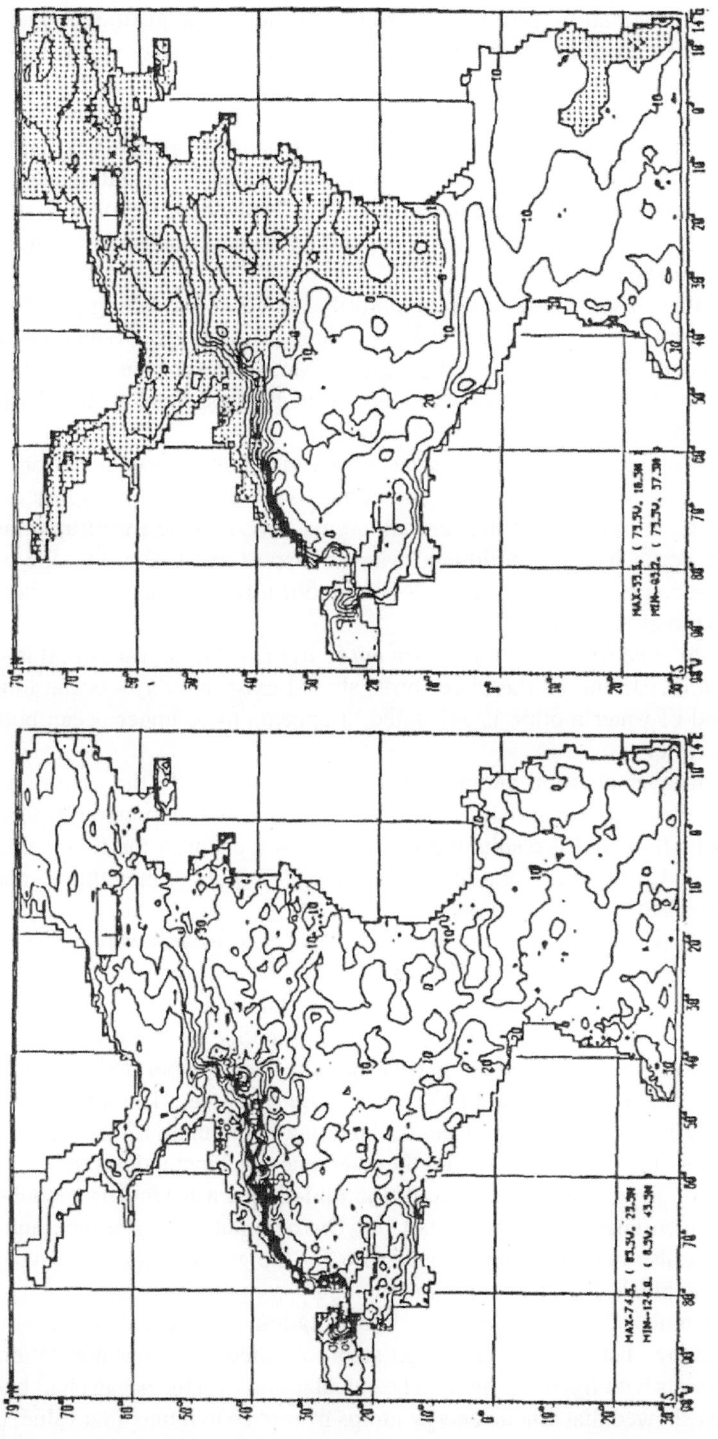

Fig. 3.2 An example of the SSH (cm) transformation from **(a)** diagnostic to **(b)** adjustment stage. Contour intervals – 10 cm; the values in shaded area are negative (Sarkisyan, 2001)

knowledge.The pragmatic approach proposed in our opinion is justified by the fact that the system of equations and boundary conditions are inaccurate themselves. With this approach initial conditions are more important; in particular, this refers to measured values of the water temperature and salinity, which are regarded by us as being most accurate. To present some results of adjustment calculations and their comparison with prognostic investigations, we will base on the initial equations in the spherical system of coordinates, like (1.1.1)–(1.1.7) of Chapter 1, or in Cartesian system, like (2.1.1)–(2.1.8) of Chapter 2, depending on the considered basins size, and relevant boundary conditions.

We begun this survey by dividing temporarily the ocean modelling papers into two directions. It is time now to see "where we are". The Ekman – type modelling with homogeneous liquid is of limited prospects. Mostly the chosen idealization is too strong to be of practical significance. Baroclinicity is essential even in shallow waters like the Azov Sea with its depth of some 8 m.

The circulation of shallow coastal zones of oceans or seas depends not only on the local wind stress but on off-coast baroclinic forces as well. The upper ocean Ekman pumping is only part of the vertical circulation. Also the Sverdrup relation balancing the ocean integral meridional water transport and the curl of the wind mass is a simple approximation. At the same time it turns out that $rot\vec{\tau}$ is not the only ocean external mechanical driving.

The wind stress actually acts in the form $rot\vec{\tau}$, $div\vec{\tau}$ and even just as $\vec{\tau}$. It is seen from equation (3.10) that all the three forms should exist in today's ocean model. Once any kind of water motion is generated, it appears to be under ocean bottom topography control.

The bottom layer pressure anomaly does not follow exactly the bottom relief isolines. This seemingly small inclination is the effect of bottom topography forcing to horizontal motion, and it results in the vertical component of the flow velocity, which is maximal at the ocean bottom and becomes 2–3 order of magnitude smaller at the ocean surface.

The β-effect plays an essential but not decisive role in westward intensification. Apart from pressing water to the western coast there is separation from the western coast; there are coastal intensifications, coastal (especially east-coastal) intense upwelling-downwelling processes. The baroclinicity of "open ocean" zonal currents, bottom topography, and the JEBAR do play a very important (and in some circumstances and places determining) role in those processes. The important role of baroclinicity in the generation of the ocean climatic circulation was understood at the beginning of the 20th century. But the reference-level method has many drawbacks listed in Chapter 2. That is why in the mid-sixties a diagnostic model was constructed which takes into account not only baroclinicity, the β-effect, and the wind stress but also the other factors necessary for the geophysical hydrodynamic model, for example, basin's geometry and boundary conditions.

The old diagnostic calculations with linear stationary models revealed many interesting features listed above. The next step presented nonlinear nonstationary diagnostic models with fixed T, S fields. The calculations with the ocean (sea) basins initially at rest showed that kinetic energy jumps from zero to it maximal value, then after some small-amplitude oscillations it reaches the steady state in a model-time of month to years, depending on the basin and model parameters.

Interestingly enough the adjustment and prognostic calculations confirmed almost all the conclusions made in the diagnostic calculations. The main exception is the drastic decrease in kinetic energy at the adjustment stage and its further slow decrease at the prognostic stage. This is a problem of parametrization of turbulence. Another problem is the use of Levitus's *T, S* fields as initial data, but these fields are already overly filtered. The diagnostic calculations played their positive role, and there are no computer or computational problems in the adjustment calculations. The last may be done with very simple parametrization of turbulence and small value of turbulent coefficients. The adjustment calculations are now done starting with absolutely the same original equations and in some cases with the same boundary conditions as the prognostic calculations. It means that both directions of ocean modelling, which were discussed at the beginning of this paper, met with each other. The formal difference is the model-time of integration.

As a result the adjustment calculations pretend to present the "observed" data while the prognostic calculations may answer to some theoretical questions and may give a forecast. The main problem of the prognostic calculation is their strong dependence on turbulence parameters.

The adjustment calculations will be necessary in the near future despite the successes of the prognostic calculations. There are at least four reasons:

1. The adjustment calculations may be done with as high resolution as reasonable, they result in more intense currents and more detailed (less filtered) *T, S* fields and present "observed" data for calibration of the prognostic models.
2. One can take two periods (each 5–15 years or so) for data averaging, make adjustment calculations for each period separately and then by comparison make a conclusion about the ocean climate variations between this periods. Examples of such investigations can be found in Ezer et al. (1995) and Levitus (1989a,b,c, 1990).
3. The adjustment calculations may serve to fill the gaps in observed data, based on statistically consistent data, instead of statistical interpolation and filtering. Undoubtedly this hydrodynamical method has advantages because it gives all thermohydrodynamical (and mutually adjusted) elements and not just Levitus's *T, S* fields. An example was given in Ivanov et al. (1997).
4. Finally, the adjustment method is used for processing single hydrographic sections. The WOCE sections are the examples. Preliminary results of such processing were given in Levitus and Sarkisyan (2001).

3.5 A Brief Analysis of Prognostic Calculations

As mentioned by K. Bryan (1969) baroclinic ocean modelling began in early 1960-ies (Sarkisyan, 1961, 1962). In those papers was shown that nonlinear density advection play the main role in ocean climatic currents generation and thermodynamic characteristics evolution. As for the $rot_z\vec{\tau}$, the effect of wind stress is essential first of all in its role in the wind-drift advection of density anomalies. It means that analytical models are too idealized and privilege should be given to numerical modelling. In those first papers the momentum balance equation were linearized.

The first calculations with simplified models already demanded tens of years of model-time for spin-up (Kochergin et al., 1972; Marchuk et al., 1973, 1976, 1979; Friedrich, 1967, 1970; Friedrich and Klimok, 1988; Klimok et al., 1986; Sündermann et al., 1982). A more sophisticated model of Bryan type with very idealized initial data and coarse resolution demanded thousands of years. That is why M. Cox noted that the calculation with higher resolution is impractical. Nowadays almost every scientist making prognostic calculations begins with Levitus's T, S fields as initial data rather than with idealized uniform stratification (Beckmann and Döscher, 1997; Bleck et al., 1992; Bleck and Smith, 1990; Chassignet et al., 1996; DYNAMO Group, 1997; Haidvogel and Beckmann, 1999; Marsh et al., 1996; Mikolajewicz et al., 1994).

As for the model-time integration the scientists' opinions and practical realizations differ widely, from tens to thousands of years (Kochergin et al., 1972; Zalesny, 1996, 1997; Demyshev et al., 1992).

Note that the JEBAR does exist implicitly in any paper where both baroclinicity and the bottom relief are taken into account. Papers of Bryan (1969) and Friedrich (1970) provide the first examples of such contributions. But this effect is practically lost by coarse resolution and/or too long model-time investigation with an overestimated mixing coefficient. The JEBAR was examined in many interesting papers and monographs (see e.g. Myers et al., 1996). Lately the essential role of this factor was once more pointed out in a paper by Friedrich and Sündermann (1998), Mellor (1999) and in series of papers of Dirk Olbers with coauthors (see for example Olbers and Eden, 2003).

Bryan (1969) and Bryan et al. (1975) pioneered the performance of prognostic calculations, relying on the complete nonlinear model until climatic characteristics were adjusted starting from an idealized initial state. The fields of temperature, salinity, or only density have been qualitatively reconstructed in these works, although in a considerably smoothed form as compared with the known charts prepared from observational data. The fields of general circulation have been also qualitatively obtained, although with significantly underestimated absolute values of the flow velocity. It has been revealed that partly in view of the underestimated flow velocity, the processes of adjustment in deep layers are very slow (millennia of model time); therefore the local derivatives have been artificially accelerated up to two thousand fold, and the local heat capacity of deep waters has been decreased (up to twenty-five fold) so that calculations may be realized under the potentialities of 1970-ies supercomputers at least with a rough horizontal resolution (3°–6°) and increased time step (up to hundredfold in deep layers). However, the point is not only, if at all, in the underestimated velocities, smoothed fields, and distorted physics of deep oceanic layers. In our opinion, a principal difficulty is the problem of parametrization of subgrid-scale processes.

The calculations performed by Bryan et al. (1975) have demonstrated that first, considerably increased values of the coefficient of horizontal viscosity ($\approx 10^9$–10^{10} cm^2 s^{-1}) are to be preassigned in models of this type and, second, computational results are very sensitive to values of the coefficient of turbulent viscosity. Taking into consideration the fact that the case in point is a linear parametrization of subgrid-scale processes with phenomenological coefficients, we arrive at the conclusion that this is the main difficulty in prognostic calculations. It is natural that the question arises of whether it is possible to decrease numerical values of the parameters and reduce the sensitivity of computational results to them through a decrease in horizontal step. Since we are going to concentrate our attention on this problem, numerous interesting studies conducted by K. Bryan with disciples and colleagues with models of this type will not be discussed here, whereas several works devoted to prognostic calculations with a comparatively high resolution will be briefly analyzed. It is the practice to call studies of such a kind eddy-resolving models. In our opinion, this name is not appropriate. First, a model does not change because of the resolution step at which calculations are performed. Second, it is not less important to resolve jet flows, fronts, and upwellings. However, there is another peculiarity in recent numerical experiments. Since K. Bryan has shown that the adjustment of the temperature and salinity of deep oceanic layers to forcings acting at the ocean surface is of a very slow nature, it has become evident that even with the most powerful supercomputers it is impossible to combine a high resolution and a complete adjustment of deep oceanic waters with a temperature and salinity being initially constant along the horizontal. Therefore in many studies T and S taken Levitus's atlas are taken as initial conditions.

Modelling of the overall World Ocean on the basis of the "primitive" equations of thermohydrodynamics at a resolution of 0.5° with 20 vertical levels is a cumbersome problem even with supercomputers like those exploited by Semtner and Chervin (1992). Thus the authors have resorted to a number of simplifications and, partly, to disputed, in our opinion, methods for the calculations to be brought to completion.

As a result, the World Ocean simulated by the authors consists of several parts in which, in essence, different systems of equations have been solved, since the method of Sarmiento and Bryan (1982) is sensitive to the value of the coefficient γ for the additional source $\gamma\,(T_1 - T)$, introduced for the simulated ocean to avoid a model climatic drift occurring in integrations over a long time. This is the main disadvantage of the calculations performed by Semtner and Chervin (1992). It is correct to mention that the above disadvantages result from difficulties which have not been surmounted by any present-day author of prognostic calculations with much higher resolution. Thus the study carried out by Semtner and Chervin (1992), despite its limitations, is an essential contribution to computational oceanology.

When performing calculations, first, the authors have prescribed a multi-year average annual forcing at the ocean surface, and then a seasonal forcing has been preassigned. Moreover, the representations of the horizontal turbulent viscosity in the form of the Laplace and the biharmonic operators have been employed. It is natural that the results of integration for the past 10 years of model time with the seasonal forcing and the biharmonic representation of the horizontal turbulent viscosity are of prime interest. Numerous charts, curves, sections of calculated characteristics, their comparisons with observational data, and their analyses are given in this voluminous work. Let us concentrate on some results to which little attention has been given by the authors, if at all. Besides, some conclusions arrived at in the paper do not receive our support. We will dwell on their criticism as well.

Let us begin with variations of the basic integral function Ψ. The calculations with the two variants (average annual and seasonal forcing) have shown that the fields of Ψ averaged over time are almost identical visually; thus the authors show only the variant with the seasonal forcing. It is mentioned that with a few exceptions, Ψ in the middle latitudes is governed by Sverdrup's dynamics. The influence of the seasonal forcing on the sea level is even less noticeable. The authors explain this circumstance by the assumption that a baroclinic compensation of pressure proceeds with depth. We do not support this conclusion. A baroclinic compensation may proceed, first, only in some regions of the real ocean (the Gulf Stream, the Canary current, and so on) and, second, at significant depths. The circulation and density gradients in deep layers slightly depend on seasonal variations and slightly influence the sea level ζ; it is the integral transport Ψ on which they have a stronger effect. A reason should be searched for in limitations of the model. We will concentrate on one of them below.

With such an insignificant standard deviation of the seasonal sea level ζ from the average annual value, the spectra of oscillations of Ψ for the overall 10 years of integration with the seasonal forcing are unusual. More interest in these spectra has been expressed by us than by the authors. Some values in our tables are inexact, since we have visually derived them from figures; however, this fact will not influence our qualitative conclusions.

The presence of a wide spectrum of oscillations with different periods in different regions (Tables 3.1 and 3.2) and nonperiodicity despite the fact that the forcing is a regular harmonic seasonal oscillation is quite natural, since an external action is followed by the influence of geometry, nonlinearity of the intraoceanic interaction, and different rates of origination and evolution of flows in upper and lower layers of the ocean. So, even if the feedback of the ocean on the atmosphere is not taken into consideration, the ocean has its own internal dynamics and thus is not passively subjected to the action of the atmosphere; in particular, it does not repeat a harmonic oscillation. This is reflected in Tables 3.1 and 3.2 for many regions of the World Ocean and presents a reliable and interesting result obtained by Semtner and Chervin (1992). A strictly seasonal behavior of the meridional velocity component in the Somali current (Table 3.2) in the upper oceanic layer (37.5 m) is clear as well.

It is also interesting that the presence of multi-year trends or oscillations which have not engaged authors' attention are visually noticeable in their figures; we have commented on them in the tables. However some other results require a critical evaluation; as follows.

1. For some reason, the overall body of the ocean in six regions of intensive currents (four of them are listed in Table 3.1) is subjected to strictly seasonal oscillations.

Table 3.1 The principal characteristics of the integral mass transport in the jetlike flows (compiled from paper by Semtner and Chervin, 1992)

Current	Fundamental period of oscillations	Limits of oscillations in Sverdrups	Average value in Sverdrups	Comments
Gulf Stream	100 days to 1 year	(+15)–(+45)	30	
Kuroshio	Above 1 year	(−20)–(+200)	100	A multi-year trend toward decreasing is noticeable
Brazilian current	1 year	(−100)–(−40)	−70	A multi-year oscillation is noticeable
Drake Passage	Half a year to 1 year	(+196)–(+204)	200	
Agulhas current	30–50 days	(−50)–(−130)	−70	
Somali current	Strictly seasonal cycle	(−30)–(+60)	20	
Oyashio	Strictly seasonal	(−10)–(−50)	−30	
Labrador current	Strictly seasonal	(−35)–(−50)	−40	A multi-year oscillation is noticeable
East Madagascar current	300–100 days	(−35)–(−65)	−45	
East Australian current	100–400 days	(−40)–(+70)	+30	Significant multi-year variations, a trend toward decreasing
North Brazilian current	Strictly seasonal	(+10)–(+30)	20–25	
Current of the Gulf of Mexico	200–300 days	(+9)–(+18)	15	

2. The Gulf Stream is "underevaluated" by the authors, since its average flow is smaller than that of the Kuroshio, Brazilian, Agulhas, or East Australian currents by a factor of 2.5–3 and is 1.5 times smaller than that of the Labrador and East Madagascar currents; it is comparable only with the flows of the Oyashio and North Brazilian currents.

The above remarks may be supplemented with the following two disadvantages: (1) the place of the Gulf Stream separation is not in agreement with observational data and (2) the global pattern of Ψ is governed by Sverdrup's dynamics, and we regard this as a disadvantage as well.

We have listed these four disadvantages in view of the fact that in our opinion, they are of a common origin. This is the vorticity equation with a subtle balance

Table 3.2 Some characteristics of the meridional component of the flow velocity at a depth of 37.5 m (compiled from paper by Semtner and Chervin, 1992)

Current	Fundamental period of oscillations	Limits of variations in cm/sec	Comments
Gulf Stream	About 50 days and more	(+50)–(+90)	A multi-year oscillation is also noticeable
Current of the Gulf of Mexico	Half a year	(−70)–(+30)	A multi-year behavior of the oscillation amplitude is noticeable
Somali current	Strictly seasonal	(−70)–(+120)	
Eastern part of The central Pacific	30 days and more	(60)–(+100)	

of terms; the solution of this equation is sensitive to the value of the coefficient of horizontal turbulent diffusion (when integrating over a long time period), and this coefficient is overestimated. As integration is performed, the gradients of temperature and salinity (density, pressure), which are partly produced by internal thermohydrodynamics of the ocean itself from preassigned initial values of T and S and partly result from the heat exchange across the ocean surface and the work of wind on the redistribution of density, are smoothed.

A smoothing of these internal gradients results, on the one hand, in their underestimation, in the suppression on the role of the JEBAR (and the vorticity equation is very sensitive to this factor) and, on the other hand, in the overestimation of the role of a direct wind action (Demin and Ibrayev, 1986; Sarkisyan, 1995). The subjection to Sverdrup's dynamics, the underestimated value of the Gulf Stream flow (the basin width as well as the value and constancy of sign of the gradient of the zonal wind component play an important role in Sverdrup's dynamics), the strictly seasonal oscillations of Ψ in some regions, and so on may be explained by this. The strictly seasonal oscillation of Ψ has a peculiarity; we reproduce a figure from the authors. Figure 3.3 shows that changes become seasonal beginning immediately in the first year of the imposition of seasonal boundary conditions, that is, the ocean has not needed much time for deep layers to come under oscillation. This means that many oceanic regions have been already subjected to the influence of wind, i.e., to Sverdrup's dynamics, over the previous 22 years of integration.

Unfortunately, the authors do not point out the reasons for their employment of a diagnostic variant of calculations in the equatorial zone. One more peculiarity may be revealed in the equatorial zone; this is an anomalous, in our opinion, intensity of circulation.

Prognostic calculations performed by Rosati and Miyakoda (1988) are of great interest. The calculations were carried out for 18 levels of the World Ocean at a horizontal step of 1° everywhere except for the tropical zone from 10 °S to 10 °N, in which a step of $(1/3)^\circ$ was used. The minimum depth of the ocean comprises 25 m, and the maximum depth is 3 km.

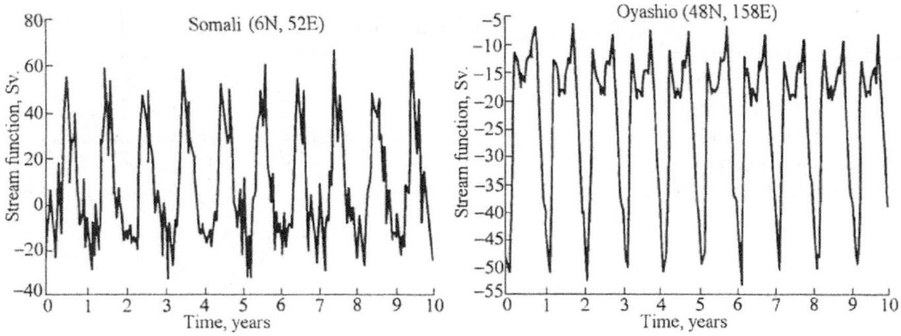

Fig. 3.3 An example of seasonal behavior of flows in the regions of the Oyashio and Somali currents (in Sverdrups) (Semtner and Chervin, 1992)

The calculations were done in the following way. At the first stage multi-year monthly average fields of the temperature T and salinity S were taken as initial data, and integration was begun from a state of rest. The calculation of adjustment (initialization), until a seasonal behavior in the upper oceanic layer is established, was then performed over 6 years of model time. Thereupon, taking the results of calculations as initial data and the meteorological information (heat fluxes, the wind, field and so on) over 1983 and 1984 as boundary conditions at the ocean surface, Rosati and Miyakoda have calculated two variants of "actual" values of oceanographic characteristics over these 2 years of the catastrophic El Niño.

The main achievement of this work is the consideration of parametrization of subgrid processes and of high-frequency atmospheric action. Thus these are precisely the factors whose influence on the sea surface temperature (SST), the structure of flows, the depth of a mixed layer, and so on is studied by the authors.

The authors could decrease the value of the eddy coefficient by 1–2 orders of magnitude off the tropical zone and in the "free ocean" without appearance of parasitic waves. This is a great progress, since a traditionally overestimated value of A_M in prognostic calculations presents a significant disadvantage. However, in the tropical zone, where the step is 3 times smaller, the coefficient increases almost up to $10^8\,\mathrm{cm}^2\,\mathrm{s}^{-1}$, which is no longer desirable. The significantly overestimated value of the coefficient of horizontal turbulent viscosity with the decreased step is a considerable limitation of the parametrization used by Rosati and Miyakoda.

In global charts the difference between the calculated and measured values of the SST is maximum in the regions of intensive currents (4°–5°), and it amounts up to 7° near the coast of Peru. The authors have made use of actual observational data and performed a comparison between the temporal behavior of the calculated and measured fields of temperature and zonal velocity at a point in the equator, at 95 °W, in the vertical section from the surface down to 200 m over 1982 and 1983 (Figs. 3.4 and 3.5). The authors give a correct qualitative description of the temporal variations; however, the calculated gradients of the characteristics are smoothed, and the flow velocity is smaller than the measured one by a factor of 3.

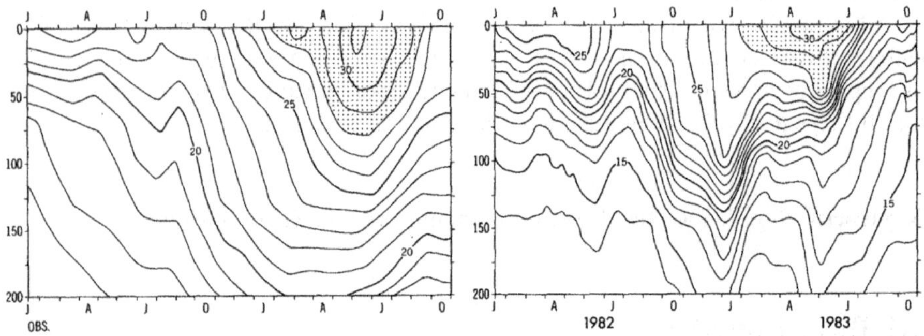

Fig. 3.4 A temporal behavior of the temperature (°C) at 95 °W at the equator over 1982–1983; the left- and right-hand fragments are prepared from the calculations by model E-12 (Rosati and Miyakoda, 1988) and observational data, respectively

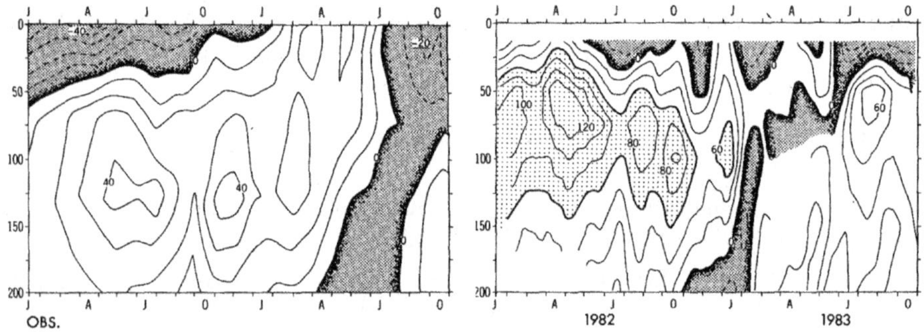

Fig. 3.5 The same as in Fig. 3.4 but for zonal velocity component (in centimeters per second). Regions of flows directed to the West are dark colored; regions where the velocity is above 60 cm/sec are light colored

Fields of T and S are taken as initial data in many studies, including both papers under review. This means that from the onset, apart from the external permanently acting factor rot$\vec{\tau}$, one more source, namely, the JEBAR, is incorporated into the right-hand side of (3.2.3). The origination, evolution, setting, and a final state of the field of Ψ depend not only on the coefficients of turbulence, steps in space, and fields of $\vec{\tau}$ and ρ but also on the length of integration interval. Whereas the initial values of T and S are the same in all three studies in question and the external force $\vec{\tau}$ at least is preassigned in a common way, the scatter in many other parameters is significant. It is known that an overestimated value of A_M is necessary for computational stability to be provided, and this leads to a smoothing of computational results. Only Rosati and Miyakoda (1988) could decrease A_M by 1–2 orders of magnitude everywhere except for the equatorial zone and some regions of intensive currents along the western coast. The steps along the vertical are adequate in all the three studies, since 18–30 levels describe the vertical structure of flows with some degree of satisfaction. The horizontal steps are still somewhat large, but this depends exclusively on potentialities of computers.

The main scatter is observed in the length of integration intervals. The adjustment to Levitus's data is performed in the course of 3–25 days and 6–32 years of model time at fluid surface areas and in the ocean, respectively. In some cases only a seasonal behavior in the upper oceanic layer is set at times, and in other cases the overall body of the ocean is significantly transformed. Furthermore, in some cases the common system of equations is solved, and in other cases an artificial source like $\gamma\,(T - T_1)$ is introduced following the method of Sarmiento and Bryan (1982). The question on the optimum length of adjustment interval remains open. It may seem that potentialities of computers are fully responsible for this as well. It is our belief that in Semtner and Chervin (1992), an integration over such long intervals is not optimum. Since in none of these three works the evolution of the JEBAR or at least of Ψ at the initial stage of calculations is discussed, we have permitted ourselves to prove this statement relying on our experience and the net present list charts of Ψ; unfortunately they present no information on the flows of the Gulf Stream and Kuroshio as well.

We have seen above that from the calculations of Semtner and Chervin the function Ψ, with a few exceptions, is characterized by Sverdrup's dynamics; besides, Table 3.5.1 demonstrates that the average mass flux of the Kuroshio exceeds that of the Gulf Stream more than by a factor of 3, which is not realistic.

Let us attempt to answer the question as to what factors and in what way should evolve for Sverdrup's dynamics to be set on (3.2.3) in consequence of integration over an unoptimally long interval.

There are two main sources, namely, the JEBAR and wind, and one main sink, namely, the horizontal viscosity, on the right-hand side of (3.2.3). An integration starts with the state of rest; thus the average flow of any current begins to grow under the influence of the two above sources and attains its maximum. However, as the length of integration interval increases, the effect of sinks becomes more and more pronounced, to be more precise, an overestimated viscosity in the right-hand side of (3.2.3) and the effects of diffusion in the equations of heat and salt transport are meant. Furthermore, unlike the wind, the JEBAR is also decreased in the average through a smoothing effect of dissipative forces. It is worth noting that the heat exchange sources specified at the upper boundary generate or intensify the gradients of density in the upper oceanic layer; however, this is shaded in the course of integration over a long interval with overestimated sinks. The diagnostic method of calculations (from the viewpoint of Sarmiento and Bryan (1982)) employed by Semtner and Chervin (1992) at the levels below 710 m has not "rescued" them in an attempt to overcome this impediment. The model drift exerts more and more considerable action on computational results. As the length of the integration interval is increased, the density gradients, which have been not intense because of the smoothness of the temperature and salinity fields taken from Levitus's atlas, are smoothed in the average; the absolute values of the flow velocity decrease as well. This results in the fact that in the right-hand side of (3.2.3) the only external permanently acting factor, which is not subjected to smoothing, namely, $\mathrm{rot}\vec{\tau}$, becomes prevailing. As for the left-hand side of this equation, the gradients of Ψ become smaller and smaller. The net result is the establishment of an almost Sverdrup balance in the bulk of the ocean, to be more precise, the following relations:

$$\beta \frac{\partial \Psi}{\partial x} \approx \frac{\mathrm{rot}\vec{\tau}_z}{\rho_0} \tag{3.5.1}$$

$$\Psi \approx \frac{1}{\beta \rho_0} \int_{x_0}^{x} \mathrm{rot}\vec{\tau}_z dx \tag{3.5.2}$$

where $x_0\,(\varphi)$ is the value of x on the eastern coast of the ocean.

From (3.5.2) it is easy to understand why the Gulf Stream flows are smaller than those of other intensive currents. A greater flow intensity at a latitude necessitates a wider ocean, a larger value of rot$\vec{\tau}$, and less number of sign alternation. It is our belief that even with a horizontal resolution as high as necessary the same point is responsible for the inaccuracy in determination of the place where the Gulf Stream is separated. As is known, the derivative of the zonal component of rot$\vec{\tau}$ with respect to latitude in the Atlantic (which is a principal part of rot rot$\vec{\tau}$) changes the sign at a significant distance to the North of Cape Hatteras. This is precisely what results in a displacement of the point of separation to the North. We propose to stop integration prior to the establishment of a Sverdrup balance, for example, when flows of many intensive currents attain their maxima.

Our review of the large-scale part of the World Ocean circulation which has been obtained in both studies will be completed by some reasonings about "the optimum length" of an integration interval. The dissipative terms appearing in the equations do not describe the physics of turbulent motion and negative viscosity properly; they are mere permanent sinks of the energy. While these terms are insignificant, an integration may be continued. However, in the course of further integration their contribution becomes considerable, i.e., the net result of integration becomes strongly dependent on those terms of the equation whose coefficients, and even analytical expressions are not determined. The stopping time, i.e., the optimum length of integration upon which a solution obtained is not distorted should be chosen depending on numerical values of the coefficients of these essentially fictitious terms.

Let us take a look at both studies specially for the results related to the equatorial oceanic zone to be analyzed. In the above consideration of the study carried out by Rosati and Miyakoda (1988) we have already pointed out some disadvantages associated precisely with this region. The authors could to decrease A_M but only outside this zone. Unfortunately, the authors do not explain why in the zone with the decreased horizontal step A_M is 1–2 orders of magnitude larger than that in the off-equatorial zone, where the step is 3 time larger.

Very early in the presentation of their strategy of numerical experiments Semtner and Chervin (1992) mentioned that in the equatorial zone 10° in width they have performed diagnostic after Sarmiento and Bryan (1982) instead of prognostic calculations. The reason of such a deviation is convincingly explained by the authors only in terms of the example of similar calculations of a deep circulation; however, they do not clarify the necessity of such calculations in the case of upper layers.

Undeniably, this is a forced step in view of the fact that, with the selected values of the parameters a computational instability would occur in the equatorial zone. Moreover, as is seen, that dynamics are more intensive in the equatorial zone as compared with other regions of the World Ocean even with such an artificial method for suppressing the model trend. A possible reason for this will be mentioned at the end of this chapter.

3.6 On the Intercalibration of Models

An intercalibration of numerical models for calculation of ocean characteristics was started by the late 1980s both in the USSR and in western countries (Demin et al., 1991; Sarkisyan and Demin, 1992). Within a year two collections of works on calibration of models appeared at a rapid pace (Gates et al., 1993; Stockdale, 1993), and one more article presenting principal results of a calibration of Soviet models was published in 1995 (Sarkisyan, 1995). The main peculiarity in the calibration of Russian models was in rigid requirements for these models; to be more precise, only complete models were admitted. Smoothing of the bottom relief, increase of the coefficients of turbulence or horizontal steps, decrease in the number of levels in height, changes in boundary conditions, and so on were not allowed. The prime objective was to test the potentialities of models in reproducing summer climatic characteristics of the Atlantic (30 °S–70 °N) without a considerable distortion of initial fields of T and S preassigned in one-degree squares. Some calculations were done to the whole World Ocean. The high requirements resulted in the fact that half the models not only did not reach the finish but also did not cope with diagnostic calculations. In the long run only models from two Soviet republics (Russia and Ukraine) passed all calibration stages. The main disadvantage of the Soviet calibration was a too short interval of integration which was dictated by extremely limited computer capacities. Let us cite some principal scientific results of Soviet diagnostic and adjustment calculations which were obtained and/or checked when calibrating models (Sarkisyan, 1995; Sarkisyan and Demin, 1992).

1. The vertical velocity component is the most important characteristic for intercalibrations of models.
2. The meridional heat transport is next in importance.
3. The sea surface height is the most stable characteristic of intercalibrations; since, in addition, it is easily checked and comparatively well studied, intercalibrations necessitate it.
4. The mass transport stream function is very sensitive to baroclinicity of deep layers and the bottom relief. The adjustment of deep-ocean hydrologic characteristics is yet an unsolved problem; thus results of intercomparisons of models with respect to the field of Ψ should not be taken for granted.
5. It is desirable to model with a space step smaller than 1° even when studying global climatic characteristics. An increase in step results in the fact that the

flow velocity and the gradients of T and S are several times underestimated and the field of w is completely distorted.

Let us take a quick look at some Western studies on intercalibration. The main distinguishing characteristic of the two western groups for intercalibration is that they have considered much more complicated problems which could not be handled with computers available to Soviet oceanologists.

Seven groups participated in an intercalibration of models for calculation of the seasonal behavior of tropical Pacific characteristics (Stockdale, 1993). Wind stress and heat flux were the same for all of them. The models differed in all other parameters so greatly that in the strict sense of the word, this undeniably very interesting action could be hardly called an intercalibration. The authors have arrived at the conclusion that in essence, none of the model is satisfactory. They mention that a part of the error in calculating the sea surface temperature (SST) could result from the inaccuracy in the preassigned climatic fields but the error is mainly due to disadvantages of the models. The principal disadvantages are as follows: the equatorial waters obtained are cold (0.5° to 2.2°) as compared with observational data; furthermore, the SST in the western and eastern parts is approximately by 1° warmer and several degrees colder, respectively. The errors are concentrated precisely in a narrow equatorial zone. Vertical meridional section of the temperature and zonal velocity demonstrate the concentration of intense currents and countercurrents near the equator. The authors note that the results depend on space steps to a lesser degree, and to a greater extent on the method of turbulence parametrization and the numerical values of the constants of turbulent diffusion and viscosity.

Essential problems with the vertical velocity component have emerged not only because of comparison difficulties but even in response to the procedure of averaging, since typical periods of the Legeckis waves are larger than the seasonal cycle. A chaotic appearance of the vertical section of w in the Kyushu model (Stockdale, 1993) is typical in this respect. Unfortunately, the authors do not discuss the degree to which computational results are dependent on initial data. All the groups but one use Levitus' atlas, although in various ways. In general, the majority of numerical experiments in the 1980s and 1990s start with Levitus' fields prescribed in the overall body of the ocean, and in many cases fields of T and S are even artificially "kept" close to initial data. This is not accidental and necessitates a special discussion when presenting computational results.

Also of interest are calibration calculations on modelling climatic characteristics of the atmosphere and ocean which have been carried out at the four large centers: the Geophysical Fluid Dynamics Laboratory (GFDL), Princeton, New Jersey, the United States; the Max Planck Institute of Meteorology (MPI), Hamburg, Germany; the American National Center for Atmospheric Research (NCAR), and the United Kingdom Meteorological Office (UKMO). An intercalibration of the first stage of experiments without growth in CO_2 is presented by Gates et al. (1993), and we will discuss this paper. The joint circulation has been integrated at these centers over a model time interval between 70 and 100 years. This has required a very long operation of supercomputers. Therefore the models run with a rough resolution; the atmospheric part has been simulated with a higher accuracy than of the oceanic part.

The resolution for the oceanic part of the model is as follows: the minimum and maximum horizontal steps comprise 2.5° and 5°, respectively, and the corresponding numbers of levels are 4 and 17. Undeniably, this rough resolution has distorted characteristics of the ocean and must influence those of the atmosphere even despite a better (approximately by a factor of 2) resolution.

The zonally averaged heat transport in the ocean is qualitatively correctly described by all the models, but quantitative differences are significant. As for the zonally averaged vertical section of the stream function, both severe quantitative and qualitative differences exist. The calculations performed at the NCAR differ greatly from the others, both in the heat transport curves and in the fields of the mass transport stream function. The calculations carried out at the other centers are qualitatively close to each another. It is likely that these distinctions partly result from a very rough vertical resolution used by this group (only four levels); however, it is more probable that NCAR did not make use of flux corrections, whose goal is the artificial compensation, of the drift of climate models. Such correction is unpromising; moreover, it conceals the reason of a model climate drift. We believe that a higher resolution would improve the results of calculations.

The other three models have common disadvantages: the meridional heat transport is underestimated and the integral global circulation has the nature of a strongly smoothed classical Sverdrup transport. The picture of the ψ-field, presented by UKMO, is the most typical example (Fig. 3.6). In particular, the water flux in the regions of the Kuroshio and Gulf Stream is several times smaller than the flows obtained by Semtner and Chervin (1992). In addition to the coarse resolution, there is another reason common to many other prognostic calculations; this deserves particular attention. We mean the optimum integration interval. The time interval of integration is limited because of difficulties in turbulence parametrization. The

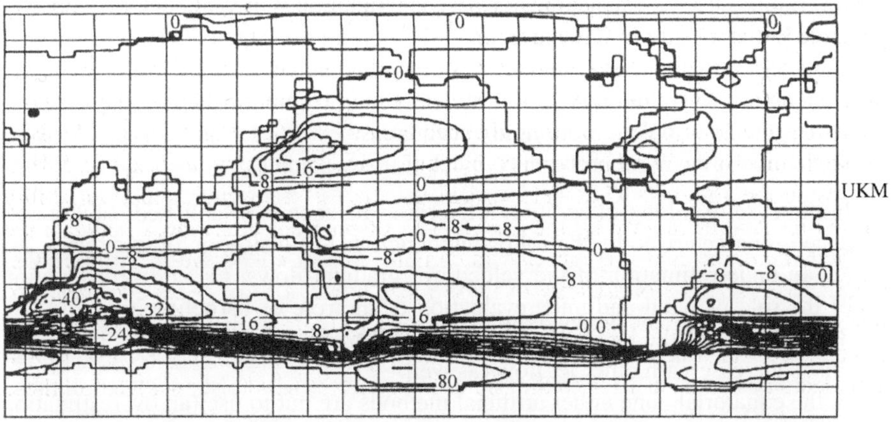

Fig. 3.6 The chart of ψ-field of UKMO. A long model-time of integration have filtered out the effect of baroclinicity and brought to Sverdrup dynamics (Sarkisyan and Sündermann, 1995)

limiting length depends on the chosen values of empirical constants, initial data, boundary conditions, resolution accuracy, and so on. We propose to terminate the calculations when the relative role of turbulence starts to increase. In the course of a long integration the effect of turbulence becomes significant at a certain time. In our opinion, this necessitates stopping of calculations even if a quasi-stationary state of the system has not been reached. If an integration is continued despite this, then a considerable smoothing of the effect of baroclinicity, an overevaluation of the role of wind, a suppression of the role of the JEBAR, the establishment of a Sverdrup balance in the field of Ψ, a displacement of the place where the Gulf Stream is separated away from the American east coast, and so on occur. Not only the three groups (GFDL, MPI, and UKMO) have these disadvantage, but also the calculations of Semtner and Chervin (1992) posed the same problems. Intercomparing the calibration calculations carried out in the USSR and the Western countries, we may state that the time interval of integration in the USSR has been too short and, on the contrary, that in the western countries has been unoptimally long. The problem of finding a criterion for the length of the integration interval is not easy; it depends not only on the above reasons but is also different, for example, for oceanic layers being distinct in depth. New numerical experiments for the elaboration of methods for performing adjustment and prognostic calculations may be necessary.

Rosati and Miyakoda (1988) failed to decrease the coefficient of horizontal mixing in the regions of intensive currents and in the vicinity of the equator. Semtner and Chervin (1992) have been forced to "restore" the results of calculations closer to Levitus' fields precisely in the vicinity of the equator. Furthermore, an unnatural intensity in the upper layer of the equatorial ocean is pronounced against the background of the general flatness of the overall computational results, which is natural in the case of a rough resolution.

It is our preposition that in all these cases a long integration of the vorticity equation is not advisable. In some studies this is an equation in Ψ obtained in an explicit form, in others the equation is represented implicitly in a difference form; however, this does not change the essence. The function Ψ is very sensitive not only to errors in the density of deep layers and the bottom relief, but also to areas where the nonlinear terms in the equations of motion become very important. The vicinity of the equator and (in the case of a sufficiently high resolution) regions of intensive boundary and/or turbulent motions are such areas.

So, let us list the main problems typical of many studies.

1. In the case of a long integration a Sverdrup balance is established; such a balance is characterized by the following deficits: (1) a distortion of integral flows; (2) an underestimation of the velocity of gradient flows; (3) an overevaluation of the role of wind and underevaluation of the role of baroclinicity; and (4) a displacement of the location where the Gulf Stream is separated; it is likely that this is the case with other jet flows as well.
2. In the equatorial zone either artificial methods are put to use (an overestimation of the turbulence coefficients, introduction of an artificial source like $\gamma\,(T_1 - T)$, and so on), or models produce an unnatural local intensification.

In our opinion, all the above and other problems are reduced to the following: (1) in all adjustment and prognostic (dealing with a long integration) calculations it is necessary to track constantly the limit of model potentialities, since it is unpromising to continue integrating once this limit is attained in view of the fact that the viscosity and diffusion begin to play an important role; and (2) dynamics in an oceanic zone near the equator in models relying on the vorticity equation for the mass transport stream function is distorted; it is likely that this is the case in view of the fact that the nonlinear terms increase in importance here.

On the whole the intercalibration, especially the multinational one, is a very progressive action. The authors of this book repeatedly discussed the above results of calibration and that resulted in our long-time collaboration (Sündermann and Sarkisyan, 1991).

References

Beckmann A. and D. Döscher, 1997. A method for improved representation of dense water spreading over topography in geopotential coordinate models. *J. Phys. Oceanogr.* 27: 581–591.

Bleck R. and L. T. Smith, 1990. A wind-driven isopycnic coordinate model of the North and equatorial Atlantic Ocean. *J. Geophys. Res.* 95: 3273–3285.

Bleck R., C. Rooth, D. Hu, and L. T. Smith, 1992. Ventilation patterns and mode water formation in wind- and thermodynamically driven isopycnic coordinate model of the North Atlantic. *J. Phys. Oceanogr.* 22: 1486–1505.

Bryan K., 1969. A numerical study of the World Ocean circulation. *J. Comput. Phys.* 4(3): 347–376.

Bryan K., S. Manabe, and R. C. Pacanowski, 1975. A global ocean-atmosphere climate model, II: The oceanic circulation. *J. Comput. Phys.* 5: 30–46.

Bulushev M. G. and A. S. Sarkisyan, 1996. Energetics of the equatorial currents adjustment processes initial period. *Izv. Ross. Akad. Nauk, Atmos. Ocean. Phys.* 32(5): 600–612 (in Russian).

Chassignet E. P., L. T. Smith, R. Bleck, and F. Q. Bryan, 1996. A model comparison: numerical simulation of the North and Equatorial Atlantic oceanic circulation in depth and isopycnic coordinates. *J. Phys. Oceanogr.* 26: 1849–1867.

Cox M. D., 1970. A mathematical model of the Indian Ocean. *Deep Sea Res.* 17: 47–75.

Demin Yu. L. and A. S. Sarkisyan, 1977. Calculation of equatorial currents. *J. Mar. Res.* 35(2): 339–356.

Demin Yu. L., H. J. Friedrich, R. A. Ibrayev, A. S. Sarkisyan, and J. Sündermann, 1990. A note on modeling the world ocean climate. *Ocean Model.* 89: 3–5.

Demin Yu. L., R. A. Ibrayev, and A. S. Sarkisyan, 1991. Calibration of circulation models and reproduction of the World Ocean climate. *Izv. Acad. Sci. USSR Atmos. Oceanic Phys. Engl. Transl.* 27: 1054–1067.

Demyshev S. G., G. K. Korotaev, and A. S. Sarkisyan, 1992, Numerical experiments of adjustment of hydrological field in the equatorial Atlantic on the basis of conservative model. *Russ. J. Numer. Anal. Math. Modelling.* 7(1): 1–24.

DYNAMO Group, 1997. Dynamics of North Atlantic models: Simulation and assimilation with high resolution models. *Ber. Inst. f. Meereskunde Kiel*, 294 pp.

Ezer T. and G. L. Mellor, 1994. Diagnostic and prognostic calculations of the North Atlantic and sea level using a sigma coordinate ocean model. *J. Geophys. Res.* 99: 14159–14171.

Ezer T., G. L. Mellor, and R. J. Greatbatch, 1995. On the interpentadal variability of the North Atlantic Ocean: Model simulated changes in transport, meridional heat flux and coastal sea level between 1955–1959 and 1970–1974. *J. Geophys. Res.* 100(C6), 10: 559–566.

Friedrich H., 1967. Numerical computations of the wind-induced mass transport in a stratified ocean. In: *Proc. Symp. Math.-Hydrodyn. Invest. Phys. Processes in the Sea*, Moscow, 1966. Mitt. Meeresk. Univ. Hamburg, 10, 134–139.

Friedrich H., 1970. Preliminary Results from a Numerical Multilayer Model for the Circulation in the North Atlantic. Dt.Hydrogr.Z. 23(4), 145–164.

Friedrich H. and J. Sündermann, 1988. On the joint effect of baroclinicity and bottom effect (JEBAR). *Izv. Ross. Akad. Nauk, Atmos. Ocean. Phys.* 34(5): 733–736 (in Russian).

Friedrich H. and V. I. Klimok, 1988. Numerical simulation of seasonal variation of the North and tropical Atlantic circulation obtained by a semidiagnostic method. *Scientific and Metodological Seminar on Ship Hydrodynamics*, Varna, 17 Session BSHC, 2.

Gates W. L., et al., 1993. An intercomparison of selected features of the control climates simulated by coupled ocean-atmosphere general circulation model. *World Climate Programme, Res. WCRP-82*. WMO/TD-No.574, 46 pp.

Haidvogel D. B. and A. Beckmann, 1999. *Numerical Ocean Circulation Modelling*. Imperial College Press, 318 pp.

Ivanov Yu. A., K. V. Lebedev, and A. S. Sarkisyan, 1997. Ocean modeling by general adjustment. *Izv. Ross. Akad. Nauk, Atmos. Ocean Phys.* 33(6): 812–818 (in Russian).

Klimok V. I., V. P. Kochergin, and H. J. Friedrich, 1986. Numerical modeling of seasonal variations of the World Ocean. *Izv. Acad. Sci. USSR Atmos. Oceanic Phys.* Engl. Transl, 22(9): 940–947.

Kochergin V. P., V. I. Klimok, and A. S. Sarkisyan, 1972. Numerical experiments to compute the density field of the Northen Atlantic Ocean. *Meteorologiya I Gidrologiya* 8: 54–61.

Levitus S., 1989a. Interpendal variability of temperature and salinity in the deep North Atlantic, 1970–1974 versus 1955–1959. *J. Geophys. Res.* 94(C11): 16125–16131.

Levitus S., 1989b. Interpendal variability of temperature and salinity at intermediate depth of the North Atlantic Ocean, 1970–1974 versus 1955–1959. *J. Geophys. Res.* 94(C5): 6091–6131.

Levitus S., 1989c. Interpendal variability of temperature and salinity in the upper 150 m of the North Atlantic Ocean, 1970–1974 versus 1955–1959. *J. Geophys. Res.* 94(C7): 9679–9685.

Levitus S., 1990. Interpendal variability of steric sea level and geopotential thickness of the North Atlantic Ocean, 1970–1974 versus 1955–1959. *J. Geophys. Res.* 95(C4): 5233–5238.

Levitus S. and A. S. Sarkisyan, 2001, Ocean dynamics characteristics obtained by synthesis of WOCE and climatic informations. *Izv. Ross. Akad. Nauk, Atmos. Ocean Phys.* 37(9): 534–546 (in Russian).

Marchuk G. I., A. S. Sarkisyan, and V. P. Kochergin, 1973. Numerical methods and results of calculation of flows in a baroclinic ocean. *Geophys. Fluid Dyn.* 5: 89–100.

Marchuk G. I., A. S. Sarkisyan, and V. P. Kochergin, 1976. Numerical methods of flows in a baroclinic ocean, Mathematical Models in Geophysics. *IASH AISH Publ.* 116: 11–17.

Marchuk G. I., V. B. Zalesny, and V. I. Kuzin, 1979. A Numerical Model of Calculation of Thermodynamic Characteristics of the World Ocean. Paper presented at Joint IOC/WMO Seminar on Oceanographic Products and the IGOSS Data Processing and Service System, Moscow. (Apr. 2–6, 1979).

Marchuk G. I., V. P. Dymnikov, and V. B. Zalesny, 1987. Mathematical Models in Geophysical Fluid Dynamics and Numerical Methods of Their Realization. Leningrad, Gidrometeoizdat (in Russian).

Marchuk G. I. and A. S. Sarkisyan, 1988. *Mathematical Modeling of Ocean Circulation*. Springer-Verlag, 262pp.

Marchuk G. I., J. Sündermann, and V. B. Zalesny, 2001. Mathematical modeling of marine and ocean currents. *Russ. J. Numer. Anal. Math. Modelling* 16(4): 331–362.

Marsh R., M. J. Roberts, R. A. Wood, and A. L. New, 1996. An Intercomparison os a bryan-cox-type ocean modeling and an isopycnic ocean model, Part II: The subtropical gure and meridional heat transport. *J. Phys. Oceanogr.* 26: 1528–1550.

Mellor G. L., 1996. *Introduction to Physical Oceanography*. AIP press, Woodbury, New York, 260 pp.

Mellor G. L., 1999. Comments on 'On the utility and disutility of JEBAR'. *J. Phys. Oceanogr.* 29(8): 2117–2118.

Mikolajewicz U., U. Cubasch, G. Hegerl, H. Hock, E. Maier-Reimer, B. D. Santer, and S. Schulz, 1994. Changes in oceanic circulation of the North Atlantic as a result of an increase in atmospheric greenhouse gas concentration. *ICES Mar. Sci. Symp.* 198: 292–296.

Myers P. G., A. F. Fanning, and A. J. Weaver, 1996, JEBAR, bottom pressure torque and Gulf Stream separation. *J. Phys. Oceanogr.* 26: 671–683.

Olbers D. and Eden C., 2003. A simplified general circulation model for a baroclinic ocean with topography. Part I: Theory, waves and wind-driven circulations. *J. Phys. Oceanogr.* 33: 2719–2737.

Rosati A. and K. Miyakoda, 1988. A general circulation model for upper ocean simulation. *J. Phys. Oceanogr.* 18(11): 1601–1626.

Sarkisyan A. S., 1961. On the role of the density advection by wind in dynamics of baroclinic ocean. *Izv. Ross. Acad. Nauk SSSR* 9: 1396–1407.

Sarkisyan A. S., 1962. On dynamics of wind-driven currents in a baroclinic ocean. *Oceanologia.* II(3): 393–409.

Sarkisyan A. S., 1977, The diagnostic calculation of a large scale oceanic circulation. *The Sea, Marine Modelling*, V. 6, New-York-London-Sydney-Toronto, pp. 363–458.

Sarkisyan A. S., 1969a, *Theory and Computation of Ocean Currents.* U.S. Dept. of Commerce and the NSF, Washington, DC.

Sarkisyan A. S., 1969b. Deficiencies of barotropic models of ocean circulation. *Izv. Acad. Nauk SSSR, Ser. Fiz. Atmos. Okeana* 5(8): 818–835 (AGU English translation).

Sarkisyan A. S. and A. A. Serebryakov, 1969. A non-stationary model of equatorial currents. *Oceanologiya.* IX(1): 87–91 (in Russian).

Sarkisyan A. S. and Yu. L. Demin, 1983. A semidiagnostic method of sea currents calculation. *Large-scale Oceanographic Experiments in the WCRP* 2(1): 201–204.

Sarkisyan A. S., 1991. *modeling of Ocean Dynamics.* Gidrometeoizdat, St. Petersburg, 296pp. (in Russian).

Sarkisyan A. S. and Yu. L. Demin, 1992. *Numerical Models and Results of Calibration Calculations of Flows in the Atlantic, Ser. Atmosphere-Ocean-Cosmos, The Section Program.* The Institute of Computational Mathematics RAS, Moscow, 285pp.

Sarkisyan A. S. and J. Sündermann, 1995. On the direction in World Ocean mathematical modeling initiated by Academian G. Marchuk. *Izv. Ross. Akad. Nauk, Atmos. Ocean Phys.* 31(3): 427–454 (in Russian).

Sarkisyan A. S. and A. N. Sidorova, 1998. Numerical modeling of physical characteristics in the neighbourhood of an isolated hydrological section in the Barents Sea. *Russ. J. Numer. Anal. Math. Modelling.* 13(6): 537–549.

Sarkisyan G. A., and V. N. Stepanov, 1999. A method for calculating physical characteristics of the Ocean from an individual hydrological section. *Izv. Ross. Akad. Nauk, Atmos. Ocean Phys.* 35(4): 550–555 (in Russian).

Sarkisyan A. S., 2001. On some milestones in Ocean Modelling History. *Russ. J. Numer. Anal. Math. Modelling.* 16(6): 497–518.

Sarmiento J. L. and K. Bryan, 1982. An ocean transport model for the North Atlantic. *J. Geophys. Res.* 87: 394–408.

Semtner J. A. and R. A. Chervin, 1992. A Ocean general circulation from a global eddy-resolving model. *J. Geophys. Res.* C4(97): 5493–5550.

Stockdale T., 1993. Intercomparison of tropical ocean GCMS. *World Climate Programme, Res. WCRP-79.* WMO/TD-No.545.

Sündermann J. and A. S. Sarkisyan, 1991. Ideal partners in physical oceanography. In *Development: Collaboration Between the Acad. Sci. USSR and the German Scientific-Research Society*, pp. 116–119.

Sündermann J., V. P. Kochergin, V. I. Klimok, V. A. Sukhorukov, and H. J. Friedrich, 1982. Mathematical Modeling of Seasonal Variations of the World Ocean General Circulation with

Consideration for the Surface Turbulent Layer. Comp. Cent. Siberian Dept. Acad. Sci. USSR, Novosibirsk, preprint no. 381, p. 21.

Zalesny V. B., 1996. Numerical simulation and analysis of the sensivity of large-scale ocean dynamics. *Russ. J. Numer. Anal. Math. Modelling*. 11(6): 421–443.

Zalesny V. B., 1997. Variability and equilibrium states of the world ocean circulation. *Russ. J. Numer. Anal. Math. Modelling* 12(6): 547–567.

Chapter 4
Synthesis of Models and Observed Data

Abstract We consider the behavior of prognostic models in the initial phase starting with observed data, i.e. time and space averaged T and S fields. Analysis of the initial stage energetics shows that a dramatic fall of kinetic energy happens during the first six hours of integration. Here and in other parts of the book we justify our opinion on the necessity to monitor the calculation energetics from the very initial stage. Further, in this chapter a method used by WOCE and other programs is presented of T, S sections data processing generating the three velocity components and enriching the observed T, S data. Then based on Kalman filtering, a model of a four-dimensional analysis of hydrophysical ocean fields is presented, its accuracy tested and applied to data from several regions of the North Atlantic.

Keywords WOCE data processing · Four-dimensional analysis · Accuracy · Energetics

4.1 Energetics of Ocean's Hydrophysical Characteristics Mutual Adjusting Processes

4.1.1 Introduction

The method of hydrophysical adjustment proposed by Sarkisyan and Demin (1983), Marchuk and Sarkisyan (1988) is an approach to the classical oceanographic problem of reconstructing the flow field by specified temperature and salinity.

The method consists of two stages. At the first stage the nonlinear system of ocean dynamic equations is integrated by specified temperature and salinity (pure diagnostic calculation).

At the second stage the equations for temperature and salinity (with relevant boundary conditions) are added and the integration is continued (adjustment calculation). The result of the first stage serve as the initial conditions for the second stage while the wind stress is acting in both stages.

The integration normally is stopped by means of some criteria (usually by time-behavior of kinetic energy) depending on the objective. The main aim of the

A.S. Sarkisyan, J.E. Sündermann, *Modelling Ocean Climate Variability*,
DOI 10.1007/978-1-4020-9208-4_4,

adaptation stage is to obtain the ocean upper baroclinic layer temperature and salinity fields approximately adjusted to the stationary system of equations.

It is assumed that the data is statistically consistent and that the resulting fields of temperature and salinity not essentially differ from the original data. The standard error of observed data serve as a limit of deviation between original *T,S* data and results of adjustment calculation. In addition the original large-scale structure of *T,S* fields pattern should be conserved.

This latter demands an additional visual control. Below some other criteria will be noted.

The robust diagnostic method of Sarmiento and Bryan (1982) is closest to the one presented above.

A short model-time of integration is characteristic of thermohydrodynamic adjustment. In mid-latitudes about a month model-time is enough for the diagnostic stage and several months for adjustment of the ocean's upper layer. In the equatorial area about twice more time is needed which is quite short anyway. Such a short model-time enables to solve the adaptation (adjustment) problem for limited open areas because the side boundary condition errors do not affect the solution essentially. The method usually is used without preliminary data filtering, because the numerical model itself makes filtering during integration. More difficult is the implementation of the adjustment processes for areas without data and for ocean deeper layers. We shall return to these problems later.

The criterion to stop the integration process, as mentioned above, is the end of the kinetic energy abrupt dropping process.

In mid-latitudes it corresponds to the period when inertial-gravitational waves, exited in the beginning of the adjustment calculation, damp. The situation of wave-like processes is more complex for narrow equatorial area and needs a special investigation. Below we present an example for the Tropical Atlantic.

4.1.2 Characteristics of the Model and Numerical Experiments

The starting system of equations used in these experiments is traditional for ocean hydrothermodynamics with the assumptions of hydrostatics and incompressibility with rigid lid at the ocean surface. See equations (1.1.1)–(1.1.7) presented in Chapter 1. To simplify the description of energy correlations we present all the equations in Cartesian coordinates and, besides we assume a linear equation of state. So we choose one equation for density instead of equations for temperature and salinity. The system of equations reads now:

$$\frac{\partial u}{\partial t} + u\frac{\partial u}{\partial x} + v\frac{\partial u}{\partial y} + w\frac{\partial u}{\partial z} - fv = -\frac{1}{\rho_0}\frac{\partial p}{\partial x} + \frac{\partial}{\partial z}K_M\frac{\partial u}{\partial z} + A_M\Delta u \tag{4.1.1}$$

$$\frac{\partial v}{\partial t} + u\frac{\partial v}{\partial x} + v\frac{\partial v}{\partial y} + w\frac{\partial v}{\partial z} + fu = -\frac{1}{\rho_0}\frac{\partial p}{\partial y} + \frac{\partial}{\partial z}K_M\frac{\partial v}{\partial z} + A_M\Delta v \tag{4.1.2}$$

$$\frac{\partial p}{\partial z} = g\rho \tag{4.1.3}$$

$$\frac{\partial u}{\partial x} + \frac{\partial v}{\partial y} + \frac{\partial w}{\partial z} = 0 \tag{4.1.4}$$

$$\frac{\partial \rho}{\partial t} + u\frac{\partial \rho}{\partial x} + v\frac{\partial \rho}{\partial y} + w\frac{\partial \rho}{\partial z} - fv = \frac{\partial}{\partial z}K_T\frac{\partial \rho}{\partial z} + A_T\Delta\rho \tag{4.1.5}$$

The boundary conditions at the sea-surface are:

$$\left(\rho_o K_M\frac{\partial u}{\partial z}, \frac{\partial v}{\partial z}\right) = (\tau_x, \tau_y) \tag{4.1.6}$$

$$w = 0 \tag{4.1.7}$$

$$\rho = \rho^{(0)} \tag{4.1.8}$$

At the bottom ($z = -H$):

$$\left(\rho_o K_M\frac{\partial u}{\partial z}, \frac{\partial v}{\partial z}\right) = (\tau_x^b, \tau_y^b) \tag{4.1.9}$$

$$\mathrm{w} = -\left(\mathrm{u_H}\frac{\partial \mathrm{H}}{\partial \mathrm{x}} + \mathrm{v_H}\frac{\partial \mathrm{H}}{\partial \mathrm{y}}\right) \tag{4.1.10}$$

and density anomaly isolation $\left(\frac{\partial \rho}{\partial n} = 0\right)$ is approved.

At side boundaries a free slip condition for the horizontal flow velocity is used and for the density anomaly – the isolation condition. At liquid side boundaries velocity and density are specified. The bottom stress components are calculated by the formulae of Weatherly (1972).

The specific peculiarity of the model is the use of sea surface topography ζ instead of mass transport stream function Ψ. This method has a number of advantages:

(1) A comparison of right-hand sides of equations (2.2.46) and (2.2.62) of Chapter 2 shows, that in the equation for Ψ the gradient of the density anomaly, existing in the JEBAR, is multiplied by z; in the oceans upper layer the density gradient is bigger and more accurate (measured or calculated) but it is multiplied by a small value of z in deeper layer, in which the density gradient is smaller and less accurate, – vise-versa. As a result it leads to suppressing the baroclinicity and overestimating the effect of bottom topography in the Ψ -pattern.
(2) There is no multiplier z in JEBAR in the equation for ζ; besides the equation contains another source of potential energy- the BARBE, which doesn't depend on bottom topography. Consequently ζ is formed mainly by the ocean's upper layer baroclinicity and is lesser dependent on bottom topography.
(3) The ζ -field represents the ocean's surface layer gradient currents, i.e. all the intense currents, which are measured partly and partly calculated by the dynamic method. So any crude error in the calculation of ζ will be easily seen and corrected.

(4) In many areas exist undercurrents, in some even a three layer structure, which twice change directions in vertical. The Ψ -field demonstrates only vertically integrated, consequently indefinite flow, the errors of which is not easily detectable.
(5) In multiple-connected basins, i.e. basins possessing n islands one has to calculate $n - 1$ Ψ-functions, while there are no problems with ζ : only one ζ field with necessary boundary conditions at islands is calculated.
(6) To use the Ψ-function one has to imply the rigid lid boundary condition at the sea surface, instead of the normal free-surface boundary condition for ζ.
(7) There are regular measurements of the sea-surface height at boundaries of many islands. Besides the measurements of ζ from space become more dense and more accurate. That is impossible for the Ψ-function.

There is seemingly one advantage of using the Ψ -function: the case of small, closed, homogeneous basin, where $\Psi \equiv 0$ at boundaries. But that is an exception because any basin deeper than some 10–20 m is baroclinic either due to river discharge or other variations via the free surface. Therefore the side-boundary conditions depend on z.

The finite-difference relations for SST used here, were obtained from finite-difference versions of the equations for momentum, continuity and hydrostatics without preliminary choice of alternative equations of (3.1.8) or (3.1.10) Chapter 3. Naturally it appeared, that the finite-difference version of the SST equation is the analog of equation (3.1.10) because the continuity equation, used for determination of the vertical component of flow velocity w contains the horizontal components divergency.

This circumstantance once more stresses the privilege of (3.1.10) compared to (3.1.8). In Chapter 3 it was shown that the wind stresses action on the ocean surface is threefold: by $\vec{\tau}$, $\text{div}\vec{\tau}$, $\text{rot}\vec{\tau}$.

The details of the numerical model used here is described below in Chapter 6 (Ibrayev, 1993). Here we just mention, that the finite-difference solution method is analogous to the Bryan's box-method (1969) with an Arakawa B-grid. The numerical scheme is conservative for a closed basin.

To form an equation for energy we consider the balance of kinetic energy for barotropic (i.e. average in vertical) and for the baroclinic (i.e. declination of flow velocity from average in vertical) part. Let us induce the following notations (Bulushev and Sarkisyan, 1996)

$$\vec{\nabla} = \left(\frac{\partial}{\partial x}, \frac{\partial}{\partial y}\right), \vec{v} = (u, v) \tag{4.1.11}$$

$$KE = \rho_0 \frac{\left(\vec{v} \cdot \vec{v}\right)}{2} \tag{4.1.12}$$

is the kinetic energy density.

To obtain the kinetic energy balance equation we sum the equation (4.1.1) multiplied by u with (4.1.2) multiplied by v. Then we have:

$$\frac{\partial KE}{\partial t} = -\rho_0\left[\left(\vec{v},\vec{\nabla}\right)\vec{v} + w\frac{\partial\vec{v}}{\partial z}\right]\vec{v} - (\vec{\nabla}p)\vec{v} + \\ + A_M\left(\Delta\vec{v}\right)\vec{v}\rho_0 + \left(\frac{\partial}{\partial z}\nu\frac{\partial\vec{v}}{\partial z}\right)\vec{v}\rho_0 \qquad (4.1.13)$$

Using the continuity equation (4.1.4), we transform the first term of the equation's (4.1.13) right-hand side by following way:

$$\rho_0\left[\left(\vec{v},\vec{\nabla}\right)\vec{v} + w\frac{\partial\vec{v}}{\partial z}\right]\vec{v} = (\vec{v},\vec{\nabla})KE + w\frac{\partial KE}{\partial z} \\ = \vec{\nabla}(KE\vec{v}) + \frac{\partial}{\partial z}(KEw) - KE\frac{\partial w}{\partial z} - KE\frac{\partial w}{\partial z} - \\ - KE(\vec{\nabla}\cdot\vec{v}) = \vec{\nabla}(KE\vec{v}) + \frac{\partial}{\partial z}(KEw) \qquad (4.1.14)$$

Then we transform the second term of equation (4.1.14) right-hand side by using (4.1.3) and (4.1.4):

$$(\vec{\nabla}p)\vec{v} = \vec{\nabla}(p\vec{v}) - p(\vec{\nabla}\cdot\vec{v}) = \vec{\nabla}(p\vec{v}) + p\frac{\partial w}{\partial z} = \\ = \vec{\nabla}(p\vec{v}) + \frac{\partial pw}{\partial z} - w\frac{\partial p}{\partial z} = \vec{\nabla}(p\vec{v}) + \frac{\partial pw}{\partial z} - \rho g w \qquad (4.1.15)$$

We introduce the whole ocean basin's averaging operator

$$<\varphi> \; = \frac{1}{v_0}\iiint \varphi dv_0 \qquad (4.1.16)$$

and implement the operator (4.1.16) to equation (4.1.15). As a result we obtain the formula for velocity of variation of volume-mean kinetic energy:

$$\left\langle\frac{\partial}{\partial t}KE\right\rangle = -\frac{1}{V}\iiint\left[\vec{\nabla}(KE\vec{V}) + \frac{\partial}{\partial z}(KEw)\right]dV \\ -\frac{1}{V}\iiint\left[\vec{\nabla}\left(P\vec{V}\right) + \frac{\partial Pw}{\partial z}\right]dV \\ + \langle\rho g w\rangle + \left\langle\mu\left(\Delta\vec{v}\right)\vec{v}\rho_0\right\rangle + \left\langle\left(\frac{\partial}{\partial z}\nu\frac{\partial v}{\partial z}\right)\vec{v}\rho_0\right\rangle \qquad (4.1.17)$$

Let us reconstruct the first term of the equation's (4.1.17) right-hand side, by using the boundary condition (4.1.7) and the formula

$$\int_V div(\vec{a})dv = -\iint_S (\vec{a}\cdot\vec{n})dS \qquad (4.1.18)$$

where $\vec{n}$ is an internal normal of the surface S. We have:

$$-\frac{1}{v}\iiint\left[\vec{\nabla}(KE\vec{v})+\frac{\partial KEw}{\partial z}\right]dv=\int_{\Gamma_1}\int KEv_n dS+\int_{\Gamma_2}\int KEv_n dS \tag{4.1.19}$$

where V_n is projection of total vector (u, v, w) to the internal normal. The first term of the tight-hand side is integration over side boundary, the second one, which is integration over ocean bottom, diminishes due to boundary condition (4.1.11). The reconstruction of the second term is analogous. The last term of equation (4.1.17) may be presented in a form:

$$\left(\frac{\partial}{\partial z}v\frac{\partial\vec{v}}{\partial z}\right)\vec{v}\rho_0=\rho_0\frac{\partial}{\partial z}v\left(\vec{v}\cdot\frac{\partial\vec{v}}{\partial z}\right)-\rho_0 v\left(\frac{\partial\vec{v}}{\partial z}\cdot\frac{\partial\vec{v}}{\partial z}\right) \tag{4.1.20}$$

Then we have

$$\left\langle\left(\frac{\partial}{\partial z}v\frac{\partial\vec{v}}{\partial z}\right)\vec{v}\rho_0\right\rangle=\frac{1}{v}\int\int\vec{\tau}\cdot\vec{v}\,dxdy-$$
$$-\frac{1}{v}\int\int\vec{\tau}^b\cdot\vec{v}\,dxdy-\left\langle\rho_0 v\left(\frac{\partial\vec{v}}{\partial z}\cdot\frac{\partial\vec{v}}{\partial z}\right)\right\rangle \tag{4.1.21}$$

where $\vec{\tau}$ is the wind stress, and $\vec{\tau}^{b}$ is the bottom stress.

Finally we come to the following formula

$$\left\langle\frac{\partial}{\partial t}KE\right\rangle=ADVKE+PEKE+WPF+WKE+DISKE \tag{4.1.22}$$

The physical meaning of items of the last formula are as follows $ADVKE = \frac{1}{V}\iint_{\Gamma_2} KEV_n dS$ – the kinetic energy flux via side boundaries.

In a closed basin $V_n = 0$ so the volume-mean kinetic energy does not change due to advection in a closed basin;

PEKE – the gravity forces work, responsible for exchange between kinetic and potential energies;

$WPF = \frac{1}{V}\int_{\Gamma_2} PV_n dS$ – the pressure forces work at the boundary, which is equal to zero also for a closed basin;

$WKE = \frac{1}{V}\iint \vec{\tau}\cdot\vec{v}\,dxdy$ – the wind stress's work at the sea surface;

DISKE – volume-mean kinetic energy dissipation due to horizontal turbulent viscosity, bottom stress and vertical viscosity.

Let us now introduce vertical averaging in operation by:

$$\hat{\varphi} = \frac{1}{H}\int\limits_{-H}^{0} \varphi dz \tag{4.1.23}$$

We denote $\varphi'' = \varphi - \hat{\varphi}$ – the deviation from average and the formula holds

$$\langle \varphi_1 \cdot \varphi_2 \rangle = \langle \hat{\varphi}_1 \cdot \hat{\varphi}_2 \rangle + \langle \varphi_1'' \cdot \varphi_2'' \rangle \tag{4.1.24}$$

Then we denote

$$KBT = \rho_0 \frac{\left\langle \hat{\vec{v}}, \hat{\vec{v}} \right\rangle}{2} \quad \text{and} \quad KBC = \rho_0 \frac{\left(\vec{v}'', \vec{v}'' \right)}{2} \tag{4.1.25}$$

the barotropic and baroclinic parts of kinetic energy respectively.

The formula for $<KBT>$ we obtain as follows: we implement the vertical averaging operation (4.1.23) to equations (4.1.1), (4.1.2), multiply them by $\hat{u}$ and $\hat{v}$ respectively, add them to each other and, finally, average the result over the whole volume.

$$\left\langle \frac{\partial}{\partial t} KBT \right\rangle = WNTKBT + WPFKBT + WKBT + DISKBT \tag{4.1.26}$$

with

$DISKBT = \left\langle \left(\mu \Delta \hat{\vec{v}} \right) \hat{\vec{v}} \rho_0 \right\rangle - \frac{1}{S_0} \iint \tau^b \cdot \hat{\vec{v}} dxdy$ – dissipation of barotropic kinetic energy;

$WKBT = \frac{1}{S} \iint \vec{\tau} \cdot \vec{v} dxdy$ – the input into KBT of the wind stress work;

$WNTKBT = -\left\langle \rho_0 \left[\widehat{(V, \nabla) V} + \widehat{w \frac{\partial V}{\partial z}} \right] \cdot \hat{\vec{v}} \right\rangle$ – the input of the nonlinear terms work;

$WPFKBT = -\langle (\hat{\nabla} P) \cdot \hat{V} \rangle$ – the input of the pressure force work. In a closed basin $WPFKBT = 0$ when bottom topography or density anomaly is constant or, in other words, when the $JEBAR = 0$.

So in these cases the pressure gradients work gives no input into the variation of barotropic part of kinetic energy.

It is easy to obtain the formula for $\langle \frac{\partial}{\partial t} KBC \rangle$ as a difference of relevant terms of formula for K and KBT by using (4.1.24)

$$\left\langle \frac{\partial}{\partial t} KBC \right\rangle = WNTKBC + WPFKBC + WKBC + DISKBC. \tag{4.1.27}$$

The meanings are analogous to those of (4.1.26):

$$
\begin{aligned}
&WKBC = WKE - WKBT = \frac{1}{V}\iint \vec{\tau}\cdot\vec{v}''dxdy\\
&DISKBC = DISKE - DISKBT\\
&WNTKBC = ADVKE - WNTKBT\\
&WPFKBC = PEKE + WPF - WPFKBT
\end{aligned}
\qquad (4.1.28)
$$

Let us make two remarks concerning to some terms of (4.1.28) for the case of a closed basin:

1. In this case $ADVKE = 0$ and, consequently, nonlinear terms only redistribute the kinetic energy between baroclinic and barotropic parts without changing the total energy.
2. As it was mentioned above, in a flat-bottom closed basin $WPFKE = WPFKBT = 0$, i.e. due to work of gravity forces the potential energy can be exchanged only with in baroclinic parts of kinetic energy.

4.1.3 Results of Adjustment Calculation

The summer season climatic data for wind stress temperature and salinity have been chosen for numerical experiments. Initial T and S fields, taken from Levitus's data, was not subjected to any filtration.

In version No. 1: $A_M = A_T = 10^8\,\mathrm{cm}^2/\mathrm{s}$ and $K_M = K_T = 10\,\mathrm{cm}^2/\mathrm{s}$.

The version No. 2 differs from version 1 by values of the turbulent diffusion coefficients only: $A_T = 5{*}10^7\,\mathrm{cm}^2/\mathrm{s}$ and $K_M = 1\,\mathrm{cm}^2/\mathrm{s}$.

The grid mesh is uniform $\delta\lambda = \delta\theta = 1^\circ$, the grid points of the velocity meridional component were at the very equator. There were 17 irregularly distributed levels in vertical, extending from 10 to 5000 m.

The region chosen for the experiment is the Equatorial Atlantic between 10.5 °S–10.5 °N and 60.5 °W–12.5 °E. Initially the ocean was at rest despite space variability of specified temperature and salinity fields.

First performed was a diagnostic integration (i.e. with "freezed" temperature and salinity fields) for 90 days of model time to generate a motion and adjust flow velocities to specified three dimensional "noisy" temperature and salinity fields. After every 24 two hourly time steps the Matsuno scheme was used to filter noises.

Then two prognostic experiments were performed to make mutual adjustment of flow velocity and *T,S* fields. These two experiments' parameters were mentioned above. The prognostic integration with the first group of parameters were continued only for 38 days, while with smaller turbulent coefficients up to 158 days.

One may argue that five months of prognostic integration is too short for baroclinic adjustment. We shall see below that even one month of model time prognostic calculations appear to be enough for conclusions. Moreover, the character of the

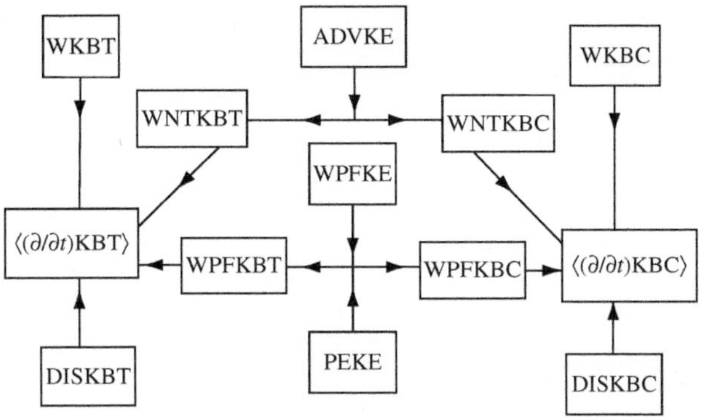

Fig. 4.1 The kinetic energy general balance scheme (Bulushev and Sarkisyan, 1996)

kinetic energy balance at the very beginning of adjustment is essentially different from the one at the so called "steady state" (the ocean is never in steady state). Let us construct the scheme of kinetic energy balance, based on above differential relations, before discussing the obtained results. The analysis showed, that the finite-difference approximations are good analogs of space differentials. Consequently the finite-difference balance of kinetic energy may be schematically shown in a form presented in Fig. 4.1.

The calculations showed, that in experiment No. 1 the turbulence parameters are overestimated, which leads to overly filtration, so we shall discuss mainly the results of the experiment No. 2.

The sea surface height both after diagnostic and adjustment calculations are shown on Fig. 4.2. The result of the diagnostic calculation is "noisy" – as a the result of sub-grid scale disturbances, and/or nonsynchronic processes and other kinds of errors.

These small scale structures diminish after adjustment calculation, the result is now more regularized, the noise is filtered. But filtration is not the aim of calculation, it is not the main result either. The main result is the equatorial undercurrent which was not apparent in the diagnostic version and was revealed clearly after adjustment. The adjustment means nonlinear thermohydrodynamical interpolation which is the only realistic method in comparison of any scholastic mathematical interpolation.

The adjustment process, when integration is done with reasonable model-time (not shorter, not longer) and stopped by some criteria (depending on the models quality and the goal seeking in the experiment) corrects to some extent the specified initial *T,S* fields.

Let us now discuss the physical processes which led to the above mentioned variations of the hydrophysical fields during adjustment. To do it we analyze the schemes of the kinetic energy balance. In Fig. 4.3 we demonstrate the energy balance after 90 days of diagnostic adjustment of flow velocities to the fixed *T,S* fields and in Fig. 4.4 after only 6 hours of adjustment. The arrows directed into the boxes mean a positive sign of term and the ones, directed out of the boxes – a negative sign.

The absolute values of the kinetic energy differentials are 10^{-7} J/m^3 s and the value of energy itself – 10^{-1} J/m^3.

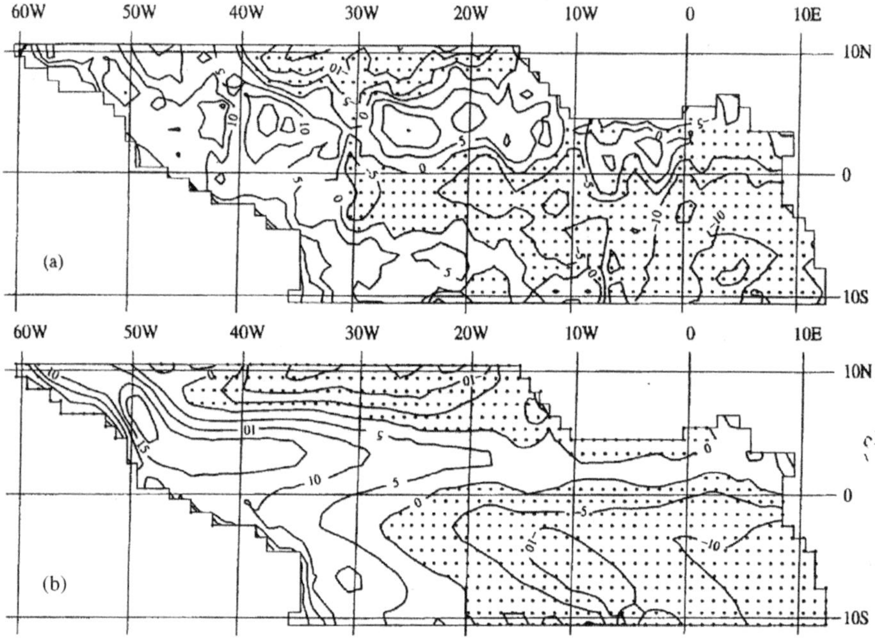

Fig. 4.2 The sea surface height (cm) in diagnostic (**a**) and adjustment (**b**) calculations (Bulushev and Sarkisyan, 1996)

The Fig. 4.5 shows kinetic energy balance after 38 days of integration. As it seen on Fig. 4.3, the main positive input into both baroclinic and barotropic parts of kinetic energy is obtained due to work of pressure gradients and it is balanced by viscosity. The effect of wind stress is positive also but it accounts for 10–15% only compared to work of pressure gradients. The advection of kinetic energy via boundaries is small compared to *WNTKBT* and *WNTKBC*, consequently, the main baroclinic energy pumping into the barotropic part is being done due to nonlinearity. For better understanding, the Figs. 4.3–4.5 should be compared with the Fig. 4.1.

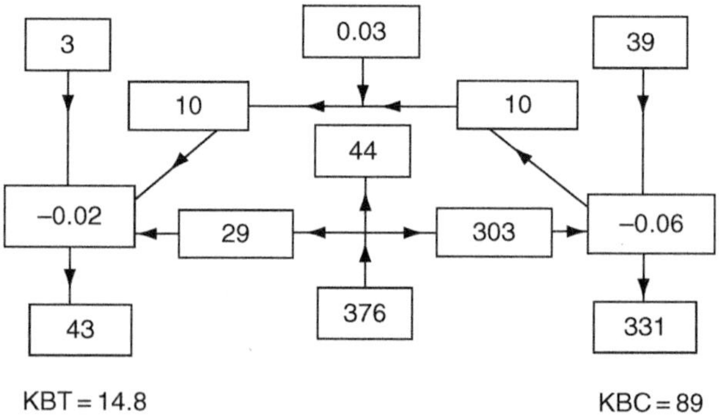

Fig. 4.3 The kinetic energy balance scheme after 90 days' diagnostic calculations to generate the flow velocity and adjust it to the "frozen" T, S-fields, basin's geometry, equations of motion and boundary conditions (Bulushev and Sarkisyan, 1996)

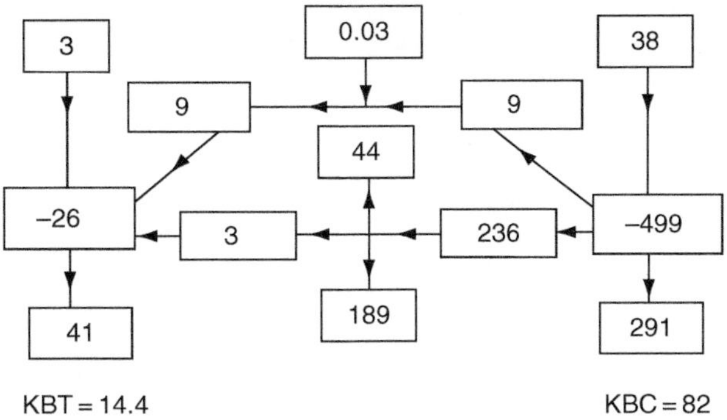

Fig. 4.4 The energy balance scheme after the first 6 hours of prognostic calculations for mutual adjustment of flow velocity and T, S fields (Bulushev and Sarkisyan, 1996)

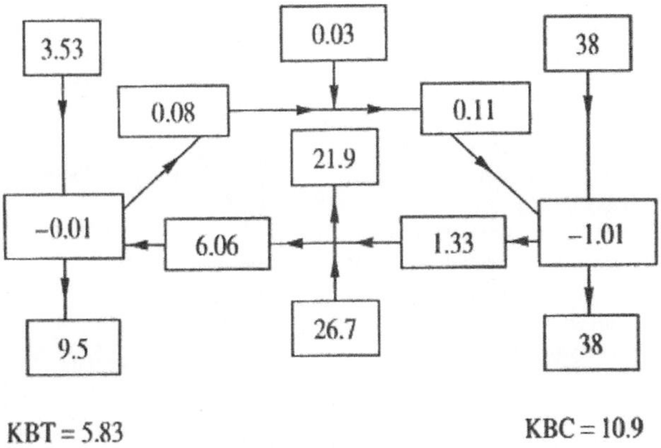

Fig. 4.5 The energy balance scheme after 38 days of adjustment calculations (Bulushev and Sarkisyan, 1996)

The energy balance obtained in the diagnostic stage of the flow velocity calculation, formally corresponds to the stationary solution of the momentum equations with fixed density field; but actually it represents an unnatural condition when the flow velocity does not affect on temperature and salinity. That is why a dramatic variation of the energy balance happens during the first six model-hours of prognostic adjustment process. The scheme in Fig. 4.4 shows that drastic variations occur only in the terms responsible for the pressure gradient work. The maximal variation is in *WPFKBC*, which has almost the same value as in the Fig. 4.3 but with opposite sign. It means that a pumping of kinetic energy (mainly its baroclininc part) into potential one already begun.

The Fig. 4.5 shows that the role of the pressure gradients work decreases essentially compared to diagnostic calculations. The main positive input now is given by the wind stress and as a main sink of energy serve the horizontal and vertical turbulence. The baroclinic part of kinetic energy decreases more than the barotropic one.

The time-series of terms, which constitute the equations for the baroclinic and barotropic parts of the kinetic energy are presented in Figs. 4.6 and 4.7. The flow velocity zonal component at $z = 75$ m of the very equator shows a spreading of perturbation of 4° to 6° horizontal scale with the velocity which is close

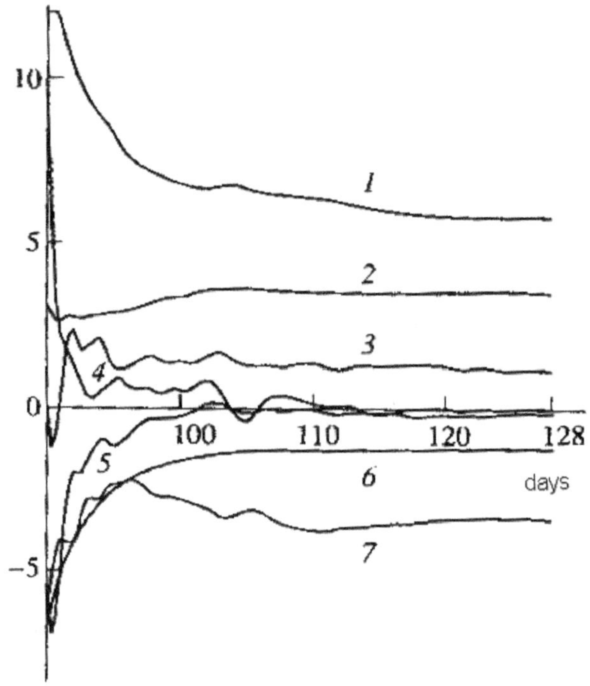

Fig. 4.6 The time behavior of different items of the kinetic energy barotropic part (experiment No. 1) after 90 days of diagnostic calculation followed by 38 days of adaptation. The notations: The barotropic part of kinetic energy (1), the kinetic energy barotropic part variation due to the work of wind stress (2) (in 10^{-6} erg), the variation of the pressure force work (3) (in 5×10^{-6} erg), the time differential of the kinetic energy barotropic part (4) (in 5×10^{-6} erg), the time variation of the work of nonlinear terms (5) (in 5×10^{-6} erg), the time variation of horizontal viscosity (6) (in 5×10^{-6} erg), the time variation of vertical viscosity (7) (in 5×10^{-6} erg) (Bulushev and Sarkisyan, 1996)

to phase velocity of gravitational waves of the first and second baroclinic mode. The time scale of perturbations is about 1–2 days. As for an analogous diagram for the extraequatorial zone they are of local character during the first 10 days. A dramatic decreasing of kinetic energy occurs during 10 days (the baroclinic part almost for an order of magnitude and the barotropic one – about two times) i.e. the motion became more barotropic. During that period the rapid and small-scale oscillations, which are connected with inertial gravitational waves, diminish and causes a rapid decreasing of kinetic energy. Afterward both barotropic and baroclinic energy decreasing process as themselves itself decelerate for an order of magnitude.

Consequently the second period of the adjustment process occurs. In this second period the time variation of barotropic kinetic energy is sign changeable and to the end of adjustment its absolute value becomes smaller by three orders of magnitude. The exchange between potential and kinetic energy is not intense any more.

The oscillation processes analysis shows, that after the first 10 days of adjustment Yanai, Kelvin and Rossby waves are much more evident. To identify these processes an analysis of fields of pressure and flow velocity horizontal components was made, the velocity of oscillations was studied as well as the wave's meridional and vertical structure and polarizational relations. The Kelvin waves of some 10° lengths with phase velocities of 70–80 cm/s are clearly seen, they decrease not essentially to the

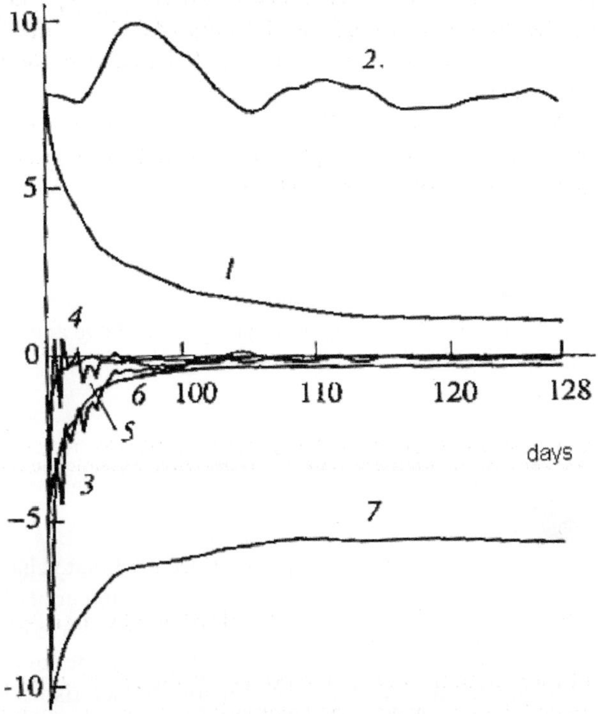

Fig. 4.7 Same as in Fig. 4.6 but for kinetic energy baroclinic part. The notations: The baroclinic part of kinetic energy (1), the kinetic energy baroclinic part variation due to the work of wind stress (2) (in 10^{-6} erg), the variation of the pressure force work (3) (in 5×10^{-6} erg), the time differential of the kinetic energy baroclinic part (4) (in 5×10^{-6} erg), the time variation of the work of nonlinear terms (5) (in 5×10^{-6} erg), the time variation of horizontal viscosity (6) (in 5×10^{-6} erg), the time variation of vertical viscosity (7) (in 5×10^{-6} erg) (Bulushev and Sarkisyan, 1996)

end of adjustment. Their characteristics are close to the third baroclinic mode; their amplitude at the equator are about 4–5 cm. Oscillations, excited by the Yanai wave with wave number close to zero and structures corresponding to the second baroclinic mode are seen, both in the upper and lower layers as well as in the central and western parts of the basin. Their amplitude decreases 2 to 5 times to the end of adjustment process. The most striking process of the ocean's upper layer adjustment is the forming of the equatorial undercurrent which happens when the Kelvin wave of first baroclinic mode has past.

The structure of the Kelvin wave is symmetric relative to the equator. It is absent in the flow velocity field but evidently seen in the pressure field. The integration in experiment No. 2 was continued some extra 120 days to examine whether the structure of the ocean upper layer is in steady or in intermediate stage of some nonstationary process.

The results showed, that in the upper layers the process of adjustment practically has finished within the first month but in deeper layers intense nonstationary processes are continuing.

Besides, during these 120 days of integration further smoothing of the original T, S fields occur and the thermocline becomes eroded.

From a physical point of view it can be concluded that the second period of integration for adjustment was stopped in due time. Theoretically mathematical criterions may be used to finish the integration. For example:

$$\frac{\partial}{\partial t}KE < \sigma_1(z) > /N_1(z) \ll \varepsilon_1$$
$$\frac{\partial T}{\partial t} < \sigma_2(z) > /N_2(z) \ll \varepsilon_2$$
$$\frac{\partial T}{\partial t} < \sigma_3(z) > /N_3(z) \ll \varepsilon_3$$

where σ_i are the space dispersions of corresponding characteristics at the depth z and N_i may be chosen experimentally. The above analysis shows that some practical experience of the modeler is necessary.

4.2 Ocean Modelling Experiments by General Adjustment (OMEGA)

The direct prognostic calculation of ocean climate characteristics might be the best method of ocean modelling, but there are a number of obstacles which have not been overcome yet. Some of them have been listed above in Chapters 2 and 3, the others will be discussed below when describing a number of ocean models. The alternative is the diagnostic calculation of climatic characteristics by specified temperature and salinity. This method proved to be a useful tool but it demands the T and S fields specified in every grid point and statistically homogeneous.

Usually the observed data is unevenly distributed. Interpolation helps in case of statistically significant observed T, S data but in many parts of the World Ocean the data is scarce. In some areas there is no T, S data at all. In these areas any kind of simple or even mathematically very sophisticated methods of interpolation or extrapolation will result in a scholastic information which is far from ocean thermohydrodynamics data. The above mentioned standard prognostic calculation which starts from constructed initial T, S field will "cut the pics and fill the holes". The longer the integration, the stronger filtration happens. That distorts the initially reliable part of the T, S field.

The question arises: is it possible to improve the less reliable information by neighboring statistically significant one but to avoid the strong filtration of the data. For this aim we propose the OMEGA procedure which is the following.

1. Make a diagnostic calculation to prepare initial values of all thermohydrodynamic characteristics.
2. Mark the areas, layers or even separate points with reliable T, S values and after some n time steps of prognostic calculation replace the calculated values of T and S by them.
3. Make another n time steps of prognostic calculation and again replace the achieved data by initial reliable T, S values, etc. After some m calculations the less reliable data in internal and surrounding grid points will be corrected through

the influence of the reliable data. With some prescribed accuracy they will reach a steady state.

4. Finally some n prognostic time steps should be done to get a mutual adjustment of the initially reliable T, S data with the ones corrected (or even created) by these procedures.

So the OMEGA procedure is a nonlinear thermohydrodynamical interpolation and, for the surrounding area, extrapolation of reliable (statistically significant) observed T, S data.

This procedure corrects the scholastically interpolated or extrapolated unreliable data, creates appropriate flow velocities and other dynamic characteristics, and adjusts the initial data to the basin's geometry and to the ocean surface boundary conditions with minimal variation of reliable information. We describe very briefly below a numerical experiments (Ivanov et al., 1997) which proved this procedure to be a useful tool for synthesis of modelling and limited observed data.

The North Atlantic was chosen for the numerical experiment. Figure 4.8 shows the temperature field at 200 m after 180 days of integration by the OMEGA method for two small testing areas; one of them is chosen at a site with small gradient current, the other- vise-versa. The adjustment calculations were done for standard hydrological levels, by using a modern 3 D-PE numerical model (Ibraev, 1993) and

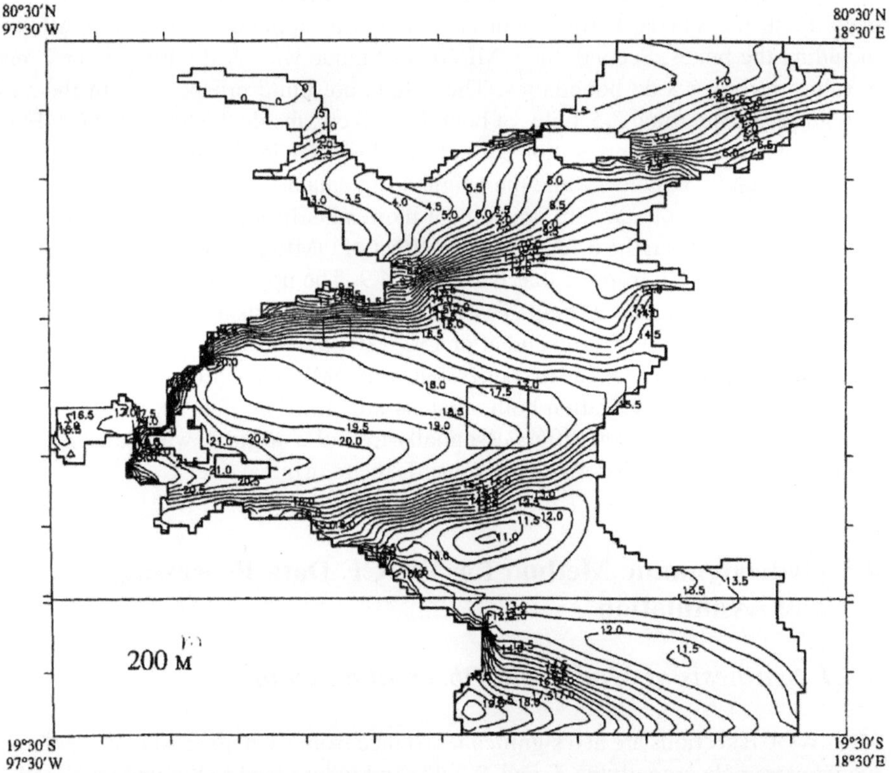

Fig. 4.8 The North Atlantic temperature (°C) at $z = 200$ m after 180 days' integration by the OMEGA method. The locations of both testing areas are shown (Ivanov et al., 1997)

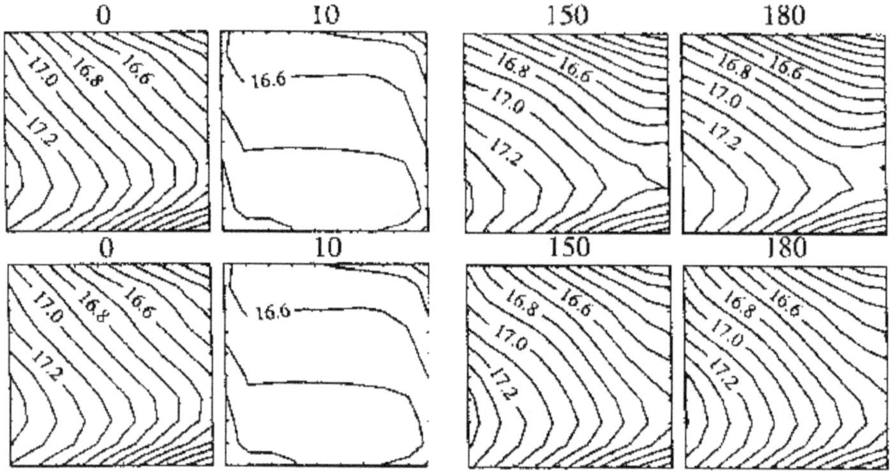

Fig. 4.9 The evolution of temperature (°C) at 250 m of the high gradient box (days). The upper parts of the figures correspond to the standard method of integration and the lower parts to the integration by the OMEGA method (Ivanov et al., 1997)

Levitus multi-year mean T,S data. In these two boxes the data was distorted by very strong filtering. Afterward two calculations have been done to regenerate the data gaps. First, the standard adjustment procedure was applied to the North Atlantic including the boxes. Second, the OMEGA technique was used with regular T and S values at boxes side boundaries. The results are quite prospective: in the case of regular regenerating T,S-fields at boundary layers, inside the boxes the OMEGA technique in a 6 month integration corrects the distorted data getting them very close to the original T,S values, while the neighboring data do not change essentially.

As for the case of standard adjustment the results inside and outside the boxes essentially differ from the original. So the effect of OMEGA is obvious. We demonstrate it only for one example that is the Fig. 4.9. The upper curves of boxes show, respectively, the original T fields (with "0" on top), the distorted T field after 10 days of model time integration and after 150, 180 days.

The lower curves are as the first ones, but with OMEGA technique of integration.

After 10 days of integration both versions are alike but after six months the result of standard adjustment differs essentially from the original, while the OMEGA method regenerate the distorted field towards the to original one.

4.3 Hydrodynamic Method for WOCE Data Processing and Assimilation

4.3.1 Synthesis of Model and Observation Data

Most WOCE sections are at a significant distance from each other, and interpolation would extremely smooth the T and S fields and would lead to the underestimation of the velocity of gradient currents. The WOCE section data added in this way to a vast body of climatic data will not have a noticeable effect over a large part of the World Ocean. In Sarkisyan and Stepanov (1999), Sarkisyan and Levitus (1999), a technique is proposed to solve the problem of obtaining the most complete information about current dynamics from an individual hydrological section without using the dynamic (reference level) method.

The technique consists of the following stages.

(1) With the climatic T and S fields (Conkright et al., 1994) used as initial data and with a nonlinear model of ocean hydrodynamics, a short term diagnostic computation (usually for a few months of model time) is performed to recover the current velocity and some other physical characteristics. Further, a shorter term diagnostic computation (for about one to two months) is carried out to obtain all climatic characteristics adjusted to each other to, as well as to the equations, bottom topography and boundary conditions.
(2) The computed T and S are replaced by the WOCE section data and are "frozen" during longer term integration, causing the surrounding data to adjust to the inlaid section. This procedure is completed when the hydrodynamic characteristics remain virtually unchanged in the region nearest to the section.
(3) The section data are "unfrozen", and a prognostic computation (for about two months) is performed to achieve mutual adjustment.

The above procedure is referred to as model-data synthesis or amalgamation. As a result of amalgamation, one can obtain all dynamically adjusted physical characteristics (u, v, w, ζ, T, S and ρ) both at the section and in its surrounding region (1 to 3 degrees from the section). The initial T and S data at the section become slightly smoothed but undistorted.

4.3.2 Results of Experiments

Two experiments were conducted.

The first experiment was carried out in the following way. The climatic T and S data and the current fields obtained after adaptation were regarded as "synoptic" data. Further, these data were strongly smoothed and were regarded as "climatic" information. The difference between the "synoptic" and "climatic" fields served as an error field. One of the sections in the climatic field was replaced by the "synoptic" T and S data, and the above amalgamation procedure was applied to assimilate and process the inlaid data. The computations have shown that the errors decrease by a factor of two to three at the points located around the inlaid section.

The second experiment was carried out using an actual synoptic section within in the Barents Sea. The following conclusions can be given:

(1) The proposed method does correct the climatic T and S fields adjacent to the section, bringing them closer to the "synoptic" data.
(2) The inlaid section only affects several nearest grid points (2°–3°) (Sarkisyan and Sidorova, 1998).

It follows from the second conclusion that it is impossible to produce realistic T and S fields or other hydrodynamic characteristics for the world ocean by using the WOCE section data alone.

Sarkisyan and Levitus (1999) carried out a further experiment with some advancements:

(1) A more complete numerical model (Ibraev, 1993) and a real WOCE section are used.
(2) The chosen section (24.5 °N in the North Atlantic) was used in different independent studies, which allows a comparative analysis of results.
(3) A more detailed analysis of computation results was performed.

The region was chosen in the North Atlantic, climatic hydrological fields for the summer season were taken from the climatic atlas, and wind stresses were taken from the National Center for Environmental Prediction of the National Center for Atmospheric Research (NCEP/NCAR) (Kalnay et al., 1996). Fixed values of T and S are specified at the surface, and insulation conditions are imposed at the side boundaries and at the bottom.

The model is based on the conventional system of nonstationary three-dimensional equations written in spherical coordinates in the Boussinesq approximation with molecular viscosity and diffusion. Nonlinear terms and the bottom topography are taken into account. The model has 1° horizontal resolution and 24 vertical levels.

4.3.3 Computation Results

The climatic distribution of T isolines at the zonal-vertical section in question (24.5 °N, Fig. 4.10a) has an anticyclonic character. Except for the region near the Bahamas, where upwelling occurs, large-scale downwelling is observed at a larger part of the section. After five-months model-time computations, the T distribution between 76.5° and 78.5 °W, is characterized by a vertical slope, which is indicative of an intense vertical circulation in the immediate vicinity of the continental slope (Fig. 4.10b). At a larger part of the section (east of 75.5 °W), T and S was replaced by the corresponding data observed at the A05 WOCE section before the computation. The original T and S fields at the WOCE section contain vertical deviations with upwellings and downwellings, which are indicative of eddies with diameters of about one hundred kilometers (Fig. 4.10c).

The horizontal-velocity field computed for a depth of 100 m and presented in Fig. 4.11 seems to be realistic, the velocity values in the Gulf Stream region amount to 40 to 60 cm s^{-1}. These values are normal for the 1° resolution model. The velocities of the equatoria] currents are overestimated, because the transition to the steady state is not completed in that region, unlike the region in question between 23° and 26 °N, where the steady state is virtually reached and the velocity is primarily directed northward.

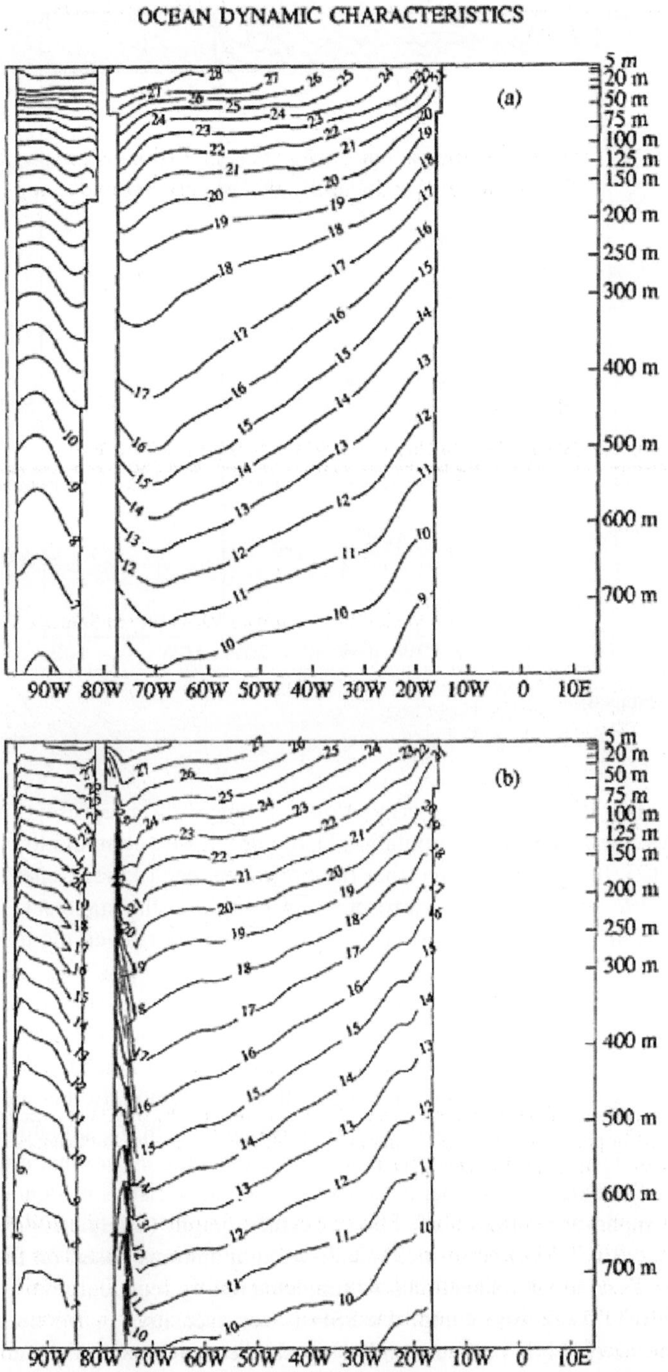

Fig. 4.10 Vertical temperature section (°C) at 24.5 °N. (**a**) Levitus climatic data; (**b**) climatic temperature after five months of adaptation; and (**c**) the WOCE AO5, 1992 section

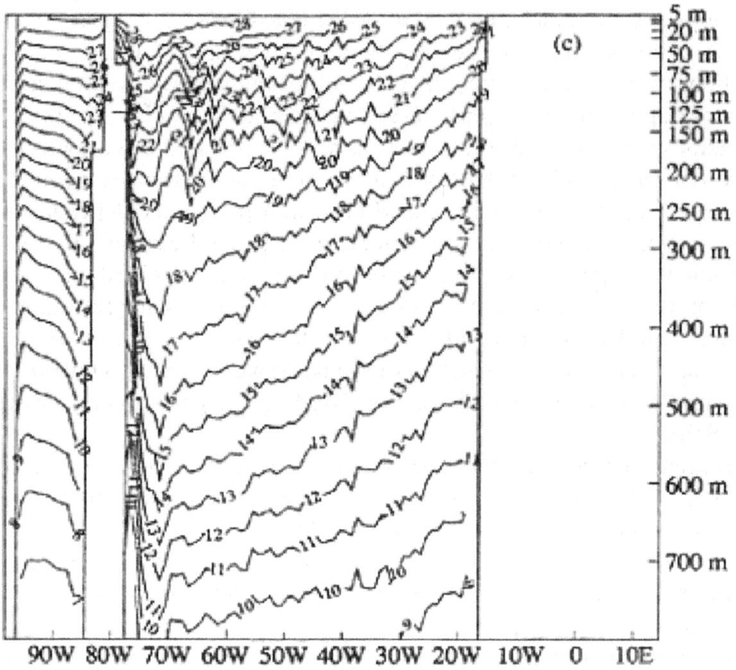

Fig. 4.10 (continued)

The variation of the kinetic energy at the section 24.5 °N represents all computation stages: the diagnostic computation D, the prognostic computation A1, and so on (Fig. 4.12). It should be noted that the portion of each kinetic-energy curve at the stage A2 is higher than the corresponding portion at the stage A1. This result demonstrates the effect of "noisy" synoptic data even after seven months of model integration.

The field of the velocity difference between computations based on an inlaid WOCE section and those performed with climatic data (Fig. 4.13a,b) shows that the effect of synoptic data is substantial in a narrow band near the section and is propagated westward and northward in the Gulf Stream region. Eddies are generated near the inlaid WOCE section and are especially intense in the Bahamas region of bottom slope. The effect of synoptic data decreases in the meridional direction and with depth. The behavior of other hydrodynamic characteristics is identical.

For a comparative analysis of the sea-surface height (SSH) at 24.5 °N values were computed using different methods. The computations based on the dynamic method were carried out for section data subjected to different degrees of smoothing. In the case of a weak smoothing, the SSH is very sensitive to the ocean's upper layer baroclinicity, while in the strongly smoothed case it is unrealistic and deviates essentially from the model result. As for the stream function (not shown here), it is strongly sensitive to variations in deeper layer density. This latter result is also noted

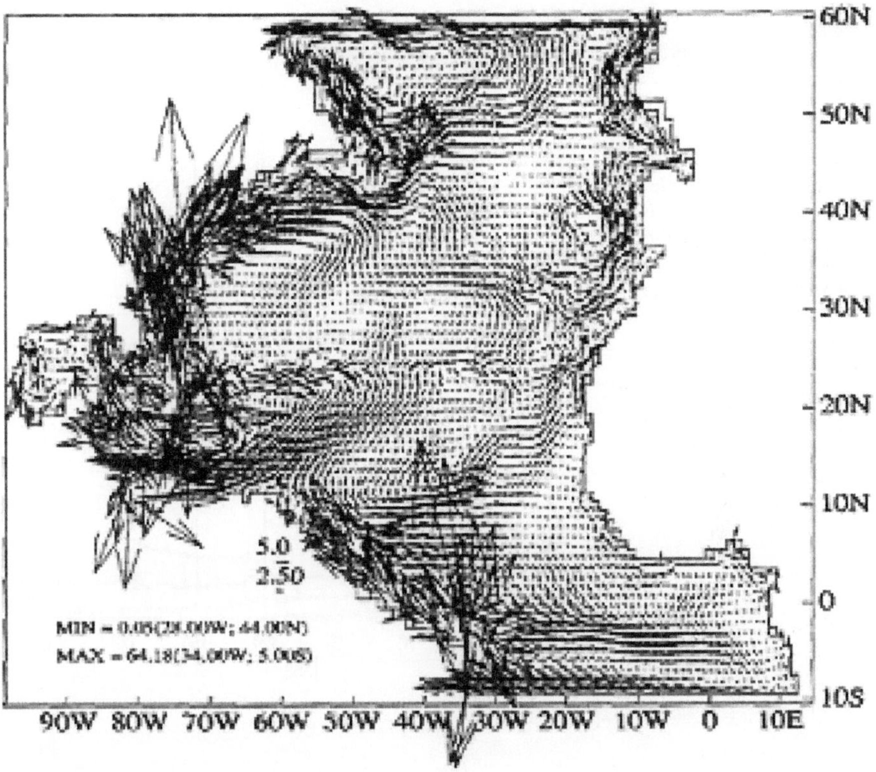

Fig. 4.11 The climatic field of the horizontal velocity at a depth of 100 m at 24,5 °N (Levitus and Sarkisyan, 2001)

in a recent work by the scientific group engaged in the CLIPPER project (Treguier et al., 1999).

Due to intense eddies and other-type spatial variations in T and S at the section, an intense mass and heat transport occurs in the meridional direction (Tables 4.1, 4.2, 4.3, and 4.4).

Table 4.2 shows that the most strong mass transport in the model occurs in sectors I and II, which is caused by the western intensification due to the JEBAR and β-effect.

The heat transport values corresponding to the water mass transport values presented in Tables 4.1 and 4.2 are given in Tables 4.3 and 4.4. It follows that the northward and southward transport values computed separately are an order of magnitude greater than the net transport. The two variants of heat transport computed by the dynamic method lead to southward heat transport, which is unrealistic.

The experiments conducted indicate that the sign and magnitude of integral mass and heat transports using the dynamic method depend on the degree of smoothing the T and S fields or on the choice of the zero-surface depth. Therefore, we do not consider the dynamic method to be substantiated.

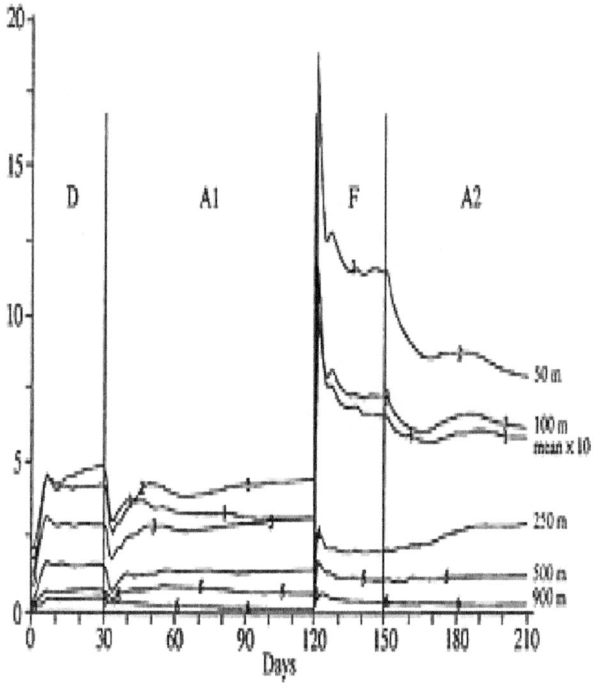

Fig. 4.12 The kinetic energy (10^{-3} m^2 s^{-2}) at different depths at the section at 24,5 °N: D corresponds to diagnostic computations for one month of model time, A1 corresponds to adaptation computations (three months), F corresponds to adaptation with the "frozen" section *T* and *S* data inlaid into the adapted climatic data at 24,5 °N (one month), and A2 corresponds to mutual adaptation of the "unfrozen" data at the section and the region surrounding it (two months). (Levitus and Sarkisyan, 2001)

4.3.4 Discussion and Conclusions

We start our discussion with a brief survey of the results presented below in Tables 4.5 and 4.6. The presented values of mass and heat transport were obtained by different methods and in different years. Nevertheless, the divergence of the estimates obtained for a narrow latitudinal belt is too large. In principle, one could consider the results of the experiments to be satisfactory, because the data presented in Tables 4.1–4.4 are close to some values listed in Tables 4.5 and 4.6. However, we have chosen another way.

Let us consider a few quotations from only two of a large number of papers listed in Tables 4.5 and 4.6.

Böning et al. (1991): "An annual signal in the region east of the Bahamas is masked by strong oscillations in water flow on a characteristic scale of about 100 days, which are due to Antilles Current instability."

Fillenbaum et al. (1997): "A time series of local heat transport (relative to 0° temperature) exhibits a strong variability within the limits ±2 PW on a characteristic scale of about 100 days." "At 26.5 °N, the annual mean transatlantic heat transport obtained by combining the Levitus climatic data and the measurements taken with anchored buoy stations located in the Florida Current region oscillates within the limits 1.44 ± 0.33 PW."

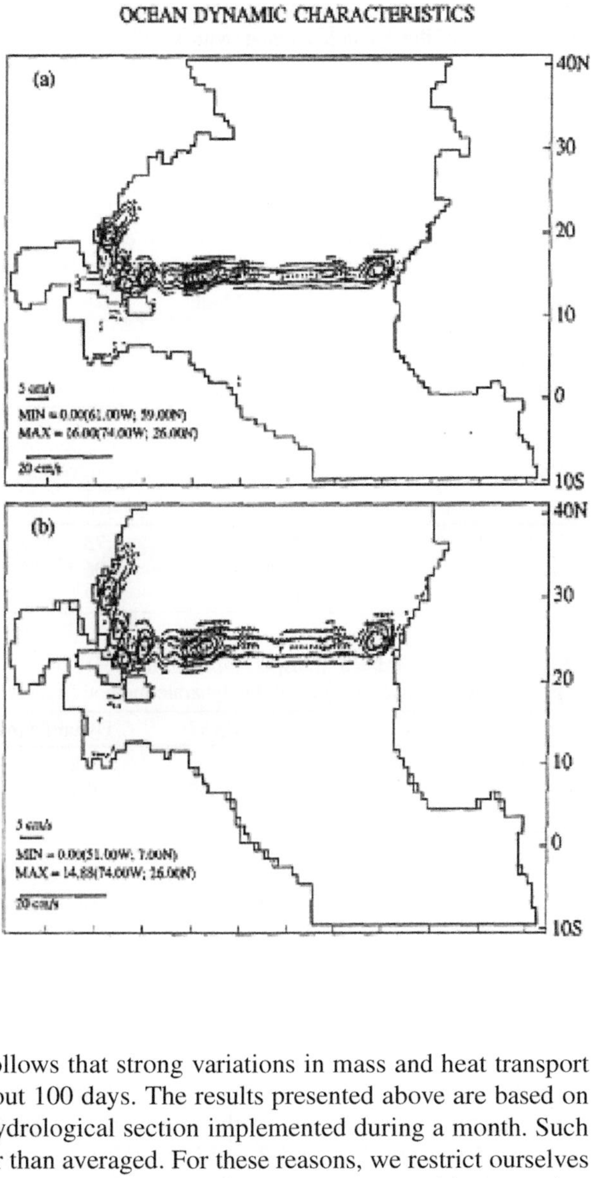

Fig. 4.13 Field of the difference between the horizontal velocities obtained in the experiments with the inlaid WOCE section and calculated by climatic data at the levels of (**a**) 10, (**b**) 100 (Levitus and Sarkisyan, 2001)

From the preceding, it follows that strong variations in mass and heat transport occur on a time scale of about 100 days. The results presented above are based on the data of a transatlantic hydrological section implemented during a month. Such data are instantaneous rather than averaged. For these reasons, we restrict ourselves to the statement that these results are only in qualitative agreement with some other results, and a greater emphasis we place on the main characteristic features and new qualitative conclusions.

As noted above, the temperature at the WOCE section (Fig. 4.10c) is characterized by a rather strong spatial variability. The question arises: What is the nature of this variability? The proposed method was used for the nonlinear hydrodynamic extrapolation of synoptic data with simultaneous amalgamation with the surrounding climatic data near the section. As a result, eddies were recovered near the section

Table 4.1 Northward and southward mass transport values a 24.5 °N computed by the model and two versions of the dynamic method (with weakly and strongly smoothed data)

Mass transport, Sv	Model	Dynamic method 1	Dynamic method 2
Northward	98.8	93.5	29.2
Southward	98.84	89.9	35.8
Total	−0.04	3.6	−6.6

Table 4.2 Northward mass transport across different sectors at the section in question computed by the model and two versions of the dynamic method (with weakly and strongly smoothed data)

Mass transport, Sv	Model	Dynamic method 1	Dynamic method 2
Sector I 78.5°−74.5 °W	55.0	17.7	8.4
Sector II 74.5°−70.5 °W	-42.5	1.0	2.6
Sector III 70.5 °W−15 °E	-8.6	−19.3	−21.8

Table 4.3 Northward and southward heat transport in petawatts (PW) at 24.5 °N computed by the model and two versions of the dynamic method (with weakly and strongly smoothed data)

Heat transport, PW	Model	Dynamic method 1	Dynamic method 2
Northward	5.59	6.73	2.57
Southward	5.06	7.06	3.33
Total	0.53	−0.32	−0.77

Table 4.4 Northward heat transport through different sectors at the section in question computed by the model and two versions of the dynamic method (with weakly and strongly smoothed data)

Heat transport, PW	Model	Dynamic method 1	Dynamic method 2
Sector I 78.5° − 74.5 °W	2.52	1.21	0.65
Sector II 74.5° − 70.5 °W	−1.56	−0.13	0.03
Sector III 70.5° − 15 °E	−0.6	−1.90	−1.94

(Fig. 4.13a,b), and these eddies are responsible for the spatial variability. We consider this result to be realistic.

A clearly defined asymmetry is observed in the spatial distribution of the effect of inlaying synoptic data (Fig. 4.13a,b). Namely: (a) the effect of synoptic data decays in the meridional direction at a distance of a few degrees from the section; of course, this effect also decays with depth because of baroclinicity; (b) in the zonal direction, the effect is propagated westward under the action of Rossby waves and other factors; in the region of the continental slope (east of 80 °W), this effect is enhanced through the JEBAR and is propagated northward by the Antilles Current and further by the Gulf Stream. This western intensification is clearly reflected in the mass and heat transport values (Figures are not shown here).

Due to eddies and other-type spatial variations, as well as to the JEBAR (at least in sectors I and II), the meridional transport in the region of the section is an order of magnitude stronger (Tables 4.1–4.4) than those computed from climatic data. This result has been revealed from direct observations in the Bahamas region (Lee et al., 1990; Rosenfeld et al., 1989). As a result of computations, the authors arrived at the conclusion that the western intensification and a high degree of variability of the

Table 4.5 Water mass southward transport by the Deep Western Boundary Current in the North Atlantic (Sv)

Authors	Mass transport, Sv	Commentary
Schmitz and McCartney (1993)	17	
Marsh et al. (1996)	15; 16.5; 25	Three different models at 26 °N
Lee et al. (1990)	33	East of the Abaco Island (Bahamas) below 800 m
Rosenfeld et al. (1989)	10	Direct measurements; strong variability from 5.3Sv in the north-ward direction in April 1985 to 34.8 Sv in the southward direction in April 1986 at 26.5 °N
Schott et al. (1993)	from 14.2 to 17.3	In the equatorial Atlantic from the northwest to the southeast
Böning et al. (1991)	6–8	

Table 4.6 Meridional heat transport values in PW estimated by different authors for the latitudinal belt from 23° to 26.5 °N in the North Atlantic

Authors	Heat transport value, Sv	Commentary
Hall and Bryden (1982)	1.2	In October 1957 at 25 °N
Roemmich and Wunsch (1985)	1.2	In August 1981 at 24 °N
Marsh et al. (1996)	0.88	In July at 23 °N (maximum)
Molinari et al. (1990)	0.69, 1.86, 1.02	In February, July and November at 26.5 °N, respectively
Treguier et al. (1999)	0.9	At 25 °N
Fillenbaum et al. (1997)	−0.8	Between Africa and the Bahamas at 26.5 °N
Böning et al. (1991)	0.44 or 0.80	At 26.5° depending on the wind field used in the model

characteristics of this region are, at least partially, due to synoptic processes in the extended zone between the Bahamas and Africa.

4.4 Four-Dimensional Analysis of Hydrophysical Ocean and Sea Fields

4.4.1 Introduction

The reconstruction of the spatiotemporal structure of ocean hydrographic parameters from available measurements is an important oceanographical problem. Experience gained from the use of modern primitive-equation numerical models has shown that predictions produced by such models reflect only the qualitative structure of hydrophysical fields of interest. Simulated fields are far from reality (especially in the case of long model runs). A more universal approach to the reconstruction of climatic (or close to actual) ocean (sea) variability is to combine numerical models with observations. Several examples of such a combination (synthesis, amalgamation) were presented above. More sophisticated versions are four-dimensional analysis schemes (Knysh, 1982; Sarkisyan et al., 1989, etc.).

In the 1970–1990s, temperature and salinity (density) measurements in certain ocean regions were the most available type of observations. In recent years, satellite altimetry has been found quite promising for reconstruction of hydrophysical fields. In both cases, four-dimensional assimilation algorithms and schemes have been developed, which made it possible to reconstruct ocean hydrophysical parameters and to obtain interesting scientific results. The goal of this paragraph is to give a brief overview of some four-dimensional analysis algorithms developed over the last 10–15 years and to discuss their validation in different world-ocean regions. The data used is also discussed here. Earlier algorithms have been applied to some hydrophysical problems stated in the framework of the "Sections," TOGA, and WOCE projects. Four-dimensional analysis procedures proposed later have been used for reconstructing ocean and sea hydrophysical parameters from altimeter data.

4.4.2 Differential Equations and the Kalman Filter

For further presentation of various four-dimensional analysis schemes, it is convenient to base on dynamic-stochastic models (DSMs) (Timchenko, 1981). The basic idea behind this approach is an extension of optimal Kalman filtering to stochastic spatiotemporal fields (Sakawa, 1972; Knysh, 1982). The basic equations and a simplified Kalman filter are presented here includings the transport-diffusion equations for heat and salt, the continuity and hydrostatic equations.

$$\frac{\partial u}{\partial t} + \frac{\partial(u \cdot u)}{\partial x} + \frac{\partial(v \cdot u)}{\partial y} + \frac{\partial(w \cdot u)}{\partial z} - fv = = -\frac{1}{\rho_0}\frac{\partial p}{\partial x} + K_M \frac{\partial^2 u}{\partial z^2} + A_M \Delta u \tag{4.4.1}$$

$$\frac{\partial v}{\partial t} + \frac{\partial(u \cdot v)}{\partial x} + \frac{\partial(v \cdot v)}{\partial y} + \frac{\partial(w \cdot v)}{\partial z} + fu = = -\frac{1}{\rho_0}\frac{\partial p}{\partial y} + K_M \frac{\partial^2 v}{\partial z^2} + A_M \Delta v \tag{4.4.2}$$

$$\frac{\partial p_1}{\partial z} = \rho_1 g \tag{4.4.3}$$

$$\frac{\partial u}{\partial x} + \frac{\partial v}{\partial y} + \frac{\partial w}{\partial z} = 0 \tag{4.4.4}$$

$$\frac{\partial T}{\partial t} + \frac{\partial(u \cdot T)}{\partial x} + \frac{\partial(v \cdot T)}{\partial y} + \frac{\partial(w \cdot T)}{\partial z} = K_T \frac{\partial^2 T}{\partial z^2} + A_T \Delta T \tag{4.4.5}$$

$$\frac{\partial S}{\partial t} + \frac{\partial(u \cdot S)}{\partial x} + \frac{\partial(v \cdot S)}{\partial y} + \frac{\partial(w \cdot S)}{\partial z} = K_S \frac{\partial^2 S}{\partial z^2} + A_S \Delta S \tag{4.4.6}$$

where ρ is a specified function of temperature (T) and salinity (S). The boundary conditions in the vertical are set as follows.

At the ocean surface $z = \zeta\,(x, y, t)$:

$$p_1 = p_a$$

$$\rho_0 K_M \frac{\partial u}{\partial z} = \tau_x,\ \rho_0 K_M \frac{\partial v}{\partial z} = \tau_y,\ w = 0,\ T = T^0\,(x, y, t)\,,\ S = S^0\,(x, y, t) \quad (4.4.7)$$

At the ocean bottom $z = -H\,(x, y)$ the no-slip condition is used for current velocity:

$$u = v = w = 0 \qquad (4.4.8)$$

For temperature and salinity we specify the conditions

$$\frac{\partial T}{\partial n_1} = \frac{\partial S}{\partial n_1} = 0 \qquad (4.4.9)$$

where n_1 is the normal to the ocean bottom.

The lateral boundaries of the domain under consideration are assumed to be vertical. At the liquid lateral boundaries, u, v, T and S are specified as functions of coordinates. For velocity, no-slip and no-flux condition are set at the solid lateral boundaries:

$$\frac{\partial u}{\partial y} = 0 \qquad v = 0 \qquad (4.4.10)$$

at the northern and southern boundaries

$$\frac{\partial v}{\partial x} = 0 \qquad u = 0 \qquad (4.4.11)$$

and at the eastern and western boundaries. The zero mass flux is specified for temperature and salinity.

$$\frac{\partial T}{\partial n} = 0 \frac{\partial S}{\partial n} = 0 \qquad (4.4.12)$$

where n is the normal to the solid lateral boundary.

The sea level elevation $\zeta = \zeta_1 + \dfrac{p_a}{\rho_o g}$ is governed by an elliptic differential equation of divergent form (Demin and Sarkisyan, 1977)

$$\Delta\zeta + \frac{1}{H}\frac{\partial H}{\partial x}\frac{\partial \zeta}{\partial x} + \frac{1}{H}\frac{\partial H}{\partial y}\frac{\partial \zeta}{\partial y} = R\,(x, y) \qquad (4.4.13)$$

with the boundary conditions of the skew-derivation type (Marchuk, 1969).

$$\alpha \frac{\partial \zeta}{\partial x} + \beta \frac{\partial \zeta}{\partial y} = r_1 (x, y) \tag{4.4.14}$$

at the boundary parallel to the y-axis and

$$\alpha \frac{\partial \zeta}{\partial y} + \beta \frac{\partial \zeta}{\partial x} = r_2 (x, y) \tag{4.4.15}$$

where $R(x, y)$, $r_1(x, y)$, and $r_2(x, y)$ are specified right-hand sides of equations. at the boundary parallel to the x-axis. Here a and b are known coefficients.

Equation (4.4.13) does not degenerate when the Coriolis parameter is zero; i.e., this equation applies to the computation of currents at the equator. The boundary conditions (4.4.14) and (4.4.15) are derived from Eqs. (4.4.1)–(4.4.4) via the boundary conditions for the velocity. In doing this, it is assumed (as adopted in theory of ocean currents) that, along with the boundary conditions for velocity, the equations of motions hold at the boundaries of the basin (if they are vertical walls).

The initial distributions of temperature and salinity (density) in the model are assumed to be known from observations.

The following notation is used in differential equations and relations (4.4.1)–(4.4.15): u, v and w are estimates (with account for temperature and salinity measurements) of the current velocity components along the x, y, and z-axes, directed east, north and upward, respectively; $\rho (x, y, z, t) = \rho_1 (x, y, z, t) - \rho (z, t)$ is the anomaly of the estimated sea-water density, ρ_1, p_1 – is the estimated pressure; p_a is the atmospheric pressure; ρ_0 is a constant density; f is the Coriolis parameter; K_M, A_M, K_T, A_T, K_S, A_S are the turbulent exchange coefficients for momentum, heat and salt; H is the ocean depth; g is acceleration of gravity; t is the time; Δ is the two-dimensional Laplacian; and τ_x, τ_y are the wind stress components.

At the time $t = t_l$ of data arrival, the temperature and salinity computed at this time by numerical solving Eqs. (4.4.5) and (4.4.6) are corrected by formulas of Knysh et al. (1988)

$$\begin{aligned} T(x, t_l^+) &= T(x, t_l^-) + \sum_{r=1}^{N} \Delta_r^T (x, t_l^-) \delta_r^T (x, t_l) \\ S(x, t_l^+) &= S(x, t_l^-) + \sum_{r=1}^{N} \Delta_r^S (x, t_l^-) \delta_r^S (x, t_l) \end{aligned} \tag{4.4.16}$$

where $\Delta_r^T (x, t_l^-)$, $\Delta_r^S (x, t_l^-)$ are the weight functions determined from temperature and salinity error autocorrelations $P_{TT}(\vec{x}, \vec{x}', t)$ and $P_{SS}(\vec{x}, \vec{x}', t)$ respectively;

$$\begin{aligned} \delta_r^T (x, t_l) &= T_i(x_r, t_l) - T(x_r, t_l) \\ \delta_r^S (x, t_l) &= S_i(x_r, t_l) - S(x_r, t_l^-) \end{aligned} \tag{4.4.17}$$

are the errors in the estimates temperature and salinity; $T_i(\vec{x}_r, t_l)$, $S_i(\vec{x}_r, t_l)$ are the temperature and salinity measurements in N points of the domain at the time $t = t_1$, $\varphi\left(t_1^+\right)$ and $\varphi\left(t_1^-\right)$, where φ is T or S, stand for the functions with and without account for the data received at $t = t_1$. At the time of data assimilation $P_{TT}(\vec{x}, \vec{x}', t)$ and $P_{SS}(\vec{x}, \vec{x}', t)$ are corrected by using expressions

$$P_{TT}(x, x', t_l^+) = P_{TT}(x, x', t_l^-) + \sum_{r=1}^{N} \Delta_r^T(x, t_l^-) P_{TT}(x_r, x', t_l^-)$$

$$P_{SS}(x, x', t_l^+) = P_{SS}(x, x', t_l^-) + \sum_{r=1}^{N} \Delta_r^S(x, t_l^-) P_{SS}(x_r, x', t_l^-) \qquad (4.4.18)$$

To reduce the dimension of $P_{TT}()$ and $P_{TT}()$ there are approximated by weighted correlations of the temperature and salinity fields themselves:

$$P_{TT}(x, x', t) \approx \sigma_T^H(x, t)\sigma_T^H(x', t) P_T^H \left|(x - x')\right| \sigma_T(x, t)\sigma_T(x', t)$$
$$P_{SS}(x, x', t) \approx \sigma_S^H(x, t)\sigma_S^H(x', t) P_S^H \left|(x - x')\right| \sigma_S(x, t)\sigma_S(x', t) \qquad (4.4.19)$$

where $\sigma^H(x, t)$ are the normalized rms derivations of the errors in the estimated temperature and salinity, $P^H\left|(x - x')\right|$ are the normalized autocorrelations for temperature and salinity and $\sigma(x, t)$ are the rms deviations in the temperature (or salinity) field.

Representation of the temperature and salinity error covariances in the form (4.4.19) were considered in Knysh et al. (1988).

The differential equation for the normalized error variance of estimated temperature is given by:

$$\frac{\partial D}{\partial t} + u\frac{\partial D}{\partial x} + v\frac{\partial D}{\partial y} + w\frac{\partial D}{\partial z} = K_T\frac{\partial^2 D}{\partial z^2} + A_T \Delta D \qquad (4.4.20)$$
$$D_T^H(\vec{x}, t) = \left[\sigma_T^H(\vec{x}, t)\right]^2$$

Here in (4.4.20) the upper and lower indices of D are dropped. Equation (4.4.20) is supplemented with the Dirichlet condition specified at the surface, lateral boundaries, and bottom. The initial variance is assumed to be known. An analogous equation and analogous boundary and initial conditions are used for the salinity error variance

$$D_S^H(\vec{x}, t) = \left[\sigma_S^H(\vec{x}, t)\right]^2$$

A more detailed description of the dynamic-stochastic model outlined above can be found in Knysh et al. (1988).

4.4.3 Synchronization and Monitoring of Hydrophysical Fields in the Newfoundland Energy Active Zone

A considerable amount of hydrographic data sufficient for estimating the spatiotemporal variability of ocean characteristics were gathered in the Newfoundland energy active zone (NEAZ) in 1981–1987. It is well known, however, that the areal measurements in the NEAZ are not synchronic, they were conducted during relatively long periods (from 10 to 28 days).

Data synchronization is a problem reasonably arising for the NEAZ, a region with complex water dynamics and high variability. A simplified DSM was used for processing hydrophysical data (22 measurement runs) (Knysh et al., 1990). The nonlinear terms and horizontal turbulent exchange were neglected in Eqs. (4.4.1) and (4.4.2) and the sea level was represented as the sum of a calculated by the dynamic formula value and the declination from it. The latter was estimated by using a relevant differential equation (Knysh et al., 1990). The temperature and salinity equations (4.4.5) and (4.4.6) remained unchanged. The vertical velocity was calculated according to Semyonov (1981).

The temperature and salinity values produced by the model were corrected by ones estimated from observations.

The numerical computations were performed for the North Atlantic region within $37^\circ - 50^\circ\ 30'$W. The grid spacing in the x- and y-directions was constant and was equal to $30'$, and 30 levels were considered in the vertical: 0, 10, 20, 30, 50, 75, 100, 125, 150, 200, 250, 300, 400, 500, 600, 700, 800, 900, 1000, 1200, 1300, 1400, 1500, 1750, 2000, 2500, 3000, 4000 and 5000 m. The time step was set equal to 1 day. The turbulent exchange of momentum coefficient and the turbulent diffusion coefficients for heat and salt were specified as

$$K_M = 100\,\mathrm{cm}^2/\mathrm{s},\ K_T = K_S = 1\,\mathrm{cm}^2/\mathrm{s},\ A_T = A_S = 10^6\mathrm{cm}^2/\mathrm{s}.$$

The wind stress components were specified by the Hellerman seasonal mean fields taken for the corresponding observation season and interpolated to the one-half-degree square grid.

The differential equation for the sea level declination was solved numerically by using a first order accurate (upwind) scheme, and differential equation for temperature and salinity were integrated by using a second-order accurate scheme. Matrix factorization in the vertical was used for solving the algebraic systems of equations corresponding to the differential equations of motions. Amplitude peaks in the annual cycle of the air and water surface-layer temperatures, quite high cyclonic and anticyclonic eddy activity in the atmosphere and the ocean, and intense atmosphere-ocean interactions were observed in the considered North Atlantic region. These processes were reflected in the structure and the spatiotemporal distributions of ocean hydrophysical parameters.

Figures 4.14 and 4.15 show the sea level maps calculated from different years and months. The topography of this integral characteristic reflects surface gradient currents.

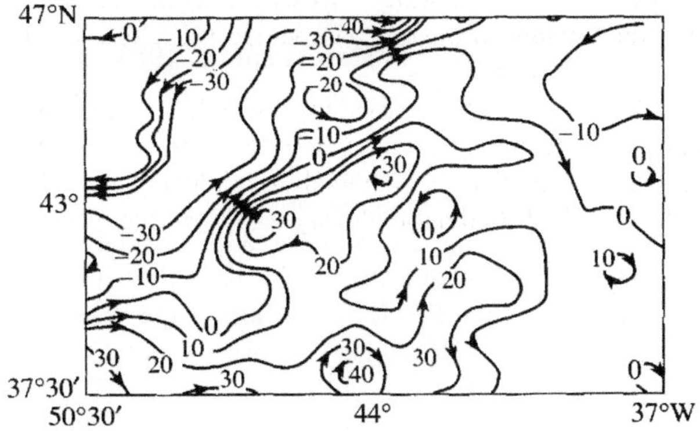

Fig. 4.14 Sea surface height (SSH) averaged over the period of *T* and *S* assimilation December 5 to 26, 1982 (2nd survey). Arrows on the isolines indicate the direction of surface gradient currents, and digits correspond to sea level values in cm (Knysh et al., 1990)

The large-scale system of currents in the NEAZ can be seen in the maps: the Gulf Stream/North Atlantic Current (NAC) (from southwest to northeast), the Slope Water Current (the northern periphery of the Gulf Stream/NAC), and an anticyclonic eddy (in the center of the zone).

In the scientific literature devoted to the thermodynamic water structure in the NEAZ, this eddy is known as a semipermanent antycyclonic eddy (SPAE).

Additionally to the large-scale currents mentioned above, one can see the southern branches of the Gulf Stream in the south of the zone. This system of large scale currents generates intense eddies of the opposite signs.

A comparison of this system of large-scale currents in the NEAZ with the bottom topography suggests that the mean position of the SPAE corresponds to the center of the Newfoundland depression. The Gulfstream/NAC turns the Southeast Newfoundland Ridge. Along converging isobaths at 3000–4500 m, its branch is directed northeast. The Labrador waters flow throw the Newfoundland valley. Let us consider the SPAE in more detail. Inspection of Fig. 4.14 showed that the SPAE stores a large amount of energy. In individual time periods, the current speed in the SPAE is increased (energy is accumulated) or the orbital velocities of water particles decrease substantially (energy is released).

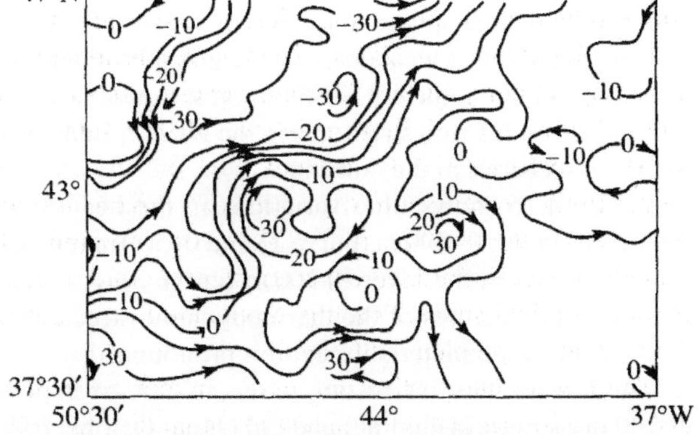

Fig. 4.15 SSH averaged over the period of *T* and *S* assimilation February 1 to 22, 1986 (19th survey). Arrows on the isolines indicate the direction of surface gradient currents, and digits correspond to sea level values in cm (Knysh et al., 1990)

An analysis of a sequence of heat content maps in the layer 0–500 m during the year from winter, 1982, to summer, 1983, and from summer to winter, 1983, suggests that the SPAE heat content decreased in the course of winter cooling from its maximum of 38.3×10^3 MJ/m^3 in October, 1982, to 35.3×10^3 MJ/m^3 in May, 1983.

A maximum amount of energy in the SPAE was observed in October, 1983 after summer overheating. The SPAE area surrounded in the heat content isoline of 35.5×10^3 MJ/m^3 was about four times as large as that observed in May, 1983.

The size of the SPAE varies during the year. In the period of intensification (roughly October–November), it covers about 250 by 150 nautical miles (Figs. 4.14 and 4.15).

The structure of the SPAE is highly variable. Nevertheless, we can indicate some characteristic structural features. In periods of strengthening, the SPAE consists of two separate eddies (e.g., December 1982, (Fig. 4.14); May–June, 1986) or a single eddy structure, whose orientation varies noticeable with time (October, 1983; February–August, 1985; February, 1986).

An analysis of vertical profiles of horizontal currents in the SPAE shows that anticyclonic water motions extend to depths of about 1000 m and deeper. The current velocities have a strongly baroclinic nature. For example, the maximum horizontal speed at the surface and depths of 200, 500, 1000 and 2000 m in December, 1984, was equal to 56.5, 45.4, 30, 18 and 9 cm/s, respectively.

One possible cause for a strengthening or weakening of the SPAE was indicated above. However, these processes cannot be considered independently of the influence exerted on them by the interactions between the SPAE and other large-scale currents, such as the NAC, the Labrador current and the Gulf Stream. In connection with this, it should be noted that the mass transport in the southern branches of the NAC decreases in periods of SPAE intensification. In such periods, these branches are much fewer than in periods of a weakened SPAE. The period of reduced SPAE strength is characterized by generating eddy structures of various types. This can easily be seen in the sea level maps obtained from measurements: in May, 1983; January and August, 1984; and April–May, 1985.

Thus, the SPAE seems to play an important role in a strengthening (or weakening) of the NAC during certain time periods and in the interactions between the ocean and the atmosphere in the NEAZ. A strengthening of the Gulf Stream and the NAC in periods of a weakened SPAE and, conversely, a weakening of these currents in periods of strengthened SPAE suggest a permanent exchange of energy between large-scale circulation structures in the NEAZ. Eddies are formed in areas of intense interactions between the currents.

Figure 4.16 shows time variations in the heat content in the order of 10^{14} MJ within the layers 0–200 m (curve 1) and 0–500 m (curve 2) and the volume-averaged kinetic energy in the layer 0–200 m (curve 3). An analysis of these curves reveals the following features of the thermodynamic structure.

First, these characteristics exhibit pronounced seasonal variability. The heat content and, especially interestingly, the volume-averaged kinetic energy in the layer 0–200 m increase in summer and fall (June–October). In the period of winter cooling

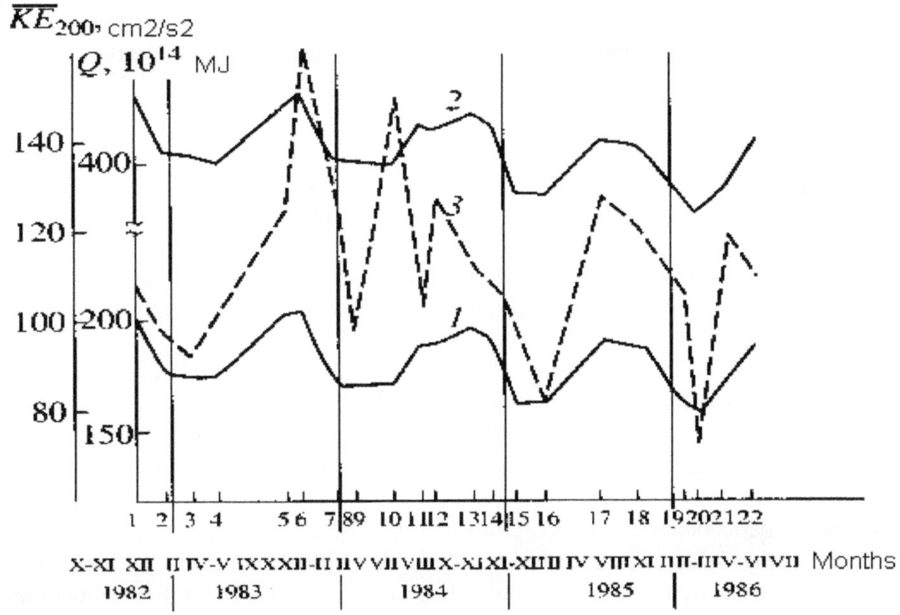

Fig. 4.16 Time variations in the heat content and the volume-averaged kinetic energy in the layer 0–200 m (Knysh and Sarkisyan, 2003)

(November–February), their values diminish considerably. Maximum values of heat content are observed in October, and minimum values occur in January–February. The vertical profiles of a horizontal averaged kinetic energy exhibit an analogous behavior down to 1000 m.

Year-to-year variations in the extreme heat contents within the layers 0–200 m and 0–500 m from 1983 to 1986 exhibit a negative trend. The same behavior is shown by the maximum values of the volume-averaged kinetic energy in the layer 0–200 m. A tendency of year-to-year decrease in the heat content was also found by Baranov et al. (1987). This phenomenon is explained by a southward shift (of unknown cause) in the NAC core during the indicated period.

Thus, the four-dimensional analysis technique developed for measurements processing has demonstrated good performance in particular physical and geographical conditions.

4.4.4 Numerical Experiments on the Estimation Accuracy of the Four-Dimensional Analysis Procedure

To estimate the degree of variability of results produced by the DSM-based four-dimensional analysis technique and to further improve of approximations, used, we have to conduct model experiments, because actual simultaneous measurements are scarce.

Obviously, numerical experiments should be as close to reality as possible. In particular, the period covered by model computation should be comparable with the period of an actual measurement run, and the computational domain should be as close to the actual physical and geographical conditions as possible.

The numerical experiments considered were conducted for the tropical Atlantic region defined by 2° 30′–12° 30′N and 15° 30′–55° 30′W. The "actual" ocean fields in these experiments were generated by the prognostic model (4.4.1)–(4.4.15). The grid spacing was 1° along the x axis and 0.5° along the y axis. The time step was $\delta t = 3\,\text{h}$. Sixteen levels were used in the vertical: 0, 20, 50,100,150, 200, 300, 400, 500, 700, 900, 1100, 1400, 2000, 3000 and 4000 m. These fields were regarded as "true". The technique we used can be described as follows.

The initial "true" hydrographic fields were obtained by optimal interpolation (to a grid domain) of nonsynchronic temperature and salinity measurements made in the layer 0–1100 m during the cruises of the research vessels "Mikhail Lomonosov" and "Akademic Vernadskii" of the Marine Hydrographic Institute, Academy of Sciences of Ukraine, in summer 1988.

The climatic (summer) temperature and salinity fields used are averages. The same climatic fields were used below the depth 1100 m. The initial currents fields were derived by performing a 12 days diagnostic run from the initial temperature and salinity fields (12 days is a period during which a steady state is achieved). Next, we made predictions of a measurement run performed in the Tropical Atlantic.

The fields obtained in each forecast day were taken as "true" fields. In the diagnostic runs of model (4.4.1–4.4.15) was set $A_M = 10^8$, $K_M = 10^2\,\text{cm}^2/\text{s}$. The prognostic computations were conducted with $A_M = A_T = A_S = 7 \times 10^7\,\text{cm}^2/\text{s}$. and $K_M = K_S = K_T = 10^2\,\text{cm}^2/\text{s}$. A climatic (summer) field was used for wind stress.

For four-dimensional analysis of hydrophysical "observations", it is reasonable to use a different, coarse model of currents, which would imitate the imperfectness of numerical models with respect to the actual ocean. In meteorology, this case is associated with the method of nonidentical cousins in contrast to the method of identical twins, in which the same model is used for truth simulation and four-dimensional analysis (Gandin and Kagan, 1976). Below are the results obtained by both methods.

In both cases, the truth was simulated as described above. In the cousin method, data were assimilated into a simplified DSM described above in Section 4.4.3. The horizontal and vertical resolutions were the same as in the simulation of "actual" fields. The model parameters were the same as in simulation of $A_T = A_S = 10^7$, $K_M = 100$, $K_S = K_T = 1\,\text{cm}^2/\text{s}$, $\Delta t = 12\,\text{h}$.

The initial gridded fields of temperature and salinity estimates were obtained by optimal interpolation of nonsynchroneous temperature and salinity "measurements" chosen from the "true" fields in the layer 0–1100 m on each day.

Measurements were assumed to be available only at some points of the regular grid. The set of these grid points imitated a network of hydrographic stations in the region. Climatic temperature and salinity fields were used are averages at depth below 1100 m. The initial fields of current velocity were obtained by diagnostic computations.

At this stage, a 36 days prognostic model run was performed and the four-dimensional analysis of ocean parameters with data assimilation during the same period. In the four-dimensional analysis, "measurements" at stations located at the points of the regular grid were selected from the "true" temperature and salinity fields given at 1 day intervals down to a depth of 1100 m. In formulae (4.4.19) the product $\sigma_T(\vec{x},t)\cdot\sigma_T(\vec{x}^{\,\prime},t)$ or $\sigma_S(\vec{x},t)\cdot\sigma_S(\vec{x}^{\,\prime},t)$ was approximated by $\sigma_T(\vec{z},t)\cdot\sigma_T(\vec{z}^{\,\prime},t)$ or $\sigma_S(\vec{z},t)\cdot\sigma_S(\vec{z}^{\,\prime},t)$.

In the assimilation of temperature (salinity) "measurements" data at a particular depth, the influence of these parameters measured at other depths were neglected. In view of what was said above, we needed to know only the dependences on horizontal coordinates in the correlation function $P^H(|\vec{x}-\vec{x}^{\,\prime}|)$. These dependences were approximated by the exponent $[-\alpha(z)(x-x')^2-\beta(z)(y-y')^2]$. The initial normalized error variance for temperature and salinity was set equal to unity. Between the times of data arrival, the variance on the boundary of the domain obeyed the law (Knysh et al., 1980)

$$D[x,(t_1+m\Delta t)^+] = D\left(x,t_1^-\right)+\left\{D\left[x,\left(t_1+(m-1)\,\Delta t^+\right)\right]\right.$$
$$\left.-D\left(x,t_1^-\right)\right\}\exp\left(-\lambda_1\cdot\Delta t\right)$$

where $m=1,2$ is the time step index and $\lambda_1=0.05$ is a parameter determining the temporal behavior of the covariance. The values of $\lambda_1,\ \alpha(z),\ \beta(z)$ were estimated from actual measurements of temperature and salinity.

In four-dimensional analysis, the errors of estimated fields – sea-level (pressure), current velocity components, temperature and salinity decrease with time at all depths where temperature and salinity data were assimilated (Fig. 4.17). For example, after all measurements have been assimilated, the horizontally averaged absolute values of the errors in the recovered temperature field on day 36 decreased with respect to their values on day 1 by 67, 47, 49, 50, 31, 33 and 40% at depths of 20, 100, 150, 200, 300, 500 and 1100 m, respectively. Figure 4.18 gives a more vivid illustration of the errors in the recovered dimensional analysis. It can be seen that the temperature error is maximal at 100 m and decreases sharply at deeper levels.

The temporal behavior of the horizontally averaged errors in the recovered salinity is qualitatively the same as for the errors for recovered temperature.

The salinity errors decrease with depth and range within 0.01–0.018% in the layer 0–1100 m at the end of the analysis interval. At depths where no data is available, the errors increase with time, ranging within 0.008–0.019%.

The behavior of the errors in the recovered horizontal components of current velocity is similar to the behavior of the errors of the temperature and salinity errors. In the upper 200-m ocean layer, the layer-averaged errors in the horizontal velocity on day 36 decrease by 53% with respect to these errors on day 1. An analysis of the temporal behavior of the volume-averaged kinetic energy shows that the energy

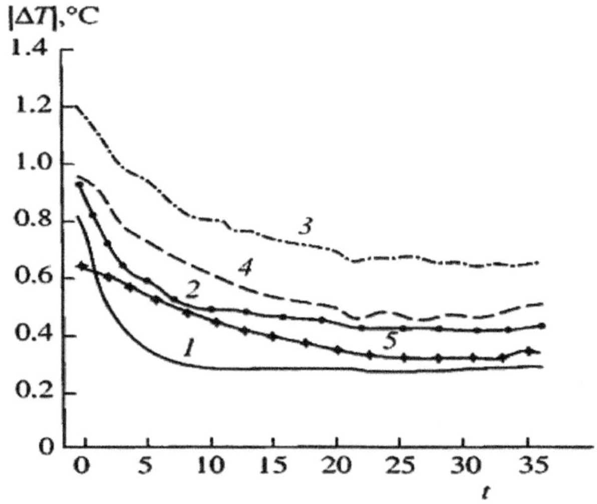

Fig. 4.17 Horizontally averaged absolute values of the errors in the model temperature vs. time at depths of (1) 20, (2) 50, (3) 100, (4) 150, and (5) 200 m in the "cousin's" method (Knysh and Sarkisyan, 2003)

levels corresponding to the "true" hydrophysical fields on day 36 are close to those computed with data assimilation.

In the "identical twin" method, the same model (4.4.1)–(4.4.15) was used for the simulations of "true" fields and for data assimilation. The model parameters were the same as in the cousin "nonidentical" method.

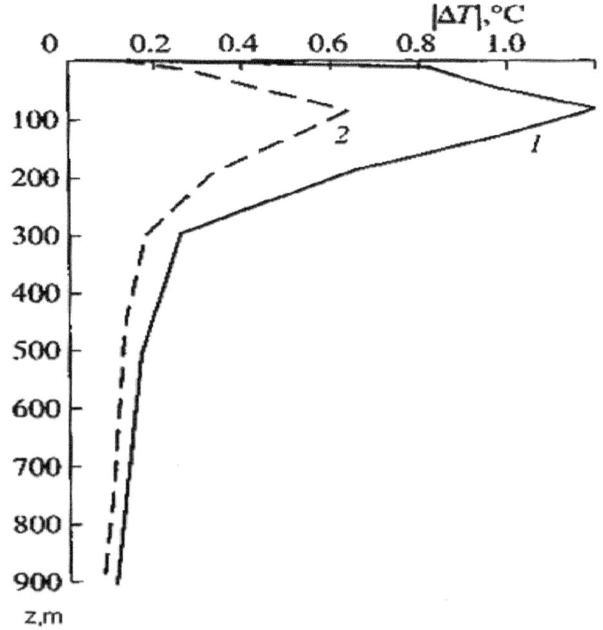

Fig. 4.18 Profiles of the horizontally averaged absolute values of the errors in the model temperature recovered on analysis day (1) 1 and (2) 36 in the nonidentical "cousin's" method. (Knysh and Sarkisyan, 2003)

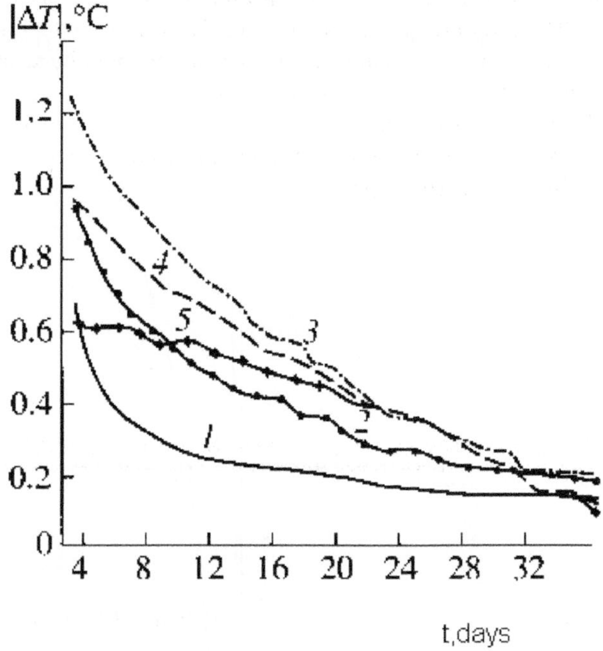

Fig. 4.19 Horizontally averaged absolute values of the errors in the model temperature vs. time at depths of (1) 20, (2) 50, (3) 100, (4) 150, and (5) 200 m in the "identical twins" method (Knysh and Sarkisyan, 2003)

The errors in the recovered sea-level, horizontal velocity, temperature and salinity in the identical twins method also decrease with time (Fig. 4.19). However, the character of temporal variability and the absolute values of these errors are substaintially different in this case. In the layer 0–1100 m, the errors in the recovered fields decrease monotonically to their minimum values at the end of the analysis interval. For example, the temperature errors range within 0.05–0.2°C at depths 20–200 m and within 0.005–0.01°C at depths of 300–1100 m. Thus, these errors are two or three times smaller than the errors in the cousin method. An analysis of the errors in the recovered salinity gives analogues results.

Figure 4.20 shows variations in the errors of the recovered temperature and salinity with depth. The errors on assimilation day 36 decrease with respect to their values on day 1 by 83, 83, 96, 96 and 88% for temperature and by 84, 90, 87, 84 and 88% for salinity at depths of 20, 50, 100, 150 and 200 m, respectively.

A comparison of similar computations performed with the same model and with different models shows that the errors of the recovered hydrophysical fields in the twin method are much smaller than those in the cousin method. This is an essential drawback of model experiments on estimating the accuracy of any four-dimensional analysis procedure.

The behavior of our hydrodynamic ocean model differs from the behavior of the actual ocean. As a result, the errors in the recovered fields derived in numerical experiments with the model turned out to be underestimated.

Nevertheless, numerical experiments performed with dynamic-stochastic models are quite useful for the study of various issues of four-dimensional analysis, such as

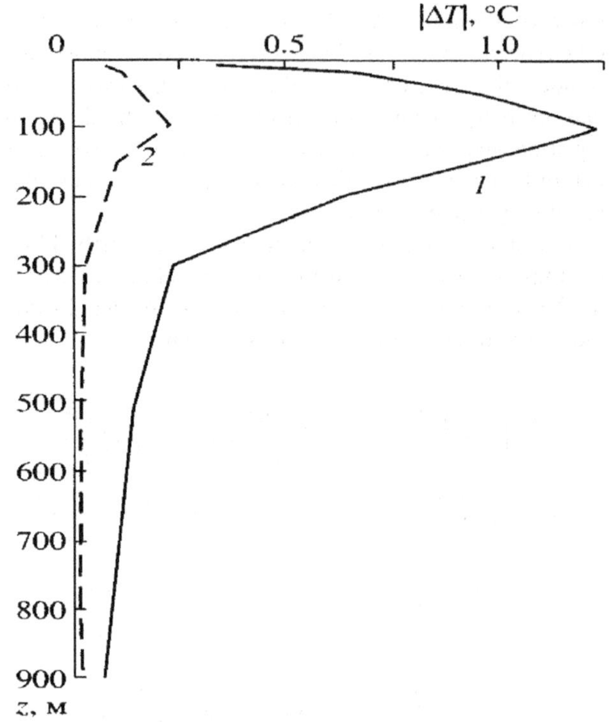

Fig. 4.20 Profiles of the horizontally averaged absolute values of the errors in the model temperature recovered on analysis day 1 (the curve 1) and day 36 (the curve 2) in the "identical twins" method (Knysh and Sarkisyan, 2003)

the estimation of its reliability and accuracy, the influence of one type of data on model reconstruction of other fields, etc.

These experiments also show that the results produced by the DSM-based four-dimensional analysis procedure are more accurate than those yielded by dynamic and diagnostic methods widely used in oceanography for calculating currents from observed temperature and salinity fields, with the fact that these observations are nonsynchronic being entirely neglected.

4.4.5 Four-Dimensional Analysis of Hydrophysical Fields in the Tropical Atlantic

The four dimensional analysis procedure proposed above was applied also to processing the tropical ocean data. Data assimilation was based on the dynamic-stochastic model (4.4.1)–(4.4.20).

In the four-dimensional analysis, the temperature and salinity data measured in the Tropical Atlantic during the 37th cruise of the research vessel "Academic Vernadskyi" (June, 28 to July 15, 1988), was used.

The initial temperature and salinity fields were obtained by the optimal interpolation (to a grid domain) of nonsynchronic temperature and salinity measurements conducted in the layer 1–1100 m over the region. Climatic (summer) fields were used at depths below 1100 m.

The grid domain, the horizontal and vertical resolutions, and the time step were the same as in the numerical experiments described in the previous section. The initial current fields were obtained in a diagnostic run from the initial temperature and salinity fields. The turbulent exchange coefficients for momentum, heat and salt and the boundary conditions were the same as in the "cousin" method (See Section 4.4.4).

In the numerical solution of the differential equations (4.4.1) and (4.4.2), conditions (4.4.10) and (4.4.11) (the same as in the "cousin" method) were used in the lateral rigid boundary. On the liquid boundary, the horizontal velocity was computed from the geostrophic relations and was specified so that the fluid flow through the boundary was zero.

As shown above, a practical implantation of the correlation algorithm for four-dimensional analysis requires information on the statistical structure of hydrophysical fields. In some previous studies it was assumed that the temperature and salinity fields (unlike the errors in estimates) are homogeneous and isotropic with respect to the correlation function.

An analysis and estimation of the statistical structure and disturbances in the temperature and salinity fields over the tropical North Atlantic has shown that the isotropic assumption is not always valid.

To verify isotropy of the temperature and salinity fields, their autocorrelations were calculated in different directions every 10 degrees. The observations made in summer and fall of 1986 were used for this propose, because they were more data in comparison to 1988's survey. The spatial discretization choosen in the calculation was the minimal distance (equal to $30'$) between the stations in the region.

Figure 4.21 shows the correlation distances for temperature, with the average calculated as the arithmetic mean. The left half-plane is not presented because of the symmetry. Dots indicate observation stations (the distance between the station is equal to 1° along the Ox-axis and $30'$ along the Oy-axis).

It can be seen that the spatial statistical structure of temperature fields is nonisotropic with respect to the autocorrelation function. The autocorrelation function for temperature is not invariant with respect to the zonal and meridional directions. The correlation distances for T in the zonal direction are much longer than those in the meridional direction. This feature can be seen at all depths of measurements. The correlation distance varies with direction roughly according to the ellipse law. Figure 4.21 shows that the semimajor axis of the correlation ellipse is slightly tilted relative to the zonal direction. Analogous results were obtained for the salinity correlation distances.

The nonisotropy of temperature and salinity fields with respect to correlations testifies to differences between large-scale oceanic processes in these directions. Indeed, the large-scale circulation in the active layer of the tropical Atlantic is characterized primarily by zonal currents. The meridional gradients of T and S are much greater than their zonal gradients. To make allowance for the anisotropy of temperature and salinity fields, a correlation ellipse was introduced into the four-dimensional analysis algorithm at each depth:

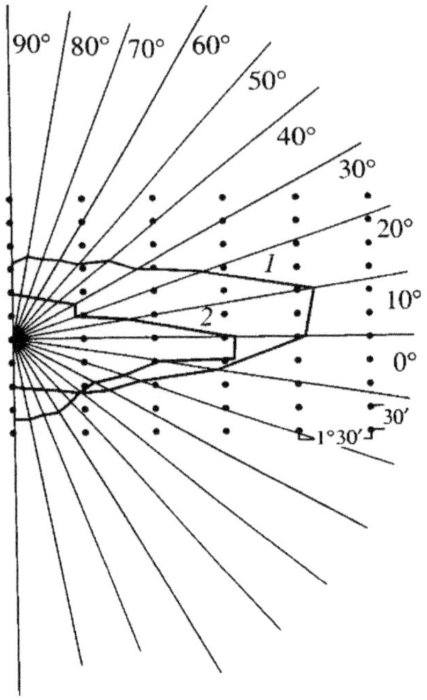

Fig. 4.21 Temperature correlation distances calculated for depths of 50 m (the curve 1) and 800 m (the curve 2) (Knysh and Sarkisyan, 2003)

$$P(\Delta r) = \exp\left[-\alpha_1(x - x')^2 - \beta_1(y - y')^2\right]$$

where α_1 and β_1 are the coefficients corresponding to approximations in the zonal and meridional directions. The coefficients α_1 and β_1 were determined by the least squares method. The tilt of the semimajor axis was neglected in the approximation of the autocorrelations.

The four-dimensional analysis was performed for 81 days. All temperature and salinity data were initially ordered in time, grouped within the assimilation period (set equal to 3 days), and averaged over depth according to the vertical grid spacing chosen in the numerical model. An analysis of the resulting sea level map (Fig. 4.22) shows that a major element of the large-scale water circulation in the region is an eastward jet current, which is identified with the Atlantic Equatorial Countercurrent (AECC). Judged from the surface gradient currents, a major part of this jet is generated by the eastward turn of the Guayana Current within 50°–55 °W and 6°–8 °N. An anticyclonic eddy structure with a maximum sea level elevation of 17 cm can be seen in the turn area south of the AECC. North of this area, the AECC is fed by the North Equatorial Current. An area of low sea levels is located north of the AECC.

The maximum difference in sea level values is 37 cm. Cyclonic gyres are traced in the area of low sea levels. Intense cyclonic and anticyclonic eddies can be seen south of the AECC. A characteristic feature of the water circulation in the region during the summer of 1988 is an abrupt southward turn of the AECC extending up to the equator (from the measurements made south of 2° 15′N). This can easily be seen in the map of horizontal currents (Fig. 4.23).

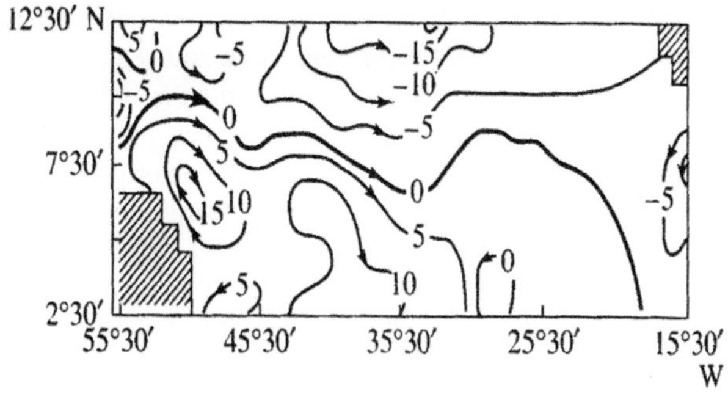

Fig. 4.22 SSH (in cm) produced by the primitive equation model with data assimilation (Knysh and Sarkisyan, 2003)

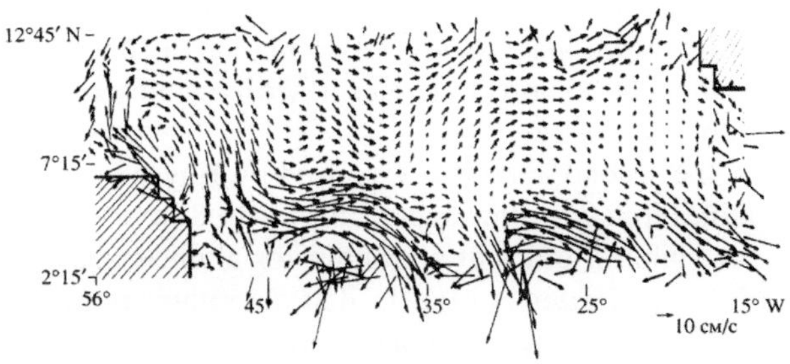

Fig. 4.23 Currents field at a depth of 100 m produced by the primitive equation model with data assimilation (Knysh and Sarkisyan, 2003)

It is well known that data assimilation into a primitive equation model may lead to the generation of spurious small-scale waves in a further model run. Such waves may substantially distort the analysis results and may even lead to numerical instability. The results of four-dimensional analysis with the primitive equation model (4.4.1)–(4.4.15) showed that this effect did occur. For this reason, the numerical procedure for four-dimensional analysis was designed so that the Matsuno scheme was applied after every assimilation cycle (Matsuno, 1966). This prevented the generation of spurious small-scale waves in the model. The results described above can be viewed as a certain technique for the four-dimensional analysis of hydrophysical fields based on dynamic-stochastic ocean models. The outputs of this technique are simultaneous fields of hydrophysical parameters – seawater temperature, salinity (density), sea level, and current velocity. These parameters can be used for calculating their anomalies and important integral ocean characteristics, such as heat content, heat transport by currents, etc.

4.4.6 Algorithm for Altimeter Data Assimilation: Numerical Experiments and Some Reconstructions of Hydrophysical Fields in the Black Sea from TOPEX/POSEIDON Altimetry

In the last decade, much attention has been given to the development of techniques for satellite altimetry assimilation (Fukumori et al., 1993; Holland and Malanotte-Rizzoli, 1989; Mellor and Ezer, 1991; Pinardi et al., 1995; Verron, 1990). This is explained by the improvement of altimeter data accuracy and by the regular arrival of altimeter data covering large ocean and sea areas.

One major difficulty in altimeter data assimilation is that the sea level is a two-dimensional space function. Thus, a key point of any altimeter data assimilation algorithm is the method used for data extrapolation to depth. There are, at least, two approaches. The first consists in using the model itself for data extrapolation. It was shown by Malanotte-Rizzoli et al. (1989) that this approach imposes stringent conditions on the model and requires relatively long model runs for recovering baroclinic information from barotropic data. In the second approach, some statistics and other prior information are used, along with a model, for extrapolating altimeter data to depth. The need for extrapolation in rigid-lid primitive equation models used for data assimilation was justified mathematically by Pinardi et al. (1995). In particular, it was shown that the direct assimilation of sea level data (without extrapolation to depth) does not affect the subsurface dynamics, because the barotropic velocity field is divergence-free.

The second approach was used by Knysh et al. (1996). It should be noted that studies following this approach are based (to a varying degree) on Kalman filtering. The simplest efficient algorithm for altimeter data assimilation was suggested by Mellor and Ezer (1991). In that algorithm, sea level anomalies (deviations from annual means) are previously converted into temperature and salinity (density) data. This is done by using the model climate and surface–subsurface correlations, for example, correlations between the sea level anomaly and temperature or salinity. Next, these temperature and salinity data are assimilated into the model by optimal interpolation at the arrival times of altimeter data with (in 0.5, 1, 3, and 6 days).

The basic differences of the altimetry assimilation method proposed by Knysh et al. (1996) from that used by Mellor and Ezer (1991) are listed below.

(i) Neither the model climate nor sea level anomalies are used. The differences between observed and simulated sea levels are transformed into residuals of temperature, salinity, and horizontal velocity.
(ii) The horizontal velocity components are directly corrected along with temperature and salinity (density).
(iii) An improved data assimilation scheme is continuous in time in the same sense as a numerical solution can be viewed as continuous in time. For this purpose, relevant sources are introduced into the right-hand sides of the heat and salinity (density) diffusion-transport equations and the equations of motion.

The last circumstance makes it possible to obviate the adjustment problem associated with the generation of spurious high-frequency inertia-gravity waves at assimilation times (see Section 4.4.5). Moreover, it will be shown below that discrete-time assimilation leads to a stepwise (temporal) behavior of the solution and, as a result, to a similar behavior of the errors. Continuous assimilation makes possible to avoid this effect.

For numerical experiments, we used model (4.4.1–4.4.15) but written in spherical coordinates (Ibraev, 1993). Equations (4.4.5) and (4.4.6) were replaced by a diffusion – transport equation for density. This model is best suited for solving the problem under study, because the model involves an equation for sea level. The rigid lid approximation on the surface was used for deriving this equation. Our numerical experiments were conducted for the Black Sea and the tropical North Atlantic.

Let us describe the numerical results obtained for the Black Sea. The model parameters used for the Black Sea included a 20 km horizontal grid spacing and 21 levels in the vertical: 5, 10, 20, 30, 50, 70, 90, 120, 150, 200, 250, 300, 400, 600, 800, 2000 m. The time step was 30 min.

The "true" hydrophysical fields were obtained in an adaptive 49 day model run from a summer climatic density field. The results corresponding to the last 25 days were used in the numerical experiments.

The runs with altimeter data assimilation were initialized with the hydrophysical (first-guess) fields obtained in an adaptive 26 day model run from a smoothed climatic density field. The "observation" data were assumed to be accurate and to arrive at each grid point of the computational domain at given time intervals.

In the framework of this setting, several numerical experiments were conducted aimed at estimating the effectiveness of the altimetry assimilation procedure described above. First an adaptive run without allowance for altimeter data was carried out (experiment 1). Then was an adaptive run was conducted with insertion of altimeter data every 10 days (experiment 2). In the data insertion procedure, the computed sea levels were only replaced by the "true" ones on model days 1, 11, and 21.

The assimilation procedure in the next experiments involved data insertion combined with simultaneous correction of other hydrophysical fields (ρ, u, v). A simplified procedure was applied, in which the sea level and density error correlations $P_{\zeta\zeta}(\ldots)$ and $P_{\rho\zeta}(\ldots)$ used in formulas of type (4.4.16) were time invariant and were not corrected. It was also assumed that $P_{\zeta\zeta}(\ldots)$ and $P_{\rho\zeta}(\ldots)$ are equal to the correlations of the sea level and density fields themselves, which in turn were computed from the initial first-guess fields.

Thus, experiment 3 was an adaptive run, with altimeter data insertion and density correction [by a formula similar to (4.4.16)] performed at 10 day intervals. Experiment 4 was analogous to experiment 3, but, in addition to density, the horizontal current velocity components u and v were also corrected by the formulae

$$\begin{aligned} u(x,t_l^+) = {} & u(x,t_l^-) - \frac{g}{a_0 l}\frac{\partial}{\partial\varphi}\sum_{r=1}^{N}\Delta_r^{1\zeta}(x,y,t_l^-)\delta\zeta(x_r,y_r,t_l) - \\ & - \frac{g}{a_0 l}\frac{\partial}{\partial\varphi}\sum_{r=1}^{N}\int_0^z \Delta_r^{2\zeta}(x,t_l^-)d\mu\,\delta\zeta(x_r,y_r,t_l) \qquad (4.4.21) \\ v(x,t_l^+) = {} & v(x,t_l^-) + \frac{g}{a_0 l\cos\varphi}\frac{\partial}{\partial\lambda}\sum_{r=1}^{N}\Delta_r^{1\zeta}(x,y,t_l^-)\delta\zeta(x_r,y_r,t_l) + \\ & + \frac{g}{a_0 l\cos\varphi}\frac{\partial}{\partial\lambda}\sum_{r=1}^{N}\int_0^z \Delta_r^{2\zeta}(x,t_l^-)d\mu\,\delta\zeta(x_r,y_r,t_l) \end{aligned}$$

Here, $\vec{x} = (\lambda, \varphi, z)$, φ is latitude, λ is longitude, z is depth, and a_0 is the Earth's radius. Note that formulae (4.4.22) were derived by using the geostrophic relations.

In experiment 5, the sensitivity of numerical results to the frequency of data arrivals was analyzed. This experiment was similar to the preceding one, but altimeter data were assumed to arrive two times more frequently, i.e., every 5 days.

Experiment 6 was aimed at studying the sensitivity of numerical results to the accuracy of the model error correlations. This experiment was analogous to experiment 4, except that $P_{\zeta\zeta}(\ldots)$ and $P_{\rho\zeta}(\ldots)$ used in formulae of type (4.4.16) and (4.4.22) were calculated from the errors in the recovered density and sea level available at the initial time.

To avoid the effect of inertia oscillations in the computed errors in the recovered hydrophysical fields, 17 hour moving averages were calculated for all fields prior to their analysis.

Figure 4.24 shows that the replacement of the computed sea level field by the "true" one without correcting other hydrophysical fields (curve 2) only slightly reduces the error in the recovered field as compared with the adaptive run (curve 1). More exactly, the error is reduced by 11% at the end of analysis. Data assimilation combined with density correction gives a substantial reduction (by 67%) in the error (curve 3). An additional correction of u and v further reduces the sea level error by 5% (curve 4). A twofold increase in the frequency of data arrivals (additional data assimilations on days 6 and 16) considerably reduces the error at the initial stage (curve 5). However, after data assimilations on days 11 and 21, this difference reduces significantly, and curves 4 and 5 become rather close. In the case of using more accurate model error correlations (curve 6), the error immediately after assimilation is greater than that in experiment 4, but it becomes smaller than in experiment 4 in four days after assimilation (curve 6 lies below curve 4).

Altimeter data assimilation made it possible to improve the density and velocity fields only in the upper 600 m ocean layer. Moreover, in the case of data assimilation with density and velocity correction, the errors in the recovered density and velocity fields in the layer 0–200 m were reduced considerably as compared with the mere replacement of the computed sea level by the "true" one. For example, the mean

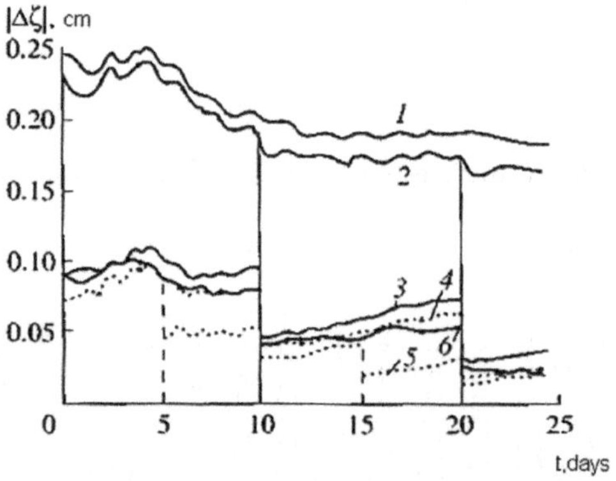

Fig. 4.24 Spatially averaged absolute errors of the SSH field recovered in different experiments. Digits indicate the experiment numbers (Knysh and Sarkisyan, 2003)

absolute errors in the recovered u and v fields in the upper 120 m layer were reduced by only 15% in the case of altimeter data insertion, while an additional correction of density and velocity further reduced these errors by about 45%. It should be noted that the twofold increase in the frequency of data arrivals had a small effect on the errors in the recovered velocity and density.

The horizontally averaged absolute values of the errors in the recovered density also decreased with time. The use of more accurate error correlations additionally reduced the errors in the recovered density.

The numerical experiments conducted suggest the following conclusion. The altimeter data assimilation procedure involving data insertion, combined with correction of density and velocity, makes it possible to considerably improve the accuracy of the recovered fields. The largest effect is achieved when density correction is involved in the procedure. To conclude this section, we present some results of TOPEX/POSEIDON satellite altimetry assimilation into a general circulation model of the Black Sea.

These results were obtained by Korotaev et al. (1998), which is an important study for practical applications. Altimeter data assimilation was based on the nonlinear numerical model described in Demyshev and Korotaev (1992). The dynamic sea level of the Black Sea for subsequent assimilation was recovered from the TOPEX/POSEIDON satellite altimetry.

The dynamic sea level was assimilated into the model at every time step. To this aim, the dynamic sea level data were interpolated in time by using temporal harmonic expansions.

The difference $\delta\zeta = \zeta^a - \zeta$ between the altimeter-based and simulated sea levels was optimally interpolated to a horizontal grid (Gandin and Kagan, 1976). The algorithm for altimeter data assimilation can be described as follows. The right-hand side of the salt diffusion-transport equation was supplemented with the assimilation source $Q = \tau^{-1} k_{s\zeta} \delta\zeta$, where τ is the relaxation time and $K_{S\zeta}$ is the coefficient for transforming sea level residuals into salinity residuals. The covariance $P_{S\zeta}$ between the salinity errors at different depths and the sea level errors, as well as the sea level error variance (σ_ζ^2), was estimated from climatic monthly mean fields of salinity and sea level obtained in adaptive model runs. The coefficient K was taken as a function of the vertical coordinate and time, and the relaxation time was set equal to 10 days. Altimeter data were processed according to the technique described by Korotaev et al. (1998), and were assimilated into the model from mid-November, 1992, to April, 1993. The climatic fields from Staneva and Stanev (1998) were used to specify wind stress and heat fluxes at the sea surface. The salt fluxes were specified to be proportional to the difference between the climatic and computed salinities.

Figures 4.25a–c show the sea level topography of the Black Sea for April derived from (a) climatic data, (b) CoMSBlack'93 hydrography, (c) TOPEX/POSEIDON altimeter data assimilated into the model and (d) by optimal interpolation of TOPEX/POSEIDON altimeter data. The sea level fields in Fig. 4.25a and b were obtained in 10 day adaptive runs. It can be seen that the climatic sea level (Fig. 4.25a) is smoother than the sea level fields displayed in Fig. 4.25b and c. The sea level fields derived from the CoMSBlack'93 data and TOPEX/POSEIDON altimetry

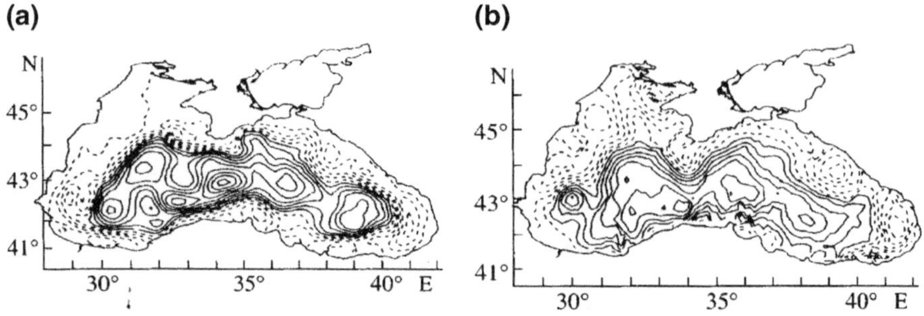

Fig. 4.25 Sea surface height for April derived from (**a**) climatic data, (**b**) CoMSBlack'93 hydrography, (**c**) TOPEX/POSEIDON altimetry assimilated into the model, and (**d**) optimal interpolation of the TOPEX/POSEIDON data. The contour interval is 2 cm. The areas of negative SSH values are shown by *solid lines* (Knysh and Sarkisyan, 2003)

assimilation are well correlated with each other and reproduce the basic features of the Black Sea general circulation.

Figure 4.25 shows the sea level field obtained by optimal interpolation of the TOPEX/POSEIDON data for April, 1993, with the climatic sea level (Fig. 4.25a) used as a reference field. The role of the model in data assimilation can be seen by comparing the sea level fields displayed in Fig. 4.25c and d and by comparing these fields with the sea level presented in Fig. 4.25b. Figure 4.26 shows the map of current velocities at a depth of 100 m at the end of the integration period. Additionally to a general cyclonic vorticity and anticyclonic eddies at the periphery, one can see cyclonic and anticyclonic mesoscale structures within the main Black Sea current. These results agree well with existing views about the Black Sea general circulation. The current speed at this depth in the main Black Sea current reaches 25–30 cm/s, which also agrees with observations.

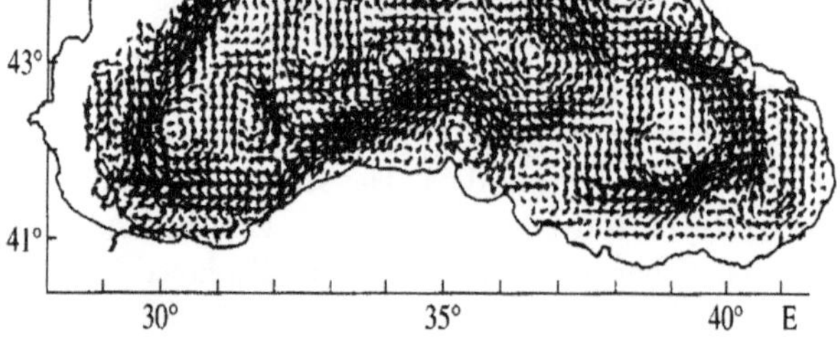

Fig. 4.26 Current velocities at a depth of 100 m at the end of the assimilation period for the TOPEX/POSEIDON data. The vector in the upper corner of the figure corresponds to a current speed of 10 cm/s (Knysh and Sarkisyan, 2003)

Experience gained from assimilation of climatic temperature and salinity distributions (Knysh et al., 2001) and the TOPEX/POSEIDON altimetry into primitive equation numerical models suggests the following important conclusion. Time-continuous assimilation in the form of sources introduced into the right-hand sides of the corresponding equations is preferred. In this case, the forecast errors are smaller than those in discrete assimilation. The form of the sources is naturally determined by the Kalman filter. A disadvantage of this approach is that observations have to be interpolated in time and space.

References

Baranov E. I., A. V. Kolinko, and V. S. Regentovskii, 1987. Climatic and anomalous oceanographic conditions of the Newfoundland energy-active zone and their relation to the atmospheric circulation. *Itogi Nauki i Tekhniki, Ser. Atmosfera, Okean, Kosmos. Programma "Razrezy"*, Moscow: VINITI 8: 82–91 (in Russian).

Bryan K., 1969. A numerical method for the study of the circulation of the world ocean. *J. Comp. Phys.* 4: 347–376.

Böning C. W., R. Döscher, and R. G. Budich, 1991. Seasonal transport variation in the Western Subtropical North Atlantic: Experiments with an Eddy resolving model. *J. Phys. Oceanogr.*, 21: 1271–1289.

Bulushev M. G. and A. S. Sarkisyan, 1996. Energetics of the equatorial current adjustments initial stage. *Izv. Acad. Sci. Atmos. Ocean Phys.* 32(5): 600–612.

Conkright M. E., S. Levitus, and T. P. Boyer, 1994. *World Ocean Atlas*. Vol. 3: Salinity, Vol. 4: Temperature. Washington, DC: U.S. Dept. Commerce.

Demin Yu. L. and A. S. Sarkisyan, 1977. Calculation of equatorial currents. *J. Mar. Res.* 35: 339–356.

Demyshev S. G. and G. K. Korotaev, 1992. *Numerical Energy-Balanced Model of Baroclinic Currents in an Ocean with a Rough Bottom on a Grid C* (Numerical Models and Results of Calibration Calculations of Currents in the Atlantic). Moscow, 163–231.

Fillenbaum E. R., T. N. Lee, W. E. Johns, and R. J. Zantopp, 1997. Meridional heat transport variability at 26.5 °N in the North Atlantic. *J. Phys. Oceanogr.* 27: 153–174.

Fukumori I., J. Benveniste, C. Wunsch, and D. B. Haidvogel, 1993. Assimilation of sea-surface topography into an ocean circulation model using a steady-state smoother. *J. Phys. Oceanogr.* 23: 1831–1855.

Gandin L. S. and R. L. Kagan, 1976. *Statisticheskie metody interpolyatsii meteorologicheskikh dannyk (Statistical Methods of Interpolating Meteorological Fields)*. Leningrad: Gidrometeoizdat (in Russian).

Hall M. and H. Bryden, 1982. Direct estimates and mechanism of ocean heat transport. *Deep-Sea Res.* 29: 339–359.

Holland W. R. and P. Malanotte-Rizzoli, 1989. Assimilation of altimeter data into an ocean circulation model: Space versus time resolution studies. *J. Phys. Oceanogr.* 19: 1507–1534.

Ibraev R. A., 1993. Reconstruction of the climatic characteristics of the Gulf stream. *Izv. Akad. Nauk, Fiz. Atmos. Okeana* 29(6): 803–814.

Ivanov Yu. A., K. V. Lebedev, and A. S. Sarkisyan, 1997. Ocean modeling by general adjustment. *Izv. Ross. Akad. Nauk, Atmos. Ocean Phys.* 33(6): 812–818 (in Russian).

Kalnay E., M. Kanamitsu, R. Kister, W. Collins., D. Deaven, L. Gandin, M. Iredel, S. Saha, G. White, J. Wool, Y. Zhu, M. Chellian, W. Ebisuzaki, W. Higgins, J. Janoviak, K. C. Mo, C. Ropelewski, A. Leetmaa, R. Reynolds, and R. Jenne, 1996. The NCEP/NCAR 40-year reanalysis project. *Bull. Am. Meteorol. Soc* 77: 437–471.

Knysh V. V., V. A. Moiseenko, A. S. Sarkisyan, and I. E. Timchenko, 1980. Combined use of measurements on hydrophysical ocean fields in a four-dimensional analysis. *Dokl. Akad. Nauk SSSR* 252(4): 832–836.

Knysh V. V., 1982. Multi-element four-dimensional analysis of the main hydrophysical fields of the ocean. *Izv. Akad. Nauk SSSR, Fiz. Atmos. Okeana* 18(4): 391–398.

Knysh V. V., V. A. Moiseenko, and V. V. Chernov, 1988. Some results of a four-dimensional analysis of hydrophysical fields of the Tropical Atlantic. *Izv. Akad. Nauk SSSR, Fiz. Atmos. Okeana* 24(7): 744–752.

Knysh V. V., V. A. Moiseenko, and V. V. Chernov, 1990. Monitoring hydrophysical fields of the newfoundland energy-active zone on the basis of a four-dimensional analysis of 1982–1986 observational data. *Itogi Nauki i Tekhniki, Ser. Atmosfera, Okean, Kosmos–Prograirana "Razrezy"*. Moscow: VINITI, 13: 154–167 (in Russian).

Knysh V. V., O. A. Saenko, and A. S. Sarkisyan, 1996. Method of assimilation of altimeter data and its test in the tropical North Atlantic. *Russ. J. Numer. Anal. Math. Model.* 11(5): 333–409.

Knysh V. V., S. G. Demyshev, G. K. Korotayev, and A. S. Sarkisyan, 2001. Four-dimensional climate of seasonal Black Sea circulation. *Russ. J. Numer. Anal. Math. Model.* 16(5): 409–426.

Knysh V. V. and A. S. Sarkisyan, 2003. Four dimensional analysis of Hydrophysical ocean and sea fields. Numerical experiments and reconstructions. *Izv. Acad. Sci. Atmos. Ocean Phys.* 39(6): 816–833.

Korotaev G. K., O. A. Saenko, and Ch. D. Koblinski, S. G. Demyshev, and V. V. Knysh, 1998. Accuracy estimation, methods, and some results of TOPEX/POSEIDON data assimilation in a model of the Black Sea general circulation. *Issled. Zemli Kosmosa* 3: 3–17.

Lee T. N., W. Johns, K. Schott, and R. Zantopp, 1990. Western boundary current structure and variability East of Abaco, Bahamas at 26.5 °N. *J. Phys. Oceanogr.* 20: 446–466.

Malanotte-Rizzoli P., R. E. Young, and D. B. Haidvogel, 1989. Initialization and data assimilation experiments with primitive equation model. *Dyn. Atmos. Oceans* 13: 349–378.

Marchuk G. I., 1969. Numerical solution of the poincare problem for the ocean circulation. *Dokl. Akad. Nauk SSSR* 165(5): 1041–1044.

Marchuk G. I. and A. S. Sarkisyan, 1988. *Mathematical Modeling of Ocean Circulation*. Springer-Verlag, 262pp.

Marsh R., M. J. Roberts, R. A. Wood, and A. L. New., 1996. An intercomparison of a Bryan-Cox-type ocean model and an isopycnic ocean model. Part II: The subtropical gyre and meridional heat transport. *J. Phys. Oceanogr.* 26: 1528–1550.

Matsuno T., 1966. Numerical integrations of the primitive equations by a simulated backward difference method. *J. Meteorol. Soc. Jpn.* Ser. 2, 44: 76–84.

Mellor G. L. and T. Ezer, 1991. A gulf stream model and an altimetry assimilation scheme. *J. Geophys. Res.* 96: 8779–8795.

Molinari R. L., E. Lohns, and I. F. Festa, 1990. The annual cycle of meridional heat flux in the Atlantic Ocean at 26.5 °N. *J. Geophys. Res.* 20: 476–482.

Pinardi N., A. Rosati and R. C. Pacanowski, 1995. The sea surface pressure formulation of rigid lid models. Implications for altimeter data assimilation studies. *J. Mar. Syst.* 6: 109–119.

Roemmich D. and C. Wunsch, 1985. Two transatlantic sections: Meridional circulation and heat flux in the sub tropical North Atlantic. *Deep-Sea Res.* 32: 619–664.

Rosenfeld L. K., R. L. Molinari, and K. D. Leaman, 1989. Observed and modeled annual cycle of transport in the straits of Florida and East of Abaco Island, the Bahama (26.5 °N). *J. Geophys. Res.* C 94(4): 4867–4878.

Sakawa Y., 1972. Optimal filtering in linear distributed parameter system. *Int. J. Contr.* 16(1): 115–127.

Sarkisyan A. S. and Ju. L. Demin, 1983. A semidiagnostic method of sea currents calculation. *Large-Scale Oceanographic Experiments in the WCRP* 2(1): 210–214.

Sarkisyan A. S., E. V. Semenov, and V. V. Efimov, 1989. Numerical model of a four-dimensional analysis of on thermohaline measurements. *Izv. Akad. Nauk SSSR, Fiz. Atmos. Okeana* 25(1): 53–63.

Sarkisyan A. S. and A. N. Sidorova, 1998. Numerical modeling of physical characteristics in the neighborhood of an isolated hydrologic section in the Barents Sea. *Russ. J. Numer. Anal. Math. Modell.* 13(6): 537–549.

Sarkisyan A. S. and S. Levitus, 1999. Ocean climate characteristics as obtained by amalgamating WOCE-Levitus hydrographic data. *EOS Trans. Am. Geophys. Union* 80(49): OS210-03.

Sarkisyan G. A. and V. N. Stepanov, 1999. Method for calculating physical characteristics of the ocean from an individual hydrologic section. *Izv. Akad. Nauk, Fiz. Atmos. Okeana*, 35(6): 550–555; *Izv. Atmos. Ocean. Phys. (Engl. Transl.)*, 35(6): 501–505.

Sarmiento J. L. and K. Bryan, 1982. An ocean transport model for the North Atlantic. *J. Geophys. Res.* 87: 394–408.

Schmitz W. J. and M. S. McCartney, 1993. On the North Atlantic circulation. *Rev. Geophys.* 31: 29–49.

Schott F. A., J. Fischer, J. Reppin, and U. Send, 1993. On mean and seasonal currents and transports at the western boundary of the Equatorial Atlantic. *J. Geophys. Res.* 98: 14353–14368.

Semyonov E. V., 1981. Calculation of vertical motions in numerical models of water circulation. *Okeanologiya (Moscow)* 21(3): 433–434.

Staneva J. V. and E. V. Stanev, 1998. Oceanic response to atmospheric forcing derived from different climatic data sets: Intercomparison study for the Black Sea. *Oceanol. Acta* 21(3): 383–417.

Timchenko I. E., 1981. *Dynamic-Stochastic Models of the State of the Ocean.* Kiev: Naukova Dumka (in Russian).

Treguier A. M., T. Reynaud, T. Pichevin, B. Barnier, J. M. Molines, A. de Miranda, C. Messanger, J. O. Beismann, G. Madec, N. Grima, M. Imbard, C. Le Provost, 1999. The CLIPPER project: High resolution modeling of the Atlantic. *Int. WOCE Newsl.* 36: 3–5.

Verron J., 1990. Altimeter data assimilation into an ocean circulation model: Sensitivity to orbital parameters. *J. Geophys. Res.* 95: 11443–11459.

Weatherly G. L., 1972. A study of the bottom boundary layer of the Florida current. *J. Phys. Oceanogr.* 2: 54–72.

Chapter 5
Modelling of Climate Variability in Selected Ocean Basins

Abstract Specific ocean variability modelling is discussed, in reference to selected ocean basins, namely the Arctic-North Atlantic system, the North Atlantic, the Pacific, and the Indian Ocean. For the first of the basins problems of the sea-ice and snow thermodynamics, ice thickness vertical multilayer distribution and the dynamics of ice-water phase transition modelling are considered. Besides, the Arctic is strongly connected with the North Atlantic and depends on the surrounding river discharge. The effect of rivers is two-fold: ocean water volume variation and possible release of contaminants. A paragraph devoted to critical analysis of advantages and drawbacks of the North Atlantic high resolution modelling is following. Concerning the North Pacific circulation and contaminated water propagation a short model-time of integration with high resolution (0.12°) enabled to keep the results of calculations at the realistic stage. An analysis of very successful modelling of the Kuroshio extension and the Kuroshio itself by high resolution and adequate model-time of integration is also presented in this paragraph. The last section of the Chapter contains the Indian Ocean descriptive hydrometeorology and thermo-hydrodynamics modelling.

Keywords Arctic Ocean · North Atlantic Ocean · Indian Ocean Dipole · Kuroshio · Seasonal Variability · Tracers

5.1 Numerical Simulation of in the Arctic–North Atlantic System

5.1.1 Introduction

A specific feature of the arctic part of the World Ocean is the presence of ice cover that is permanent over its substantial area. The ice itself is a complex system and affects actively heat and water exchange processes between the atmosphere, the land, and the World Ocean. The Arctic Ocean that accounts for 5% of the World Ocean area and 1.5% of its volume contributes 10% of the total fresh water inflow to the World Ocean (Ivanov, 1976, 2001). The fresh water flow into the north Atlantic from the Arctic basin happens in the form of ice or surface waters through the Fram Strait

A.S. Sarkisyan, J.E. Sündermann, *Modelling Ocean Climate Variability*,
DOI 10.1007/978-1-4020-9208-4_5,

and the straits of the Canadian Arctic Islands. Main components of the "conveyor belt" (Broecker, 1991) are deep convection in the Labrador Sea and the Greenland Sea, which generates deep Atlantic water that propagates across the bottom layers of the Atlantic Ocean to the south as a part of the global hydrological cycle. A variation in the fresh water flow from the Arctic basin can disturb the thermohaline structure of the North Atlantic and the World Ocean. In view of this, the study of heat flows and fresh water transport and transformation in the Arctic is necessary for understanding global climate changes (Fletcher, 1970).

Further ice is a main vehicle for transporting dissolved and suspended substances in the ocean. These substances may be aerosols, particulate matter, or contaminants. The high level of contamination of the Arctic region with heavy metals, oil hydrocarbons, radio nuclides, and other industrial by-products has led to the investigations of anthropogenic sources of contamination and the process of its transport. At present it is generally acknowledged that one of the main mechanisms of the contamination of the Arctic is atmospheric transport. Another potential contamination source is rivers inflow into the Arctic Ocean from industrial regions.

The contaminant transport in the Arctic Ocean is in accordance with the system of ice drift and currents in the basin, where the main elements are the Transpolar Drift crossing the Arctic Ocean from the Bering Straits to the shores of Greenland and the anticyclonic circulation in the Beaufort Sea (Treshnikov and Baranov, 1972).

The role of the Arctic Ocean within the global climate can be studied by numerical simulation with multicomponent coupled models.

The description of thermodynamics of sea ice and snow is based on locally one-dimensional models of different complexity, different formulations of the ice distribution over its thickness, different numbers of ice-thickness levels, different parameterizations of albedo and penetrative radiation. For example, for ice uncovered by snow, the heat budget is described by:

$$\rho_i c_i \frac{\partial T_i}{\partial t} = -\frac{\partial}{\partial z}\left(k_i \frac{\partial T_i}{\partial z}\right) - \frac{\partial}{\partial z}\left[I_0 \exp(-\lambda_i z)\right], z \in \left[0, -h_{1,}\right] \tag{5.1.1}$$

where ρ_i is the sea-ice density (assumed to be constant), $c_i(T_i; S_i)$ is the heat capacity of ice as a function of its temperature T_i and salinity S_i, $k_i(T_i; S_i)$ is the heat conduction of ice, I_0 is the flux of penetrative solar radiation with vertical scale of extinction λ_i, z is the vertical coordinate directed upward from the ice-snow surface, where $z = 0$. The problem for snow cover is solved similarly, assuming that no penetration of solar radiation into ice occurs.

On the surfaces of ice and snow, heat fluxes are given; at the snow-ice boundary, the continuity condition is taken for temperature and heat fluxes; at the lower boundary of ice, the temperature is set equal to that of seawater freezing at a given salinity. Otherwise, the heat surplus is consumed to the melting of snow or ice, which makes a positive flow of freshwater into the ocean. At the lower boundary of ice, both melting and freezing can occur, leading to the formation of a corresponding flow of freshwater or salinity.

Phase transitions change the mass of ice and snow. At present, most climate models include simple sea-ice models (Polyakov, 2001); however, some groups have turned to the more complex model of NCAR (IPCC, 2001). This is a multi-layer model accounting for the salinity distribution by ice thickness and the dependence of thermodynamic ice parameters on its temperature and salinity. All modern models incorporate snow cover on ice. The numerical features of the models are governed by the number of levels in ice, allowing for the temperature profile and the method accounting ice salinity. From the computational viewpoint, sea-ice thermodynamics presents no considerable problems. A major remaining difficulty is the appropriate parameterization of physical processes.

5.1.2 Sea-Ice Dynamics

It is supposed that the ice cover is a two-dimensional medium and can be described at each point by an ensemble of different-thickness ices (or, by a function of ice distribution by thickness); here, each gradation of ice thickness h_k can be described by its mass m_k and concentration A_k. Accordingly, the mass of snow covering the ice of gradation k can be introduced. The velocity of sea-ice drift u_i is assumed to be the same for all gradations of thickness. Then the sea-ice motion is described by

$$m\frac{\partial \vec{u}_i}{\partial t} + mf\vec{u}_i \times \vec{u}_i = -mg\vec{\nabla}\varsigma + \vec{\tau}_a + \vec{\tau}_\omega + \vec{F} \tag{5.1.2}$$

Here, m is the total mass of ice and snow, $\vec{\tau}_a$ is the wind stress, f is the Coriolis parameter, $\vec{\nabla}\varsigma$ is the ocean-surface gradient, $\vec{F}$ is the force induced by the sea-ice rheology (i.e., stresses emerging on the ice cover due to its motion and different thickness and concentration of ice).

Sea ice models describe melting and freezing (thermodynamics) and movement and deformation of the ice cover (dynamics) in response to atmospheric and oceanic forces. Model simulations reveal the importance of sea ice dynamics to the climate system. For example, in comparing simulations of Arctic sea ice volume with submarine gathered data, Hilmer and Lemke (2000) concluded that better models of sea ice dynamics are required to improve understanding of the interannual variability and discern the true climate signal. In this paper, we focus on sea ice stresses resulting from deformation of the ice cover.

The sea ice cover is formed from floes that are of approximately convex polygonal shape, have lateral dimension of between 100 m and 5 km, and are several meters thick. Sea ice deformation occurs through the relative motion of sea ice floes: under local divergence, sea ice floes can separate to form linear regions of open water called leads; in local shear, floes can slide along their common edges; and in local convergence, floes can break up to form linear regions of rubble piled above and below the ice cover along floe edges known as pressure ridges. Typically floe-scale deformation involves ridging and sliding or opening and sliding, but all types of deformation are usually present in any region containing a collection of floes.

During pressure ridging, the ice cover first breaks in flexure into blocks and the ridging stress is determined by the work required to move the ice blocks against gravity and ridging friction to form a pressure ridge. When floes slide past each other, the sliding stress is determined by friction between the floe edges. The frictional stress is normally taken to be independent of the rate of deformation so that the cumulative sea ice stress is independent of strain-rate magnitude. Since the work done in forming pressure ridges or sliding of floes past each other is irreversible, sea ice rheological behavior is considered to be plastic.

Since ridging redistributes ice among different thicknesses whereas sliding does not, the redistribution of ice thickness depends upon the relative amount of sea ice deformation realized through ridging as compared with sliding (and opening). The resultant sea ice stress used in a model of sea ice dynamics also depends upon the relative amounts of ridging, sliding, and opening as the stresses involved in these types of deformation differ. The relative amounts of ridging and sliding depend upon the type of deformation of the ice cover; for example, more sliding of floes past each other occurs in pure shear than in pure divergence.

Most sophisticated sea ice codes treat the ice within a grid cell to be distributed between several thickness categories and implement a parameterization for the ridging work fraction introduced by Flato and Hibler (1995). A particular feature of this parameterization is that it implies the absence of sliding friction in pure convergence, while discrete simulations (Hopkins, 1996; Ukita and Moriz, 2000) show that the irregular shape of floes leads to significant interfloe sliding in this case.

Sea ice models require an expression relating sea ice strength (maximum stress in compression) to the thickness of the ice that is being deformed. Most sea ice models either set the ice strength to be proportional to the potential energy change in forming a pressure ridge, in which case the strength is proportional to ice thickness squared, or treat the strength to be proportional to ice thickness, adopting a parameterization due to Hibler (1979). One of version of the sea-ice ridging to sliding deformation stress force $\vec{F}$ contents in the paper of Iakovlev (1998).

A large number of studies describe the results of simulating the circulation, ice generation and transformation in the Arctic basin and the correlation of these processes with those in the North Atlantic. Simulated results based on three-dimensional models are given, for example, by Hakkinen and Mellor (1992), Hibler and Bryan (1987), Iakovlev (1998), Karcher and Oberhuber (2002), Kuzin et al. (2006), Golubeva and Platov (2007), Maslowski et al. (2004), Polyakov and Timokhov (1995), Proshutinsky and Johnson (1997), Riedling and Preller (1991), Ryabchenko et al. (2003) etc. We present below the results of simulation using a version of a coupled ocean model for the Arctic-North Atlantic system (Kuzin et al., 2006; Golubeva and Platov, 2007).

The dynamic equations are solved by Kuzin et al. (2006) using the method of dividing into the barotropic and baroclinic modes (see Chapter 1); the barotropic part represented as the solution of the equation for the stream function. The discretization of the principal equations of the model is done by combining the finite element method and the method of splitting with respect to physical processes and geometric directions (Marchuk et al., 2001; Zalesny and Kuzin, 1995). For the time domain a combination of explicit and semi-explicit approaches is used. The advective part of the space operator is approximated by a numerical first-order scheme with dosed system viscosity. The processes occurring in the upper mixed layer are parametrisized depending on the Richardson criterion and the layer of vertically homogeneous distribution of temperature, salinity and velocity components.

To simulate the Arctic–North Atlantic interaction a combined grid for the coupled ocean-ice model was used by Kuzin et al. (2006). The region under consideration includes the northern part of the Atlantic Ocean starting from latitude 20 °S. The inclusion of the tropic zone contributes to an adequate formation of the western boundary currents, i.e., the Gulf Stream and its extension, the North Atlantic Current. At 65 °N, the grid of the spherical coordinate system adapted for the Atlantic merges with another orthogonal grid of higher resolution, which is obtained by a rotation of the spherical coordinate system and the reprojection of its semisphere on the region north of 65 °N (Murray, 1996). The maximum resolution is attained in the circumpolar region and amounts to 35 km. On the average, the nodes of the computational grid in the Arctic Ocean region are set at a distance of about 50 km. The grid resolution for the North Atlantic is taken equal to 1°. In the vertical the grid of the North Atlantic has 33 horizontal levels, more refined near the sea surface where the resolution is 10 m.

The simulated region comprises the most significant straits between the Canadian Arctic Islands. The minimum depth of the shelf zone is taken to be 50 m.

The initial temperature and salinity distribution was obtained from PHC (Polar Science Center Hydrographic Climatology) climatic data (Steele et al., 2001), which comprises monthly three-dimensional temperature and salinity fields in the layer up to 1 km depth and seasonal (winter, summer) and annual mean data for the total range of depths. The winter distribution was used as the initial condition.

The fresh water inflow from rivers were based on available data of the mean seasonal run-off of the main 13 rivers in the Arctic region, which amounts to 8.6 km^3/day and 23 rivers in the Equatorial and North Atlantic (about 29.7 km^3/day) derived from River Discharge Database (Völösmarty et al., 1998).

The conditions at the surface of the ocean-ice system include: latent and explicit heat flows, incoming solar and long-wave radiation, long-wave surface radiation, wind friction, and fresh water flux caused by precipitation.

One of the most complete sets of atmospheric data is provided by NCEP/NCAR (National Center for Environmental Prediction, National Center for Atmospheric Research) reanalysis. These data were used for the Arctic area of the system. For the Atlantic area south of the 60 °N, the flows obtained by these data were coupled with climate data on heat and wind friction flows (Trenberth et al., 1989).

The numerical experiment of the ocean-ice interaction model was carried out for the period 1948–2002. We describe and analyse below some results of the calculations. The ice field changes more dynamically than the thermohaline structure of the deep ocean. Therefore, it can be restored in a shorter time by the model itself.

Following the recommendations of the AOMIP (Arctic Ocean Model Intercomparison Project) project, the simulation of the initial ice field was implemented using a preliminary numerical experiment from 1948 to 1954. The thickness and compactness distributions for all ice categories, which were obtained in the preliminary experiment, have been used to prescribe initial conditions for the main experiment conducted from 1948 to 2002. The initial water and ice velocity components were also obtained in the preliminary experiment starting from zero values.

The most important forcing for ocean and ice dynamics is caused by surface wind friction. The most typical feature of the wind field is the anticyclonic vorticity generated over the Arctic Ocean region.

The seasonal wind field variability is mostly related to the formation of the Siberian anticyclone in winter and its attenuation in summer, which in the long run intensifies the anticyclonic wind field vorticity in winter (see Fig. 5.1a) and leads to its attenuation or even substitution by cyclonic vorticity in summer (see Fig. 5.1b).

Beside the seasonal wind field variability, there is also interannual variability caused on Azores mainly by the fluctuations of pressure differences between the Iceland minimum and the Siberian maximum, the so called North Atlantic Oscillation (NAO). The periodicity of these fluctuations is 5–10 years and they also intensify or decrease the anticyclonic wind field vorticity in the central Arctic area. There were warm periods during the mid and late 1950s, and in the early 1990s, and a cold period from 1974 to the end of 1980s. As it will be shown later all these

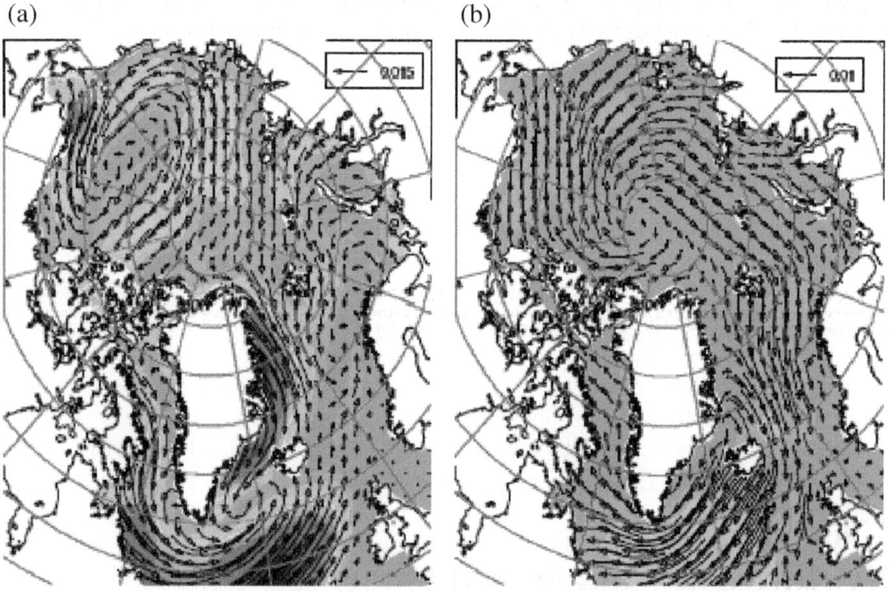

Fig. 5.1 (**a**) Wind friction pattern averaged over the whole simulation period of 1948–2002, (**b**) wind friction restored by the 3d-mode eigenfunction in EOF analysis of a wind series. The samples of arrows in the right upper corner correspond to wind friction values 0.015 and 0.01 N/m^2. The background darkening corresponds to the friction vector modulus, darker areas indicate larger values of the modulus (Kuzin et al., 2006) (See also Plate 1 in the Color Plate Section on page 339)

periods can be simulated by the ICMMG (Institute of Computational Mathematics and Mathematical Geophysics) model (Kuzin et al., 2006; Golubeva and Platov, 2007).

The general pattern of the ice circulation in the Arctic Ocean is such that the ice generated mostly in marginal seas either at once, or having done some cycle within the ocean basin is transported by the Greenland Current to warmer waters in the North Atlantic, where it melts. The calculations show that the newly-formed ice is mostly located in marginal seas.

The old ice is located further on the way from marginal seas to the basin center and drifts to the south by the dominant currents and winds in this region. The oldest pack ice is located in the basin center and its motion is circular, which produces clogging in the regions of the islands and the northern shores of Greenland and the Canadian Arctic Islands. The persistent stationary ice distribution pattern is due to the mechanical ice drift through the Fram Strait and ice melting from below in the case of a warmer water inflow under ice. The latter often occurs during anticyclonic circulation failures when intense water and ice counter drifts are generated.

The annual variability of the ice boundary is depicted in Fig. 5.2, where the ice boundaries for each month in the period 1948–1960 are shown. As seen, the greatest

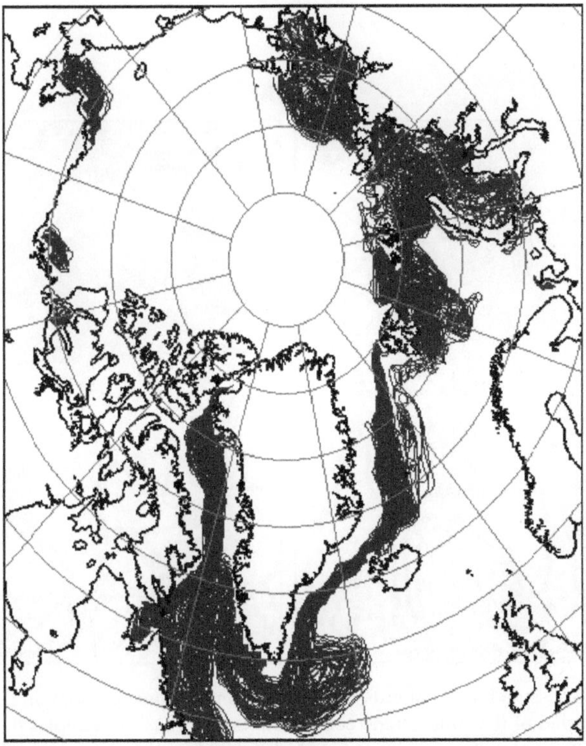

Fig. 5.2 Monthly positions of the ice edge boundary in 1948–1960 (Kuzin et al., 2006) (See also Plate 2 in the Color Plate Section on page 340)

variability zone comprises the basins of the Kara Sea, the Laptev Sea, the northern area of the Barents Sea, the Baffin Sea, the Chuchi Sea, the delta of the MacKenzie river, and the site of the contact of the cold Greenland Current with the Gulf Stream.

The interannual changes of the area of the ice surface are about 8–9% of the total ice area. The variability of ice drift through the Fram Strait is about 30% of its mean value of about 4.8 km^3/day.

The greatest variability among the simulated variables is seen in the ice drift system, including both seasonal and interannual variability. The winter drift pattern (see Fig. 5.3a) is more regular than the summer one. It includes the dominant anticyclonic cycle in the central area of the Arctic Ocean and the ice drift through the Fram Strait to the North Atlantic. The main ice mass corresponds to the anticyclonic cycle zone. This ice is the most long-living since due to its rotation it persists in the region, where the process of new ice formation dominates over ice melting.

In summer, the anticyclonic cycle is broken (see Fig. 5.3b) due to warmer waters which are found in the center of the thick ice area. This is evident from the fact that the relatively thin ice zone up to 1.5 m is generated in this region. The thickness of the main part of ice changes insignificantly. Kuzin et al. (2006) emphasize that throughout the year ice constantly drifts along the eastern shores of Greenland. This ice reaching the warm Atlantic waters melts and thus contributes an additional portion of fresh water to these regions.

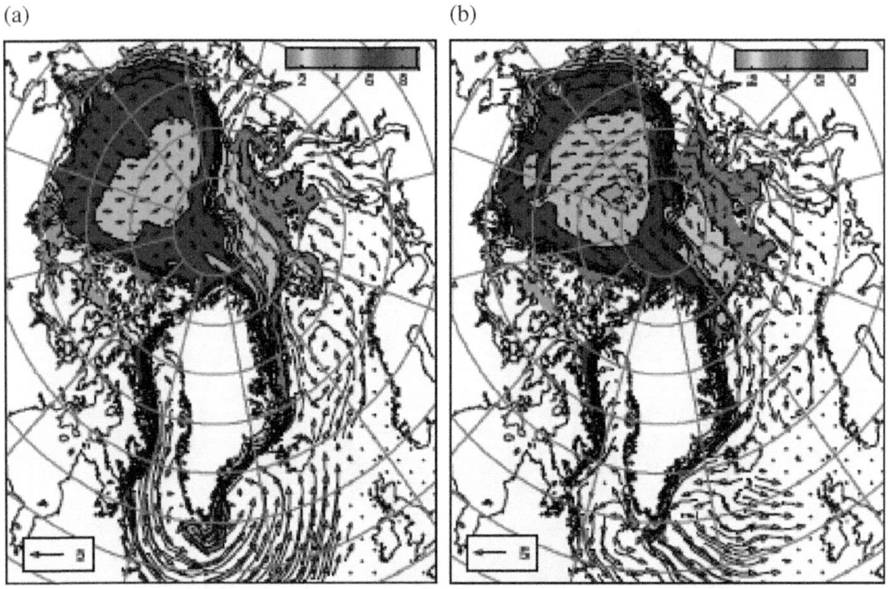

Fig. 5.3 Ice thickness and drift patterns for a typical anticyclonic circulation (**a**) and for the circulation in the period of the anticyclonic component failure (**b**). The samples of arrows in the left lower corner correspond to the velocity of 5 cm/s. Ice thickness (m) is graduated by darkness in the figures (Kuzin et al., 2006) (See also Plate 3 in the Color Plate Section on page 341)

5.1.3 Simulation of the Coupled Arctic Ocean-North Atlantic Circulation System

The saline and warm Atlantic water normally penetrates into the North European basin through the Faeroes-Shetland Strait and the Faeroes-Iceland Strait. In the region of 70 °N the flow divides into two branches. The Nordkapp Current is eastward directed and transports the Atlantic waters to the Barents Sea, where they are intensely cooled and freshened due to their contact with the coastal waters and convective mixing and propagate further into the region of the Kara Sea and the Nansen basin. The second branch of the divided Atlantic waters is directed northward along the western area of Spitsbergen. It is the West Spitsbergen Current that is regarded as the main heat source in the Arctic basin. The Arctic Ocean, in turn, transports cold and fresher water to the Atlantic Ocean, which controls the generation of bottom waters in the whole World Ocean.

The numerical experiment carried out for the period of 1948–2002 simulates the current system and 3D distribution of temperature and salinity fields. The North Atlantic is one of the best studied World Ocean regions. Its dynamics has been studied by measurement data and reproduced by numerical simulation using models of different degrees of complexity. Each numerical model, reproduces some typical features of the upper 1000 m, circulation in the North Atlantic, namely the anticyclonic cycle in the mid latitudes with the most intense west boundary current in the upper 800 to 1000 m layer the deeper layer countercurrent that transports waters from north to south, and the bottom layer northward current.

Much less is known about the Arctic Ocean circulation because this region of the World Ocean is not easily accessible. In accordance with changing atmospheric conditions on the ocean surface, the numerical experiment results in a nonstationary pattern of currents in the upper ocean layer. The surface oceanic circulation (see Fig. 5.4) shows a transport over the Arctic Ocean from the Chuchi coast to the eastern area of Greenland, where the southward-directed flow causes water transport by the Transpolar drift and the East Greenland Current from the Arctic to the Atlantic Ocean through the Fram Strait. The narrow strip of the current from the Bering Straits runs along the American coast and closes in a cyclone. The branch of Atlantic water from the Barents Sea is well seen.

The second branch of Atlantic water is northward directed in accordance with the West-Spitsbergen Current. The computational results suggest that in the surface layer the current in the Fram Strait region turns to west (see Fig. 5.4). A branch directed westward down the continental slope of the Nansen basin is seen in the computational domain only below 200 m. In the first numerical experiment the obtained model flow propagating through the Fram Strait is much weaker than the second branch of Atlantic water penetrating through the Barents Sea. However, it is reported by Ivanov (2001) that in some researchers' opinion, the role of the Fram Strait in the Atlantic water propagation is overestimated. The gradually deepening Atlantic water running down the continental slope is called the Arctic Boundary Current (Woodgate et al., 2001). The warm Atlantic waters are transported along

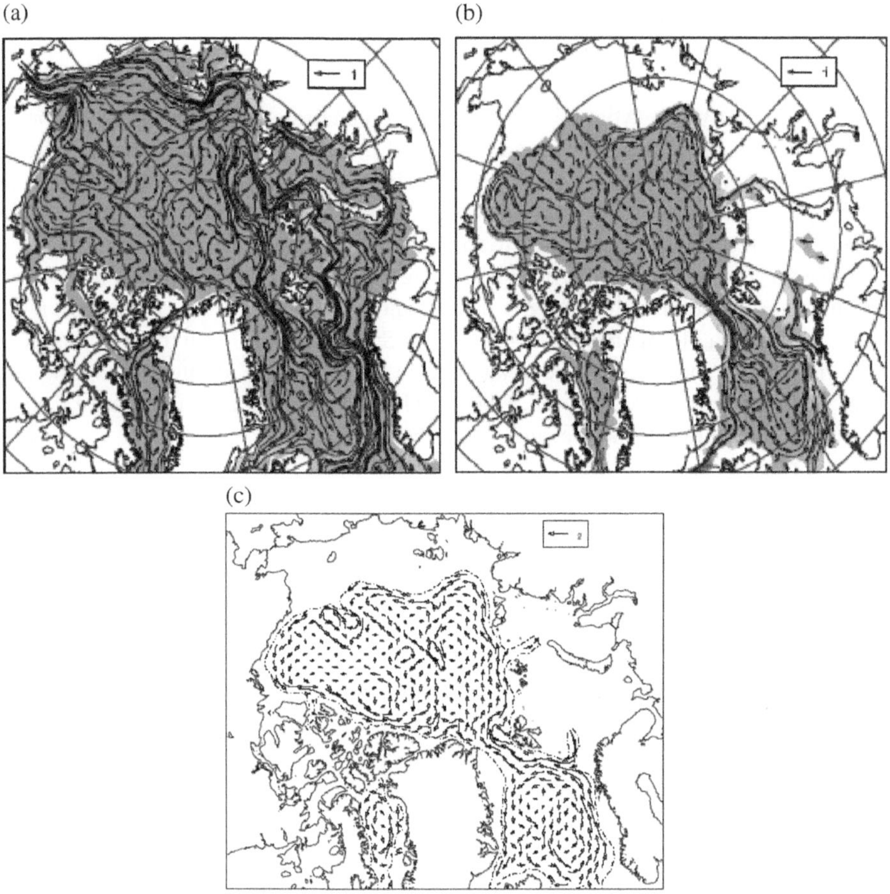

Fig. 5.4 (**a**) Surface currents field. The samples of arrows in the right upper corner correspond to the velocity of 1 cm/s. The background dark color corresponds to velocity value; (**b**) currents at a depth of 250 m. The darker shades indicate higher velocity. (**c**) Circulation at 400 m depth obtained in experiment with Neptune parameterization, shows a steady, cyclonic, topography-steered current for the entire period of investigation (Kuzin et al., 2006) (See also Plate 4 in the Color Plate Section on page 342)

the southern boundary of the Nansen basin to the Lomonosov Ridge, where the current divides into two parts: one turns north and moves along the ridge and the other continues to move down the continental slope. Modelling of Atlantic water circulation was improved by Golubeva and Platov (2007) by the use of the "Neptune" parametrization (Holloway, 1992) for the horizontal viscosity. A realistic simulation of the Atlantic water propagation without use of the "Neptune" parametrization was achieved by decreasing of the viscosity coefficients and application of a third order advection scheme.

According to oceanographic field data for the period of 1950–1990 (Ivanov, 2001), Atlantic waters in the West Arctic occupy the layer from 150 to 800 m. In the vertical distribution of the thermohaline characteristics of model fields, the cold and fresh homogeneous layer is well pronounced. Below this layer there is a region with strong temperature and salinity gradients without separation of the upper and lower haloclines. The abrupt gradient region is followed by a layer of Atlantic water, which is located at various depths depending on the region. The Atlantic water core in the Eurasian basin is at a depth of 150–200 m, whereas in the Canadian basin this depth is 700 m.

The analysis of the experiment showed that the thermohaline structure of the Arctic Ocean fields may be significantly modified by various factors. This variability can be caused by processes occurring directly in the Arctic atmosphere or by external factors. The periods of the warming and cooling in the Arctic are also well described by the model in the experiments of Golubeva and Platov (2007). Warm water from Fram Strait is transported by the boundary current along the continental slope and reaches Franz Josef Land (see Fig. 5.5). One can speculate that this happens because of the Barents Sea branch contribution, which is cooler than the Fram Strait branch due to heat loss in the shallow Barents Sea. The less intensive warm signal spreads farther to the Lomonosov Ridge and then splits into two branches, following the circulation pattern.

5.1.4 Numerical Simulation of Tracer Propagation

As mentioned above, the Arctic ice is the main vehicle of the transport of suspended materials in the Arctic. It includes various aerosols, precipitation material of cryogenic origin, including biological substances transported by rivers and radioactive contamination. Therefore, studying the pathways of migrating materials by ice and water in the Arctic basin and its drift to the North Atlantic is a topical problem of numerical simulation.

As an example, the propagation of a neutral impurity initially located in the ice cover of the Yenisei delta is considered below. The initial instant occurred in the winter season and the seasonal variations of input characteristics for the model were chosen by mean climatic distributions. The calculations were carried out for ten years of model time. The model results are given in Fig. 5.6. The impurity transport by ice (Fig. 5.6a) and the upper oceanic layer currents (Fig. 5.6b) is represented as the distribution of the impurity in one and four years. The shaded domain corresponds to regions where the concentration exceeds 10% of the maximum for ice and 1% of the maximum for water. The corresponding maximum values are indicated in the figure captions. The isolines are given on a logarithmic scale. The minimal isoline corresponds to 10^{-5} of the maximum quantity. Generally, the large-scale ice drift facilitates the impurity drift towards the North Pole and further, in accordance with the transarctic transport towards the North Atlantic. In accordance with thermodynamic processes in spring and in summer, the ice melts on its way and the impurity partially penetrates into water. Afterwards, the impurity that has neutral buoyancy is also transported in water by currents without sinking. In the results of the numerical experiment, a considerable part of the impurity is transported to the eastern shores of Greenland both by the drifting ice system and the ocean currents.

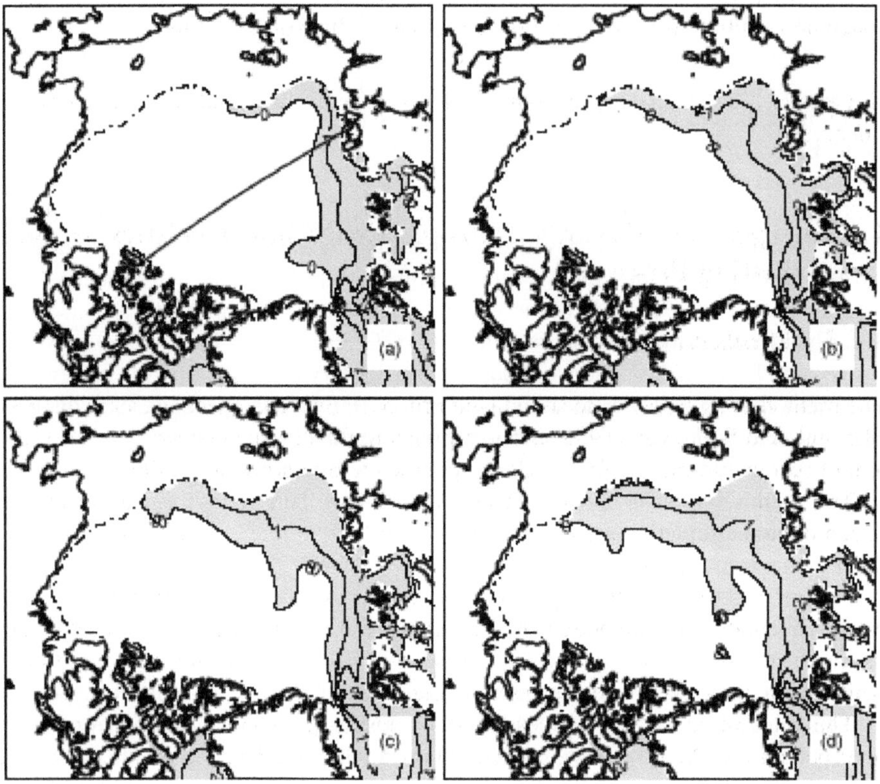

Fig. 5.5 Warming of the Atlantic layer. Temperature distribution at 200 m depth (**a**) in 1990, (**b**) in 1992, (**c**) in 1994, and (**d**) in 1996

According to the model estimates, in two years the impure ice reaches the Fram Strait and in four years the Denmark Strait. Once on the shores of Greenland, the contaminated oceanic ice is a source of impurity penetrating into water for its further propagation in the Atlantic circulation system for several more years until it disappears completely. According to the results obtained for the chosen source a much smaller concentration is observed in the Canadian basin. However, being involved in the anticyclonic circulation, the contaminated ice persists in the central Arctic area with impurity concentration virtually constant for a long time.

5.1.5 Propagation of Fresh Water from Siberian Rivers and Pacific Waters of the Bering Straits

The global hydrological cycle in the atmosphere and in the ocean is of vital importance for specifying the climate condition on the Earth. The Arctic Ocean contributes to the World Ocean 10% of its total fresh water volume (Ivanov, 1976).

The change in the fresh water flux from the Arctic basin can disturb in-depth convection conditions and, as a result, the thermohaline circulation in the North Atlantic and the global hydrological cycle. A substantial source of fresh water transported to the upper layer of the Arctic Ocean is river run-off. The main contribution is made by the Siberian rivers that are responsible for the majority of the fresh water volume compared to other rivers. Though the water inflow through the Bering Straits is significant and amounts to about 0.8 Sv, these waters, however, are only slightly fresher (31–32 psu) than the Arctic waters (33 psu). Thus, the role of the Siberian rivers in the formation of the Arctic–North Atlantic climate system is critical. For the adequate simulation of both the regional Arctic climate and the global processes it is necessary to investigate the fresh water balance and its transformation in the Arctic basin and in the North Atlantic.

The simulation results of fresh water propagation from the Ob, the Yenisei, and the Lena rivers and the waters of the Bering Straits is presented in a paper of Kuzin et al. (2006). The fresh water propagation from each of these sources was identified by parallel computation of the corresponding neutral tracer. Each tracer had a single concentration source in the delta and followed the pathway of water propagation according to the circulation processes. The tracer concentration equations were analogous to the salinity equations. The tracer concentration on the surface varied with the fresh water inflow from the atmosphere and vertical mixing was extended to the field of this tracer. On the lateral boundaries the concentration also varied with the inflows of other rivers or from the Bering Straits.

The calculations were carried out for 37 years of model time using the 1965–2002 data. Fig. 5.6 shows the patterns of the tracer propagation from the sources in the Ob delta and Fig. 5.7 – from the Bering Straits. The values are given on a logarithmic scale. For the Ob river the main part is concentrated in the central Arctic area with the tracer carried by the Transpolar current to the eastern shores of Greenland.

For the Pacific waters penetrating through the Bering Straits the largest tracer concentration is at the Canadian coast, in the Beaufort Sea, and in the seas of East Siberia. However, the impact of the Transpolar drift in this case is seen as well in the transport of part of the tracer to Greenland but more symmetrically with respect to the eastern and western shores. Furthermore, the fresh water tracer drifts to the Fram Strait and is transported to the Greenland Sea. When reaching the southern area of the Greenland Sea, part of this water sinks up to depths of 3000 m. At these depths, part of the water propagates south along the Central Atlantic Ridge. The other part following the Gulf Stream Countercurrent propagates south-west. Fig. 5.8 shows the tracer concentration distribution identifying the Ob run-off in the North Atlantic at the latitude 30 °N, 37 years after the onset of emission. The greatest concentration is achieved in the region west of the Central Atlantic Ridge and the Gulf Stream Countercurrent at depths below 1500 m. Fig. 5.10 gives the vertical cross-section of the tracer distribution along the section indicated in Fig. 5.9 by a solid line. Transported together with Arctic water, the tracer travels south in the deep layer with its maximum below 1500 m. For Pacific waters the pattern of the distribution over the Atlantic Ridge is analogous. In the upper layers a somewhat different pattern is observed. At a depth of 300 m the main concentrations are found in the region

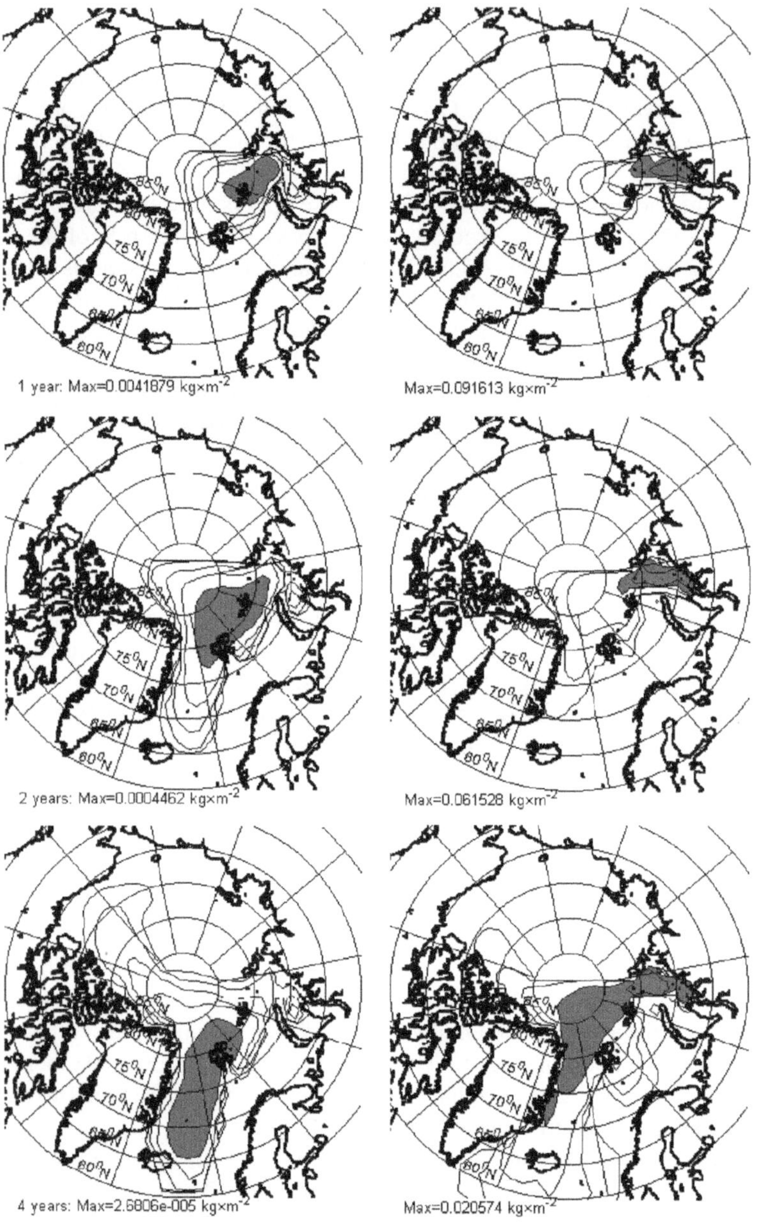

Fig. 5.6 Impurity transport by ice (**a**) and by the upper ocean layer currents as a function of time (**b**). The impurity source is located in the Yenisei delta. The shaded area corresponds to the regions where the concentration of impurity exceeds 10% of the maximum for ice and 1% of the maximum for water. The isolines are given on a logarithmic scale. The minimum isoline corresponds to 10^{-5} of the maximum quantity (Kuzin et al., 2006)

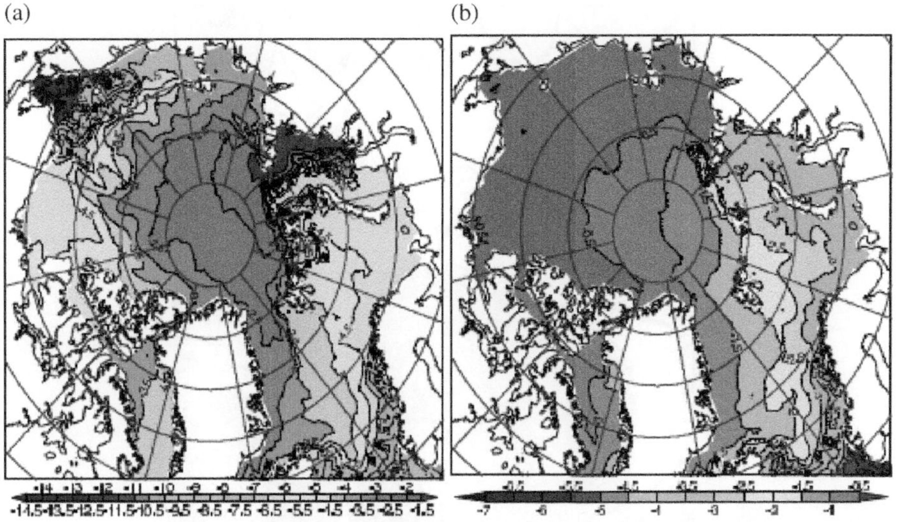

Fig. 5.7 Passive tracer distribution 37 years after the onset of emission: (**a**) the Ob, (**b**) Pacific water of the Bering Straits. The isolines are the decimal logarithms of the corresponding tracer concentration (Kuzin et al., 2006) (See also Plate 5 in the Color Plate Section on page 343)

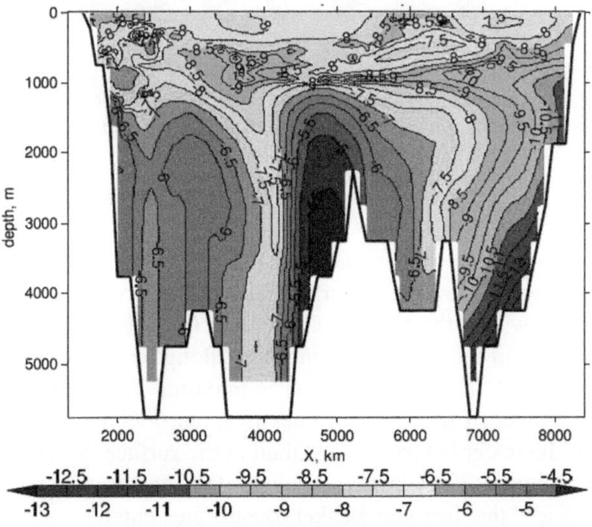

Fig. 5.8 The vertical cross-section of the tracer concentration field of the Ob River in the North Atlantic along the parallel 30 °N 37 years after the onset of emission. Left–west, right–east. The isolines are the decimal logarithms of concentration (Kuzin et al., 2006) (See also Plate 6 in the Color Plate Section on page 344)

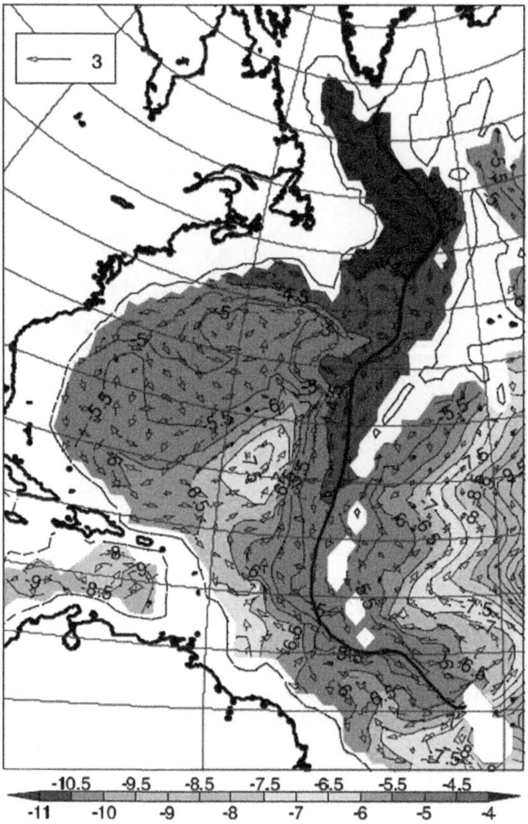

Fig. 5.9 The tracer propagation of the Ob run-off in the North Atlantic at a depth of 3 km 37 years after the onset of emission. The arrows indicate the direction and velocity of the current (the *arrow* corresponding to the velocity 3 cm/s is indicated in the *left upper corner*) (Kuzin et al., 2006) (See also Plate 7 in the Color Plate Section on page 345)

south of Greenland and in the Labrador Sea (see Fig. 5.11). At the same time, part of the tracer follows the reverse eastern branch of the North Atlantic Current to the south and further within the Canary Current in the subtropic hydrological cycle. The vertical section of the tracer concentration along the line indicated in Fig. 5.11 is shown in Fig. 5.12. The concentration is maximal at a depth of 300 m and the extension of the layer of maximum values is about 6000 km. At a depth of 500 m the main tail of the tracer is further west than in the surface layers. At the same time the tracer in the subtropic hydrological cycle does not propagate uniformly but is shaped as "tongues" that become weaker toward the center.

Thus, the fresh water tracer in the model propagates not only within the Arctic basin, but drifts to the North Atlantic and propagates in the subtropic hydrological cycle. The direct drift of the tracer to the south occurs only in the deep layers, which can be regarded as bottom water propagation.

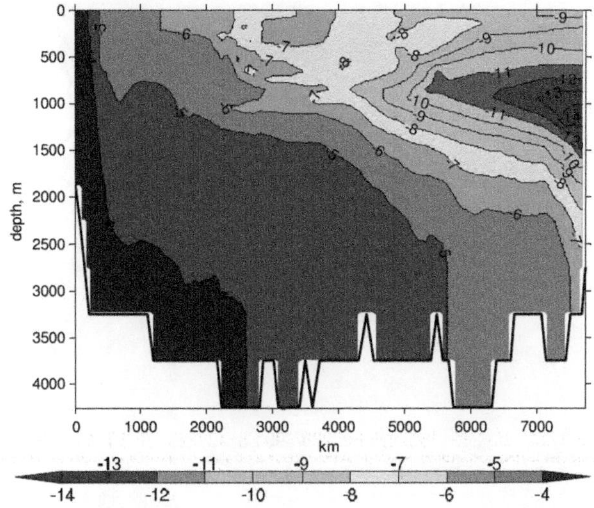

Fig. 5.10 The cross-section of the tracer concentration field corresponding to the Ob run-off 37 years after the onset of emission along the approximate trajectory of its propagation, which is indicated in Fig. 5.9 by a *solid line* (Kuzin et al., 2006) (See also Plate 8 in the Color Plate Section on page 346)

5.1.6 Conclusions

The numerical integration of the coupled ocean-ice model in the Arctic–North Atlantic region reproduces the system of large-scale circulation and annual variability of this part of the World Ocean. On the basis of their research Kuzin et al. (2006) and Golubeva and Platov (2007) made the following conclusions:

Using the ice thermodynamic model is of crucial importance when simulating the Arctic basin and Siberian seas circulation.

In the coupled simulation the greatest variability is seen in the ice drift pattern, whereas the oceanic circulation is more stable.

On the whole, the simulated results quantitatively are in agreement with other authors' computational results obtained by the AOMIP project models.

The propagation process of an idealized impurity initially located in the ice cover of the Kara Sea near the Yenisei delta has been simulated. Two impurity propagation trajectories were identified:

(a) involvement in the anticyclonic cycle of the Canadian basin and transport to the central Arctic basin area by the ice drift system;
(b) ice drift to the shores of Greenland with the gradual transition of the impurity from ice to water in the spring-summer season and further propagation by the surface current system and its involvement in the subpolar cycle of the North Atlantic;
(c) the fresh water tracer in the model propagates not only within the Arctic basin, but drifts to the North Atlantic and propagates in the subtropic hydrological cycle.

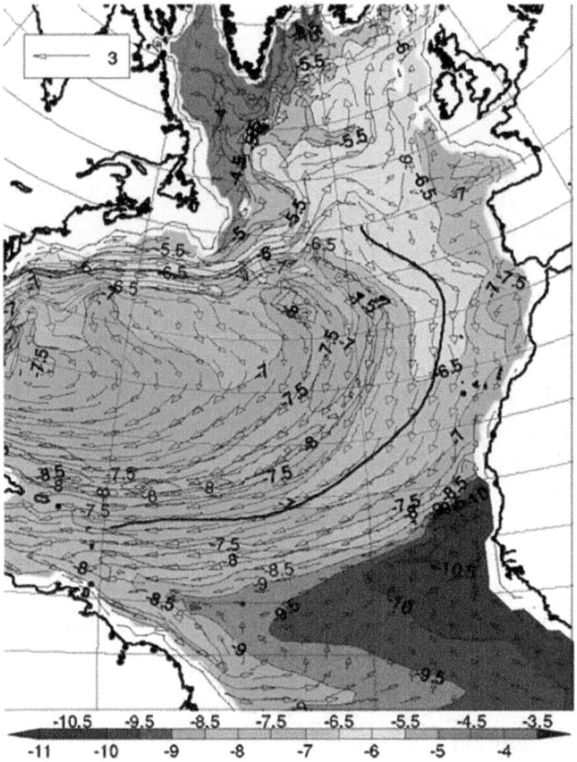

Fig. 5.11 The tracer propagation in the Ob run-off in the North Atlantic at a depth of 300 m 37 years after the onset of emission. The arrows indicate the direction and velocity of the current (the arrow corresponding to the velocity 3 cm/s is indicated in the left upper corner) (Kuzin et al., 2006) (See also Plate 9 in the Color Plate Section on page 347)

The direct tracer drift to the south occurs in the deep layers only, which can be treated as bottom water propagation.

5.1.7 Latest Results of Arctic Ocean Modelling and Intercomparison Project (AOMIP)

Recently sixteen papers on the AOMIP results were published by Journal of Geophysical Research (JGR); afterwards they were reprinted and published jointly also by the American Geophysical Union (AGU). In the following we will review a few of the papers which in our mind characterize the state of art of Arctic Ocean modelling.

Martin and Gerdes (2007) made a comparison of sea ice drift results from different sea ice-ocean coupled models and the validation with observational data in the period 1979–2001. The models all took part in the Arctic Ocean Model

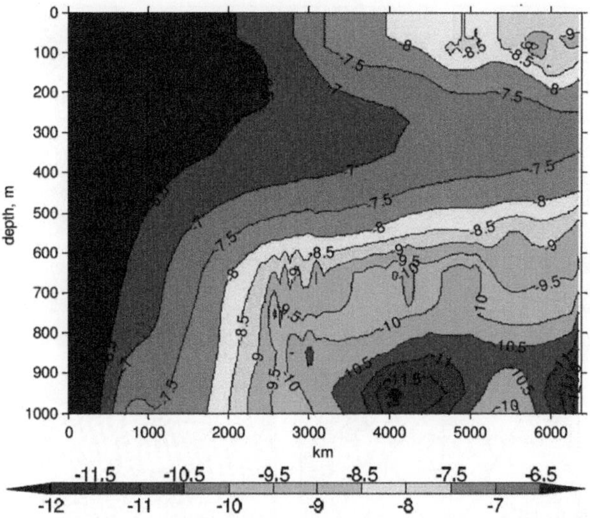

Fig. 5.12 The cross-section of the tracer concentration field corresponding to the Ob run-off 37 years after the onset of emission along the approximate trajectory of its propagation, which is indicated in Fig. 5.11 by a *solid line* (Kuzin et al., 2006) (See also Plate 10 in the Color Plate Section on page 348)

Intercomparison Project (AOMIP) and the observations are mainly based on satellite imagery. Concerning speed distributions, one class of models has a mode at drift speeds of around 3 cm/s and a short tail toward higher speeds. Another class shows a more even frequency distribution with large probability of drift speeds of 10 to 20 cm/s. Observations clearly agree better with the first class of model results.

The authors list several reasons of differences: the wind stress forcing, sea ice model and sea ice-ocean coupling characteristics which are not alike in the models considered. In the Fig. 5.13 we demonstrate one of the authors result.

Based on a three-dimensional coupled ocean/ice model, Holloway and Proshutinsky (2007) have studied the role of tides in Arctic ocean/ice climate. A three dimensional coupled ocean/ice model, intended for long-term Arctic climate studies, is extended to include tidal effects. From stored output data of an Arctic tides model, the authors introduced parametrizations for (1) enhanced ocean mixing associated with tides and (2) the role of tides fracturing and mobilizing sea ice. Results show tides enhancing loss of heat from Atlantic waters. The impact of tides on sea ice is more subtle as thinning due to enhanced ocean heat flux competes with net ice growth during rapid openings and closings of tidal leads. The model results are compared with an ensemble of nine models within AOMIP. There is a tendency for models to accumulate excessive Arctic Ocean heat throughout the intercomparison period 1950–2000 which is contrary to observations. Tidally induced ventilation of ocean heat reduces this discrepancy. We reproduce below two of their figures (Figs. 5.14 and 5.15). They give an impression about drastically different sizes of polynyas and about the leads shaped by tidal currents.

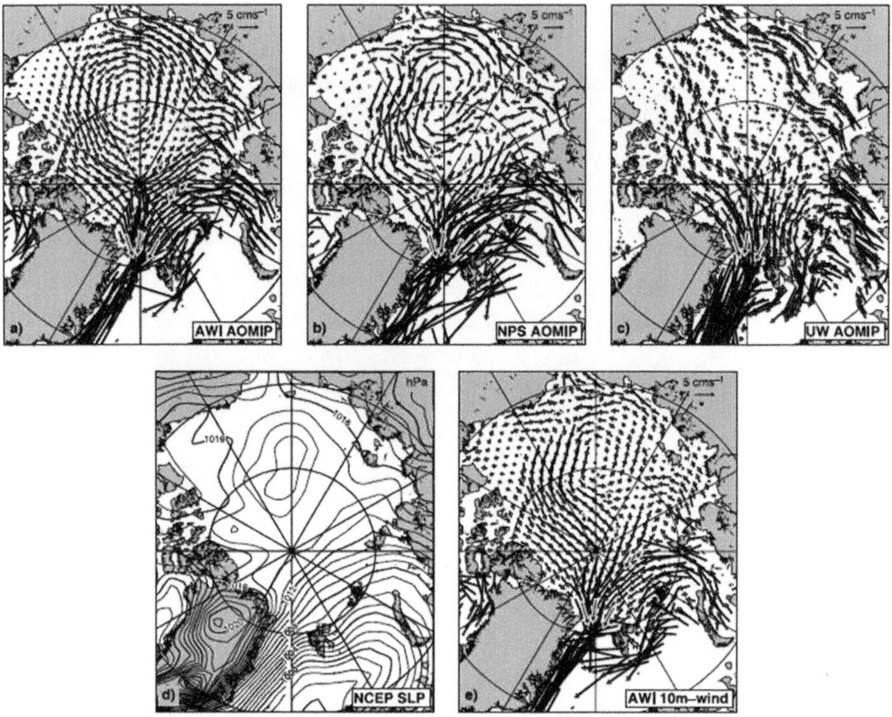

Fig. 5.13 Sea ice drift velocity averaged over the winter (November–April) of 1994/1995 from AOMIP models (**a**) AWI, (**b**) NPS, and (**c**) UW and (**d**) the corresponding NCEP/NCAR reanalysis sea level pressure field. (**e**) Sea ice drift of this winter from the AWI 10-m wind experiment. Figures a, b, c and e include IABP buoy drift vectors in bold black. For clarity reasons the horizontal resolution of modeled data is individually reduced and changing spatial coverage is due to particular ice extent

Out of sixteen papers only one was using the variational data assimilation technique to reconstruct fields of climatological data that is the paper of Panteleev et al. (2007). The authors considered the Kara Sea summer circulation. Their model satisfies the dynamical and kinematic constraints of quasi-stationary primitive equation ocean circulation. The reconstructed circulations agree well with the measurements. They are characterized by inflows of 0.63, 0.8, 0.51 Sv through Kara gate and 1.18, 1.1, 1.12 Sv between Novaya Zemlya and Franz Josef Land, for mean climatologic conditions, positive and negative AO indexes, respectively. The major regions of water outflow for these regimes are the St. Anna Trough (1.17, 1.21, 1.34 Sv) and Vilkitsky/Shokalsky Straits (0.52, 0.7, 0.51 Sv). The optimized velocity pattern for the mean climatological summer reveals a strong anticyclonic circulation in the central part of the Kara Sea (Region of Fresh Water Inflow, ROFI zone) and is confirmed by ADCP surveys and laboratory modelling. This circulation is well pronounced for both high and low AO phases, but in the positive AO phase it is shifted

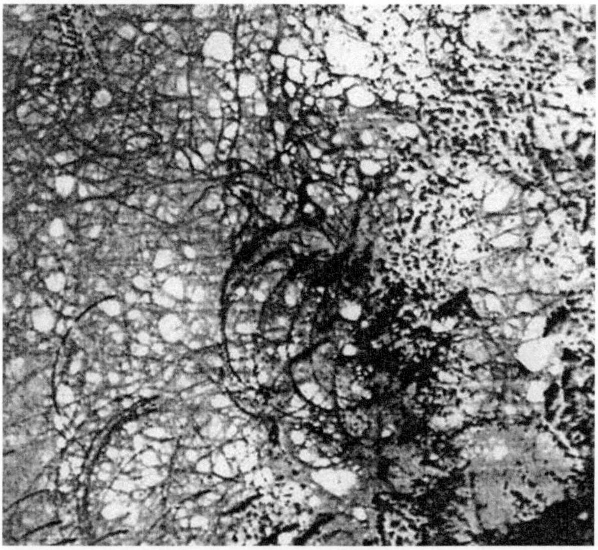

Fig. 5.14 Satellite image of ice cover in the vicinity of Spitsbergen on June 1, 1988, from Dmitriev et al. (1991) with permission from Polar research. Elliptically shaped leads are formed behind grounded icebergs as sea ice is driven by tidal currents

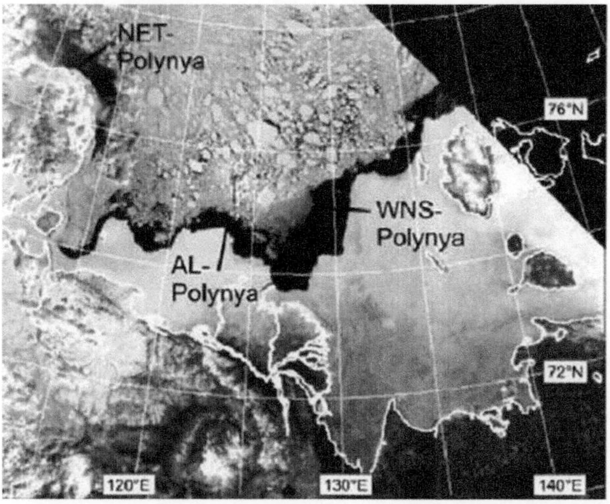

Fig. 5.15 NOAA AVHRR image (visible channel) of the Laptev Sea polynyas (Great Siberian Polynya) on 3 June 1995: (Panteleev et al., 2007). The main flaw polynyas as parts of the Great Siberian Polynya are: Northeastern Taimyr Polynya and East Severnaya Zemlya Polynya

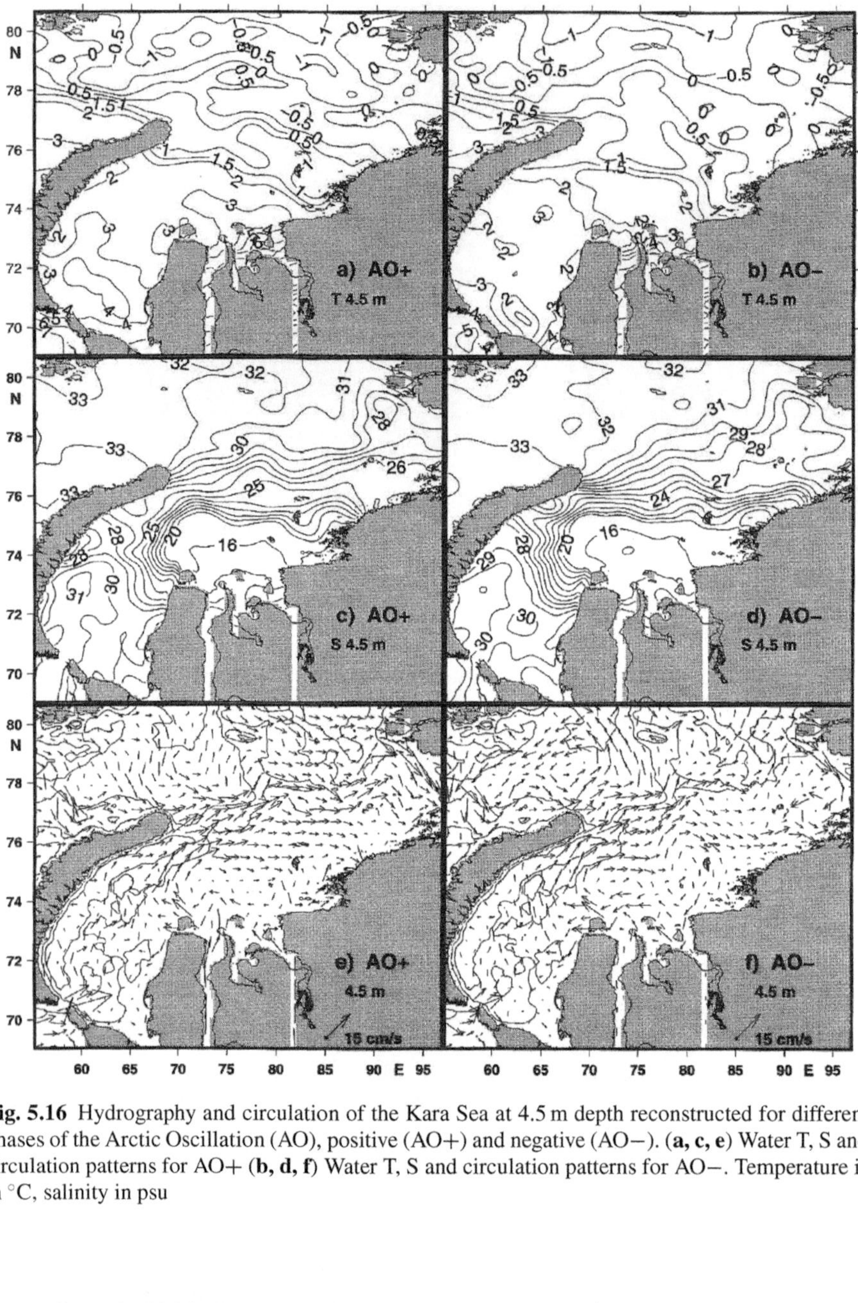

Fig. 5.16 Hydrography and circulation of the Kara Sea at 4.5 m depth reconstructed for different phases of the Arctic Oscillation (AO), positive (AO+) and negative (AO−). (**a, c, e**) Water T, S and circulation patterns for AO+ (**b, d, f**) Water T, S and circulation patterns for AO−. Temperature is in °C, salinity in psu

approximately 200 km west relatively to its climatological position. We present one out of their numerous figures (Fig. 5.16).

Using data assimilation in our mind is very prospective for studying ocean climate and especially for better understanding the extremely complicated ocean/ice thermohydrodynamics of the Arctic Ocean. At the end of this book we shall present some comments on the state of art of ocean climate variation modelling studies, including AOMIP.

5.2 The Gulf Stream

5.2.1 Modelling the Gulf Stream in the DYNAMO Experiment

Among studies performed at a grid step of about 0.3°, the most interesting is certainly the DYNAMO experiment. (DYNAMO Group, 1997 in references to Chapter 3). Here, we discuss only one aspect: how to explain the evolution of parameters from the calculation onset to the so-called steady state after 12–16 years of the model run. It would be very interesting to see the evolution of the kinetic energy KE of a few upper oceanic layers, but this information is not included in the publication. The only time plot provided is the curve of peak meridional mass overturning. Based on this a rough qualitative construct was made demonstrated in Fig. 5.17 (by four main points of each curve). This figure shows that the process did not reach a steady state, simply the rate of decrease in mass overturning slowed down. It would be very strange if such a steady state did occur. The upper layers might initiate a very slow evolution of KE; however, as the characteristic evolution time for temperature (T) and salinity (S) in lower layers is known to be several orders of magnitude longer than in the upper layers, the so-called steady state for parameters averaged over the whole body of the ocean is a very conditional notion. But the essence is in another point: we believe that the experimental resu lts of DYNAMO were better than assumed by the very authors. For example, we claim that, at the initial stage of the calculation, the Gulf Stream separation near Cape Hatteras did occur for

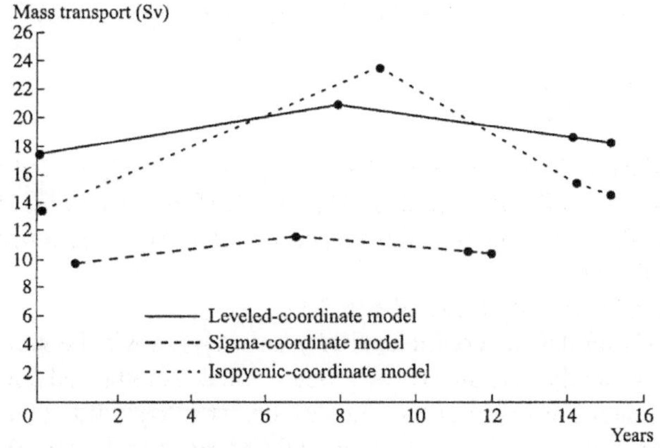

Fig. 5.17 Schematic evolution of peak mass meridional overturning (Sv), constructed from Fig. 3.1 of the DYNAMO Group experiment (1997). It can be seen that 7–9 years of model integration time yield a decreasing trend for the magnitude of overturning

all three models, but, in the course of the calculation, the authors missed this fact. Indeed, let us consider the calculations at such a rough resolution as 1° (Demin and Ibrayev, 1992). What is the reason that Fig. 3.2, demonstrated above in Chapter 3, shows the separation near Cape Hatteras? Simply because the information about this separation existed in the initial T and S fields, and a short-term integration had not yet managed to destroy this information. Despite the "feeding" performed in the DYNAMO experiment through the boundary conditions, the long-term integration results in a dissipation of the initial potential energy, smoothing of the T and S fields, and/or other undesirable consequences. The integration time must be limited by the quality of a model, and any attempt to reach a steady state with a low-quality model (for example, at a coarse resolution, failed of accurate parametrization of turbulence, etc.) would lead to unrealistic results.

Of course, the authors are right in concluding that it is necessary to use a higher resolution. This was clearly shown in Bleck et al. (1995). However, we are not sure that the results of this work would be very good at a similar long-term integration and coarse resolution as in the DYNAMO experiment. In our opinion, one needs not only to reduce the grid size, but also to monitor the evolution of parameters in the course of integration.

Reaching a steady state (in general, the ocean has no such state) is not an end in itself. The rate of mutual adaptation between the parameters and realistic results depends on stopping the integration at the right time. Considering the prediction details of the Gulf Stream separation near Cape Hatteras, we extend the problem by pointing to the very reasons for this separation and its localization. To address these questions correctly, one must have an understanding of how the Gulf Stream develops. The prevailing view among oceanographers is that the western intensification (or, in other words, the "suppression" of currents to the west coast) is driven by the β-effect. Assume that this is the case, but what about the separation? If it is known that the β-effect holds at all latitudes, including that of Cape Hatteras, why does this effect fail to prevent the separation? To understand this, we recall the fact that there are two centers of atmospheric action over the North Atlantic: the subtropical anticyclone and subpolar cyclone. The "free" (i.e., outside the boundary layers) ocean involves the same centers of action, except that they are asymmetric (shifted westward). But is the β-effect the single factor responsible for this asymmetry? Our answer is negative, and we demonstrate this through a qualitative analysis of two equations. The first of them was formulated as early as 1974 (Sarkisyan, 1974). In a simplified and dimensionless form, this equation is

$$\left(\frac{\beta H}{l} - \bar{u}\right)\frac{\partial \zeta_1}{\partial x} = a_1 Q' + \varepsilon_1 rot\vec{\tau}_z \underbrace{- \frac{\beta}{l\rho_0}\int\limits_0^H (H-z)\frac{\partial \rho}{\partial x}dz}_{BARBE} \tag{5.2.1}$$

where a_1 and ε_1 are dimensionless parameters, Q' is a dimensionless function of the heat flux, ζ_1 is the ocean level induced by the density anomaly, and $\bar{u}\frac{\partial \zeta_1}{\partial x}$ is the advection of the density anomaly by a zonal velocity of the trade-wind current in the "free" ocean. BARBE is baroclinic β-effect. The remaining notations are standard.

Let us analyze the expression in brackets on the left-hand side of Eq. (5.2.1). At southern latitudes, $\bar{u} < 0$; i.e., the trade-wind current has the same sign as the β-effect. Both these factors jointly "transport" the baroclinic-fluid mass westward, but obstacles such as the Greater Antilles and the North American continent itself emerge on this pathway. It is the existence of meridionally directed continents in the ocean (unlike the atmosphere) that transforms the subtropical anticyclone and creates asymmetry. In fact, the currents at southern latitudes in the free ocean have a northwestward rather than westward direction. Partially, it is these currents that generate the Antilles Current and the Gulf Stream.

At midlatitudes (30°–40°N), the ocean includes a west-to-east transport identical to that in the atmosphere; i.e. $\bar{u} > 0$, baroclinic-fluid advection is directed opposite to the β-effect, it overrides this effect, and it separates the Gulf Stream from the west coast.

Now, we write the simplified equation of sea-surface height

$$\frac{1}{2\alpha}\Delta\zeta - \left(\frac{\partial H}{\partial y} - \frac{H\beta}{l}\right)\frac{\partial \zeta}{\partial x} + \frac{\partial H}{\partial x}\frac{\partial \zeta}{\partial y} =$$

$$= \frac{1}{\rho_0 g} rot_z \vec{\tau} - \underbrace{\int_0^H J(H,\rho)dz}_{JEBAR} - \underbrace{\frac{\beta}{l\rho_0}\int_0^H (H-z)\frac{\partial \rho}{\partial x}dz}_{BARBE} \qquad (5.2.2)$$

Here, we are interested only in the left-hand side of Eq. (5.2.2), namely, the multipliers at $\frac{\partial \zeta}{\partial x}$ and $\frac{\partial \zeta}{\partial y}$.

Oceans are known to narrow down and shallow on the average when moving from midlatitudes to the north. Therefore, we have $\frac{\partial H}{\partial y} < 0$ normally, and the bottom topography assists in the western intensification not less than the β-effect, because $\frac{\partial H}{\partial y}$ is of the same order of magnitude as $\frac{H\beta}{l}$. For the same reason, one can speak about the "pressing" of the currents to the continental slope, even after the Gulf Stream leaves Cape Hatteras, because $\frac{\partial H}{\partial x} > 0$. Due to the important role played by the shallowing of the ocean bottom at the continental slope, one can also talk about a coastal intensification, apart from the western one.

Thus, Cape Hatteras alone has no effect. Gulf Stream formation, northward motion, and separation as well as the further northeastward motion of the North Atlantic current are controlled by large-scale processes outside the boundary layer: sea-water baroclinicity, and the planetary continental slope. For example, if the gradients of T and S are suppressed, the Gulf Stream will separate at a place significantly north of Cape Hatteras. This is the case when model deficiencies (large horizontal grid steps, increased values of turbulence coefficients, etc.) and/or inadequately long-term integrations lead to a further smoothing of the initially smooth T and S fields.

This was just the case with the DYNAMO experiment. For an integration time of only six months (Fig. 5.18a), the Gulf Stream separated near Cape Hatteras (had the authors presented the results for a shorter integration time, the separation point would be determined more exactly). However, a four-year integration time (Fig. 5.18b) yields a separation point located considerably more northward, and the core of strong surface currents turns out to be linked to gradients of the bottom topography. Therefore, in the four-year integration case, the model managed to distort the initial T and S fields and generate unrealistic alternative fields because of the fact that a grid size of 0.3° is coarse for Gulf Stream modelling (flow velocity, water mass flux, etc.) in general, and the separation points – in particular, if the model integration time is inadequately long.

Clearly, this also occurs in the most extreme case (in the model of a homogeneous ocean when the Gulf Stream separation point is much norther). A gradual decrease in the grid size (Bleck et al., 1995) led to an increase in the T and S gradients; i.e., the Gulf Stream belt narrowed down and a geographically correct point of stream separation was obtained. Without high-gradient baroclinicity, both the correctness of the separation point and the very existence of the Gulf Stream may be under question. High-resolution numerical calculations almost ignore this fact, relying merely on a small horizontal step size. In our opinion, this approach is erroneous. If the horizontal and vertical planetary gradients of T and S are decreased by one or two orders of magnitudes, any decrease in the grid mesh can neither assist in the formation of the intense Gulf Stream nor lead to a correct point of separation. The example demonstrated above in Fig. 5.18 shows that a correct point of separation could have been obtained in the DYNAMO experiment if the integration time had been chosen appropriately and the turbulence coefficients had been minimized accordingly. As to the β-effect, its role was erroneously exaggerated in those early works where the variability of T, S, and H was either disregarded or highly reduced.

5.2.2 Crossing the Rubicon in Modelling the North Atlantic Dynamics

There has been a number of interesting recent works on high-resolution modelling of the Gulf Stream dynamics. It is the small horizontal grid mesh that characterizes the studies considered by us below. The authors of these studies include the grid size (varying from 1/10° to 1/64°) in titles of their papers, and for good reason.

The necessity to turn to higher resolutions was first evidenced by Bleck et al. in 1995, where it was shown that a gradual decrease in the horizontal grid size 0.9° $\rightarrow$ 0.225° $\rightarrow$ 0.08° (all other model parameters being the same) narrows down the flow belt, leading to the formation of meanders and eddies, and the Gulf Stream separation precisely near Cape Hatteras. The model integration time was taken to be sufficiently small and despite a small area of the northwestern Atlantic

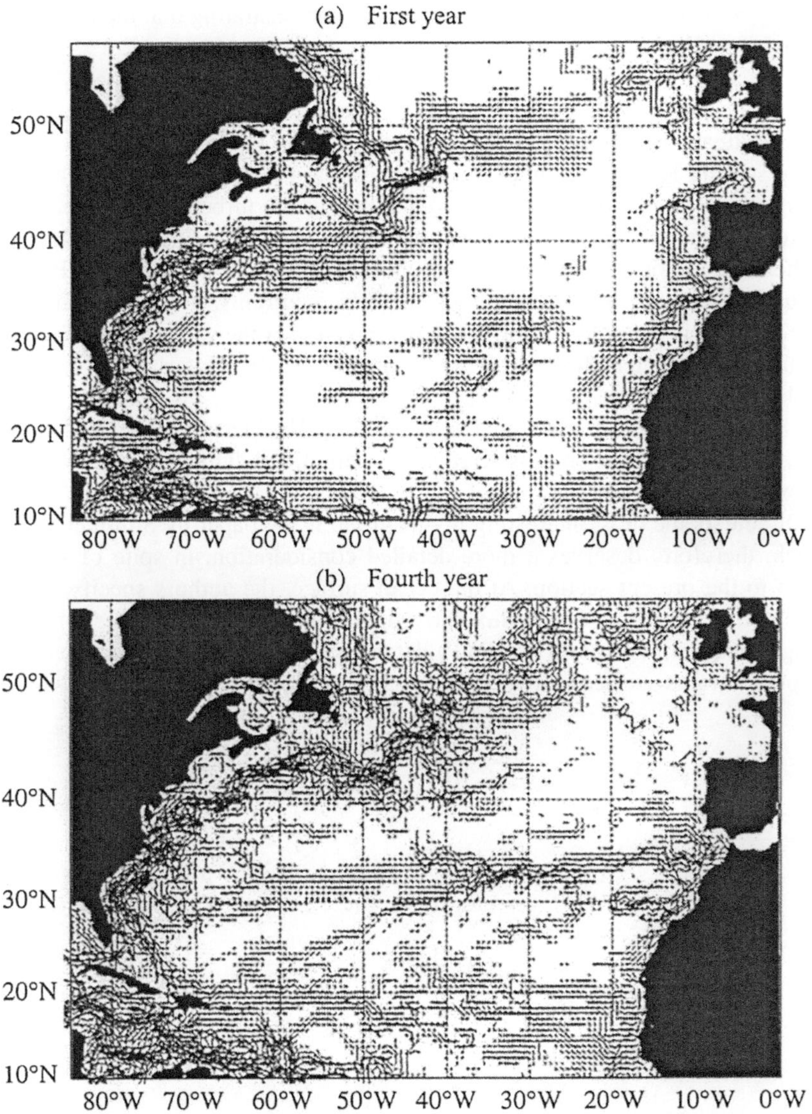

Fig. 5.18 Sea-surface currents calculated by the DYNAMO experiment with an integration time of (**a**) 6 month and (**b**) 4 years. Figure 3.7a is characterized by a partial preservation of baroclinicity of the initial fields and the Gulf Stream separation point, while, in Fig. 3.7b, likely T and S fields have already been smoothed; therefore, the extension of the Gulf Stream and the North Atlantic Current are largely controlled by the bottom topography

was covered, the authors succeeded in clearly demonstrating that their results had a requisite quality.

Longer time calculations (six years) were carried out by Paiva et al. (1999). The map of the sea-surface height obtained by these authors (Fig. 5.19) shows that an intense stream is formed and that the Gulf Stream separation point was correct. The resulting KE curve (Fig. 5.20) indicates that the process has oscillations with a period of a month; however, the given values of the turbulence coefficients make its trend divergent. This is the reason why there was little sense in conducting any further calculations. Note that, among these works, only the paper by Paiva et al. includes the time evolution of the KE of surface currents. In the remaining works, the KE curve is either less informative (averaged over the whole body of the ocean) or missing completely. It would be interesting to give the KE curves for several more levels.

The highest resolution is found in the work by Hurlburt and Hogan (2000); however, the physics of the process is reduced to simplified dynamics, while the number of levels is small.

Of course, the most interesting in this group is the paper by Smith et al. (2000), which, therefore, deserves a more detailed consideration, in spite of space limitation to the present section. At the ocean surface, the authors specify a seasonal variation of the climatic heat flux derived from the European Center for Medium-Range Weather Forecasts (ECMWF) database, and the salinity is formed through adaptation to its surface climatic value. As to the wind field, during the whole time

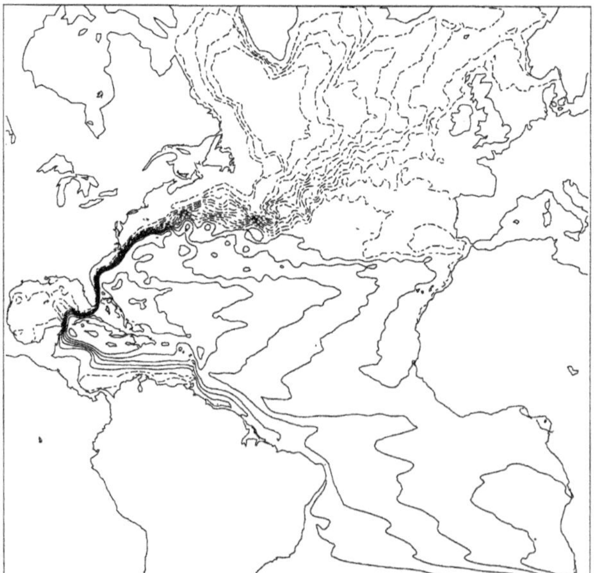

Fig. 5.19 Model mean surface height. Solid lines indicate positive values broken lines negative ones. Contour interval is 10 cm (Paiva et al., 1999)

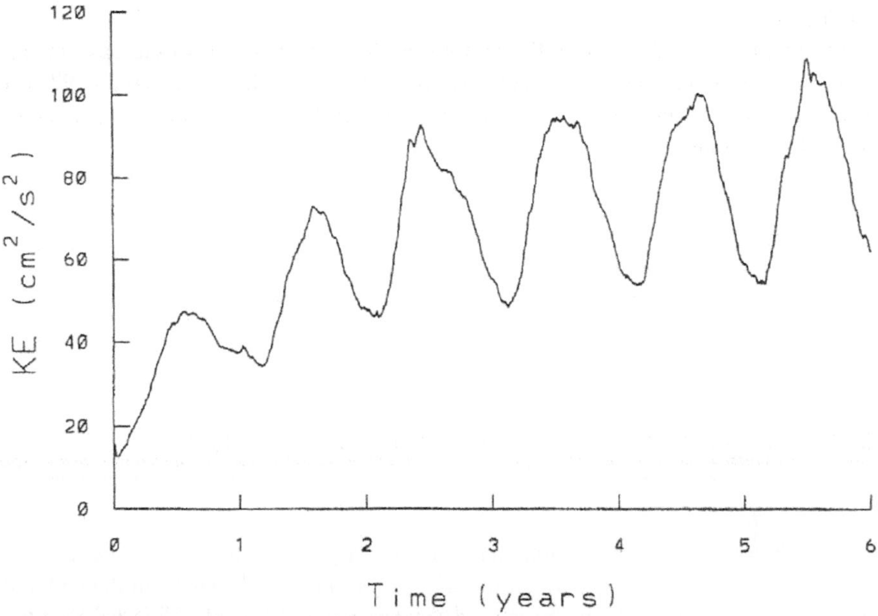

Fig. 5.20 Time evolution of the model surface kinetic energy per unit mass (Paiva et al., 1999)

of integration, interpolated values of daily ECMWF data were used. For the first five years, the integration is performed twice for initialization (more precisely, for adaptation of at least the upper layer of the ocean to the boundary conditions given at the surface). Unfortunately, no analysis or monitoring is provided for the evolution of T and S fields in the upper layer of the ocean. Also, there is no study of sensitivity of upper-layer parameters to the impact of boundary conditions on the ocean. Because of this, the results obtained can hardly be linked to specific temporal data as the authors think.

First, let us consider the achievements of this work. These include a quite realistic mass flux of the Gulf Stream, the increase in this flux from 90 Sv near Cape Hatteras to 150 Sv at 55 °W, realistic vertical profiles of flow velocity, a correct place of the Gulf Stream separation, the formation of meanders and eddies, a high-level eddy KE in the water area under study, etc. This experiment was conducted within the Parallel Ocean Program (POP) and has some advantages over novel high-resolution experiments conducted by the Miami Isopycnic Coordinate Model (MICOM) (Chassignet and Garaffo, 2001; Chassignet et al., 2001). In particular, the MICOM with a resolution of $1/12°$ yields a peak Gulf Stream velocity of 150 cm/s, while the POP yields 190 cm/s. However, at present, it is the MICOM experiments that can be compared to the POP. So, let us consider the latter model in more detail.

The calculations were performed practically for the same North Atlantic area, using a 20 year model run. Figure 5.21 presents the sea-surface topography obtained by either model. The horizontal grid sizes ($1/10°$ and $1/12°$) and integration times

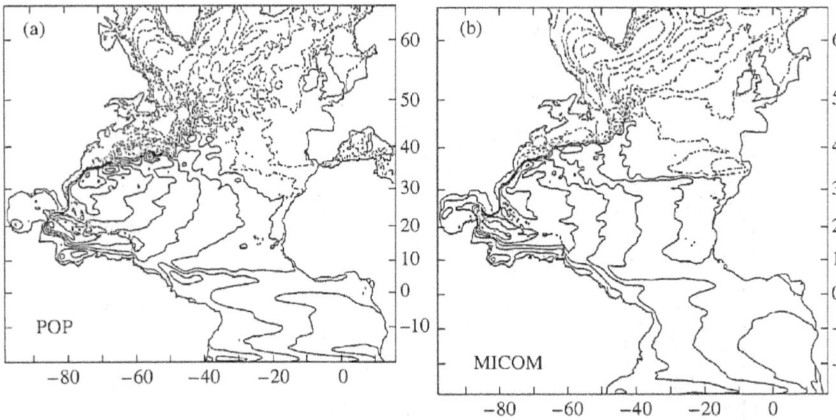

Fig. 5.21 Comparison of the ζ fields between the (**a**) POP and (**b**) MICOM experiments. Both models succeed in predicting a correct separation point; however, some differences appear later. Contour intervals – 10 cm (Smith et al., 2001)

(16 and 20 years) are comparable. The distinctive features are the following. This version of the MICOM experiment includes a forcing source based on the COADS data. In the POP model, the turbulence is represented by biharmonic operators, while the MICOM employs a combination of harmonic and biharmonic operators. Although the Gulf Stream separation is correct in both experiments, there is a substantial difference in predicting its trajectory, intensity, and variability. In the POP, the Gulf Stream moves after the separation in a more zonal direction, thus deviating by 1°–1.5° south of the correct direction. In the MICOM, the separation trajectory is realistic early in its development, but, later, it deviates by 1°–2° north of the correct direction, and the amplitude of the seasonal core oscillation is abnormally large.

It is of high interest to study the sensitivity of both models to the values of the turbulence coefficients. If one tries to reduce the diffusivity and viscosity values (in front of the biharmonic operators) by a factor of three, the Gulf Stream separation in both models will occur south of Cape Hatteras. An increase in these values may correct the situation in the POP experiment. The same is also true for the MICOM experiment; however, an accompanying stable eddy appears north of Cape Hatteras (Fig. 5.22). By the way, such an eddy (warm and anticyclonic) can also be found in other experiments with a biharmonic viscosity, due to the selectivity of the operator (DYNAMO Group, 1997). Increased values of the coefficient in front of the biharmonic viscosity operator reduce the eddy intensity, while increased values of the coefficient in front of the harmonic operator impair the large-scale circulation. Because of this, the MICOM experiment has to adopt an optimal combination of the coefficients.

There was also another interesting experiment performed with the MICOM model. Here, the modelling results were compared for different data (namely, COADS and ECMWF) on boundary conditions at the ocean surface. The ECWMF-based results were more reliable, while the COADS-based experiments led to the fact that the amplitude of seasonal oscillations of the Gulf Stream core and the

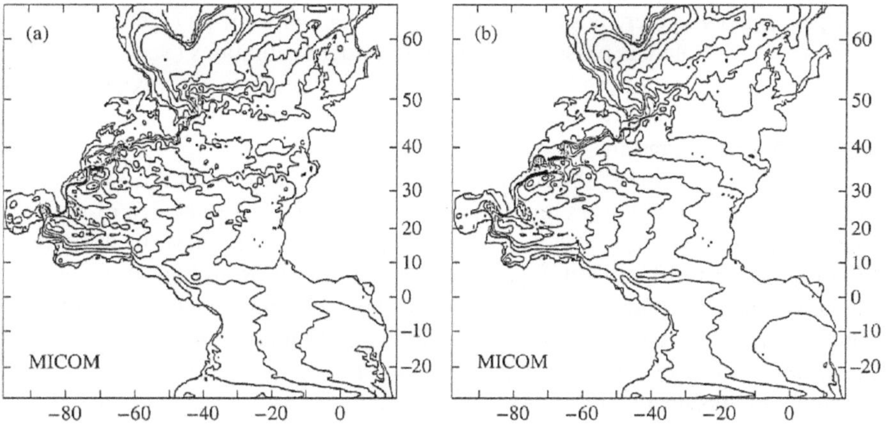

Fig. 5.22 ζ field in the MICOM experiments: (**a**) the value of the biharmonic viscosity is kept unchanged, and (**b**) with value doubled. (Contour intervals – 10 cm). Fig. (a), shows that the Gulf Stream separation is south of Cape Hatteras, while Fig. (b), is free of this error, with an abnormally stable eddy appearing north of Cape Hatteras (Smith et. al., 2001)

mixed-layer depth in winter turned out to be overestimated and the proper stream in an area east of the New England Ridge is shifted northward by $1°–2°$.

In our opinion, there is an aspect of Gulf Stream modelling that has been addressed too widely: the Gulf Stream separation point near Cape Hatteras. The reason is that the separation point, like the Gulf Stream's other properties, could not be realistically predicted with the help of coarse-resolution models for longer integration times. Had the initial T and S fields been sufficiently realistic and high-gradient, the separation point and Gulf Stream parameters would be correctly obtained without present-day advanced models, using such a primitive formula as

$$\zeta = -\frac{1}{\rho_0} \int_{-H}^{0} \rho dz \tag{5.2.3}$$

This means that all depends on the quality of the initial T and S fields and how correctly the model treats these fields. A coarse resolution leads to unreliable approximations and large numerical values of viscosity, while a long-term integration leads to the dissipation of T and S, we have demonstrated in Chapter 3 the North Atlantic surface height obtained from the initial T and S fields, first diagnostically and, then, after a short-term adaptation, with coarse horizontal grid size of $1°$. Both the Gulf Stream and its separation point are realistic in this figure. Smith et al. (2000) point out that their model density is not much different from the initial climatic field. Nevertheless, it is this small difference that is responsible for the strong Gulf Stream, the geographically correct point of its separation, and many other realistic results, which can be explained by nonlinear thermohydrodymamic interpolation and extrapolation, a high-resolution adaptation, and a sufficiently small integration time.

Up to this point, there has been no Russian model with such a high resolution. In addition, much time has passed from the last model calibrations (Sarkisyan, 1995). In view of this, the author of this paper has initiated research on solving these two problems. To this end, two different 3DPEM models were used. The first is a z-coordinate model, and the second is a σ-coordinate model (Diansky et al., 2002). The model area is clearly the North Atlantic, the grid resolution is $1/12^{\circ}$, and the number of levels (for the preliminary test calculations) is 10 and 15, respectively. The initial and boundary conditions in the calculations are close to those of the aforementioned POP and MICOM models, which will allow us both to calibrate the two models and to compare the results obtained by different models. We note that the early stage of calculations is promising and the preliminary results on both models are qualitatively close to each other. In view of this, we present only two figures reflecting the results of calculations conducted by N. A. Diansky by σ-model.

Figure 5.23 demonstrates the flow field at the level $z = 10\,\mathrm{m}$, after only five days of the model run. This is an almost diagnostic calculation of currents, with all advantages and disadvantages of the initial T and S fields provided by the group of Levitus (1994). An advantage is that the Florida Current and the Gulf Stream have sufficiently high velocities as well the Gulf Stream separation near Cape Hatteras. Figure 5.24 demonstrates the results of a two-month adaptation. The calculations have smoothed currents (particularly, at southern latitudes) and a commencing trend of eddy formation in the North Atlantic Current; however, the Gulf Stream separation point is shifted northward, because the model west-to-east transport has not yet managed to form.

Correct increased values of velocities (and, consequently, the mass flux) of the Gulf Stream, a correct point of its separation at the initial stage of integration, and the subsequent northward shift in the separation point (along with other problems)

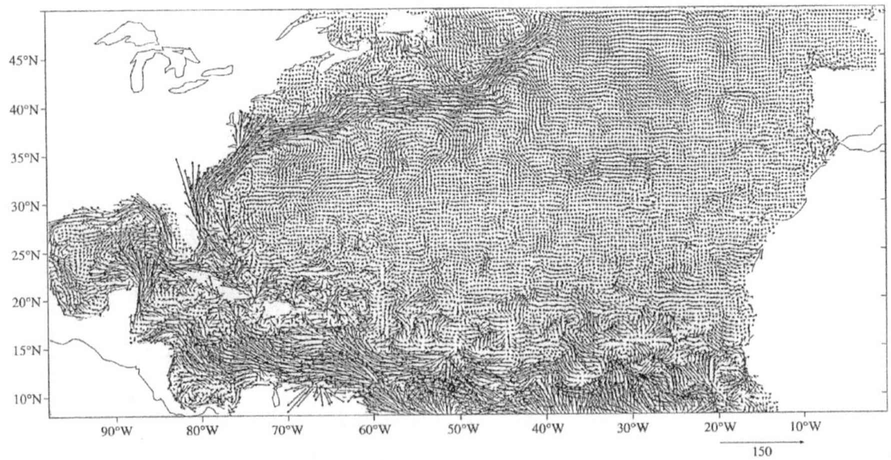

Fig. 5.23 Flow field in the North Atlantic at the level $z = 10\,\mathrm{m}$ after 5 days of model integration time by the σ-model (Diansky, personal communication)

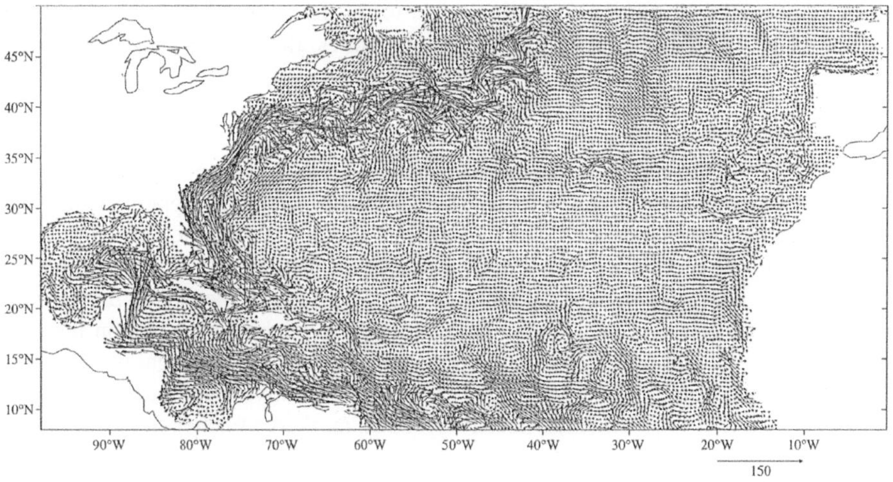

Fig. 5.24 Flow field in the North Atlantic at the level $z = 10$ m after 60 days of model integration time by the σ-model (Diansky, personal communication)

are common to all models listed above. Depending on the quality of models and the values of parameters, these problems may appear sooner in some models and later in others, but they cannot be avoided. The authors of some papers (Smith et al., 2000; Chassignet and Garaffo, 2001) managed to "shift the separation point back" to Cape Hatteras and to correct different shortcomings only due to a special treatment of the turbulence coefficients. The question is whether this is the only way of solving the problem.

Now, we briefly touch upon the problem of numerical convergence (more precisely, nonconvergence). The authors of paper Smith et al. (2000) have repeatedly stressed that no convergence can be demonstrated without turning to a higher resolution. This simply means that, if they continue the calculations, the model will diverge. In our opinion, there is no question about this. An analysis of their calculations clearly demonstrates that the results crucially depend on the turbulence parameters used. In this case, a long-term integration leads to either a steady state with distinctly unrealistic and highly filtered results or to a numerical divergence, because the energy cascade from small- to large-scale processes is parameterized rather than by deterministic equations. Therefore, some criteria are needed to stop the calculations timely. The authors are right in noting that their approach lies not very far from the Levitus climatology; yet, then, they ought to have realized how far to go, or at least, based on the calculated fields, they might have shown the evolution of T, S, ρ, KE, and other characteristics of the upper layer of the ocean.

Works of Bleck et al. (1995), Paiva et al. (1999), Smith et al. (2000), Chassignet and Garaffo (2001) did cross the Rubicon: they have shown that ocean-dynamics models must have a horizontal grid step no more than $0.1°$ and a number of levels greater than or equal to 20. However, having "the curtain lifted," we have not yet seen beyond the Rubicon what we expected. It turned out that, even with such a small horizontal resolution that allows one to resolve some eddies, the first baroclinic mode of Rossby waves, and the frontal zones, the models failed to describe energy transfer from small- to large-scale processes. The results have depended on the turbulence coefficients very heavily. What do these two best (among the aforementioned five high-resolution studies) works (Smith et al., 2000; Chassignet and Garaffo, 2001) describe? It can be seen from their calculations that the resulting Gu1f Stream characteristics are realistic. It seems likely that the upper oceanic layer of a small thickness was reconstructed against the initial climatic data, but, unfortunately, it was not demonstrated how and to what degree the T and S fields had been changed. It is likewise clear that, in the whole body of the ocean, the variation in these fields was not substantial. Thus, merely a "technical" Rubicon has been crossed with the aid of modem computers and advances in numerical mathematics. Modelling the climatic characteristics of the whole body of the ocean as of now remains a serious problem, and these two studies marked a further step in achieving this goal.

5.3 Pacific Ocean Circulation and Tracer Transport Modelling

5.3.1 Water Circulation Modelling

As for the Atlantic the Pacific Ocean modelling may be divided into two types: global or basin-wide large-scale characteristics modelling by coarse resolution with long model-time integration; selected small basins modelling by high resolution and short model-time of integration.

In the following we discuss some representative modelling approaches. We start with the paper of Losch and Heimbach (2007). Their model is global but with very crude resolution, $4^\circ \times 4^\circ$ and only 15 levels in vertical. The authors started the integration with the ocean being at rest and with the specified climatic T, S fields (Levitus and Boyer, 1994; Levitus et al., 1994) and spun up by running the model for as much as 2000 years of model time integration for investigation of direct and adjoint (Marchuk and Zalesny, 1993; Marchuk, 1995; Marchuk et al., 2001, 2003, 2005; Wunsch, 1996) sensitivities of ocean circulation to bottom topography. The model results are naturally very sensitive to bottom topography despite the crude presentation of the topography itself. The authors themselves admit that "the study is in part a proof of concept" and that "more experiments are needed to explore the sensitivities at higher resolution". Moreover, for such a long time integration it can hardly be judged what are effects of turbulent and/or numerical viscosities and what is real advection of physical b characteristics.

Besides of modelling, the analysis of observational data also raises fundamental problems. Harrison and Carson (2007) in the very title of their paper did ask – "Is the World Ocean Warming?" Based on the last half of the 20-th century observed data analysis they came to conclusion: "there is large spatial variability of 51 years trends in the upper ocean, with some regions showing cooling in excess of 3 °C, and others warming of similar magnitude".

We add that for a substantiated conclusion on centennial trends of the upper ocean, the deeper layers should be included, but the data is much less and the modelling results are much more uncertain there. We shall return to the problem at the end of this book.

We turn now to the papers with resolutions of 1° or less than one degree and tens of years of model-time integration.

We constantly pay attention to model's resolution and model-time of integration because they define the physical characteristics size, climatic significance and, in our mind, the model results quality as well. Luo and Yamagata (2003) studied the 1988–89 warming event in the Northern North Pacific based on the Modular Ocean Model (MOM2) of the Geophysical Fluid Dynamics Laboratory (GFDL) (see Pacanowski, 1996).

The domain considered covers the North Pacific from 30 °S to 65 °N. The horizontal grid mesh was irregular $(1/2)^\circ$ to $(1/4)^\circ$, with 30 irregular vertical levels.

The surface data was derived from NCEP-NCAR reanalysis archive. Both the surface wind stress and heat flux were specified. The model was spun up from ocean in state of rest by Levitus climatology (Levitus et al., 1994) for 15 yr model time then was forced by NCEP-NCAR daily reanalysis of wind stress for the period from 1979 to 1997. The model reproduces realistic upper-ocean temperature changes in comparison with observed data. Authors came to the conclusion that the mean horizontal geostrophic advection of anomalous temperature is the main contributor to the heat content increase at around 1988–1989, and surface heat flux forcing is responsible for increasing mixed layer temperature.

Authors note that near the western boundary the model heat content has a strong warm bias due to unrealistic northward overshooting of the mean Kuroshio system. We discussed already in paragraph 5.2 the Gulf Stream overshooting in the DYNAMO and other experiments, the same cause is with the Kuroshio – i.e. too long model-time of integration. In other words, the model has been "exploited" longer (totally almost 35 yrs) than the grid mesh and other parameters allow for a realistic description of processes. As a result the model passed the realistic Kuroshio separation point and the overshooting happened.

Niiler et al. (2003) have analyzed Lagrangian data from 657 SVP drifters, CNES/AVISO (Archiving, Validation and Interpretation of Satellite Oceanographic data) satellite altimeter sea level anomaly and Levitus et al. (1994) hydrographic data to develop maps of the mean and eddy circulation in the northwestern Pacific. The data used indicated two prominent meanders in the Kuroshio Extension (KE) jet, three anticyclonic recirculation eddies south or southeast of Kuroshio, a "snaky" frontal zones in the area of 30–40 °N, 140 °W to 180, etc. The AVISO currents appeared to correlate at 0.8 with drifter geostrophic velocity. We demonstrate in Fig. 5.25 one picture of this fundamental analysis.

Hosoda et al. (2004) made a set of numerical experiments to investigate interdecadal variations in the North Pacific subsurface temperature using the GFDL

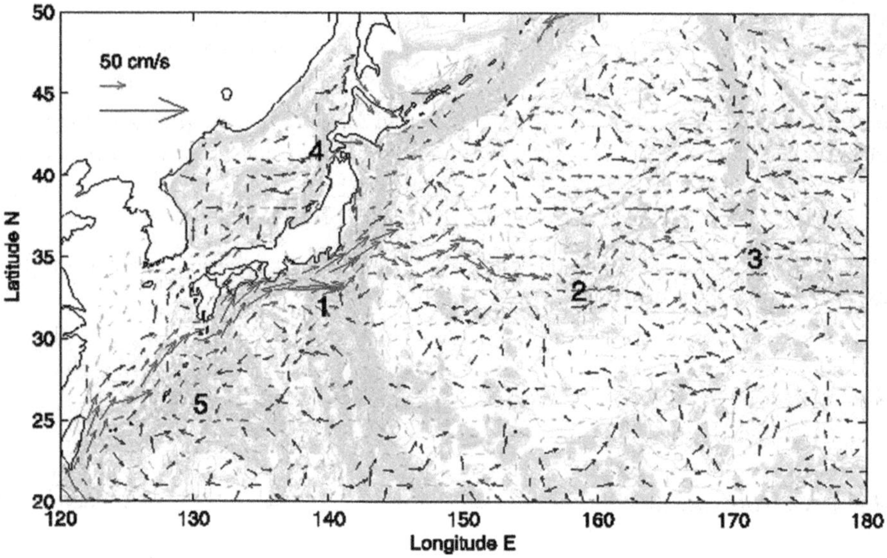

Fig. 5.25 One-degree average drifter velocities. Large velocity vectors (> 10 cm/s) are red and rescaled. Vectors, whose length is smaller than standard error (computed using 15 days decorrelation time), are light red and blue. Light green contour interval of bottom topography is 250 m. Numbers denote locations of the Izu Ridge (1), Shatsky Rise (2), Emperor Seamounts (3), Tsugaru Strait (4), and Daito Island (5) (Niiler et al., 2003)

modular ocean model, MOM 1.1. The resolution was $1° \times 1°$ with 32 vertical levels. The experiments design is common: preliminary spin up, which in this work was 40 years of model-time, and thereafter a set of experiments with specified interannual variations of the surface forcings. The area considered scopes 15 °S to 65 °N in meridional direction and practically coast-to-coast- in zonal (120E°–120 °W) direction.

The authors focus their attention to a special form of water mass which they, and some other scientists, call central Mode Water (CMW), a water mass with characteristic low potential vorticity which forms in the Kuroshio-Oyshio extension (KOE) area.

The name "Mode Water" needs explanation. The Kuroshio carries warm water and a large amount of heat release to the atmosphere in winter; that leads to development of vertically homogenized water mass which in higher latitudes enters the pycnocline if the mixed layer depth (MLD) shoals along its path. As a result this water mass is insulated from direct contact with the atmosphere as a thermostat or pycnostat. These water masses are considered to be essential for understanding the interannual to interdecadal variabilities because they retain the memory of the atmospheric conditions.

The main results of this study are as follows. Associated with the unusually large atmospheric changes after mid-1970s climate regime shift, the upper thermocline cooled as negative SST anomalies subducted and advected southward. Besides the CMW path shifted anomalously eastward due to anomalous Ekman pumping. The bottom of the winter mixed layer waters in KOE is critical to the formation of CMW. Authors showed also that coarser resolution ($3^\circ \times 3^\circ$) modelling yields unrealistic results.

To investigate formation and circulation of the North Pacific Mode Waters, Tsujino and Jasuda (2004) used the Meteorological Research Institute Community Ocean Model (MRI.COM) and ran it with high resolution $(1/4)^\circ \times (1/6)^\circ$, 48 vertical levels and corresponding precise bottom topography. The model is driven by monthly mean climatology of wind stress and the heat flux field derived from 1979 to 1993 of reanalysis conducted by ECMWF. The authors ran the model for 25 years of model time and outputs from the last five years are used for the analysis. Due to comparatively high resolution authors succeeded to reproduce climatological fronts and strong currents in the Kuroshio-Oyashio extension and all the three separate regions where the winter mixed layer reaches 300 m. The latter produce three major peaks in annuals subduction of further interest is the comparison of the total mass transport stream function with the Sverdrup one. This has been done for the whole North Pacific's (north of 5 °S) and for a comparatively smaller area of the western part. We reproduce the latter on Fig. 5.26 from which one can see the frontal zones, meanders and eddies in one of them and nonreal zonal isolines in another. In case of coarse resolution and inadequately long model time integration the JEBAR effect would be overly filtered and "Sverdrupization" would happen.

Based on Massachusets Institute of Technology (MIT) ocean general circulation model, Fukumori et al. (2004) investigated the variability of Niño-3 water, which is located at the eastern equatorial Pacific Ocean surface within 5 °S × 5 °N, $150^\circ - 90\,^\circ$W. The model comprises the whole Pacific Ocean with 1° resolution and gradually reduced the grid mesh within 10° of the equatorial zone. The author's large-scale circulation is of limited significance due to crude resolution and long integration time resulting practically in Sverdrup balance. So the sea-surface elevation across the Kuroshio is small and across the Gulf Stream even smaller. Much more interesting are the results of the Niño-3 modelling both of direct and adjoint calculations. Simulated passive tracers and their adjoint enabled to explicitly identify where the tracer-tagged water mass goes and from where it comes.

Their results contrast to the hypothesized subtropical cell, because the circulation appeared to be an open circuit with water moving to the western regions of the two hemispheres (subtropical gyres) and to Indian Ocean instead of returning to its origin.

The KE frontal zone, meanders and mesoscale eddies' decadal variability is clearly seen in Fig. 5.27 reproduced here from a paper of Qiu and Shuiming (2005).

Authors used for their analysis twelve years of sea surface height (SSH) data from multiple satellite altimeters. Data showed steady growing weak of the KE jet from 1993 to 1996 followed by gradual strengthening after 1997. The same decadal modulation is seen in the strength of the KE recirculation gyre. The decadal variability represented above in Fig. 5.27 shows that modelers necessarily should pass to a grid mesh smaller than 0.1° for investigating the oceans large-scale processes variability.

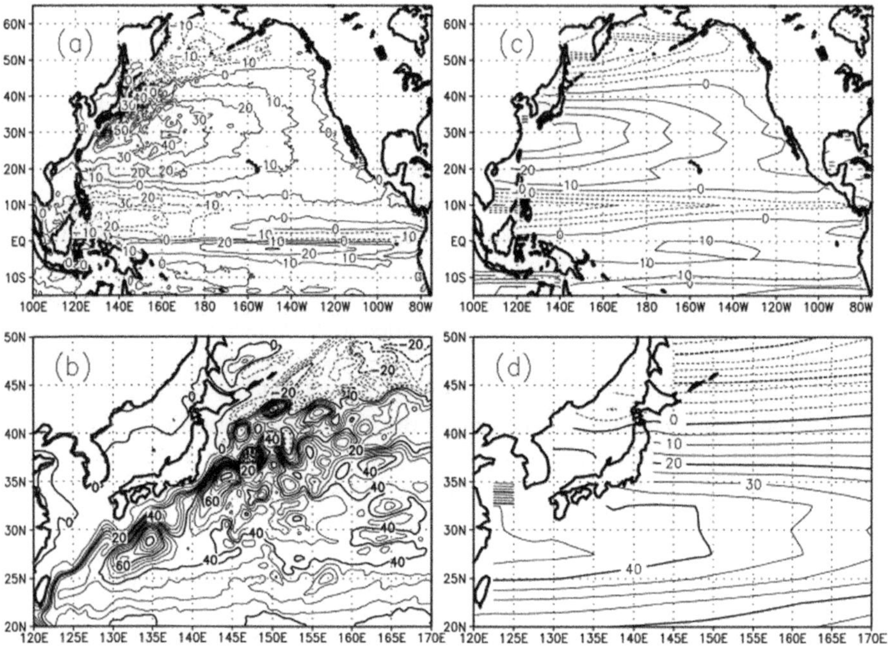

Fig. 5.26 Vertically integrated mass transport streamfunction (Sv) for (**a**) the whole model domain and (**b**) the western North Pacific, and Sverdrup streamfunction (Sv) calculated from the annual mean wind stress derived from ECMWF reanalysis for (**c**) the whole model domain and (**d**) the western North Pacific (Tsujino and Jasuda, 2004)

Guo et al. (2003) in a triply nested ocean model have examined the effect of horizontal resolution on JEBAR, see Fig. 5.28. The horizontal grid size was decreased from $(1/2)^{\circ}$ in the Nest 1 to $(1/18)^{\circ}$ in the NEST3, and the main attention was directed to area of the latter. The Princeton Ocean Model (POM- 3DPEM) with the level – 2, 5 Mellor-Yamada turbulent closure was used and sigma coordinates in the vertical (Blumberg and Mellor, 1987).

After spin up for 20 years, the model was forced by monthly wind stress and monthly heat fluxes for period 1991 to the end of 1998. The results obtained for the hind cast period was compared with observed data. Some of the main results of modelling are as follows. The Kuroshio path seasonal meridional migration, flow velocity, vertical structure, variability of sea level as well as reproduction of JEBAR become closer to observations as the model resolution increases from $(1/2)^{\circ}$ to $(1/18)^{\circ}$.

Authors point out that "the summer intensification of JEBAR ... yields a strong offshore volume transport across the shelf break. In winter, the weakened JEBAR, combined with the increased wind stress curl, decreases the offshore volume transport."

One interesting analysis made by authors deserve detailed discussion and additional comments. That concerns to values of the vertically integrated vorticity

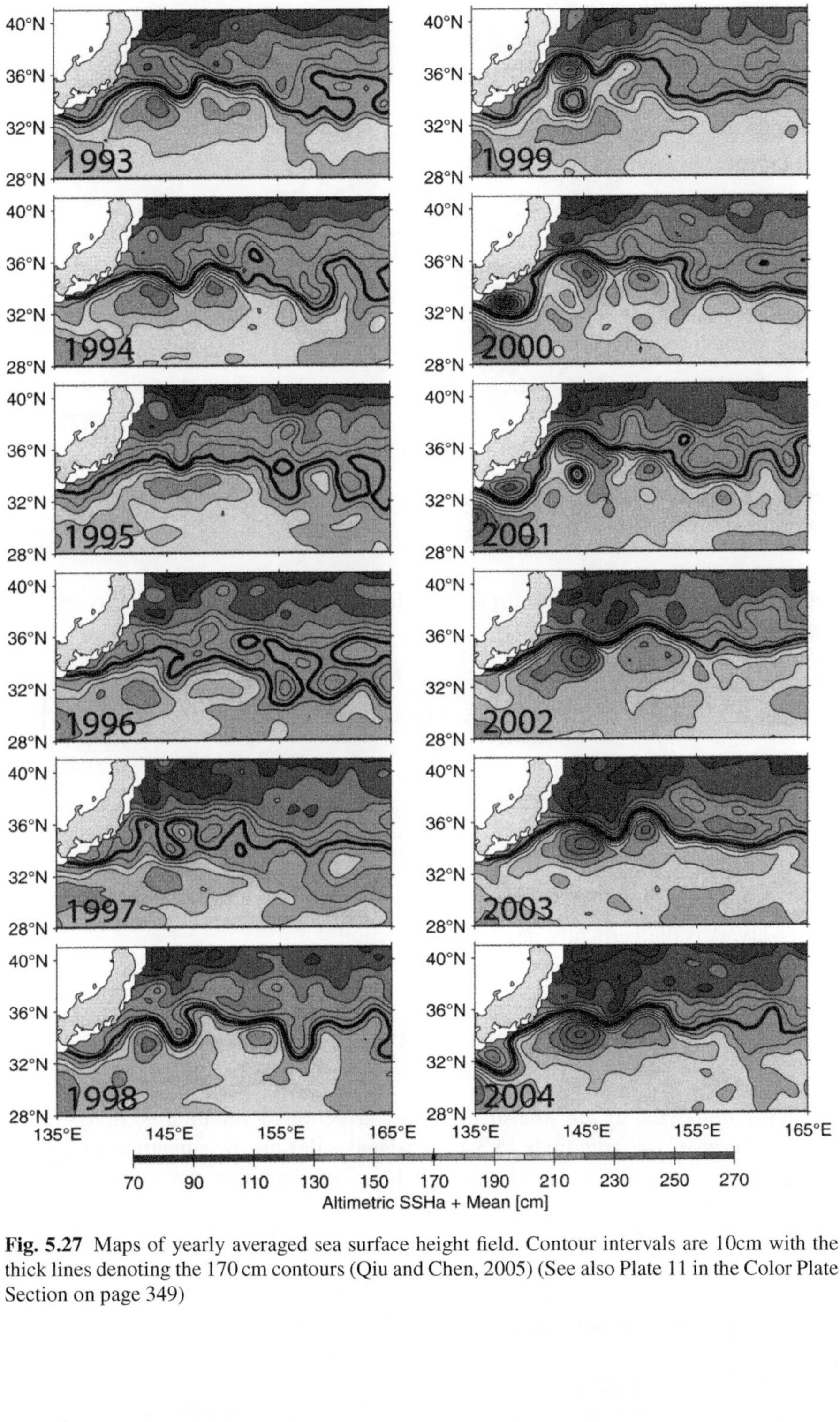

Fig. 5.27 Maps of yearly averaged sea surface height field. Contour intervals are 10cm with the thick lines denoting the 170 cm contours (Qiu and Chen, 2005) (See also Plate 11 in the Color Plate Section on page 349)

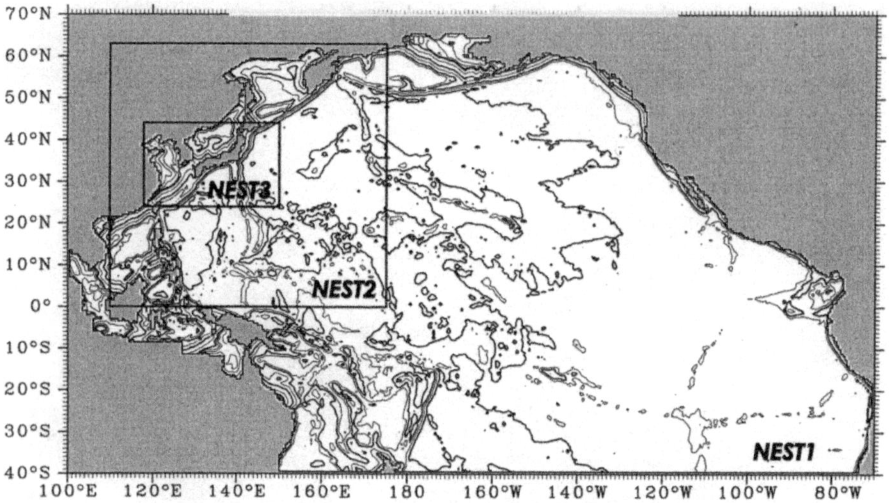

Fig. 5.28 Domain of triply nested models: NESTl, NEST2, and NEST3 (Guo et al., 2003). The isolines are depth contours

equation terms averaged over a small box from NEST3 area. The vorticity equation reads (see Chapter 2):

$$\frac{\partial}{\partial t} rot_z \left(\frac{\vec{M}}{H} \right) + \vec{M} \cdot \vec{\nabla} \left(\frac{f}{H} \right) = J\left(\chi, H^{-1}\right) + rot_z \left(\frac{\vec{\tau}_b}{\rho_0 H} \right) + rot_z \left(\frac{\vec{D}}{H} \right)$$
$$- rot_z \left(\frac{\vec{A}}{H} \right) - rot_z \left(\frac{\vec{\tau}_a}{\rho_0 H} \right) \quad (5.3.1)$$

where $\vec{M}$ is the horizontal mass transport, $\vec{D}$ the vertically integrated horizontal viscosity, $\vec{A}$ is the vertically integrated advection.
$\vec{\tau}_a$ is the wind stress, $\vec{\tau}_b$ the bottom stress,

$$J(\chi, H^{-1}) \text{ is the JEBAR with } \chi = \frac{g}{\rho_0} \int_{-H}^{0} \rho z dz$$

It is emphasized by the "that JEBAR and advection of potential vorticity (i.e. the second term of the equations (5.3.1) left-hand side) are two major contributions to the vorticity balance in this area". They form together the bottom pressure torque (BPT).

This can be seen in Fig. 5.29. Actually it was shown by many scientists that these two terms are the major terms practically in any area of the World Ocean and that JEBAR is the main source of energy for the stationary ocean currents (see Sarkisyan, 2006). Figure 5.29 shows how exactly the advection follows the JEBAR in seasonal

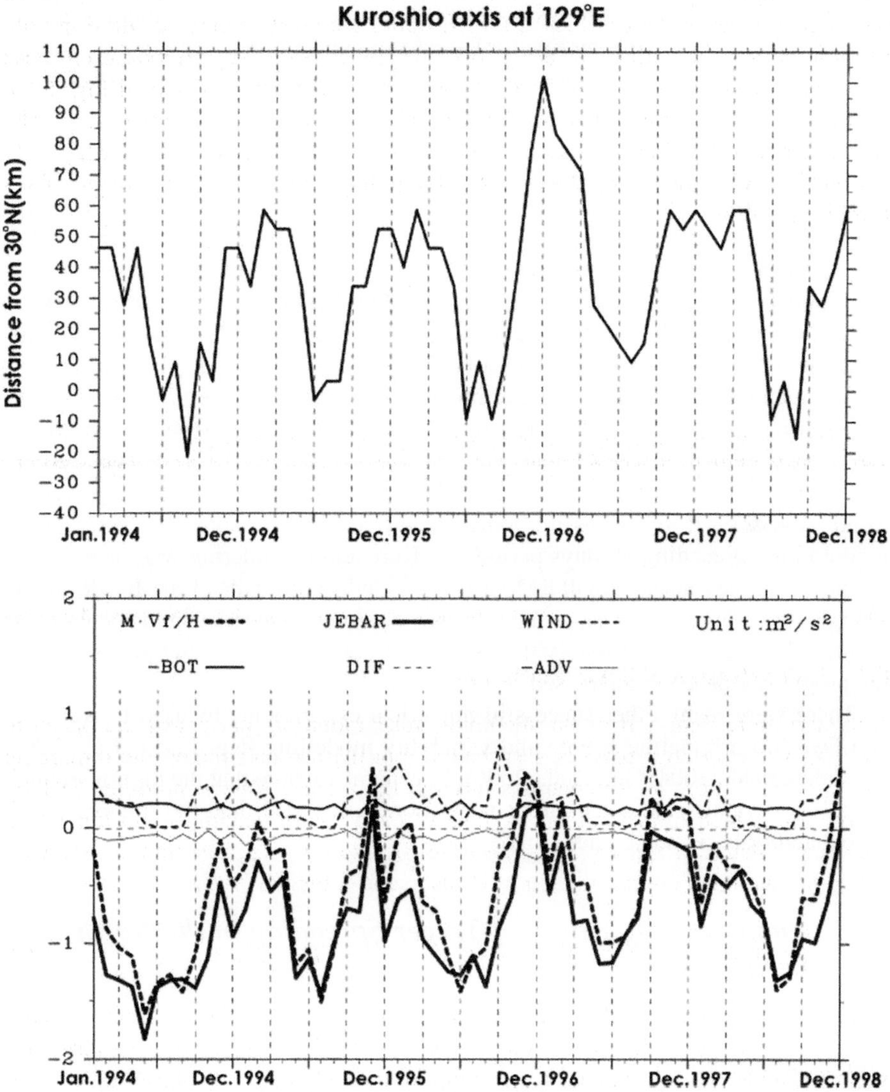

Fig. 5.29 (*top*) Seasonal variation of the Kuroshio veering latitude at 129 °E; values are monthly averaged ones for the years 1994–91, The Kuroshio axis is defined at the latitude where maximum surface speeds occur (shifts from the mean latitude of the Kuroshio axis at 30 °N are measured in km), (*bottom*) Seasonal variations of six terms in equation (5.3.1); monthly averaged results from NESTS 3 hindcast run for the years 1994–98 are used to calculate these terms (Guo et al., 2003)

variations during many years. In some moments the advection exceeds the JEBAR and that correlates very well with the increasing of the wind stress.

Our conclusion is that the BPT is as essential (and as small) as all the other terms of equation (5.3.1) are. For the first time that was shown by Gill and Bryan (1971).

Only the splitting of BPT into two terms provides the JEBAR as the major source of energy. As an analog one may say, BPT is an atom and the JEBAR is atomic energy. The BPT is the small difference of two big values and this difference is shown in Fig. 5.29. Another analog is $\frac{\partial w}{\partial z}$ in comparison with the sum $\frac{\partial u}{\partial x} + \frac{\partial v}{\partial y}$ in the equation of continuity.

Recently the JEBAR was examined with respect to energetic by Sakamoto and Umetsu (2006). They showed that the role of JEBAR is to transfer energy between the barotropic and baroclinic fields. In another paper, based on the same POM, Miyazawa et al. (2005) made essential success in forecasting the Kuroshio meandering for 60 days. A high resolution model (spatial grid of $(1/12)^{\circ}$ and 35 sigma levels) was imbedded into the basin-wide lower-resolution model ($((1/4)^{\circ}$ and 21 levels). The inner domain covers the northeast Pacific (12 °N–56 °N), 117 °E–180°) and its lateral boundary conditions are determined from the basin-wide model using one-way nesting (Guo et al., 2003). The low-resolution model after 10 years spin-up was driven by 6-hourly surface forcings covering from 1 January 1991 to 30 November 1999. The high resolution model was driven same way but from 8 August 1996 to 30 November 1999. By assimilating sea surface height anomaly into the ocean model in the preceding 40 days period, the Kuroshio meandering was successfully predicted (in the sense of small RMS error) 2 months before the event. In addition to a single trajectory forecast experiment, ensemble forecasts were conducted using 10 perturbed initial states. Possibly this is the first comprehensive attempt for solving the prognostic problem addressed.

There were many other successful numerical experiments focused to Kuroshio and Oyashio origination source and variability modelling (e.g. Kono et al., 2004; Mitsudera et al., 2004; Guo et al., 2006). In general we note that the high-resolution models for small-scale ocean areas were more successful than the low-resolution global and/or basin-scale ones.

5.3.2 Numerical Modelling of Tracer Spreading in the North Pacific Ocean

Potential sources of North Pacific waters contamination are wrecks of ships transporting used radioactive fuel to Vladivostok from Petropavlovsk-Kamchatsky. Accidents may happen due to weather hazards, technical problems or terrorist's action. Even in case of negligible contamination the very declaration of it would result in a great loss of fisheries industry in the North American west coast because of normal human panic. Nuclear contamination of coastal waters is possible also due to accidents in a coastal state institutions, factories as well.

The radionuclids are simple passive tracers, which do not play any role in the ocean/sea hydrodynamics. So, in view of modellers, after ocean thermohydrodynamic characteristics calculation, it is enough to solve one more equation, like the equation for salinity, to monitor the potentially contaminated waters spreading. Moreover, it is possible to find the location of origination of water mass presently flowing at a certain spot.

In the following results will be presented of a thermohydrodynamic model (Marchuk et al., 2001; Sarkisyan and Zalesny, 2000) completed by a radionuclide spreading model (Antipov et al., 2006). It is a 3DPEM σ-model in spherical coordinates with hydrostatic and Boussinesqu approximation. Flow velocity, potential temperature, salinity and sea surface height are prognostic variables in the model. The relation of Bryden et al. (1999) serves as equation of state. The splitting up procedure is used for solving the system of equations on a C-grid (Mesinger and Arakawa, 1976).

To prepare the heat and salt fluxes at the ocean surface, as well as the wind stress, the data of OMIP (Ocean Model and Intercomparison Project) and the NCEP reanalysis data was used. There are seven liquid segments at the domain's side boundaries, at which the Levitus and Boyer (1994) *T, S* data was specified. The bigger rivers discharge was taken into account too. The integration was continued only for six month, considering it to be enough for upper ocean adjustment to observed climatology. The resulted surface currents of the whole North Pacific (not shown here) correspond to well-known chart of G. Schott (1943) and other more recent data. Despite the calculation was done for the whole North Pacific's we shall pay attention to the northwest part of it only because there is the track of radionuclid transportation. The surface currents are demonstrated in Fig. 5.30. The flow is meandering of both the Kuroshio and its extension. Maximal flow velocity in the Kuroshio jet reaches 150 cm/s; the jets size is about 100km; in the area of Kuroshio Extension the flow velocity reaches 50 cm/s. These characteristics, as well as the Kuroshio separation point, its bifurcation following the separation, existence of eddies are in good agreement with the results of Mitsudera et al. (2004). Besides,

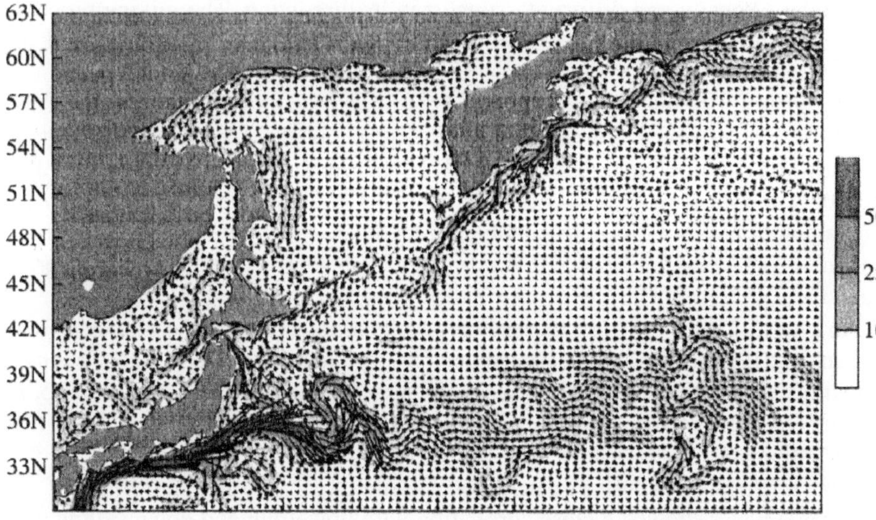

Fig. 5.30 Multiyear mean flow velocity of the Pacific Ocean surface layer (cm/s). The velocity scale is shown on the right-hand side colored diagram (Antipov et al., 2006)

the calculated East-Sakhalin current corresponds to the data of direct flow velocity measurements, performed during the Japan-American-Russian expedition (Oshima et al., 2004).

Assuming the calculated flow velocity field to be in stationary state, an equation for passive tracer transport was integrated for one year model time. As concerns the wreck source at the very initial moment a contaminated water volume of 100 m $\times$ 100 m $\times$ 5 m and a certain radioactive elements concentration were assumed. But the grid mesh is much bigger, so the question is how much time is needed for this small volume to spread up to one square of four grid points. Analytical estimates with horizontal mixing coefficient equal to $10^6\,cm^2/s$ showed that the spreading time is less than 1 day, so it may be considered to be instantaneous. For pure diffusion it needs $\sim$ 10 days for spreading to 10 km distance.

As for the concentration, it equals to the value of the small volume (100 m $\times$ 100 m$\times$5 m) divided by the volume of one grid-cell with minimal thickness (because of the σ-coordinate model the distance between levels depends on the local depth).

Two levels of radionuclids concentration have been considered: level 1 equals $1\,BK/m^3$ which about for 30% exceeds the normal background radiation; the level 2 equals $400\,BK/m^3$, which a little exceeds the upper limit of permissible concentration.

The dissolving and spreading stage began from four grid-points contaminated as highly dangerous as 400 Bk. Any concentration which exceeds 1 Bk is considered to be dangerous and consequently belongs to category 2. Authors made numerical experiments on hypothetical wreckage of ships in several points on the ships route from Petropavlovsk-Kamchatsky to Vladivostok. We have chosen two out of them. Figure 5.31 shows the spreading of contamination released at Petropavlovsk-Kamchatsky. The main cause of the contamination spreading is the water circulation while the viscosity and diffusivity support in its dissolving. The mixing processes and especially the water upwelling and downwellings generate penetration into deeper layers.

Figure 5.32 shows that only low concentration of radionuclids reaches as deep as 500 m but detailed examination showed that actually in a small area with intense downwelling the contamination of level 2 reached to that depth within 15 days, and only then it dissolved in a few weeks. This and other experiments show that contamination dissolves quite rapidly and attenuates without making any harm to Japanese or American waters if the accident happens before the ships enter the Japan sea.

In Figs. 5.33 and 5.34 one can sea the results of the contaminated waters spreading in case if hypothetical release happens in the Vladivostok. The highly contaminated waters in this case reach the Japan coasts without essential damping. So, the danger may come to the Japanese coastal waters mainly from the Japan Sea or from the Vladivostok.

One further aspect deserves attention. The analysis of vertical movement of contaminated water showed a close correlation with upwelling and downwelling, with cyclonic and anticyclonic eddies. So a three dimensional flow field should be analyzed when examining the contaminated waters spreading. For example, calculation

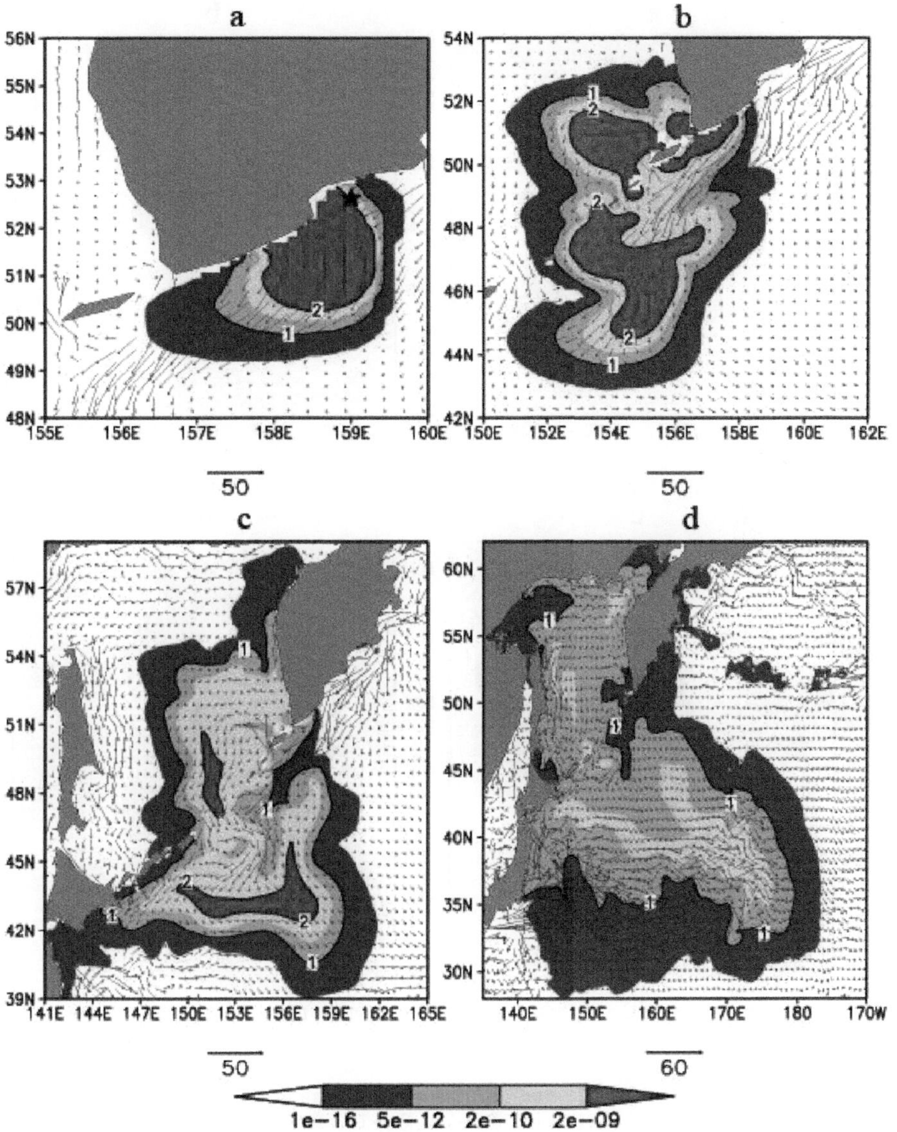

Fig. 5.31 A spreading of "contaminated" water in the ocean surface layer from a hypothetical source, located near Petropavlovsk-Kamchatsky. The plots show the following selected time-moments after starting the calculations: (**a**) 5 days, (**b**) 1,5 month, (**c**) four months, (**d**) one year. The flow velocities are in cm/s. Numbers 1 and 2 characterize the danger level of "radioactivity" (Antipov et al., 2006) (See also Plate 12 in the Color Plate Section on page 350)

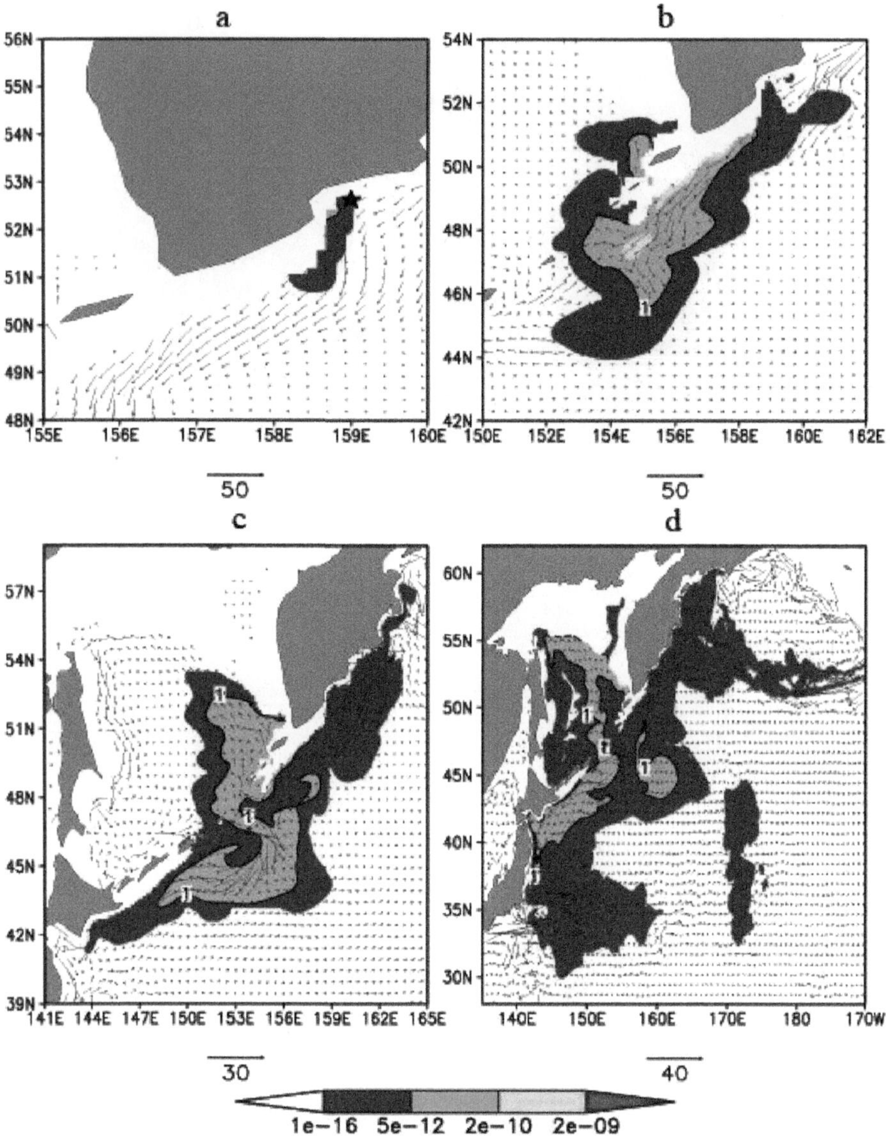

Fig. 5.32 The same as in Fig. 5.31 but for the depth 500 m. The *asterisk* shows the location of hypothetic source of radioactivity (Antipov et al., 2006) (See also Plate 13 in the Color Plate Section on page 351)

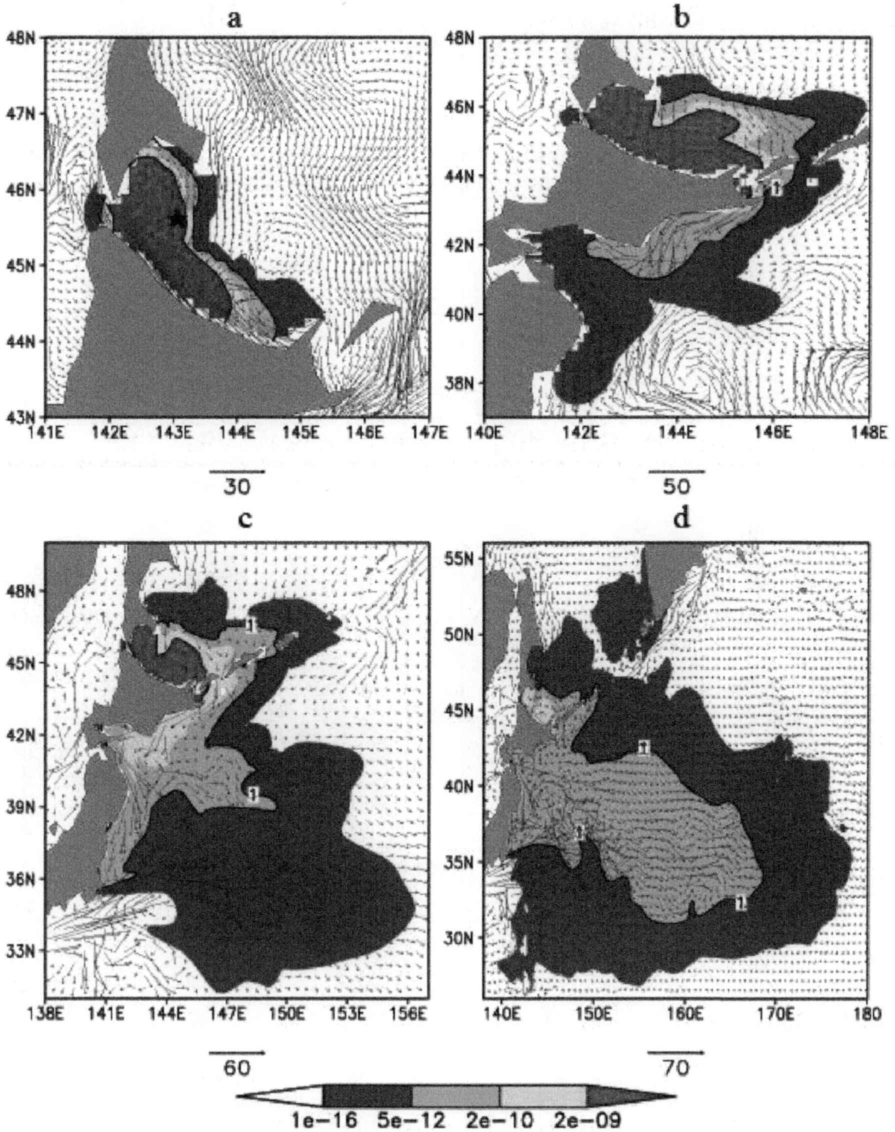

Fig. 5.33 The same as in Fig. 5.31 but for the case when the source of radioactivity is near the city Vladivostok (Antipov et al., 2006) (See also Plate 14 in the Color Plate Section on page 352)

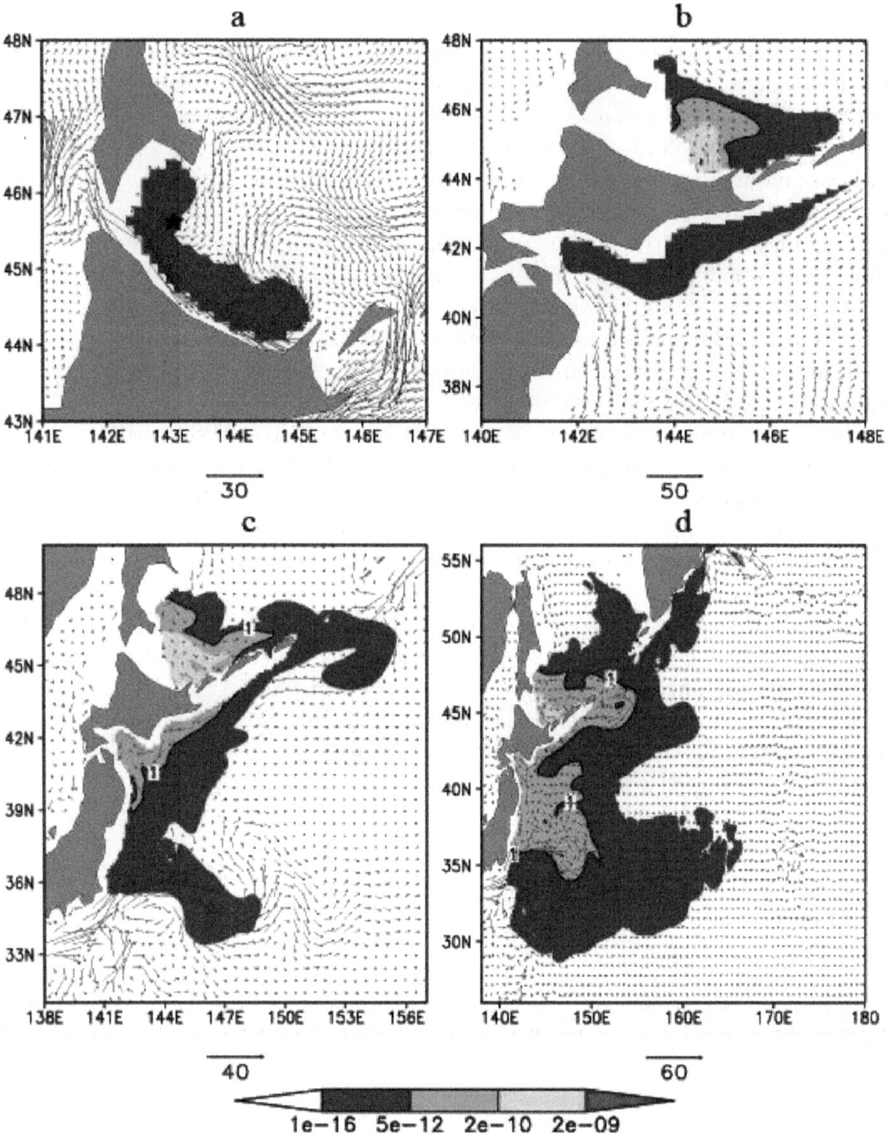

Fig. 5.34 The same as in Fig. 5.33 but for the depth 500 m (Antipov et al., 2006) (See also Plate 15 in the Color Plate Section on page 353)

of hypothetic contamination generated in the Laperouse Bay showed an isolated spot of essential concentration which appeared near southwestern coast of Kamchatka peninsula. It turned out to be contaminated upwelling water which was transported to the area by deeper layer horizontal movements.

5.4 The Indian Ocean

5.4.1 The Indian Ocean Descriptive Hydrometeorology

The Indian Ocean descriptive hydrometeorogy is very well presented in two big papers of Shankar et al. (2002) and Tomczak and Godfrey (2003). We use the second of these papers for a short introduction.

The Indian Ocean, is in several respects very different from the Atlantic and the Pacific Ocean. The most striking difference is the seasonal reversal of the monsoon winds and its effects on the ocean currents. The absence of a temperate and polar region north of the equator is another peculiarity with far-reaching consequences for the circulation and hydrology.

The Indian Ocean is the smallest of all oceans (including the Southern Ocean). The only large shelf area is the Northwest Australian Shelf, a region of strong tidal dissipation.

Three mediterranean seas influence the hydrographic properties of Indian Ocean water masses. The Persian Gulf is the smallest of the three; with a mean depth of 25 m, a maximum depth of only 90 m. The Red Sea is a very deep basin with maximum depths around 2740 m and a mean depth near 490 m; its sill depth is about 110 m. The Australasian Mediterranean Sea, a series of very deep basins with depths exceeding 7400 m, communicates with the Indian Ocean through various passages between the Indonesian islands where the depth is in the range 1100–1500 m.

Monsoonal climate dominates the northern Indian Ocean. Annual mean distributions of atmospheric and oceanic parameters are therefore of only limited use. Instead we define two mean states and discuss both separately.

The wind over the northern Indian Ocean represents the Trades, but because of its seasonality it is known as the Northeast Monsoon.

The Northeast or Winter Monsoon determines the climate of the northern Indian Ocean during the northern hemisphere winter (December–March). It is characterized by high pressure over the Asian land mass and northeasterly winds over the tropics and northern subtropics. The situation resembles the annual mean wind circulation over the Pacific Ocean, except that the Intertropical Convergence Zone (ITCZ) and the Doldrums are located south of the equator (near 5 °S) rather than north. Since most of the air pressure gradient is retained behind the Tibetian Plateau, air pressure gradients over the ocean are small. This protects the ocean from the full force of the winds blowing off the Mongolian high pressure region and results in a wind of moderate strength, comparable to the Pacific Northeast Trades which are also relatively weak at this time of year. The wind also carries dry air, and the Winter Monsoon season is the dry season for most of southern Asia.

The situation in the southern hemisphere is dominated by the pressure gradient between the tropical low and the subtropical high pressure belts. The Southeast Trades are rather uniform and somewhat stronger than in the Pacific Ocean.

The Southwest or Summer Monsoon determines the climate of the northern Indian Ocean during the northern hemisphere summer (June–September). A deep heat low develops over northern Arabia and Pakistan. The Australian heat low of the southern summer is replaced by a centre of high pressure, while the atmospheric high north of the Kerguelen Islands is shifted westward towards southern Africa. Whereas during the winter monsoon season the north-south pressure gradient from Arabia to Madagascar barely exceeds 6 hPa, there is now a gradient of 22 hPa acting in the opposite direction. As a result the winds in the northern Indian Ocean reverse completely and are no longer like the Trades anywhere.

The Southwest Monsoon, as this wind is called in the northern hemisphere, is the continuation of the southern hemisphere Trades, which between 10 and 20 °S are stronger, during this time of year, than anywhere else in the world. The Southwest Monsoon skirts the low over Pakistan to deposit rain on the Himalayas, thus bringing with it the monsoon rains and floods that are so crucial to Asian agriculture.

The distribution of hydrological properties in the Indian Ocean is much less affected by the seasonal monsoon cycle than the near-surface current field. Direct monsoonal influence is restricted to the surface mixed layer and the western boundary currents. The two most important factors which make Indian Ocean hydrology different from the hydrology of the other oceans are the closure of the Indian Ocean in the northern subtropics and the blocking effect of the equatorial current system for the spreading of water masses in the thermocline.

The most striking characteristic of the rainfall distribution over the Indian Ocean is the anomalous difference between the eastern and western regions in the north. Annual mean precipitation varies between 10 cm per year in the west (on the Arabian coast) and 300 cm per year or more in the east (near Sumatra and over the Andaman Sea). This is the reverse of the situation usually encountered in the subtropics, where the Trades bring dry continental air out over the sea in the east and rain to the western coast. The normal situation of little rain in the east and high rainfall in the west prevails in the southern Indian Ocean; Western Australia receives less than 50 cm per year but Madagascar some 200 cm per year. As a result, contours of equal precipitation show more or less zonal orientation in the south but meridional orientation in the north.

As in the other oceans, sea surface salinity follows the P-E distribution outside the polar and subpolar regions closely. The P-E minimum near 30 °S is reflected by a SSS maximum. The decrease of SSS values further south continues into the Southern Ocean, reflecting freshwater supply from melting Antarctic ice.

However, the lowest surface salinities are found in the northern subtropics, where they reach values of 33 psu and below on annual mean; during the Summer Monsoon season, surface salinity in the inner Andaman Sea is below 25 psu.

Antarctic Bottom Water (AABW) fills the Indian Ocean below approximately 3800 m depth. By the time it leaves the Circumpolar Current its properties correspond to those of Antarctic Circumpolar Water (potential temperature 0.3 °C, salinity 34.7 psu).

The situation is not really much different from that in the Atlantic Ocean, where the water at the ocean floor is usually called Antarctic Bottom Water (AABW) The distribution of potential temperature below 4000 m indicates two entry points. Entry into the Madagascar Basin has been well documented and occurs through gaps in the Southwest Indian Ridge near 30 °S, 56–59 °E. The flow gradually finds its way across to the Madagascar continental slope, where it forms a deep western boundary current. In a zonal temperature section it is seen as a steep rise of the deep isotherms against the slope, consistent with northward geostrophic movement in which the speed increases with depth. In the east, AABW enters the South Australia Basin via the Australian-Antarctic Discordance, a region in the interoceanic ridge system with multiple fractures near 50 °S, 124 °E south of Australia. Having filled the depths of the Great Australian Bight it moves west and then north into the Perth, Wharton, and North Australia Basins, forming a western boundary current along the Ninety East Ridge. Flow of AABW through the Mozambique Basin is blocked by Mozambique Strait; nevertheless, AABW recirculation in the basin must be swift, since observations of bottom currents in the 4500 m deep channel between the Agulhas Plateau and the African shelf gave average northward speeds underneath the Agulhas Current of 20 cm/c.

This water eventually enters the Arabian Basin, where it must disappear through gradual upwelling into the overlying Deep Water. AABW from the eastern path proceeds into the Mid-Indian Ocean Basin, flowing over deep saddles in the Ninety East Ridge near 10 °S and 5 °S, and turns south; the differences in temperature and oxygen across the Ninety East Ridge below 3000 m testify for the different ages of the waters on either side of the ridge.

The depth range from 3800 m upward to about 1500–2000 m and above is occupied by Indian Deep Water (IDW). Based on water mass properties the transition from Bottom to Deep Water is gradual and some authors refuse to use the terms Bottom and Deep Water, referring to lower and upper deep water instead. This may appear logical since in the southern Indian Ocean both water masses have been observed to move northward together.

A look at meridional hydrological sections proves, however, that the distinction between Deep and Bottom Water is justified. Indian Deep Water is characterized by a salinity maximum in the southern hemisphere exceeding 34.8 psu in the west and reaching 34.75 psu in the east. It occupies the depth range 2000–3800 m north of 45 °S and comes to within 500 m of the surface further south.

We continue with brief qualitative description of some of the Indian ocean currents. The western boundary current begins east of Madagascar, where the ship drift currents (Fig. 5.35) show a separation of the South Equatorial Current into a northern and southern branch. The contribution of the northern branch to the circulation in the southern hemisphere is the Mozambique Current; it is maintained throughout the year. The contribution to the circulation in the northern hemisphere ceases during the Northeast Monsoon season. The southern branch feeds the East Madagascar Current. The current field of this small but well-defined western boundary current reaches to the 2000 m level. Below 3100 m some 4–5 Sv are carried northward, with little movement in-between. The contribution of the Mozambique Current to the Agulhas Current is comparatively small.

South of 30 °S the flow continues as the Agulhas Current, one of the strongest currents of the world ocean. In contrast to other western boundary currents it shows little seasonal variation. Mean speeds are of 160 cm/c throughout the year, and peak

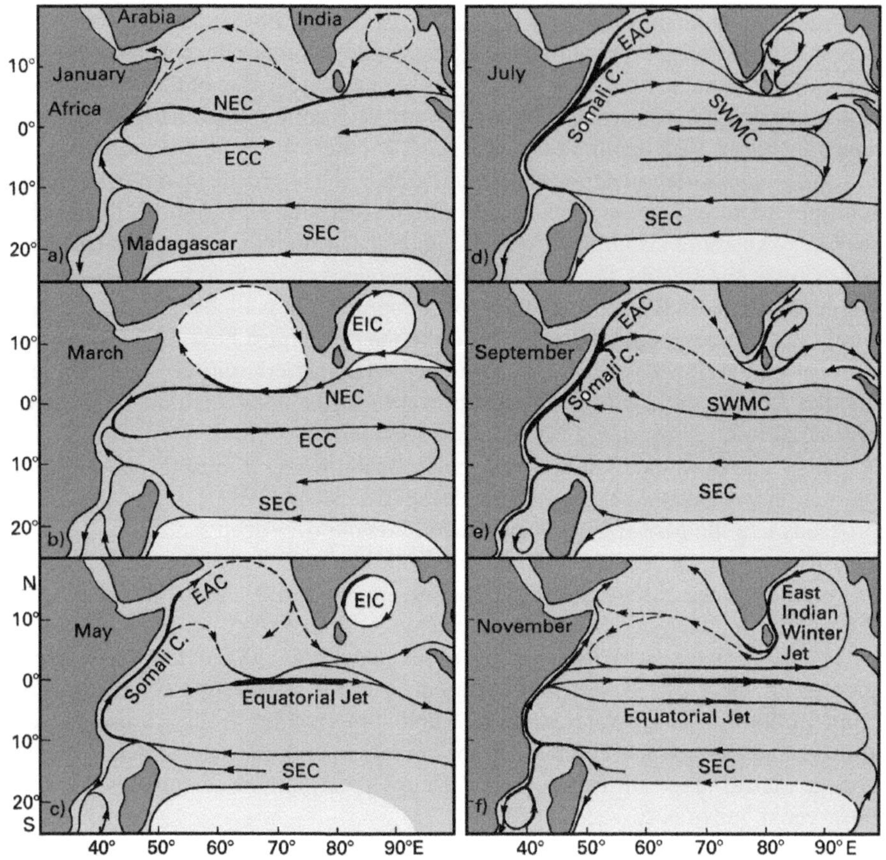

Fig. 5.35 Surface currents in the northern Indian Ocean for six different months as derived from ship drift data. SEC: South Equatorial Current, NEC: North Equatorial Current, ECC: Equatorial Countercurrent, SWMC: Southwest Monsoon Current, EAC: East Arabian Current, EIC: East Indian Current (Cutler and Swallow, 1984) (See also Plate 16 in the Color Plate Section on page 354)

speeds exceed 250 cm/c in most months. Transport estimates from observations give 70 Sv near 31 °S and an increase of 6 Sv for every 100 km (as in the Gulf Stream). On approaching the shallow Agulhas Bank near 35 °S it carries 95–135 Sv. The current occasionally floods the bank, lowering the inshore temperatures by several degrees. This upwelling is a result of the thermocline slope across the current and not related to the wind.

North of 10 °S the East African Coastal or Zanzibar Current flows northward, fed by the northern branch of the South Equatorial Current. During the Northeast Monsoon season it runs against light winds and is opposed by the southward flowing Somali Current. The point where northward and southward flows meet at the surface moves from 1 °N at the beginning of the season to 4 °S during the February peak, when the Zanzibar Current is at its weakest. Soon after the peak it starts moving north again, reaching the equator by early April. Throughout this period the current feeds into the Equatorial Countercurrent. Below the surface flow the Zanzibar Current flows northward across the equator at all times, taking the form of an undercurrent under the southward flowing Somali Current during the Northwest Monsoon season. During the Southwest Monsoon season the strength of the Zanzibar Current increases considerably; INDEX observations from April/May 1979 show it with speeds of 200 cm/s and a transport of 15 Sv. The current now feeds the northward flowing Somali Current; some consider it part of the Somali Current during these months.Southward flow in the Somali Current during the Northeast Monsoon is limited to the region south of 10 °N. It first occurs in early December south of 5 °N and expands rapidly to 10 °N in January with velocities of 70–100 cm/c. In March the southward flow contracts again to 4 °N, until the surface flow reverses in April. During the Southwest Monsoon the Somali Current develops into an intense jet with extreme velocities; INDEX observations gave surface speeds of 200 cm/s for mid-May and 350 cm/s and more for June.

South of 5 °N the Somali Current is extremely shallow; below 150 m depth southward flow is maintained throughout the year. Further north the jet deepens, eventually embracing the permanent thermocline. The current structure on the equator is extremely complex and shows layering similar to the equatorial flow further east but oriented northward–southward.

The period of northward flow can be divided into two phases of different dynamics. During the transition in May, flow in the Equatorial Jet is eastward, and westward flow develops only slowly during the following two months. From the point of view of mass continuity there is not much need for a strong western boundary current until the monsoon reaches its peak in August, and the Somali Current is first established as a response to the wind reversal along the African coast. Winds are southerly but light in late April and May and strengthen abruptly in late May or June. This drives northward flow across the equator; but the flow turns offshore near 3 °N, and a coastal upwelling regime develops between 3 °N and 10 °N.

The Indonesian throughflow from the Pacific Ocean enters the Indian Ocean during both seasons as a narrow band of low salinity water. It is embedded in generally westward flow and therefore apparently does not develop the strong lateral shear necessary to induce much instability. It continues westward, providing the core of maximum westward flow in the equatorial current system, and can be followed over the entire width of the Indian Ocean.

The subtropical gyre of the southern hemisphere is seen with two western boundary currents, one along eastern Madagascar and one along the coast of Mozambique.

The evolution of surface currents through the seasons is shown in Fig. 5.35. The North Equatorial Current is prominent in January and March when the Northeast Monsoon is fully established. It runs as a narrow current of about 30 cm/s from Malacca Strait to southern Sri Lanka, where it bends southward and accelerates to reach 50–80 cm/s between 2 °S and 5 °N in the region between 60 °E and 75 °E. The South Equatorial Current occupies the region south of 8 °S with velocities rarely exceeding 30 cm/s. Between these westward flows runs the Equatorial Countercurrent with 50–80 cm/s in the west but getting weaker in the east; in January it does not reach beyond 70 °E, being opposed in the east by weak westward flow.

The transition from Northeast to Southwest Monsoon (Fig. 5.35c) is characterized by the intense Indian Equatorial Jet first described by Wyrtki (1973). The long-term mean distributions derived from the ship drift data show it from early April until late June with velocities of 70 cm/s or more. It is possible that in any particular year the jet appears within the three-month window April–June as a feature of shorter (one month) duration with higher peak velocities. The averaging employed with the ship drift data would spread it over the three months as a weaker feature. The jet is easily observed with drifting buoys since the current converges at the equator, keeping drifting objects trapped near its core. Away from the equator the current speed falls off to less than 20 cm/s at 3 °S or 3 °N.

When the Southwest Monsoon is fully established during July and September, the entire region north of 5 °S is dominated by the eastward flow of the Southwest Monsoon Current, the only exception being a narrow strip along the equator in July to which we shall return in a moment. Velocities in the Southwest Monsoon Current are generally close to 20–30 cm/s, but an acceleration of 0.5–1.0 m s^{-2} occurs south and southeast of Sri Lanka. The South Equatorial Current expands slightly towards north, reaching 6 °S in September. The transition before the onset of the Northeast Monsoon (Fig. 5.35f) is again characterized by the Equatorial Jet. Concentrating all eastward flow in a 600 km wide band along the equator it reaches its peak in November with velocities of 100–130 cm/s and disappears in early January, when the annual cycle is repeated.

Several layers of alternate flow direction were found, with velocities reaching 12 cm/s at 4000 m depth. The uppermost layer of westward flow near 200 m depth is commonly referred to as the Equatorial Undercurrent; but it is obvious that westward transport occurs at much greater depths as well. During the Southwest Monsoon of 1976 mean flow at moorings in the area 52–58 °E was westward with a northward component into the Somali basin at 500 m and 1500 m depth.

The oscillations of the undercurrent known from the Pacific and Atlantic Oceans occur in the Indian Ocean as well. At 200 m depth they were seen from August 1979 until January 1980 (in a record which covered the period March 1979–June 1980). They were of 1300 km wavelength and 20 cm/s meridional amplitude and shifted the undercurrent axis back and forth by about 150 km either side of the equator. Drifting buoy data for the period 1979–1982 indicate that at the peak of the Southwest Monsoon in August and September the oscillations reach into the surface layer with meridional flow amplitudes up to 80 cm/s.

Seasonal reversal of the currents dominates the two major subdivisions of the northern Indian Ocean as well, but the opposing flows occupy periods of very different length, and the transitions are less well defined than in the equatorial zone. Weak westward flow, an extension of the North Equatorial Current with velocities rarely exceeding 0.2 m/s, dominates in the Arabian Sea at the peak of the Northeast Monsoon season. Northwestward flow along the western Indian shelf begins as early as November (Fig. 5.35f) and persists into January, with a width of some 400 km and a depth of about 200 m in the south (10 °N), getting narrower and deeper as it flows along the continental slope. This current flows against the prevailing Northeast Monsoon and thus cannot be wind driven. The East Indian Winter Jet supplies fresh, low density water from the Bay of Bengal at this time, while on the north Indian coast cold continental winds result in cooling and convective overturn (Shetye et al., 1993). Westward flow prevails south of 15 °N and west of 65 °E until late April, while in the remaining area currents are less and less well defined and change gradually into the weak anticyclonic pattern of Fig. 5.35b. The Somali Current responds quickly to the onset of the Southwest Monsoon in April; northward flow develops, strengthening the pattern in the west. By mid-May (Fig. 5.35c) the East Arabian Current is fully established with velocities of 50–80 cm/s. At the same time the anti-cyclonic pattern breaks up from the east where the flow joins the Equatorial Jet around southern India and Sri Lanka. Moderate eastward flow, an extension of the Somali and Southwest Monsoon Current dominates the region during the next 4–5 months. During its peak in June and July it reaches 30 cm/s and more but weakens rapidly in October when the second occurrence of the Equatorial Jet concentrates most eastward transport in the equatorial zone and outflow from the Bay of Bengal begins to oppose eastward flow around Sri Lanka. By mid-November currents are again diffuse; south of 15 °N they are weak but already westward. General westward flow is again established by early December.

A notable feature of the Arabian Sea circulation is the occurrence of strong coastal upwelling in the East Arabian Current. As in other coastal upwelling regions it owes its existence to an offshore transport direction in the Ekman layer (the Southwest Monsoon blowing parallel to the coast with the coast on its left). Positive $rot(\frac{\vec{\tau}}{f})$ over a 400 km wide strip along the coast adds to the upwelling through Ekman suction.

The circulation in the Bay of Bengal is characterized by anticyclonic flow during most months and strong cyclonic flow during November (Shetye et al., 1993). In January currents are weak and variable. In the west, the East Indian Current strengthens as the Northeast Monsoon becomes stronger, exceeding 50 cm/s in March (Fig. 5.35b) and remaining strong (70–100 cm/s) until May/June. Throughout this time the current runs into the wind, apparently as an extension of the North Equatorial Current.

An indication of a current reversal in the west is seen in September (Fig. 5.35e). Currents are consistently southwestward and strong (50 cm/s and more) north of 15 °N, and close to the shelf southwestward flow prevails. Complete reversal of the East Indian Current into the East Indian Winter Jet is not achieved until late October, when water from the Equatorial Jet enters the Bay in the east and a cyclonic circulation is established. The East Indian Winter Jet is a powerful western boundary current with velocities consistently above 10 cm/s. It follows the topography south of Sri Lanka and feeds its water into the Arabian Sea. Very little exchange occurs with the Equatorial Jet south of Sri Lanka; currents in the separation zone between the two jets (near 3 °N) are weak and variable. The East Indian Winter Jet fades away from the north in late December, its southern part merging with the developing North Equatorial Current.

The eastern boundary of the tropical Indian Ocean is different from the Atlantic and Pacific eastern boundaries in several respects. First, the mean temperature and salinity stratification is less developed than off Peru and West Africa, so when upwelling occurs it generally does not generate marked surface cooling. Secondly, the monsoon wind reversal is so complete along this coast that the annual mean alongshore wind stresses here are close to zero, from the northern Bay of Bengal south to Java and the upwelling in one season is counteracted by downwelling in the following season. Thirdly, twice a year (around May and November) the Equatorial Jet feeds warm water towards Sumatra, generating a pulse of current that flows poleward in both hemispheres. The seasonal cycle thus has strong semiannual as well as annual components, and is quite complicated. The strongest westward currents along the southern coast of Java, the seasonally reversing South Java Current, occur in August, when the monsoon winds are easterly and the Equatorial Jet is inactive. Surface cooling occurs off South Java at this time. This is also the time when the sea level difference from Java to Australia is largest, implying maximum strength in the Indonesian throughflow and suggesting that at least some of the water for the South Java Current is then supplied from the Pacific Ocean.

The dynamics of the eastern boundary current along the western Australian coast, known as the Leeuwin Current, are very unusual and require some explanation. In the Pacific and Atlantic Oceans equatorward winds along the eastern boundary produce coastal upwelling, an equatorward surface flow, and a poleward undercurrent. In the Indian Ocean, annual mean winds along Western Australia do blow towards the equator, but at the surface a vigorous poleward flow runs against the wind, and the undercurrent is equatorward. The reason is that eastern boundary currents are driven by the combined effects of alongshore winds and alongshore pressure gradients in the upper ocean.

5.4.2 Indian Ocean Climate Variability Modelling

We discuss in the following a group of ten papers, devoted to various aspects of the Indian Ocean modelling.

Marchant et al. (2006) reviewed the causes of the Indian Ocean Dipole (IOD), how it develops within the Indian Ocean, the relationships with ENSO and evaluated current research initiatives focused on characterizing and constraining the IOD. The IOD represents the zonal dipole structure of various coupled ocean-atmosphere parameters such as SST, surface pressure, ongoing longwave radiation and sea surface height anomalies (Saji et al., 1999; Yamagata et al., 2003, 2004a,b). The IOD occurs inter-annually in the tropical parts of the Indian Ocean. Typical of ocean-driven climate oscillations, the IOD experiences, a "positive" and a "negative" phase (Fig. 5.36). During a positive IOD event the SST drops in the south-eastern part of the Indian Ocean: off the northern coast of Australian, the eastern coast of Japan and Indonesia, counteracted by SST rises in the western equatorial Indian Ocean, off the

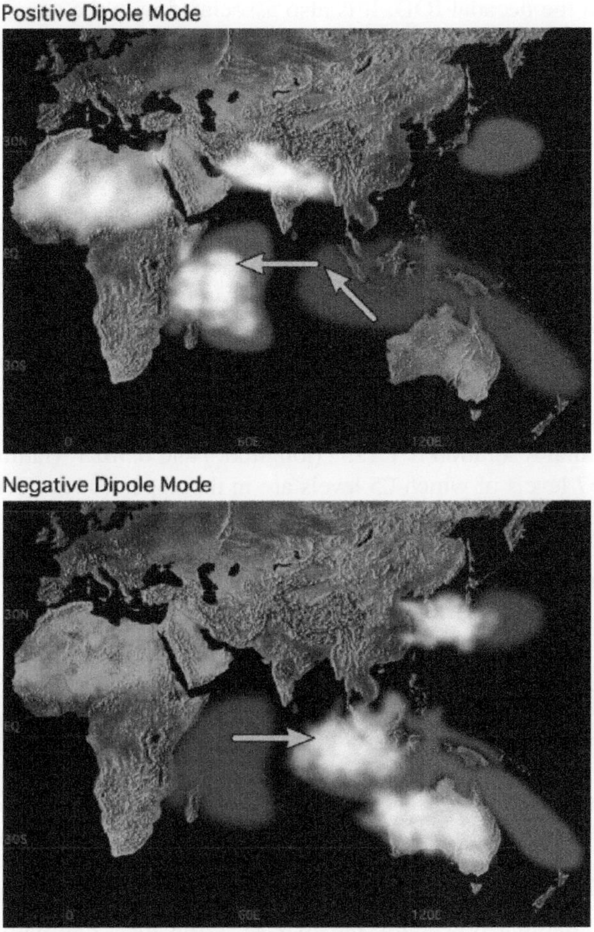

Fig. 5.36 Schematic diagram of sea surface temperature anomalies (*red shading* denoting warming; *blue* cooling) during a positive Indian Ocean Dipole (IOD) event (*top*). White patches indicate increased convective activity with arrows indicating wind direction. The negative IOD (*bottom*) which is, in effect, the reversal of the positive IOD – complete with increased convective activity over Australia, Indonesia and Japan (Saji et al., 1999) (See also Plate 17 in the Color Plate Section on page 355)

eastern coast of Africa, from the northern edge of Somalia. Inverse conditions exist during a negative IOD event.

Using the Simple Ocean Data Assimilation (SODA), NCEP/NCAR reanalysis and the GISST datasets from 1950–1999, and an atmosphere-ocean coupled general circulation model, Ashok et al. (2004) explored the possible existence of decadal (IOD) variability for the first time. They found that there are strong decadal IOD events, and that the time series of the decadal IOD and decadal ENSO indices are not well correlated. The simulated decadal signal of the IOD index is highly correlated with the 20 °C isotherm depth anomaly, indicating that ocean dynamics is involved in the decadal IOD. It is also associated with the zonal wind anomaly. They suggest that the decadal IOD in the tropics is interpreted as decadal modulation of the interannual IOD events.

The Interannual Indian Ocean Subtropical dipole (IOSD) event in the subtropical southern Indian Ocean was studied by Suzuki et al. (2004) using a coupled general circulation model (CGCM) to derive a scenario describing its complete life cycle for the first time.

The ocean component (OGCM) of the CGCM is based on the Modular Ocean Model 2.2 (Pacanowski, 1996), which was developed at the Geophysical Fluid Dynamics Laboratory/National Oceanic and Atmospheric Administration (GFDL/NOAA). This OGCM resolves the global oceans, except for the region north of 70 °N, the North Sea, the Mediterranean Sea, the Black Sea, and the Caspian Sea. Because it does not include a sea-ice model, temperature and salinity at the northern and southern boundaries in the CGCM are bound to the climatological values. The model horizontal resolution is 1.125° (longitude) and 0.5625° (latitude). In the vertical, it has 37 levels of which 25 levels are in the upper 400 m. Horizontal mixing is parameterized using a biharmonic scheme, in which the coefficients of horizontal eddy viscosity and diffusivity are fixed at $5 \times 10^{20}\,\text{cm}^4/\text{s}$. The vertical mixing parameterization depends on the Richardson number (Pacanowski and Philander, 1981).

Before coupling the atmospheric component (AGCM) with the OGCM, the individual parts are run in stand-alone mode to obtain appropriate initial conditions. The AGCM is integrated for nine months from 1 April 1988 to 1 January 1989 using the climatological SST as the boundary condition. The OGCM is spun up for 10 years from a state of rest using surface forcing derived from the climatological monthly mean wind stress and the surface temperature and salinity. The annual mean temperature and salinity fields are used at all levels as the initial condition at the beginning of the spin-up.

After obtaining the required initial conditions, the AGCM and OGCM are coupled. In the coupling procedure, the AGCM is run for a day with a time step of 300 s, while the daily averaged wind stress, surface heat flux, and precipitation minus evaporation values are used to force the OGCM. The OGCM is run for a day with a time step of 900 s. Daily averaged SST values are computed to force the AGCM for the next day.

The CGCM is integrated for 59 years. Data sets excluding the first 10 years are used in the analysis.

The positive (negative) IOD is characterized by an anomalous warm (cool) sea surface temperature (SST) in the southwestern region of the Indian Ocean and by an anomalous cool (warm) SST in the southeastern region.

The positive event brings about enhanced precipitation in the southeastern Africa during the peak phase. Composite pictures for the positive and negative IOD clarify that an anomalous latent heat flux is the dominant factor in its formation. This flux anomaly is caused by an anomaly in the climatological wind field, which is associated with a pressure anomaly in the central region of the southern Indian Ocean. Since the flux anomaly starts during austral fall in the year previous to the event peak and develops for the next 9 months, air-sea interaction must play an active role in the formation of the IOD. The reason the peak of the IOD is locked to the austral summer is that the latent heat flux influences the sea surface temperature most efficiently in the austral summer when the depth of the surface mixed layer is shallowest. Vinayachandran et al. (2005) studied bifurcation of the East Indian Coastal Current (EICC) by using OGCM MOM3 developed at GFDL.

The model domain is global and has high horizontal and vertical resolutions with a grid spacing of $1/10°$ in both longitude and latitude. There are 54 levels in the vertical with 13 levels in the upper 100m. The model is forced for 50 years using climatological winds, heat and freshwater fluxes from NCEP/NCAR reanalysis (Kalnay, 1996).

The model salinity, in addition, is restored to climatological monthly mean values with a time scale of 6 days. Results from the 50th year of the model run for the Bay of Bengal was used for the study.

The East India Coastal Current (EICC) flows equatorward during October–December carrying low salinity water from the Bay of Bengal en route. Using results from a high resolution ocean general circulation model, satellite altimeter data, Argo float profiles and ocean color images the authors show that the EICC bifurcates east of Sri Lanka. One part continues along the coast of Sri Lanka but the major part of the EICC, called by authors as the East Sri Lanka Jet (ESLJ) flows eastward into the Bay of Bengal. As a result of this bifurcation, there is offshore transport of chlorophyll a rich low salinity water from the coast of Sri Lanka. Altimeter data from 1993 to 2004 show that the bifurcation occurred every year except during the Indian Ocean Dipole (IOD) years of 1994 and 1997. The bifurcation occurs when an anticyclonic eddy that propagates westward ahead of a downwelling Rossby wave front impinges on the Sri Lanka coast. This new finding suggests that the main route of the low salinity water from the Bay of Bengal into the southeastern Arabian Sea may not be along the coast around Sri Lanka but through the Winter Monsoon Current. A series of seminal papers on the Indian ocean climate variation modelling were published during 2006 and 2007 by the "Journal of Climate". The twentieth-century simulations using by 17 coupled ocean-atmosphere general circulation models (CGCMs) submitted to the Intergovernmental Panel on Climate Change's Fourth Assessment Report (IPCC AR4) were evaluated by Saji et al. (2006) for their skill in reproducing the observed modes of Indian Ocean (IO) climate variability. Most models successfully capture the IO's delayed, basin wide warming response a few months after El Nino-Southern Oscillation (ENSO) peaks in the Pacific, but ENSO's oceanic teleconnection into the IO, by coastal waves through the Indonesian archipelago, is poorly simulated in these models, with significant shifts in the turning latitude of radiating Rossby waves. In observations, ENSO forces, by the atmospheric bridge mechanism, strong ocean Rossby waves that induce anomalies of SST, atmospheric convection, and tropical cyclones in a thermocline dome over the southwestern tropical IO. While the southwestern IO thermocline dome is simulated in nearly all of the models, this ocean Rossby wave response to ENSO is present only in a few of the models examined, suggesting difficulties in simulating ENSO's teleconnection in surface wind.

A majority of the models display an equatorial zonal mode of the Bjerknes feedback with spatial structures and seasonally similar to the Indian Ocean dipole (IOD) in observations. This success appears to be due to their skills in simulating the mean state of the equatorial IO. Corroborating the role of the Bjerknes feedback in the IOD, the thermocline depth, SST, precipitation, and zonal wind are mutually positively correlated in these models, as in observations. The IOD-ENSO correlation during boreal fall ranges from −0.43 to 0.74 in the different models, suggesting that ENSO is one, but not the only, trigger for the IOD.

In summary, CGCMs are only partially successful in simulating the rich spectrum of SST variability in the Indian Ocean. At large, most models reproduce the SST variations that arise from local air-sea interaction (IOD), but many of them fail to adequately simulate the responses to ENSO. This deficiency, as well as the divergence of the models result is demonstrated in the Fig. 5.37. The implications of this deficiency are significant, both for the prediction of present as well as future climate. ENSO is an important forcing of Indian Ocean variability, accounting for a significant fraction of SST and rainfall variability in the Indian Ocean region. It is also one of the important triggers for the IOD. In a changing climate, ENSO may alter its characteristics (Timmermann et al., 1999) and hence its impact on Indian Ocean climate through altered tele-connection processes. Thus, it is of prime importance for CGCMs to simulate SST variability and its climate impacts realistically in both the Pacific and Indian Oceans, as these models are used to project climate changes and their regional distributions. The possibility that Indian Ocean SST variations exert influences on ENSO and beyond, as suggested by recent studies (Saji and Yamagata, 2003a,b; Giannini et al., 2003; Annamalai et al., 2005), adds more importance in their simulation in climate models.

An known as the Scale Interaction Experiment Frontier version 1 (SINTEX-Fl) model was used by Behera et al. (2006a,b) to understand the intrinsic variability of the Indian Ocean dipole (IOD).

The SINTEX-F1 adapted to run on the Earth Simulator is an upgraded version of the SINTEX model described in Gualdi et al. (2003). The atmosphere model has a spectral representation which is roughly equivalent to a horizontal grid mesh of $1° \times 1°$. The ocean model adopts the Arakawa C grid with a finite mesh of $2° \times 0.5°$ cosine (latitude); the meridional grid resolution increases toward the equator with a grid length of 0.5° in the equatorial region. The SINTEX-Fl has shown remarkable skill in reproducing variabilities in the Indian and Pacific Oceans (Yamagata et al., 2004b). The monthly data from the last 200 year of the total 220 year model results are used in the analysis.

In addition to the 220 ∼ yr-long control experiment the authors performed an additional model experiment to suppress ENSO variability by decoupling the tropical Pacific. In the decoupling procedure climatological SSTs between 25 °N and 25 °S, instead of evolving interannual SSTs, derived from the last 200 yr result of the control experiment are supplied to the atmospheric model as the lower boundary condition. This effectively suppresses the ENSO evolution in the coupled model while allowing the independent evolution of IOD. The 70 yr decoupled experiment is named the noENSO experiment.

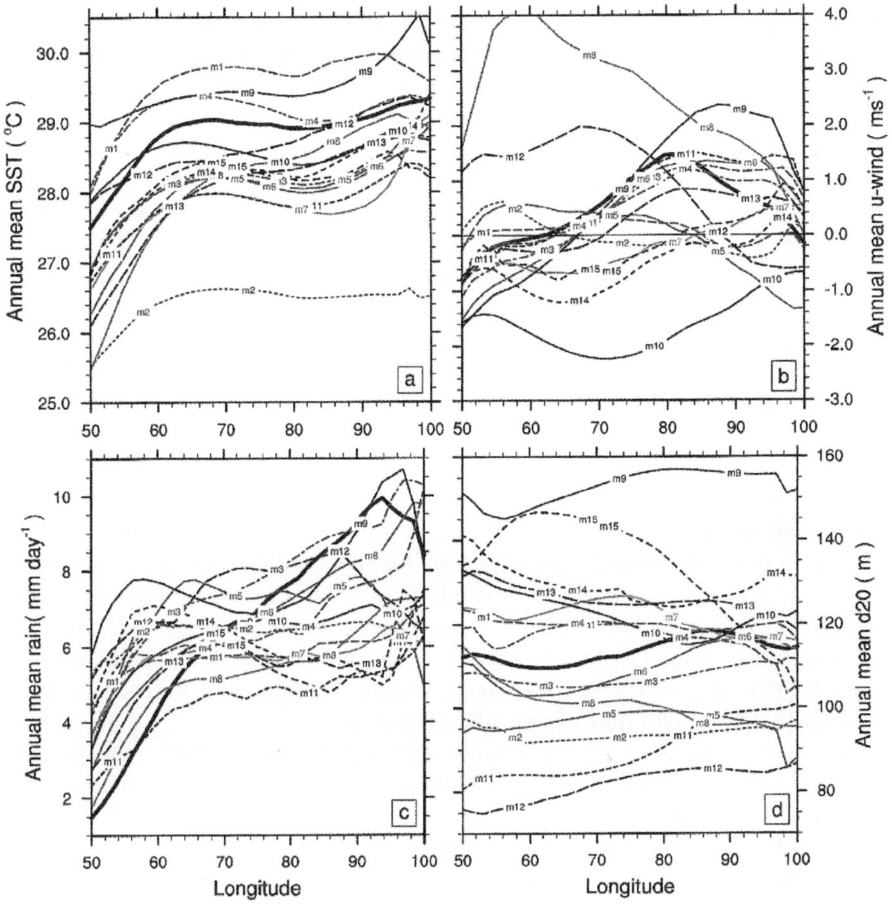

Fig. 5.37 Annual-mean equatorial climatologies for (**a**) SST, (**b**) zonal wind, (**c**) rain, and (**d**) D_{20} in climate simulations and observations (*black curve*). For the observational curves, SODA is used for SST and D_{20}, CMAP for rain, and ERA40 for zonal winds. Model curves are colored in rainbow colors from red to violet in descending order of NinÕ-3 index variance (Saji et al., 2006) (See also Plate 18 in the Color Plate Section on page 356)

In addition to a globally coupled control experiment, a Pacific decoupled noENSO experiment has been conducted. In the latter, the El Nino-Southern Oscillation (ENSO) variability is suppressed by decoupling the tropical Pacific Ocean from the atmosphere. The ocean-atmosphere conditions related to the IOD are realistically simulated by both experiments including the characteristic east-west dipole in SST anomalies. This demonstrates that the dipole mode in the Indian Ocean is mainly determined by intrinsic processes within the basin. In the EOF analysis of SST anomalies from the noENSO experiment, the IOD takes the dominant seat instead of the basinwide monopole mode. Even the coupled feedback among anomalies of upper-ocean heat content, SST, wind, and Walker circulation over the Indian Ocean is reproduced.

As in the observation, IOD peaks in boreal fall for both model experiments. In the absence of ENSO variability the interannual IOD variability is dominantly biennial. The ENSO variability is found to affect the periodicity, strength, and formation processes of the IOD in years of co-occurrences. The amplitudes of SST anomalies in the western pole of co-occurring IODs are aided by dynamical and thermodynamical modifications related to the ENSO-induced wind variability. Anomalous latent heat flux and vertical heat convergence associated with the modified Walker circulation contribute to the alteration of western anomalies. It is found that 42% of IOD events affected by changes in the Walker circulation are related to the tropical Pacific variabilities including ENSO. The formation is delayed until boreal summer for those IODs, which otherwise form in boreal spring as in the noENSO experiment.

In general, seasonal characteristic of SST, thermocline, wind, and rainfall in the Indian Ocean (Hastenrath et al., 1993) are captured well by the SINTEX-F1 simulation (see Behera et al., 2006a,b). Similar seasonal variabilities also are found in the noENSO experiment results. The southeasterlies in boreal summer cross the equator and turn into southwesterlies on the western Indian Ocean. As a result of these winds, the thermocline is shallow in the west compared to that in the east where the warm pool is prominent. Nevertheless, the thermocline rises in the east during boreal fall when the seasonal southeasterly trade winds cross the equator in the east, which is prominent in the non-ENSO experiment results. These seasonal conditions of the eastern Indian Ocean introduce the seasonal phase-locking characteristic of the IOD; the equatorial westerlies reverse to easterlies along with the rise in seasonal thermocline in the east during a positive IOD event.

Much more informative is the Journal of Climate V20, N13, 2007 – a big volume, which is twice as much as our book and exclusively devoted to the Indian Ocean climate variability examining. Owing to the restricted volume of this book we briefly describe here only three out of dozens of them.

The Indian Ocean zonal dipole as mentioned above as mentioned above is a mode of variability in sea surface temperature that seriously connected the climate of many nations around the Indian Ocean rim, as well as the global climate system. It has been the subject of increasing research, and sometimes of scientific debate concerning its existence/nonexistence and dependence/independence on/from the El Nino-Southern Oscillation. Much of the debate occurred because scientists did not agree on what years are the El Nino or La Nina years, not to mention the newly defined years of the positive or negative dipole. A method that identifies when the positive or negative extrema of the El Nino-Southern Oscillation and Indian Ocean dipole occur is proposed in a paper of Meyers et al. (2007), and this method is used to classify each year from 1876 to 1999. The method is statistical in nature, but has a strong basis on the oceanic physical mechanisms that control the variability of the near-equatorial Indo-Pacific basin.

An index of he upwelling in the SST anomaly in the area spanning the equator to $10^\circ S - 0^\circ$, 90°–$110^\circ E$, called the IODE, which is the same area used for the eastern pole in the seminal paper by Saji et al. (1999).

A purely statistical approach without considering the physical basis is not likely to work for the complex relationship between ENSO and IOD. Upwelling is a critical, physical aspect of both phenomena, as discussed above, and is selected here as a basis for the classification method. Upwelling in the Pacific cold tongue is represented by the Nino-3 region and upwelling off Java-Sumatra is represented by the IODE. They are the key time series used in the method. However, ENSO and IOD are basin-scale phenomena. Typically, the western Pacific cools when the Nino-3 area warms during El-Nino, and the western Indian Ocean warms when the IODE cools during a positive episode of IOD. Two additional time series of area-averaged SST anomalies are used here to represent the basin-scale variations: NinoWP for the western Pacific and IODW for the western Indian Ocean.

Early in the study it was found that some years could not be clearly classified due to strong decadal variation; these years also must be recognized, along with the reason for their ambiguity. The sensitivity of the classification of years is tested by calculating composite maps of the Indo-Pacific sea surface temperature anomaly and the probability of below median Australian rainfall for different categories of the El Nino-Indian Ocean relationship.

The 3D pathways of the Indonesian Throughflow (ITF) in the Indian Ocean are identified by Valsäla and Ikeda (2007) using an OGCM, with a combined set of tools: (1) Lagrangian particle trajectories, (2) passive tracers, and (3) active tracers (temperature and salinity). The model used in that study is the Australian Community Ocean Model version 2.0 (ACOM). It is a primitive equation z-coordinate ocean model cast in the B-grid. ACOM adopts a hybrid vertical mixing scheme by Chen and Rothstein (1994) and the Richardson number-dependent mixing (Price et al., 1986). The hybrid scheme improves the simulation of the mixed layer deepening in the strong vertical shear region and in the high latitude due to wind stirring and convective overturning. In the southern latitudes (south of 40 °S) the effects of the ITF appear at the surface by following the convective overturning and wintertime cooling (Wajsowicz, 2002).

The model domain contains both the Indian Ocean and the Pacific Ocean from 40 °N–65 °S, 0°–290 °E with a constant resolution of $1° \times 0.5°$ (longitude by latitude) and with 25 vertical levels. The topography is selected from ETOPO5 to keep the straits reasonably realistic. Closed circulation around Australia is permitted. The model is spun up for 20 yr (sufficient for upper 500 m evolution) forced with monthly wind stresses derived from Hellerman and Rosenstein (1983).

The Lagrangian particles, without horizontal and vertical mixing, suggest that at the entrance region the surface ITF subducts along the northwestern coast of Australia and then travels across the Indian Ocean along the thermocline depths. The subsurface ITF more directly departs westward and crosses the Indian Ocean. Using the passive tracers, which are mixed vertically under convection as well as horizontally due to diffusion, the ITF is shown to undergo vigorous mixing as soon as it enters the Indian Ocean and modifies its upper T-S characteristics. Thus, the surface and subsurface ITF watermasses lose their identities.

Upon reaching the western boundary, the ITF reroutes into three distinct depth ranges: route 1 – across the Indian Ocean just to the south of the equator (200–300 m); route 2 – across the Indian Ocean to the north of the equator (100–200 m); and route 3 – upwells in the Somali region and spreads all over the surface of the

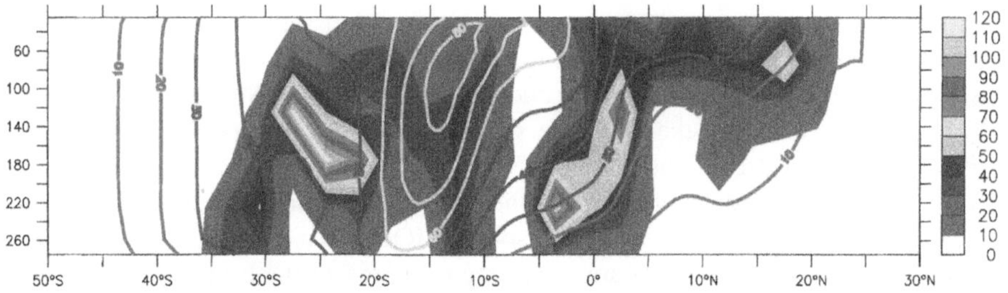

Fig. 5.38 The meridional distribution of particles and tracers along 60 °E. Shaded values are the number of particles counted in $2° \times 2° \times 50$ m bins. Contour lines represent the tracer concentrations. The equatorward slope in the existence of ITF in the northern Indian Ocean is consistently evident from both the particles and tracers (Valsäla and Ikeda, 2007) (See also Plate 19 in the Color Plate Section on page 357)

northern Indian Ocean. The scheme of 3D pathways of the ITF in the Indian Ocean is demonstrated in Fig. 5.38. The seasonality of the Somali Current is crucial to spread the ITF along route 3 during the summer monsoon (April–October) and route 2 during the winter monsoon (November–March). The basinwide spreading is responsible for a long residence time of the ITF in the Indian Ocean to be at least 20 yrs.

The effects of the ITF on the temperature and salinity are mainly accompanied with the major pathways. However, indirect effects are visible in a few spots; that is, the warm and saline feature is produced in the subsurface off the southwestern coast of Australia around 30 °S caused by the eastward surface current. This component also enhances vertical convection and warms the surface around 40 °S. The Arabian Sea high salinity water is produced extensively with the effects of the Somali upwelling, which is originally strengthened by the fresh and warm ITF.

A near-global ocean model with resolution enhanced in the southern Indian Ocean has been spin up by Murray and Reason (2001a,b) to seasonal equilibrium and then driven by NCEP-NCAR reanalysis 1 monthly mean forcings and Hadley SSTs over the period 1948–2002. The aim was to simulate changes in the subsurface properties observed in hydrographic surveys at 32 °S in the Indian Ocean in 1965, 1987, and 2002. These surveys showed a zonally averaged cooling on isopycnals corresponding to 0.5° and 0.3 °C in mode and intermediate waters between 1965 and 1987 and a warming of the mode water coupled with a continued cooling of the intermediate water between 1987 and 2002.

The model is an orthogonal curvilinear version of the GFDL model MOM. It includes a Kraus-Turner mixed layer scheme derived from that used by Chen and Rothstein (1994) and isopycnal mixing and eddy-induced transport, both of the latter using Cox (1987) numerics.

The model was spun up to a cyclic equilibrium using National Centers for Environmental Prediction-National Center for Atmospheric Research (NCEP-NCAR) Reanalysis 1 (NCEP 1) climatological monthly heat and freshwater fluxes and wind stresses (averaged for each month of the year over the period 1948–2000 inclusive) and a 30-day relaxation to climatological monthly SSTs and SSSs. World Ocean Atlas 2001 (Conkright et al., 2002) climatological monthly 10 m fields were used for the restoration of salinity but not for temperature, which, for reasons of consistency, was restored to 1948–2002 climatologically averaged SSTs from the Hadley Ice and SST (HadlSST) dataset (Rayner et al., 2003). From the equilibrium state, a number of transient simulations were run on using NCEP monthly mean forcings for the period 1948–2002.

Accelerated convergence with a tracer time step of 1 day was used during the earlier part of the equilibrium integration (Bryan, 1984). Time step inequalities were then progressively graded to equal time steps of 1.5 hours at 600 yr, after which the model was run for a further 40 yr. The solution at 640 yr was taken as the initial condition as at 1 January 1948 for the 55 yr transient synchronous simulation. During the final 40 yr preceding the transient simulation, temperature tendencies averaged over zonal regions for each level and for a particular month of the year were mostly about $5 \times 10 - 5\,^{\circ}\mathrm{C\ yr}^{-1}$, or 0.003 °C over 55 yr. This amount of drift is negligible in comparison with the interannual changes of 0.2° − 1 °C, which occurred in waters down to 1500 m during the transient simulations.

The major changes in isopycnal depth and temperature modeled in this study were confined to the mode water and were qualitatively similar to those observed but concentrated in a lower density class and in the eastern half of the section. The dominant changes here were multidecadal, with maximum temperatures on the $\sigma_\theta = 26.7\,\mathrm{kg\ m}^{-3}$ isopycnal being reached in 1968 and minimum temperatures in 1990. The simulations showed a propagation of interannual anomalies toward the section from a region of deep late winter mixed layers in the southeast Indian Ocean within a period of several years. Surface temperatures in this region were lowest in the 1960s and highest in the late 1980s. Temperatures on isopycnals showed the opposite variation, consistent with SST having the controlling effect on mixed layer density and depth. Isopycnal depths within the mode water were strongly correlated with temperature, implying a redistribution of mode water density classes, the greatest volume of mode water being produced in a higher density class ($\sigma_\theta =$ 26.8–27.0 kg m^{-3}) during the period of cooler surface forcing in the 1960s and 1970s than during the warmer period following ($\sigma_\theta = 26.6$–$26.8\,\mathrm{kg\ m}^{-3}$).

In spite of some biases resulting from errors in boundary conditions, finite resolution, and model physics, the model simulations have reproduced the important features present in the observed sections and in the 1965–87 and 1987–2002 section differences. In both the model and observations, the largest subsurface changes were in the SAMW (Sub Antarctic Mode Water) which showed a cooling and deepening of isopycnals during the first period and a warming and shallowing during the second. In the model, the changes in this water mass were unrealistically concentrated on the eastern side of the section and became peaked in a lighter density class than observed on the western side; however, the patterns of the changes were the same in the two periods (except for the change in its sign after 1987), as also was the case in the observations.

The model allows us to relate these changes to the subsurface variations of which they formed a part and to the surface variations forcing them. Because of the need for data reconstruction, limited reliance can be placed on year-to-year variations in SST datasets in the South Indian Ocean; however, we can have more faith in the longer-period variations. During the period 1948–2003, observed and modeled SSTs included a multidecadal oscillation with a period of 50–60 yr and interannual variations with periods of less than 10 yr.

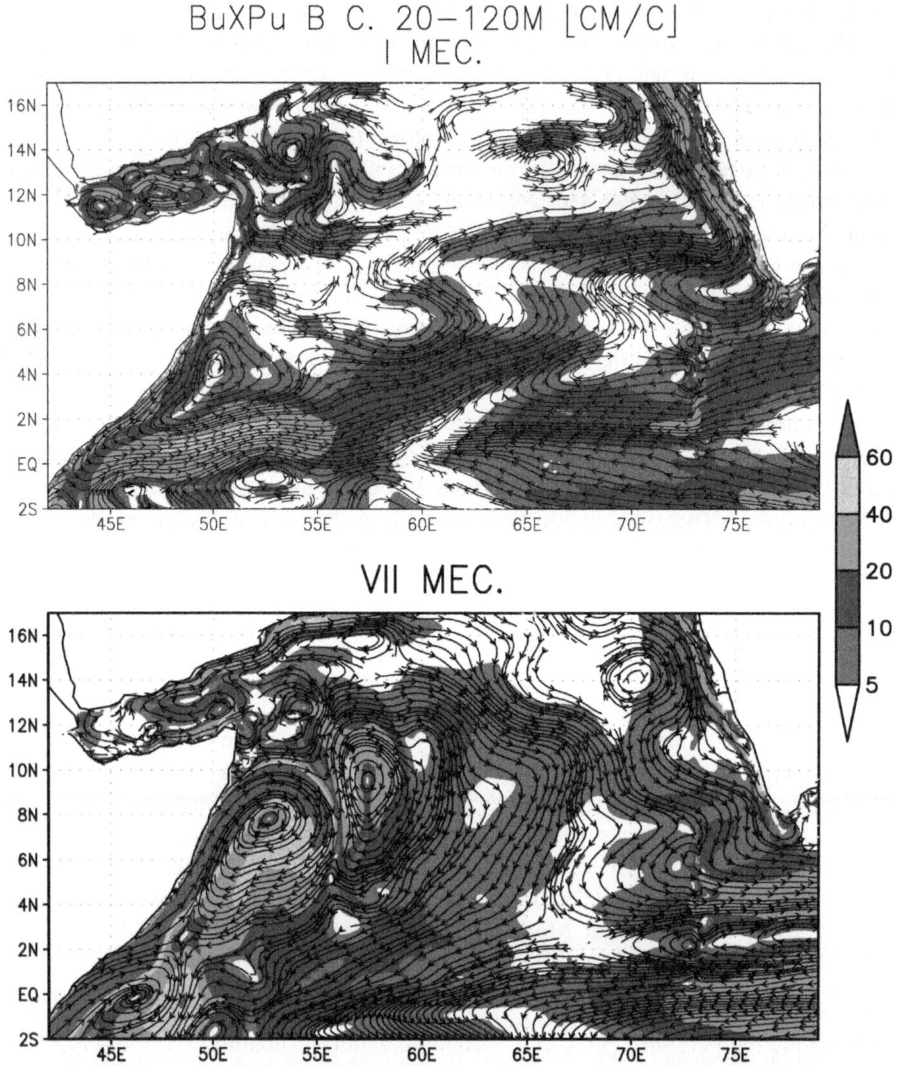

Fig. 5.39 Mean currents for (*top*) January and (*bottom*) July in the surface 10 m layer of the Indian Ocean. The streamlines of the currents are shown. The velocities (cm/s) are shown with gradations of *gray color*. The *color scale* is shown on the right (Diansky et al., 2006) (See also Plate 20 in the Color Plate Section on page 358)

One more interesting paper was published recently on the monsoon circulation in the Indian Ocean simulated with a σ-coordinate ocean model developed at the Institute of Numerical Mathematics of the Russian Academy of sciences (Diansky et al., 2006). The model has a horizontal resolution of $(1/8)^\circ \times (1/12)^\circ$ and contains 21σ – layers of uneven thickness. Realistic bottom topography and land geometry are used. The numerical experiments were carried out for 15 years starting from the Levitus climatology for January and monthly mean climatic atmospheric forcing from the NCEP reanalysis data. The annual cycle of the surface and subsurface currents and temperature and salinity fields were analyzed. The model reproduces well the Summer Monsoon and the Winter Monsoon currents and their time evolution and spatial structures. The Somali Current is adequately modeled. During the Summer Monsoon, the velocities of the current exceed 2 m/s, while the total mass transport is approximately 70 Sv. The model results show that a reversal of the Somali Current from the northern direction in the summer to the southern direction in the winter is accompanied by the generation of anticyclonic eddies, which drift westward owing to the β-effect and dissipate either near the Somali shore or in the Gulf of Aden. The monsoon variability of the equatorial surface current and equatorial subsurface countercurrent system are analyzed. It is shown that these currents are generated predominantly by the zonal component of wind stress, in which the half-year harmonic dominates. This leads to the fact that the equatorial surface current also changes its direction with a half-year periodicity almost in phase with the wind. The oppositely directed subsurface compensational countercurrent of Bengal due to the riverine runoff, make an important contribution to the circulation. This effect manifests itself especially strongly in the summer during the peak of the Ganges River runoff, which transports fresh turbid waters. The principal features of the large-scale quasi-stationary gyre structure of the Indian Ocean such as the Great Whirl, Socotra high, and Laccadive high and low are simulated. One of the results is presented on the Fig. 5.39.

References

Annamalai, H., S.-P. Xie, J.-P. McCreary, and R. Murtugudde, 2005. Impact of Indian Ocean sea surface temperature on developing El Niño. *J. Climate* 18: 302–319.

Antipov S. B., N. A. Dianskii, and A. B. Gusev, 2006. The peculiarity the radioactive contaminated waters spreading in the northwestern part of the Pacific Ocean. *Izv. Ross. Akad. Nauk. Energetics* 6: 52–70.

Ashok K., W.-L. Chan, T. Motoi, and T. Yamagata, 2004. *Decadal Variability of the Indian Ocean Dipole*. 31: L24207, doi:10.1029/2004GL021345.

Behera, S. K., J. J. Luo, S. Mason, S. A. Rao, and H. Sakuma, T. Yamagata, 2006a. A CGCM Study on the Interaction between IOD and ENSO. *J. Climate* 19: 1688–1705.

Behera, S. K., J.-J. Luo, S. Masson, P. Delecluse, S. Gualdi, A. Navarra, and T. Yamagata, 2006b. Paramount impact of the Indian Ocean dipole on the East African short rains: A CGCM study. *J. Climate* 18: 4514–4530.

Bleck R., S. Dean, M. O'Keefe and A. A. Sawday, 1995. Comparison of data-parallel and message-passing version of the Miami Isopycnic Coodinate Ocean Model (MICOM). *Parallel Comput.* 21: 1695–1720.

Blumberg, A. F., and G. L. Mellor, 1987. A description of a three-dimensional coastal ocean circulation model. In: Heaps N. (ed.), *Three-Dimensional Coastal Ocean Models*. Coastal and Estuarine Series, 4, Amer. Geophys. Union, 1–16.

Broecker W. S., 1991. The great ocean conveyer. *Oceanography* 4(2): 79–89.

Bryan, K., 1984. Accelerating the convergence to equilibrium of ocean-climate models. *J. Phys. Oceanogr.* 14: 666–673.

Bryden D., S. San, and R. Bleck, 1999. A new approximation of the equation of state for seawater, suitable for numerical ocean models. *J. Geoph. Res.* 104(C1): 1537–1540.

Chassignet E. P. and Z. D. Garaffo, 2001. From stirring to mixing in a Stratified Ocean. In: Muller P. and Henderson D. (eds.), *Proceeding of "Aha Hulico" a Hawaiian Winter Workshop*. University of Hawaii, 37–41.

Chassignet E. P., Z. D. Garrafo, R. D. Smith, and H. E. Hurlburt, 2001. High resolution Gulf Stream Modelling. http://citeseerx.ist.psu.edu/viewdoc/doi=10.1.1.38.1574

Chen, D., and L. M. Rothstein, 1994. A hybrid vertical mixing scheme and its application to tropical ocean models. *J. Phys. Oceanogr.* 24: 2156–2179.

Conkright, M. E., R. A. Locarnini, H. E. Garcia, T. D. O'Brien, T. P. Boyer, C. Stephens, and J. I. Antonov, 2002. *World Ocean Atlas 2001*. Objective analyses, data statistics, and figures: CD-ROM documentation. National Oceanographic Data Center, Silver Spring, MD, 17pp.

Cox, M. D., 1987. Isopycnal diffusion in a z-coordinate ocean model. *Ocean Model.* 74: 1–5.

Cutler, A. N. and J. C. Swallow, 1984. *Surface Currents of the Indian Ocean*. Institute of Oceanographic Sciences Report 187, 8pp. & 36 charts.

Demin Yu. L. and R. A. Ibrayev, 1992. Model of ocean dynamics. In: Sarkisyan A. S. and Demin Yu. L. (eds.), *Numerical Models and Results of Calibration Calculatiuons of Currents in the Atlantic Ocean*. IVM RAN, Moscow, (in Russian), 42–95.

Diansky N. A., A. V. Bagno, and V. B. Zalesny, 2002. Sigma model of global ocean circulation and its sensitivity to variations in wind stress. *Izv. Acad. Nauk USSR, Fiz. Atmos. Okeana*. 38: 537–556.

Diansky N. A., V. B. Zalesny, S. N. Moshonkin, and A. S. Rusakov, 2006. High resolution simulation of the Indian Ocean Monsoon Current Variability. *Oceanology* 46(5): 650–671.

Dmitriev, N. E., A. Y. Proshutinsky, T. B. Loyning, and T. Vinjl, 1991. Tidal ice dynamics in the area of Svalbard and Frans Josef Land. *Polar Res.* 9: 193–205.

Flato, G. M. and W. D. Hibler, 1995. Ridging and strength in modeling the thickness distribution of arctic sea ice. *Geophys. Res.* 100: 18 611–18 626.

Fletcher J. O., 1970. Polar ice and the global climate machine. *Bull. Atomic Scientists* 40–47.

Fukumori, I., T. Lee, B. Cheng, and D. Menemenlis, 2004. The origin, pathway, and destination of Niño-3 water estimated by a simulated passive tracer and its adjoint. *J. Phys. Oceanogr.* 34: 582–604.

Giannini, A., R. Saravanan, and P. Chang, 2003. Oceanic forcing of Sahel rainfall on interannual to interdecadal time scales. *Science* 302: 1027–1030.

Gill A. E. and K. Bryan, 1971. Effects of geometry on the circulation of a three-dimensional southern-hemisphere ocean model. *Deep-sea Res.* 18: 685–721.

Golubeva E. N. and G. A. Platov, 2007. On improving the simulation of Atlantic Water circulation in the Arctic Ocean. *J. Geophys. Res.* 112: C04S05, doi: 10.1029/2006JC003734.

Gualdi, S., E. Guilyardi, A. Navarra, S. Masina, and P. Delecluse, 2003. The interannual variability in the tropical Indian Ocean as simulated by a CGCM. *Climate Dyn.* 20: 567–582.

Guo X., H. Hukuda, Y. Miyazawa, and T. Yamagata, 2003. A Triply nested Ocean Model for Simulating the Kuroshio-Roles of Horizontal resolution on JEBAR. *J. Phys. Oceanogr.* 33: 146–169.

Guo X., H. Hukuda, Y. Miyazawa, and T. Yamagata, 2006. The Kuroshio Onshore Intrusion along the shelf break of the East China Sea: The origin of the Tsushima warm current. *J. Phys. Oceanogr.* 36: 2205–2231.

Hakkinen S. and G. L. Mellor, 1992. Modelling the seasonal variability of the coupled ice-ocean system. *J. Geophys. Res.* 97(20): 285–304.

Harrison D. E. and M. Carson, 2007. Is the Word Ocean Warming? Upper-Ocean Temperature Trends: 1950–2000. *J. Phys. Oceanogr.* 37: 174–187.

Hastenrath, S., A. Nicklis, and L. Greischar, 1993. Atmospheric-hydrospheric mechanisms of climate anamalies in the western equatorial Indian Ocean. *J. Geophys. Res.* 98(C11): 20219–20235.

Hellerman, S. and M. J. Rosenstein, 1983. Normal monthly windstress over the world ocean with error estimates. *J. Phys. Oceanogr.* 13: 1093–1104.

Hibler W. D. and K. Bryan, 1987. A diagnostic ice-ocean model. *J. Phys. Oceanogr.* 17: 987–1015.

Hibler W. D., 1979. A dynamic thermodynamic sea ice model. *J. Phys. Oceanog.* 9: 815–846.

Hilmer, M. and P. Lemke, 2000. On the decrease of Arctic sea ice volume. *Geophys. Res. Lett.* 27: 3751–3754.

Hopkins, M. A., 1996. On the mesoscale interaction of lead ice and floes. *Geophys. Res.* 101: 18 315–18 326.

Holloway. G., 1992. Representing topographic stress for large-scale ocean models. *J. Phys. Oceanogr.* 22: 1033–1046.

Holloway. G. and A. Proshutinsky, 2007. Role of tides in Arctic ocean/ice model. *J. Geophys. Res.* 112: C04S06.doi: 10.1029/2006JC003643.

Hosoda S., S.-P. Xie, K. Takeuchi, and M. Nonaka, 2004. Interdecadal temperature variations in the North Pacific central mode water simulated by an OGCM. *J. Oceanogr.* 60: 865–877.

Hurlburt H. E. and P. J. Hogan, 2000. Impact of $1/8°$ to $1/64°$ resolution on Gulf Stream model-data comparisons in basin-scale Subtropical Atlantic Ocean Models. *Dyn. Atmos. Oceans.* 32: 283–330.

Iakovlev N. G., 1998. Numerical model of the general circulation of the Arctic ocean. A new version and preliminary calculation results. *Russ. J. Numer. Anal. Math. Modell.* 13(6): 465–478.

IPCC, 2001. *Climate Change: The Scientific Basis*. Contribution of Working Group 1 to the Third Assessment report of the IPCC, Cambridge University Press, 994pp.

Ivanov V. V., 1976. Water balance and water resources of the Arctic region. *Proc. AARI.* 323: 4–24 (in Russian).

Ivanov V. V., 2001. Atlantic waters in the western Arctic. In: Lisitsin A. P., Vinogradov M. E., and Romankevich E. A. (eds.), *Experience of System Oceanologic Investigations in the Arctic*, 77–91.

Kalnay, E., 1996. The NCEP/NCAR 40-year reanalysis project. *Bull. Am. Meteolorol. Soc.* 77: 437–471.

Karcher, M. J. and J. M. Oberhuber, 2002. Pathways and modification of the upper and intermediate waters of the Arctic Ocean. *J. Geophys. Res.* 107: C63049, doi: 10.1029/2002JC000530.

Kono T., M. Foreman, P. Chandler, and M. Kashiwai, 2004. Coastal Oyashio south of Hokkaido, Japan. *J. Phys. Oceanogr.* 34: 1477–1494.

Kuzin, V. I., E. N. Golubeva and G. A. Platov, 2006. Modelling of hydrophysical characteristics of the Arctic – North Atlantic system. In: *Fundamental Investigations of Oceans and Seas*, book 1, Moscow, Nauka, 166–188.

Levitus, S. and T. Boyer, 1994. *World Ocean Atlas 1994, Vol. 4: Temperature*. NOAA Atlas NESDIS 4, 117pp.

Levitus, S., R. Burgett, and T. Boyer, 1994. *World Ocean Atlas 1994, Vol. 3: Salinity*. NOAA Atlas NESDIS 3, 99pp.

Losch M. and P. Heimbach, 2007. Adjoint sensitivity of ocean General Circulation Model to bottom topography. *J. Phys. Oceanogr.* 37: 377–393.

Luo J. J. and T. Yamagata, 2003. A model study on 1988–89 warming event in the Northern Pacific. *J. Geophys. Res.* 33: 1815–1827.

Marchant R., C. Mumbi, S. Behera, and T. Yamagata, 2006. The Indian Ocean dipole – the unsung driver of climatic variability in East Africa. *J. Compilation*, Blackwell Publishing Ltd, Afr. J. Ecol., 45: 4–6.

Marchuk G. I., 1995. *Adjoint Equations and Analysis of Complex Systems*. Kluwer, Dordrecht-Boston-London, The Netherlands, 468pp.

Marchuk G. I., A. S. Rusakov, V. B. Zalesny, and N. A. Diansky, 2005. Splitting numerical technique with application to the high resolution simulation of the Indian Ocean circulation. *Pure Appl. Geophys.* 162: 1407–1429.

Marchuk G. I., J. Schröter and V. B. Zalesny, 2003. Numerical study of the global ocean equilibrium circulation. Russ. J. Numer. Anal. Math. Modelling., V 18, N 4, 307–335.

Marchuk G. I., J. Sündermann, and V. B. Zalesny, 2001. Mathematical modelling of marine and ocean currents. *Russ. J. Numer. Anal. Math. Modell.* 16(4): 279–362.

Marchuk G. I. V. B. and Zalesny, 1993. A numerical technique for geophysical data assimilation problems using Pontryagin's principle and splitting-up method. *Russ. J. Numer. Anal. Math. Modell.* 8(4): 311–326.

Martin, T., amd R. Gerdes, 2007. Sea ice drift variability in Arctic ocean model intercomparison project models and observations. *J. Geophys. Res.* 112: C04S10, doi: 10.1029/2006JC003617.

Maslowski W., D. Marble, W. Walczowski, U. Schauer, J. L. Clement, and A. J. Semtner, 2004. On the climatological mass, heat, and salt transports through the Barents Sea and Fram Strait from a pan-Arctic coupled ice-ocean model simulation. *J. Geophys. Res.* 109: C03032, doi: 10.1029/2001JC001039.

Meyers, G., P. McIntosh, L. Pigot, and M. Pook, 2007. The years of El Niño, la Niña and interactions with the tropical Indian Ocean. *J. Climate* 20: 2872–2880.

Mesinger F. and A. Arakawa, 1976. *Numerical Methods Used in Atmospheric Models.* Vol. I JOC, GARP Publication Series, 17. Geneva, World Meteorological Organization, C. P., 5, CH-1211, Geneva, 64pp.

Mitsudera H., B. Taguchi, Y. Yoshikawa, H. Nakamura, T. Waseda, and T. Qu, 2004. Numerical study on the Oyashio water pathways in the Kuroshio-Oyashio confluence. *J. Phys. Oceanogr.* 34: 1174–1195.

Miyazawa, Y., S. Yamane, X. Guo, and T. Yamagata, 2005. Ensemble forecast of the Kuroshio meandering. *J. Geophys. Res.* 110: C10026.

Murray R. J., 1996. Explicit generation of orthgonal grids for ocean models. *J. Comp. Phys.* 126: 251–273.

Murray, R. J., and C. J. C. Reason, 2001a. A curvilinear version of the Bryan-Cox-Semtner Ocean Model and its representation of the Arctic circulation. *J. Comput. Phys.* 171: 1–46.

Murray, R. J., and C. J. C. Reason, 2001b. A curvilinear ocean model using a grid regionally compressed in the South Indian Ocean. *J. Phys.Oceanogr.* 31: 2809–2823.

Niiler, P. P., N. A. Maximenko, G. G. Panteleev, T. Yamagata, and D. B. Olson, 2003. Near-surface dynamical structure of the Kuroshio Extension. *J. Geophys. Res.* 108(C6): 24-1-24-19.

Oshima, K. L., et al., 2004. Sverdrup balance and the cyclonic gyre in the sea of Okhotsk. *J. Phys. Oceanogr.* 24: 513–525.

Pacanowski, R., 1996. *MOM2 Documentation, User's Guide and Reference Manual.* GFDL Ocean Tech. Rep. 3.1, NOAA/Geophysical Fluid Dynamics Laboratory, 260pp.

Pacanowski, R. C., and S. G. H. Philander, 1981. Parameterization of vertical mixing in numerical models of tropical oceans. *J. Phys. Oceanogr.* 11: 1443–1451.

Paiva A. M., J. T. Hargrove, E. P. Chassignet, and R. Bleck, 1999. Turbulent behavior of a fine mesh (1/12 degree) numerical simulation of the North Atlantic. *J. Marine. Sys.* 21: 307–320.

Panteleev G., A. Proshutinsky, M. Kulakov, D. A. Nechaev, and W. Maslowski, 2007. Investigation of the summer Kara Sea circulation employing a variational data assimilation technigue. *J. Geophys. Res.* 112: C04S15, doi: 10.1029/2006JC003728.

Polyakov I., 2001. An eddy parameterization based on maximum entropy production with application to modelling of the Arctic Ocean circulation. *J. Phys. Oceanogr.* 31: 2255–2270.

Polyakov I. V. and L. A. Timokhov, 1995. Thermohaline circulation of the Arctic Ocean. *Dokl. Akad. Nauk, Oceanol.* 342(2): 254–258.

Proshutinsky A. Y. and M. A. Johnson, 1997. Two circulation regimes of the wind-driven Arctic Ocean. *J. Geophys. Res.* 102(C6): 12493–12514.

Price, J. F., R. A. Weller, and R. Y. Thompson, 1986. Diurnal cycling: Observation and models of the upper ocean response to diurnal heating, cooling and wind mixing. *J. Geophys. Res.* 91: 8411–8427.

Qiu B. and S. Chen, 2005. Variability of the Kuroshio Extension Jet, recirculation gyre, and mesoscale eddies on decadal time scales. *J. Phys. Oceanogr.* 35: 2090–2103.

Rayner, N. A., D. E. Parker, E. B. Horton, C. K. Folland, L. V. Alexander, D. P. Rowell, E. C. Kent, and A. Kaplan, 2003. Global analyses of sea surface temperature, sea, ice, and night marine air temperature since the late nineteenth centure. *J. Geophys. Res.* 108: 4407, doi: 10.1029/2002JD002670.

Riedling S. H. and R. H. Preller, 1991. The development of a coupled ice-ocean model for forecasting ice conditions in the Arctic. *J. Geophys. Res.* 96: 16955–16977.

Ryabchenko V. A., G. V. Alexeev, I. A. Neelov, and A. Yu. Dvornikov, 2003. Reproduction of climatic changes in the Arctic basin by the ocean and ice circulation model without climatic salinity on the oceanic surface. *Proc. AARI.* 446, *Investigation of the Climate Changes and the Ocean-Atmosphere Interaction Processes in Polar Regions*, 60–82 (in Russian).

Saji, N. H., B. N. Goswami, P. N. Vinayachandran, and T. Yamagata, 1999. A Dipole mode in the tropical Indian Ocean. *Nature* 401: 360–363.

Saji, N., H. and T. Yamagata, 2003a. Possible impacts of Indian Ocean Dipole mode events on global climate. *Climate Res.* 25: 151–169.

Saji, N., H. and T. Yamagata, 2003b. Structure of SST and surface wind variability during Indian Ocean dipole mode years: COADS observations. *J. Climate* 16: 2735–2751.

Saji N., H., S.-P. Xie, and T. Yamagata, 2006. Tropical Indian Ocean Variability in the IPCC twentieth-century climate simulations. *J. Climate* 19: 4397–4417.

Sakamoto T. and I. Umetsu, 2006. Seasonal energy cycle of wind-driven ocean circulation with particular emphasis on the role of botom topography, *Deep-Sea Research* 53: 154–168.

Sarkisyan A. S., 1974. Mechanism of the general oceanic circulation. *Izv. Aca. Nauk SSSR, Ser. Fiz. Atmos. Okeana* 10(12): 1293–1308.

Sarkisyan A. S., 1995. Analysis of model calibration results: Atlantic Ocean Climatic Circulation. *J. Marine Sys.* 6: 47–66.

Sarkisyan, A. S., 2006. Forty years of JEBAR—the finding of the joint effect of baroclinicity and bottom relief for the modeling of ocean climatic characteristics. *Izvestiya, Atmos. Oceanic Phys.* 42(5): 534–554.

Sarkisyan, A. S. and V. B. Zalesny, 2000. Splitting up method and adjoined equation method in the ocean dynamics problem. *Russ. J. Numer. Anal. Math. Modell.* 15(3–4): 333–348.

Schott, G., 1943. Weltkarte zur Übersicht der Meeresströmungen. *Ann. Hydrogr. und Marit. Meteorol.* 71.

Shankar D., P. N. Vinayachandran, A. S. Unnikrishnan, and S. R. Shetye, 2002. The monsoon currents in the north Indian Ocean. *Progr. Oceanogr.* 52(1): 63–119.

Shetye, S. R., A. D. Gouveia, S. S. C. Shenoi, D. Sundar, G. S. Michael, and G. Nampoothiri, 1993. The western boundary current of the seasonal subtropical gyre in the Bay of Bengal. *J. Geophys. Res.* 98: 945–954.

Smith R. D., M. E. Maltrud, F. O. Bryan, and M. W. Hecht, 2000. Numerical simulation of the North Atlantic Ocean at $1/10°$. *J. Phys. Oceanogr.* 30: 1532–1561.

Steele M., R. Morley, and W. Ermold, 2001. RHC: A global ocean hydrography with high quality Arctic Ocean. *J. Climate* 14: 2079–2087.

Suzuki, R., S. K. Behera, S. Iizuka, and T. Yamagata, 2004. Indian Ocean subtropical dipole simulated using a coupled general circulation model. *J. Geophys. Res.* 109: C09001, doi: 10.1029/1023JC001974.

Timmermann, A. J., J. Oberhuber, A. Bacher, M. Esch, M. Latif, and E. Roeckner, 1999. Increased El Niño frequency in a climate model forced by future greenhouse warming. *Nature* 398: 694–697.

Tomczak M. and J. S. Godfrey, 2003. *Regional Oceanography.* Butler and Tanner Ltd., Great Britain, Parts 11–13, 175–228.

Trenberth K. E., J. C. Olson, and W. G. Large, 1989. *A Global Ocean Wind Stress Climatology Based on ECMWF Analysis*. NCAR, Boulder, Colorado. NCAR/TN-338+STR, 98pp.

Treshnikov A. F. and G. I. Baranov, 1972. *Circulation of the Arctic Basin Waters*. Gidrometeoizdat, Leningrad, (in Russian), 126pp.

Tsujino H. and T. Yasuda, 2004. Formation and circulation of mode waters of the North Pacific in a high-resolution GCM. *J. Phys. Oceanogr* 34(2): 399–415.

Ukita, J. and R. E. Moriz, 2000. Geometry and the deformation of pack ice: II. Simulation with a random isotropic model and implementation in sea-ice rheology. *Ann. Glaciolog*. 31: 323–326.

Valsäla, V. K. and M. Ikeda, 2007. Pathways and effects of the Indonesian Throughflow water in the Indian Ocean using particle trajectory and tracers in an OGCM. *J. Climate* 20: 2994–3017.

Vinayachandran P. N., T. Kagimoto Y. Masumoto, P. Chauhan, S. R. Nayak, and T. Yamagata, 2005. Bifurcation of the East India Coastal Current east of Sri Lanka. *Geophys. Res. Lett.* 32: L15606, doi: 10.1029/2005GL022864.

Völösmarty C. J., B. Fekete, and B. A. Tucker, 1998. *River Discharge Database*. Version 1.1 (RivDIS v1.0 supplement), University of New Hampshire, Durham NH, USA.

Wajsowicz, R., C., 2002. Air-sea interaction over the Indian Ocean due to variations in the Indonesian throughflow. *Climate Dyn.* 18: 437–453.

Woodgate R. A., K. Aagaard, R. D. Muench, J. Gunn, G. Björk, B. Rudels, A. T. Roach, and U. Schauer, 2001. The Arctic Ocean boundary current along the Eurasian slope and the adjacent Lomonosov Ridge: Water mass properties, transports and transformations from the moored instruments. *Deep Sea Res*. 48: 1757–1792.

Wunsch C., 1996. *The Ocean Circulation Inverse Problem*. Cambridge University Press, 442pp.

Wyrtki, K., 1973. An equatorial jet in the Indian Ocean, *Science* 181: 262–264.

Yamagata, T., Y. Masumoto, J.-J. Luo, S. Masson, M. Jury, and S. A. Rao, 2004a. Coupled ocean-atmosphere variability in the tropical Indian Ocean. Earth Climate: The Ocean-Atmosphere Interaction. *Amer. Geophys. Union* 147: 189–212.

Yamagata, T., S. K. Behera, J.-J. Luo, S. Masson, M. Jury, and S. A. Rao, 2004b. Coupled ocean-atmosphere variability in the tropical Indian Ocean. Earth Climate: The Ocean-Atmosphere Interaction. *Amer. Geophys. Union* 147: 189–212.

Yamagata, T., S. K. Behera, S. A. Rao, Z. Guan, K. Ashok, and H. N. Saji, 2003. Comments on 'Dipoles, temperature gradient, and tropical climate anomalies'. *Bull. Amer. Meteorol. Soc.* 84: 1418–1422.

Zalesny V. B. and V. I. Kuzin, 1995. Numerical aspects of modeling the general oceanic circulation. *Proc. RAS. Atmosp. Ocean Phys.* 31(3): 404–418 (in Russian).

Chapter 6
Modelling Climate Variability of Selected Shelf Seas

Abstract In this chapter we shall focus our attention to several shelf seas, the characteristics and variability of each of them are very different from others: the North Sea and the White Sea, the Caspian Sea and the Black Sea. Each of these four seas has its own peculiarities.

For the North Sea the validity of simple oceanic balances are tested against high resolving numerical modelling. The decisive factors of the White Sea dynamics are strong tidal processes. A large section of this chapter is devoted to modelling seasonal variability of the Caspian Sea, the largest enclosed water body on the Earth. The sea surface heat flux components undergo very strong seasonal variations following the surrounding continental climate. Seasonal variations of the Black Sea thermo-hydrodynamical characteristics are described in the last and the largest section. Models and numerical methods describe the seasonal variability of the Black Sea internal physical characteristics as Cold Intermediate Layer, transformation of the effect of river fresh water discharge, as well as water exchange with the Mediterranean Sea. A new four-dimensional analysis-double-correction method of data assimilation is presented. We consider this type of model-data synthesis to be one of the most perspective ways for understanding the ocean/sea long-term variability.

Keywords North Sea · White Sea · Caspian Sea · Black Sea · Tidal currents · Cold intermediate layer · Seasonal variability

6.1 The North Sea

6.1.1 Introduction

The North Sea is a shallow shelf sea adjacent to the Northern Atlantic with a mean depth of 80 m (maximum water depths in the Norwegian Trench about 800 m), see Fig. 6.1.

It is characterized by a broad connection to the ocean and by strong continental impacts from North-Western Europe. This results in a substantial interplay of oceanic influences (tides, North Atlantic Oscillation NAO, North Atlantic low

A.S. Sarkisyan, J.E. Sündermann, *Modelling Ocean Climate Variability*,
DOI 10.1007/978-1-4020-9208-4_6,

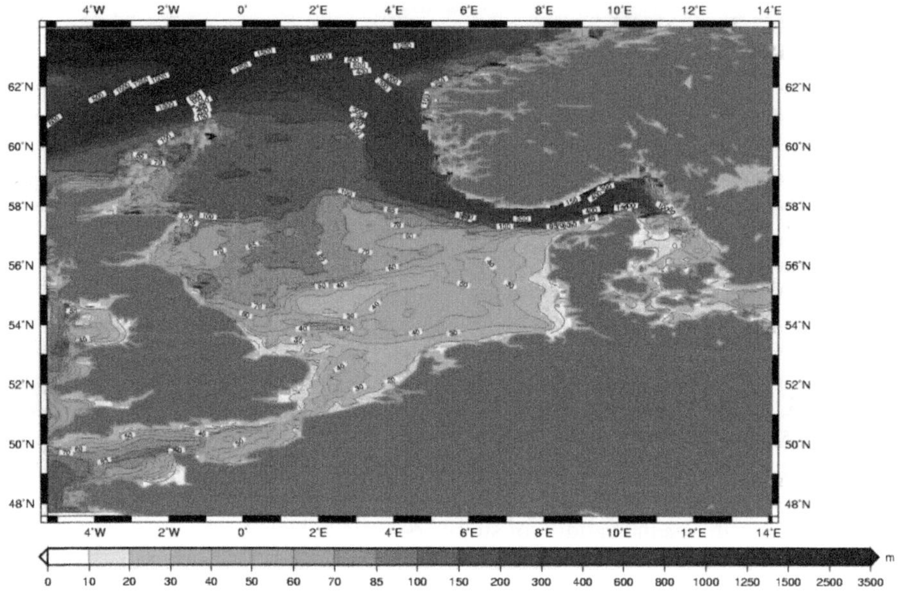

Fig. 6.1 Topography of the North Sea (m) (See also Plate 21 in the Color Plate Section on page 359)

pressure systems) and continental ones (fresh water discharge, heat flow, input of pollutants). This interaction generates a specific physical and biogeochemical regime which requires an adequate modelling concept. Oceanic circulation models cannot directly be applied to the North Sea.

The North Sea coast is a settling and economic area for about 50 Million people from nine highly developed countries. It is one of the best investigated sea regions in the world. For the present state of knowledge reference is made to Sündermann and Lenz (1983), Sündermann (1994), Charnock et al. (1994), Laane et al. (1996), Ruddick (1997), Proctor (1997), Sündermann et al. (2001), and Prandle (2000). In the following the characteristics of the North Sea and its consequences for model conception will be discussed. We are basing on the summarizing description of Pohlmann (2003).

Systematic intercomparisons of North Sea models (EU programmes NOMADS = North Sea Model Advection – Dispersion Study, NOMADS II; Proctor, 1997) did show that among modern high-resolving, nonlinear, baroclinic circulation models are only gradual differences. But, after a comprehensive validation series it had to be stated, in spite of a satisfactory general reproduction of the natural system, that for certain processes and locations still significant differences appear against field data. The main model deficits are seen presently in the parameterization of subscale processes (like turbulence, internal waves etc.) and the formulation of external forcing (like fluxes between atmosphere and ocean, lateral boundary conditions etc.). Further model improvement will be achieved less by model standardisation ("community models") than by common definition of test scenarios including quality criteria which should be passed by any existing model. Due to the societal relevance these test cases should comprise shelf sea processes such as coastal Kelvin waves, wind driven up- and downwelling, formation of fronts, density driven coastal plumes (Haidvogel and Beckmann, 1998). Regarding the complexity of the governing physics, of the initial and boundary conditions, of the natural topography, of the numerical schemes it seems not fruitful to aim at a "uniform" standard model for shelf sea dynamics (although there are "schools" worldwide existing), but to give space for the creativity of different working groups and to test the model quality by standardized criteria.

6.1.2 Special Features of North Sea Modelling

In due time there will be no universal circulation model of the world seas representing with sufficient accuracy all relevant space and time scales. Among other reasons, already the required computer capacity is beyond present possibilities. Therefore the appropriate model has to be chosen to represent and resolve the dominant processes in the region under investigation. Historically, there have been two parallel developments in circulation modelling, each mostly independent of the other. On the one hand, the simulation of the global oceanic current systems with their elementary planetary equilibriums; on the other, the mesoscalic shelf sea modelling as a superposition of many equally important processes. The North Sea forms a transition zone comprising geostrophic and Ekman regimes as well as significant deviations there from. Tides are almost strong signals which cannot be neglected.

In the following, the relevance of the fundamental oceanic forcing mechanisms (geostrophy, Ekman flow, Joint Effect of Baroclinicity and Relief JEBAR) for the North Sea will be studied. The numerical simulations base on the Hamburg Shelf-Ocean Model HAMSOM, a three-dimensional, baroclinic circulation model with free surface which operates with an implicit finite difference scheme (Backhaus, 1985).

6.1.2.1 Relevance of Geostrophy in the North Sea

Pohlmann (2003) has extracted the baroclinic part $\underline{v_g} = (u_g, v_g)$ of the geostrophic flow from the results of the complete circulation model HAMSOM. First the temperature and salinity fields T, S (x,y,z,t) have been computed on the three-dimensional model grid, next the density ρ (x,y,z,t) has been determined by the equation of state (Fofonoff and Millard, 1983) and then $\underline{v_g}$ has been calculated:

$$v_g(x, y, z, t) = \frac{g}{\rho_0 \cdot f} \cdot \frac{\partial}{\partial x}\left(\int_z^0 g \cdot \rho(x, y, z', t)\, dz'\right)$$

$$u_g(x, y, z, t) = -\frac{g}{\rho_0 \cdot f} \cdot \frac{\partial}{\partial y}\left(\int_z^0 g \cdot \rho(x, y, z', t)\, dz'\right)$$

g is the Earth's acceleration, f the Coriolis parameter and ρ_0 a reference density.

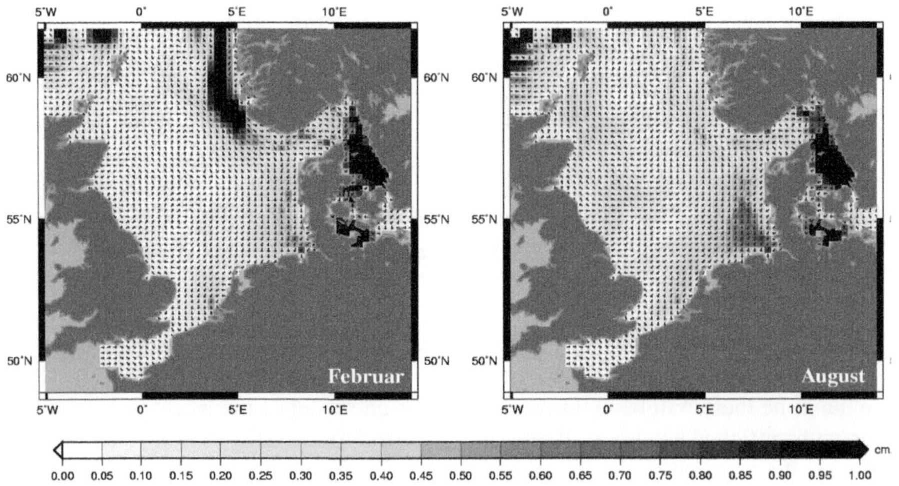

Fig. 6.2 Monthly mean of the baroclinic-geostrophic part of the flow (cm/s) in the surface layer for February and August 1991 (See also Plate 22 in the Color Plate Section on page 360)

For comparison with the total mean current field the temporally (over one month T) and vertically (over the water depth H) averaged horizontal currents have been chosen:

$$\hat{\underline{v}}_g(x, y) = \frac{1}{T} \int_0^T \int_H^0 \underline{v}_g(x, y, z', t') dz' dt'$$

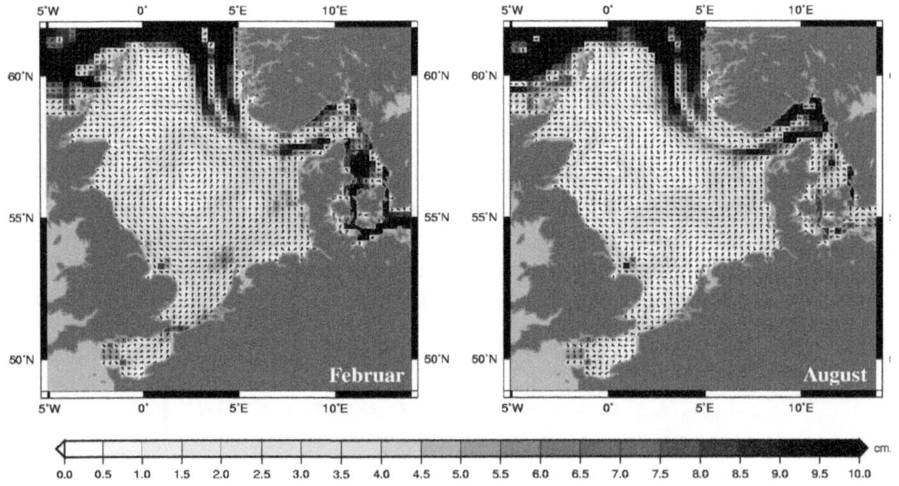

Fig. 6.3 Difference between the monthly mean of the total flow and the baroclinic-geostrophic part (cm/s) in the surface layer for February and August 1991 (See also Plate 23 in the Color Plate Section on page 361)

As examples, the Figs. 6.2 and 6.3 show the baroclinic current $\underline{\hat{v}}_g$ and the difference $\underline{\hat{v}} - \underline{\hat{v}}_g$ of the total flow and the baroclinic flow part (both monthly means)for months February and August 1991.

It is seen that the baroclinic part generally forms a basin-wide anti-cyclonic circulation which is opposite to the known cyclonic gyre. The speed is increasing towards the continental coast (Fig. 6.2). The difference plot (Fig. 6.3) shows that the baroclinic component amounts to 10% of the total flow weakening the cyclonic circulation. Computations for other years exhibit strong inter-annual fluctuations.

6.1.2.2 Relevance of the Ekman Equilibrium

The Ekman regime is characterized by the balance of the Coriolis force and the vertical exchange of momentum:

$$-f \cdot v_e(x, y, z, t) = \frac{\partial}{\partial z}\left(A_v(z) \cdot \frac{\partial u_e(x, y, z, t)}{\partial z}\right)$$

$$f \cdot u_e(x, y, z, t) = \frac{\partial}{\partial z}\left(A_v(z) \cdot \frac{\partial v_e(x, y, z, t}{\partial z}\right)$$

In the work of Pohlmann (2003) the terms on the right-hand side are again calculated by means of the complete circulation model HAMSOM. From this forcing the Ekman flow (u_e, v_e) is deduced. $A_v(z)$ is the vertical eddy coefficient and depends on depth.

Figure 6.4 shows as an example for February and August 1991 the monthly mean of the Ekman currents in 5 m depth. It has maximum values along the English coast with an onshore direction. Due to stronger winds it is higher in winter. The

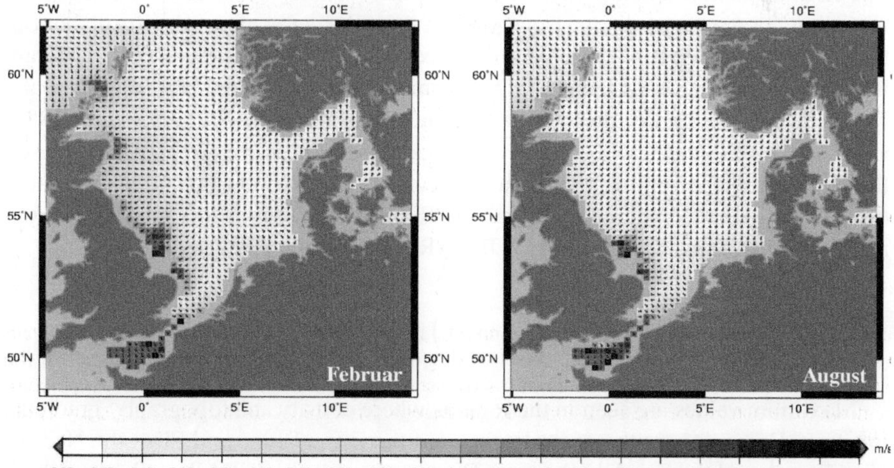

Fig. 6.4 Monthly mean of the Ekman flow in 5 m depth (m/s) for February and August 1991 (See also Plate 24 in the Color Plate Section on page 362)

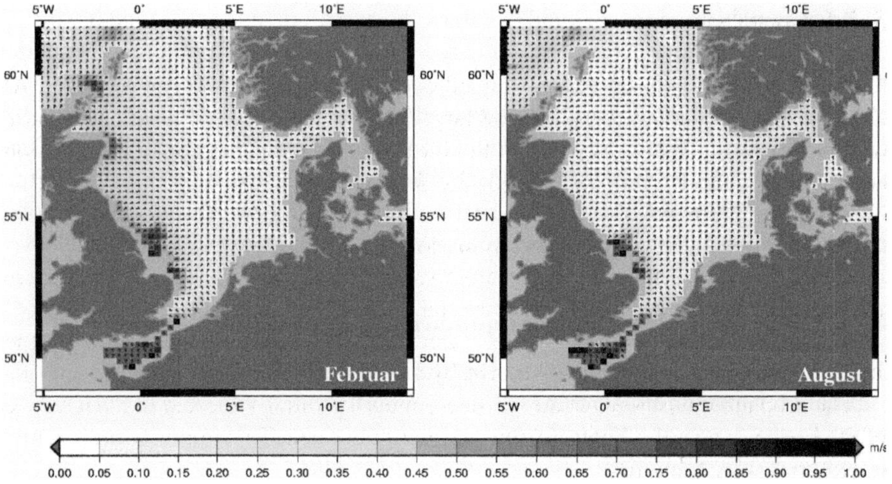

Fig. 6.5 Difference of the monthly mean of the total flow and the Ekman part (m/s) in 5 m depth for February and August 1991 (See also Plate 25 in the Color Plate Section on page 363)

difference plot (total current minus Ekman current) in Fig. 6.5 exhibits a residual flow of equal magnitude, but directed offshore (which means a compensation of the Ekman current).

Stronger currents (not Ekman balanced) are now appearing at the Norwegian coast.

6.1.2.3 Relevance of JEBAR

The JEBAR term is a component of the oceanic vorticity balance; it describes how baroclinic pressure gradients force the flow in the case of non-uniform bottom topography, see Section 2.2.3. As discussed previously, there have been different interpretations of JEBAR basing on the formal separation of the barotropic and the baroclinic motion part in the vorticity balance (Sarkisyan, 1977; Huthnance, 1984; Cane et al., 1998); they will not be discussed here. Pohlmann (2003) has analysed the vorticity balance of the North Sea for a certain time period calculating separately the β-term, the vortex stretching and the JEBAR term. From this study, Figs. 6.6 and 6.7 show the spatial structure of JEBAR for February 1991

$$J\left(\chi, \frac{1}{H}\right) = -\frac{f}{H} \cdot \left(\underline{v}_g \cdot \underline{\nabla} H\right), \text{ with } \chi \equiv \frac{g}{\rho_0} \int_{-H}^{0} z \cdot \rho \; dz$$

Maximum values are seen in the regions where density and topography gradients intersect. Examples are the outer estuaries of the rivers Rhine and Elbe, the Norwegian Trench and the Fair Isle Passage. During summer the JEBAR gradients, which are directed towards the centre of the North Sea, are enlarged due to the joint action of temperature and salinity gradients. Out of the remaining terms of the vorticity

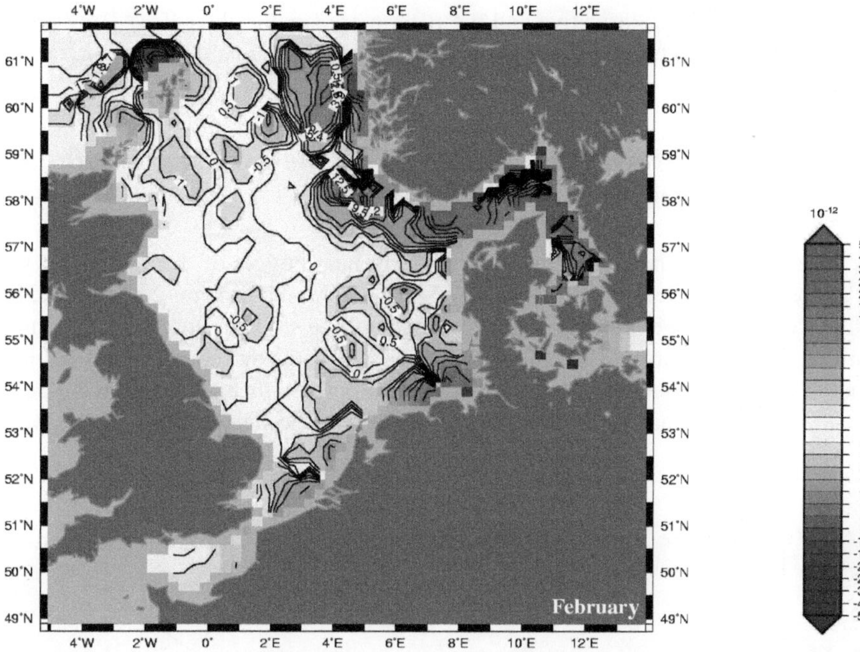

Fig. 6.6 Monthly mean of the JEBAR term ($10^{-12}s^{-1}$) for February 1991 (See also Plate 26 in the Color Plate Section on page 364)

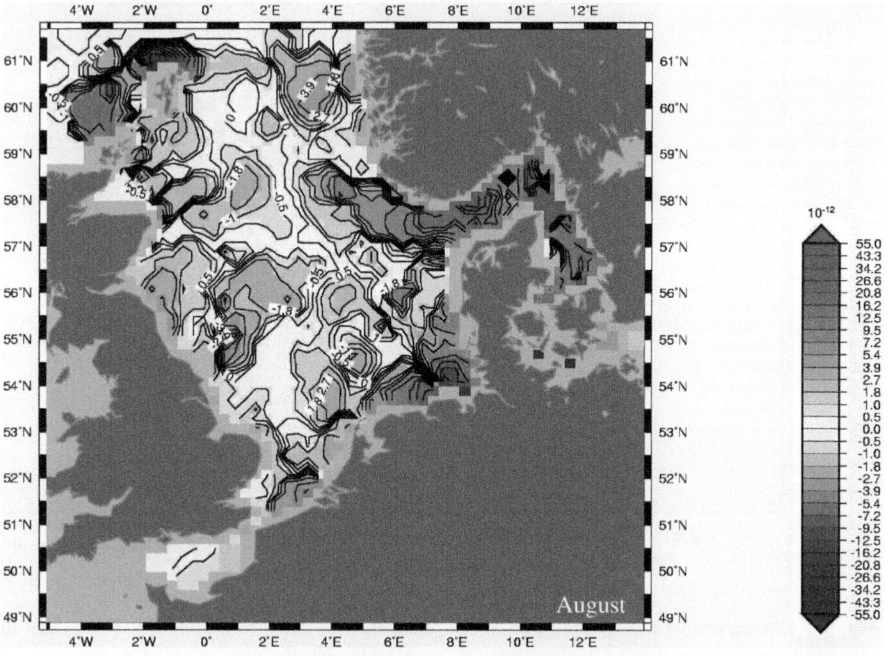

Fig. 6.7 Monthly mean of the JEBAR term ($10^{-12}s^{-1}$) for August 1991 (See also Plate 27 in the Color Plate Section on page 365)

balance, the temporal derivative and the β-term are smaller than JEBAR by one to two orders of magnitude whereas the vortex stretching is equally important.

6.1.3 Example of a Realistic Simulation: The Variable Heat Budget of the North Sea

As an example, in the following the long-term heat budget of the North Sea will be analysed using decadal simulations of HAMSOM (Pohlmann, 2003). First, the influence of wind and atmospheric heat fluxes will be studied. Surprisingly, it turns out that the correlation of maximum wind stress and maximum monthly total heat content is nearly zero. The logical expectation would be that a stronger wind deepens the upper thermal layer and enlarges therefore the heat content of the water

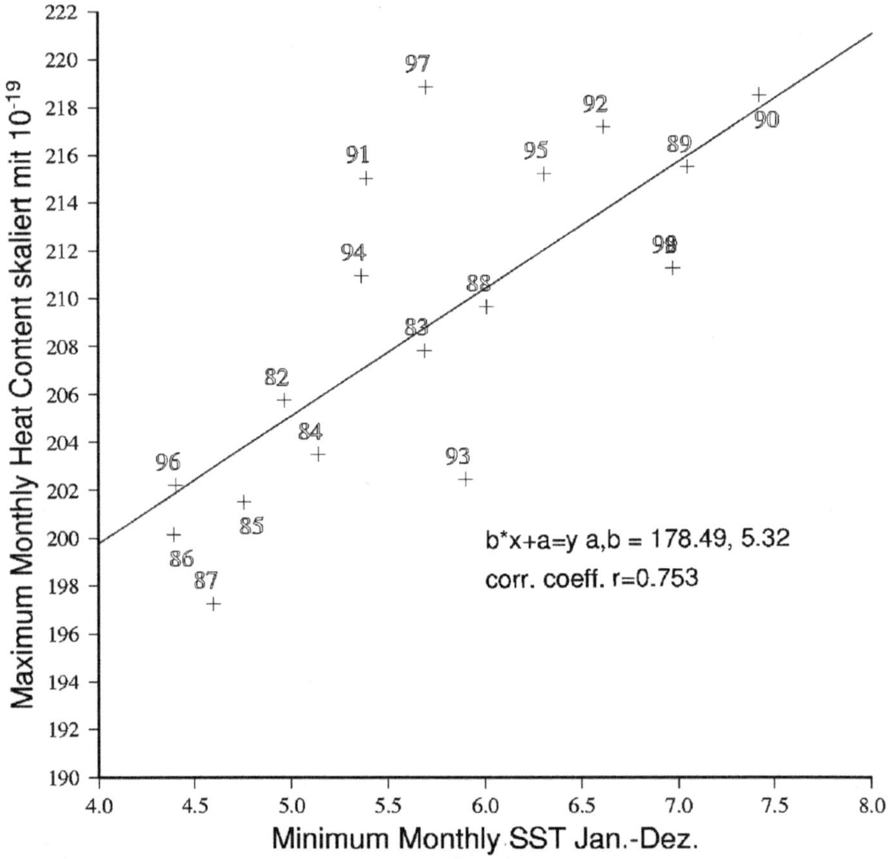

Fig. 6.8 Correlation of the minimum monthly mean surface temperature in winter (°C) and the maximum monthly mean total heat content (10^{-19} J) of the North Sea

body. The explanation is given by the negative correlation of the wind stress and the maximum sea surface temperature SST. As a fact, in the North Atlantic system a warm summer is connected with weak winds (and vice versa) which means a damping of inter-annual fluctuations in the heat content.

A clear correlation (0.75) however exists between the maximum heat content in summer and the minimum SST of the preceding winter, see Fig. 6.8.

This can be explained as follows. In winter the water column is vertically mixed resulting in an almost homogeneous temperature distribution (equal to SST). During the formation of a thermal upper layer in spring/summer the bottom water is decoupled from the actual surface processes in broad regions of the North Sea. A real interaction happens again only in the following winter. In this way the winter SST can influence the heat content in the next summer. The conservation of the winter bottom water temperature in the central and the Northern North Sea is one of the rare hydrographical phenomena with a "memory" scale of one year. Normally, typical spin-up periods (within those the preceding dynamical state is lost) amount only up to a few days in the shallow North Sea.

In the cited paper (Pohlmann, 2003) also the inter-annual variability of the North Sea's heat content has been simulated for the years 1982–1998, see Fig. 6.9.

The most striking signal is a system shift between years 1987 and 1988. Up till this event the minimum heat content in winter was rather constant at 12×10^{20} J, after that (with higher inter-annual variability) at 14×10^{20} J. There are clear indications that this shift has initiated a large-scale change within the biological species spectrum of the North Sea (Edwards and Reid, 2001). A Fourier analysis of the time series in Fig. 6.9 exhibits periods of 7–9 years correlating with modes of

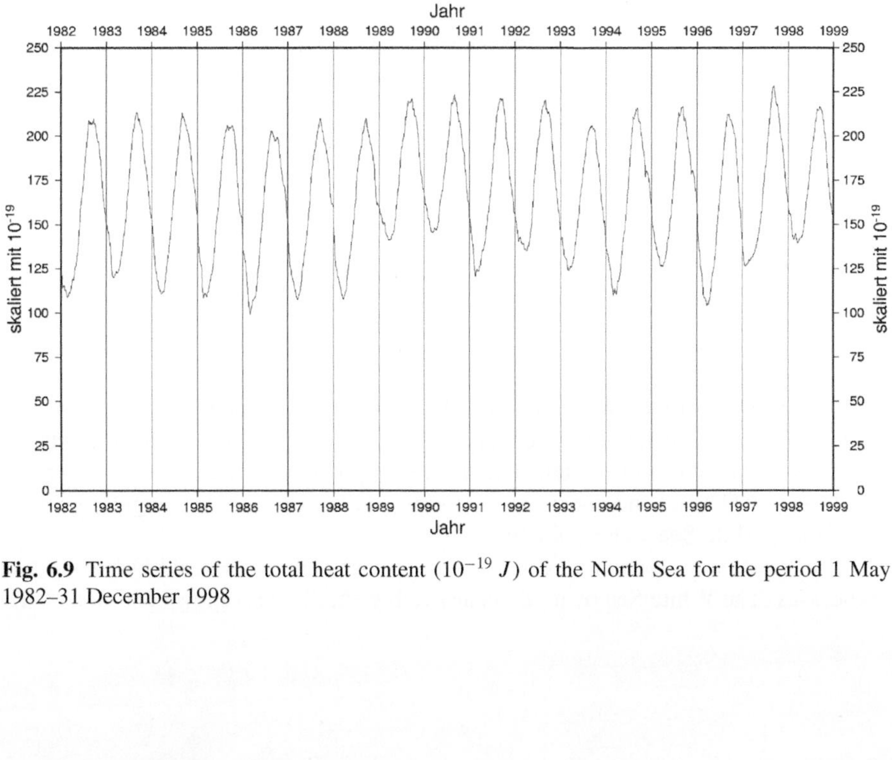

Fig. 6.9 Time series of the total heat content (10^{-19} J) of the North Sea for the period 1 May 1982–31 December 1998

the North Atlantic Oscillation NAO (Sündermann et al., 1996). Searching for the coupling mechanism it should be noticed that Becker and Pauly (1996) found high correlations (0.75) between SST and NAO in the central North Sea. This suggests a dominating role of the atmospheric heat fluxes for the heat budget in the North Sea.

6.2 The White Sea

6.2.1 Introduction

The White Sea belongs to the Arctic Ocean adjacent seas, it is connected with the World Ocean via the Barents Sea and is characterized as shelf tidal sea. The area of the White Sea equals approximately to $90,000\,km^2$, and its water volume to $6,000\,km^3$. The average depth of it is less then 70 m and the maximal one is 350 m. Its shallowest parts are the Onega Bay with an average depth of about 15 m and Dvin Bay of about 25 m. The bottom topography of the White Sea is presented on the Figs. 6.10 and 6.11.

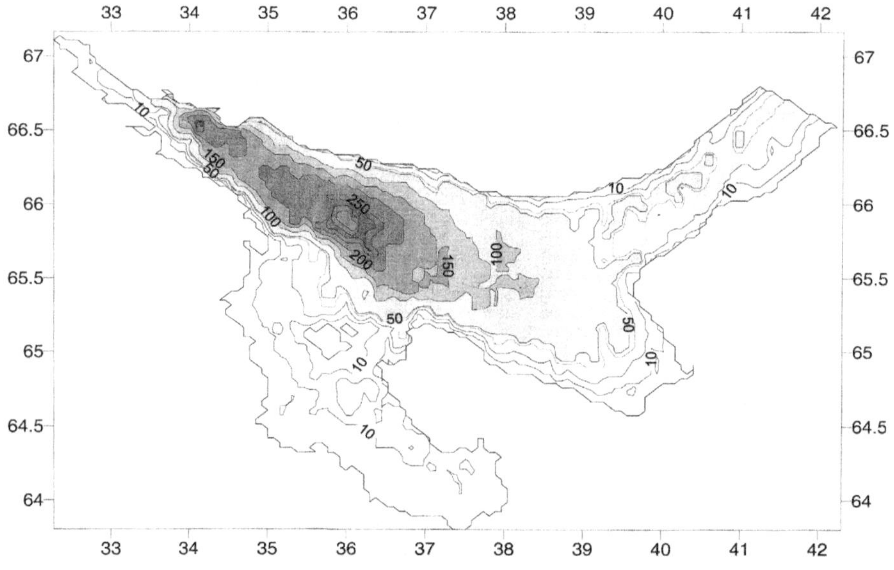

Fig. 6.10 Bottom topography of the White Sea (depth in meters)

The hydrology of the White Sea is under the strong influence of two factors acting at different parts of its side boundary layers: one originating in the Barents Sea supplies warm and saline waters and another originating in the Dvin River supplies fresh water. This second source is very essential because its discharge amounts to 4% of the White Sea water volume.

The White Sea dynamics is very nonstationary, characterizised by strong tidal processes. The White Sea own tide is negligibly small – just a few centimeters. The

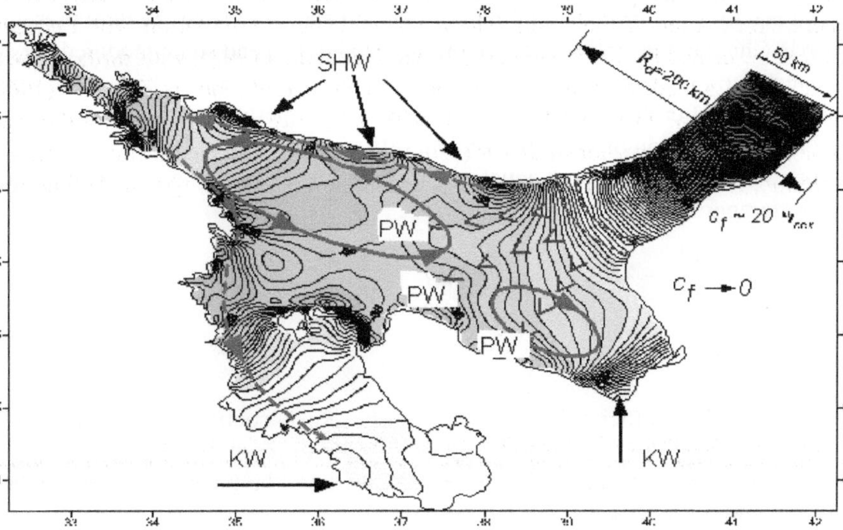

Fig. 6.11 The White Sea bottom topography and its dynamical processes schematics: R_d – tidal wave external deformation radius in the White Sea Throat; SHW – shelf waves; KW – Kelvin waves; PW – Poincare waves; C – group velocity
————— hydrological fronts at the exit of the throat;
>>>>>>>>>>>> Poincare waves radiation;
→→→→→→ Supposed alongshore cyclonic transport connected with the baroclinic effect of Poincare waves and shelf waves;
—→—→—→ Large-scale quasigeostrophic flow (Semenov, 2004) (See also Plate 28 in the Color Plate Section on page 366)

main tide comes from the Barents Sea via the narrow bay Throat. The amplitude of the incoming wave is almost 6 m; its flow velocity is up to 3 m/s. Due to the strong tides the density of relative available potential energy of the White Sea is two orders of magnitude higher than normally in the ocean. The tides main component is the lunar semi-diurnal wave M_2 while the solar tide K_1 amplitude is only some 8–10 cm. Due to nonlinear wave superposition a quarter-diurnal harmonics M_4 and MS_4 with amplitudes about 10 cm generate.

With the Throat maximal depth of 40 m the tide wave phase velocity is $c_1 \approx \sqrt{gH} \approx 20\,\text{m/s}$; the external deformation radius $a_1 = \frac{c_1}{f}$ is about 200 km compared with the cross section length of only 50 km. It means that the Earth rotation effect is filtered out when the tide wave passes via Throat to the White Sea Basin. Because the sea is shallow and located in the northern windy latitudes, the wind stress effect is very essential.

6.2.2 Y. Semenov's Model

Semenov's model is the first 3DPEM of the White Sea. Due to the abovementioned complexities the model concerns the summer season only, i.e. it does not include the ice formation, melting sublimation etc.

The equations of motion, continuity, state and statics are like the ones presented above in Chapter 1, namely the equations (1.1.1)–(1.1.4) and some modified version of the equation of state (1.1.7).

The main differences are in the equations for temperature and salinity, specifically in the formation of vertical turbulent diffusivity.

For comparison we refer to equation (1.1.5) (Chapter 1). Semenov's model equation for temperature is

$$\frac{dT}{dt} = \frac{\partial}{\partial z} K_H \frac{\partial T}{\partial z} + A_H \Delta T = \frac{1}{\Pr} \frac{\partial}{\partial z} K_M \frac{\partial T}{\partial z} + A_H \Delta T \tag{6.2.1}$$

$$\Pr = \frac{K_M}{K_H} \tag{6.2.2}$$

where K_M is the coefficient of vertical viscosity, K_H – coefficient of vertical diffusion of heat, Pr is the Prandtl number. The visible difference is that in (6.2.1) δ is absent and the Prandtl number is introduced to show the interconnection between viscosity and diffusivity.

The equation for salinity looks like (6.2.1).

In Semenov's model the coefficients K_M and K_H, as will be shown below, are not chosen empirically, but analytically determined. Another advantage concerns the boundary condition for w at the sea surface. The "rigid lid" condition is quite satisfactory for many problems of large scale ocean circulation. For some other problems the linear boundary condition $w = \frac{\partial \zeta}{\partial t}$ is satisfactory. In the present case the nonlinear boundary condition of type

$$w|_{z=\zeta} = \frac{\partial \zeta}{\partial t} + mu \frac{\partial \zeta}{\partial \lambda} + nv \frac{\partial \zeta}{\partial \theta} \tag{6.2.3}$$

was accepted to be adequate ($m = \frac{1}{R_*} \sin\theta, n = \frac{1}{R_*}$, R_* is Earth radius).

Concerning the boundary conditions at solid parts of side boundaries non-slip conditions for flow velocity and thermo-salinity isolation conditions are applied; at points of river discharges the flow velocity, T and S are specified; at the surface the wind stress, heat and salt fluxes are specified; at the bottom the flow velocity normal component equals zero, the bottom stress formally is specified but actually is being determined by the method which will be shown below.

At the White Sea entrance the boundary conditions are

$$\zeta = \zeta_S(\lambda, \theta, t), \; T = T_S(z, t), \; S = S_S(z, t) \tag{6.2.4}$$

Due to the essential role of the tidal currents a special formation of (6.2.4) was applied.

In the case of water inflow the conditions (6.2.4) are used. In the case of outflow the radiational method is used together with the Orlanski scheme (1976). If $(C_f \cdot n) > 0$, i.e. outflow (n being external normal), then the function f is determined from equation

$$\frac{\partial f}{\partial l} + C_f \frac{\partial f}{\partial n} = 0 \tag{6.2.5}$$

where $C_f = -\left(\frac{\partial \tilde{f}}{\partial t}\right) / \left(\frac{\partial \tilde{f}}{\partial n}\right)$ – is the phase velocity of pulsation's spreading along the internal normal to the boundary.

If $\left(C_f \cdot n\right) < 0$, i.e. inflow then all the three quantities (denoted by one scalar f) are defined from formulae (6.2.4).

Next, the procedure of determining the turbulence coefficients

$$K_M = -\underline{u_i{}'w'} \Big/ \left(\frac{\partial u_i}{\partial z}\right), \quad i = 1, 2 \tag{6.2.6}$$

$$K_H = -\underline{f'w'} \Big/ \left(\frac{\partial f}{\partial z}\right) \tag{6.2.7}$$

is described.

May w', $u_i{}'$, f' denote turbulent functions of velocity and a scalar quantity (temperature, density, salinity), the lower hyphen an assembly averaging. According to the semiempirical turbulence model of Launder and Spalding (1972) will be formulated in terms of the intensity of turbulent energy k and of the velocity of viscous dissipation of turbulent energy ε:

$$k = \frac{1}{2}\left(\underline{u'^2} + \underline{v'^2} + \underline{w'^2}\right) \tag{6.2.8}$$

$$\varepsilon = \underline{\upsilon\left(\frac{\partial u_i'}{\partial x_j} + \frac{\partial u_j'}{\partial x_i}\right)^2}, \qquad i, j = 1, 2 \quad i \neq j \tag{6.2.9}$$

where ν is the coefficient of kinematic viscosity.

These quantities will be calculated by simplified evolution equations. The horizontal variations were neglected compared to the vertical one. The advective transport of both the Reynolds fluxes and the stresses were neglected against dissipation and generation by work of shear and buoyancy force. With some accuracy these simplifications are applicable to the White Sea dynamics. As a result for big values of the Reynolds number the equations for k and ε evolutions may be described as:

$$\frac{dk}{dt} = \frac{\partial}{\partial z}\left(\frac{K_M}{\sigma_b}\frac{\partial k}{\partial z}\right) + P + G - \varepsilon \tag{6.2.10}$$

$$\frac{d\varepsilon}{dt} = \frac{\partial}{\partial z}\left(\frac{K_M}{\sigma_\varepsilon}\frac{\partial \varepsilon}{\partial z}\right) + C_{1\varepsilon} P \frac{\varepsilon}{k} + C_{1\varepsilon}\left(1 - C_{3\varepsilon}\right) G \frac{\varepsilon}{k} - C_{2\varepsilon}\frac{\varepsilon^2}{k} \tag{6.2.11}$$

where

$$P = -\underline{u'w'}\frac{\partial u}{\partial z} - \underline{v'w'}\frac{\partial v}{\partial z} \tag{6.2.12}$$

$$G = \frac{1}{\rho_0}g\underline{\rho' w'} \tag{6.2.13}$$

Here P is the turbulence generation due to flow velocity shear, G – the turbulence generation due to work of buoyancy forces (or the dumping of turbulence by buoyancy forces when stratification is stable).

To define K_M a number of algebraic relations was written, which was received after some transformation and simplification of equations (6.2.8)–(6.2.11): evolution of vertical velocity fluctuation $\underline{w'^2}$

$$\underline{w'^2} = \frac{2}{3}k\left(1 - \frac{1-\gamma}{C_1}\right) + \frac{kG}{\varepsilon C_1}\left\{2(1 - C_3) + \frac{2(C_3 - \gamma)}{3}\right\} \tag{6.2.14}$$

evolution of vertical density fluctuation $\underline{\rho'^2}$

$$\underline{\rho'^2} = -\frac{2Rk}{\varepsilon}\underline{w'\rho'}\frac{\partial \rho}{\partial z} \tag{6.2.15}$$

vertical turbulent density flux $\underline{w'\rho'}$

$$\underline{w'\rho'}\frac{C_{1\varphi}\varepsilon}{k} = \underline{w'^2}\frac{\partial \rho}{\partial z} + \frac{(1 - C_{1\varphi})\,g}{\rho_0}\underline{\rho'^2} \tag{6.2.16}$$

horizontal turbulent mass flux $\underline{u'\rho'}$

$$\underline{u'\rho'}\frac{C_{1\varphi}\varepsilon}{k} = -\underline{u'w'}\frac{\partial \rho}{\partial z} - (1 - C_{2\varphi})\,\underline{w'\rho'}\frac{\partial u}{\partial z} \tag{6.2.17}$$

and Reynolds stress component $\underline{u'w'}$

$$\underline{u'w'}\frac{C_1\varepsilon}{k} = -(1-\gamma)\,\underline{w'^2}\frac{\partial u}{\partial z} + \frac{(1 - C_3)\,g}{\rho_0}\underline{u'\rho'} \tag{6.2.18}$$

where R, C_1, $C_{1\varphi}$, $C_{2\varphi}$, γ, C_3 assumed to be universal constants.

The system of equations (6.2.14)–(6.2.18) after inclusion of quantities K_M, Pr and the dynamic Richardson number

$$R_f = -\frac{G}{P} \tag{6.2.19}$$

is transformed to following relations:

$$K_M = \frac{F(R_f)k^2}{\varepsilon} \tag{6.2.20}$$

$$F(R_f) = \frac{C_\mu}{1 - R_f}\left(1 - \frac{R_f}{R_{f_{cr}}}\right) \tag{6.2.21}$$

$$R_{f_{cr}} = \frac{1 - d}{3\left(\frac{1}{3}(1 - d) + \left(1 - C_{2\varphi}\right)R + d_1\right)} \tag{6.2.22}$$

$$d = \frac{1 - \gamma}{C_1}; \qquad d_1 = \frac{1}{C_1}\left((1 - C_3) + \frac{C_3 - \gamma}{3}\right);$$

$$C_\mu = \frac{2(1 - d)}{3C_{1\varphi}}$$

$$\Pr = \frac{K_M}{K_H} = \frac{R_i}{R_f} = \frac{\alpha\left(1 - \beta_1 R_f\right)}{1 - \beta_2 R_f} \tag{6.2.23}$$

$$\alpha = dC_{1\varphi}; \qquad \beta_1 = 1 + \frac{3d_1}{1 - d} + \frac{3\left(1 - C_{2\varphi}\right)\left(1 - C_3\right)}{2C_{1\varphi}(1 - d)(1 - \gamma)};$$

$$\beta_2 = \frac{1}{R_{f_{cr}}} - \frac{3(1 - C_3)}{2C_1(1 - d)} \tag{6.2.24}$$

The obtained relations are valid both for the case of stable and unstable stratifications and may be also used for describing the turbulence processes of winter convection.

The coefficients of turbulent exchange are becoming zero and the equations degenerate when the value of R_f tends to $R_{f_{cr}}$. To avoid this, the lower limits of values K_M, k, ε, Pr were specified:

$$k = k_0,\ \varepsilon = \varepsilon_0,\ K_M = K_{M_0},\ \Pr = \Pr_0 \tag{6.2.25}$$

The following boundary conditions asymptotically were formulated. The logarithmic friction layer is fulfilled, if $z_1 = [H(\lambda, \theta) - z] \to 0$:

$$\frac{\partial \vec{u}}{\partial z} = \frac{\vec{u}^*}{\kappa_1 z_1}; \qquad \vec{u} = \frac{\vec{u}^*}{\kappa_1}\ln\left(\frac{z_1}{z_0}\right) \tag{6.2.26}$$

and the friction stress obtained from the solution of equation (6.2.26) for the logarithmic layer, at the bottom layer ($z_1 = 0$) is

$$\vec{u}^*\left|\vec{u}^*\right| = -\frac{\vec{\tau}_h}{\rho} = K_M\frac{\partial \vec{u}}{\partial z} \tag{6.2.27}$$

Following the paper of Launder and Spalding (1972) the turbulent energy and the viscous dissipation velocity are presented in the form

$$k = \frac{\left|\vec{u}^*\right|^2}{C_\mu{}^{1/2}}; \qquad \varepsilon = \frac{\left|\vec{u}^*\right|^3}{\kappa_1 z_1} \tag{6.2.28}$$

In (6.2.26)–(6.2.28) $\vec{\tau}_h$ is the shear stress at the bottom, $\vec{u}$ – the horizontal flow velocity vector at the upper boundary of the logarithmic layer, $\vec{u}^*$ – the dynamical velocity, z_0 – the bottom's roughness and $\kappa_1 = 0.435$ is the Karman constant.

Finally, by expressing $\vec{u}^*$ by the near-bottom flow velocity one has

$$k = \frac{\kappa_1^2 \left|\vec{u}\right|^2}{C_\mu{}^{1/2} \left(\ln\left(z_1/z_0\right)\right)^2} \tag{6.2.29}$$

$$\varepsilon = \frac{\kappa_1{}^2 \left|\vec{u}\right|^3}{z_1 \left(\ln\left(z_1/z_0\right)\right)^3} \tag{6.2.30}$$

The boundary conditions (6.2.29), (6.2.30) should be put at some distance h from the bottom, $h >> h_\upsilon$, where $h_\upsilon \approx \upsilon/\vec{u}^*$ is the thickness of the layer where the molecular viscosity is essential (therefore the correlations (6.2.29), (6.2.30) become invalid). On the other hand h should be smaller than the thickness of the Ekman layer and the Monin-Obukhov scale L^*.

$$h << h_e \sim \left(\frac{q}{2f}\right)^{1/2}, \qquad h << L^* \approx \left(\frac{\vec{u}^{*2}}{N^2}\right)^{1/2} \tag{6.2.31}$$

where q is the viscosity coefficient, f is the Coriolis parameter and N – the Brunt-Väisälä frequency.

The wind-stress $\vec{\tau}$ is specified at the sea surface:

$$\vec{u^*}\left|\vec{u^*}\right| = -\frac{\vec{\tau}}{\rho_0}, \tag{6.2.32}$$

besides in some distance h from the free surface the relations (6.2.28) are valid.

In case of the surface waves intensive collapse, the flux of turbulent energy is specified as a boundary condition

$$K_M \frac{\partial k}{\partial z} = -M \tag{6.2.33}$$

The quantity of turbulent energy flux is expressed via wind wave characteristics and ultimately via the wind dynamical velocity $\vec{u}_\alpha{}^*$ (or via the wind velocity at 10 m height – u_{10}) (Phillips, 1985):

$$M = \left|\vec{u}_\alpha^*\right|^3 \ln\left(0.6/C_d\right) \tag{6.2.34}$$

where C_d is the sea surface resistance coefficient.

The turbulent energy dissipation velocity near the sea surface (in some distance h) is expressed in this case via the turbulence average scale l_h and the turbulent velocity of dissipation (the Kolmogorov correlation, 1942):

$$\varepsilon = \frac{C_\mu k^{3/2}}{l_h} \tag{6.2.35}$$

where the quantity of l_h is calculated by formula

$$l_h = \kappa_1 C_\mu{}^{1/4} h \tag{6.2.36}$$

Some concluding remarks to the model. As it appeared the method of calculation coefficients of viscosity and the Prandtl number (instead of empirically specifying) is connected with many simplifications which in their turn are not strongly grounded. Besides the dozen of parameters hardly may be approved as universal constants. Still the general direction of moving from the "δ-method" of instantaneous mixing to more accurate solving the vertical instability problem is prospective.

The author notes that the model was tested repeatedly and has mutually adjusted composition of parameters which took into account the so called "near-wall" effects. It is very essential that the model parametrizes the process of hydrostatical instability: buoyancy forces generate turbulent energy, enhance the coefficients of turbulent viscosity and diffusivity and diminish the hydrostatic instability.

The solution for asymptotic logarithmic layer yields transformation of energy in the whole spectral scale up to the viscous molecular sublayer. These processes appeared to be quite essential although usually they cannot be resolved explicitly in numerical models. Because of the complexity of the problems considered and the small size of the White Sea, a high space-time resolution was chosen (Semenov, 1989, 2004). In these numerical calculations the time-step was 3 min and the grid mesh 4 km.

6.2.3 The Results of Calculations

In case of the White Sea there is a big amount of data which seemingly may serve as quite satisfactory base for verification. Indeed there are tens of century-long standard hydrological sections as well as coastal hydrometeorological stations and hydrometposts.

But there exists a bias problem due to the strong tides (M_2, K_1, M_4, MS_4). The averaging of observed data for a time longer than a day would filter out these oscillating processes. It concerns not only to the flow velocities but to other characteristics as well. For example, in a paper of Semenov and Luneva (1999) was shown that the isotherms vertical displacement, excited by the internal waves of tidal period, reaches up to 20 m. So, one should be careful in interpreting the measurements made in different places and in different phases of tidal processes keeping in mind the strong vertical stratification of the White Sea.

Furthermore, the strong influence of irregular atmospheric processes complicates the averaging of observed data.

In future it may be possible to perform process-oriented measurements, the planning of which is based on results of modelling. For the time being some selected cases of measurements and/or qualitative results of data is possible to compare with the calculations.

Namely, hydrological frontal zones, the areas with vertically quasihomogeneous waters or vice-versa-the strongly stratified ones and the stable cold- and stable warm-water "pooles". To detect the frontal zones, Kravets (1987) uses the Simpson-Hunter parameter $R = \lg\left(\frac{H}{C_D|u^3|}\right)$, where H is the basin's depth, C_D – the bottom stress coefficient, u – the scale of tidal velocity. The respective stability criterium limits are $1 \leq R \leq 2$. The value $R \approx 2$ characterises the vertically stratified waters while in mixed ones $R \approx 1$.

The Semenov's calculations showed this formal mathematical criterium to be more or less indicative though the physical cause of these two kinds of water formation is more complicate: it is connected with transformation of tidal wave when it leaves the Throat and enters the Basin. The wavy character of the temperature stratification is shown along the section O-TH (Fig. 6.12) and the very location of this section is presented on Fig. 6.13.

The water temperature is mixed vertically in the Throat and strongly stratified in the Basin. As the Fig. 6.14 shows, frontal zones and eddies exist in Solovetskiy islands area their generation has advective character.

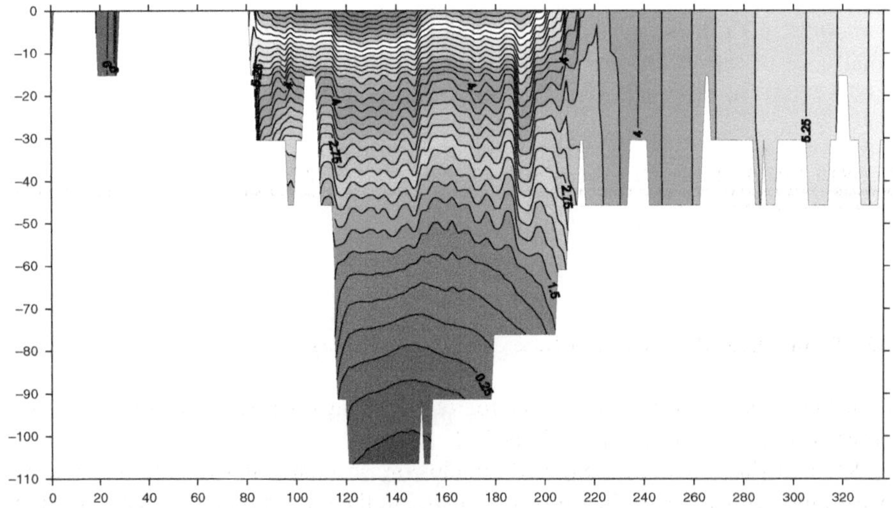

Fig. 6.12 Vertical distribution of summer temperature (°C) across the section O-TH (Onega bay to the Throat) (Semenov, 2004) (See also Plate 29 in the Color Plate Section on page 367)

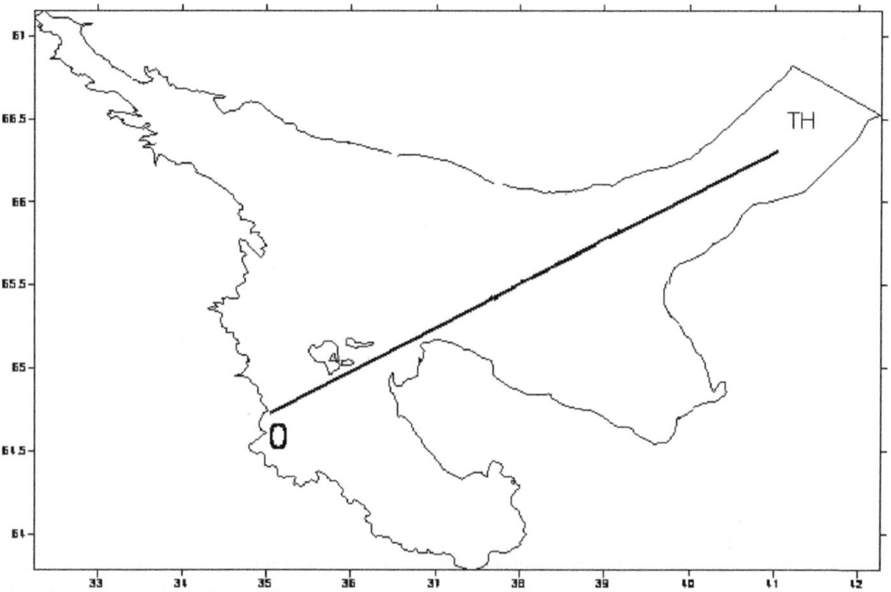

Fig. 6.13 The location of vertical section across the White Sea: O-TH-Onega bay to the Throat (Semenov, 2004)

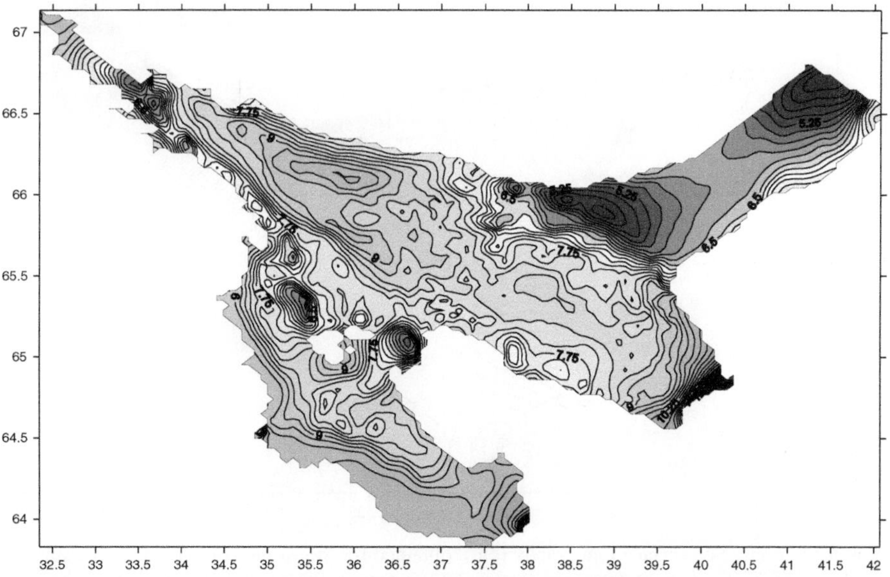

Fig. 6.14 The White Sea surface temperature (°C) in August. Both eddies and frontal zones are depicted (Semenov, 2004) (See also Plate 30 in the Color Plate Section on page 368)

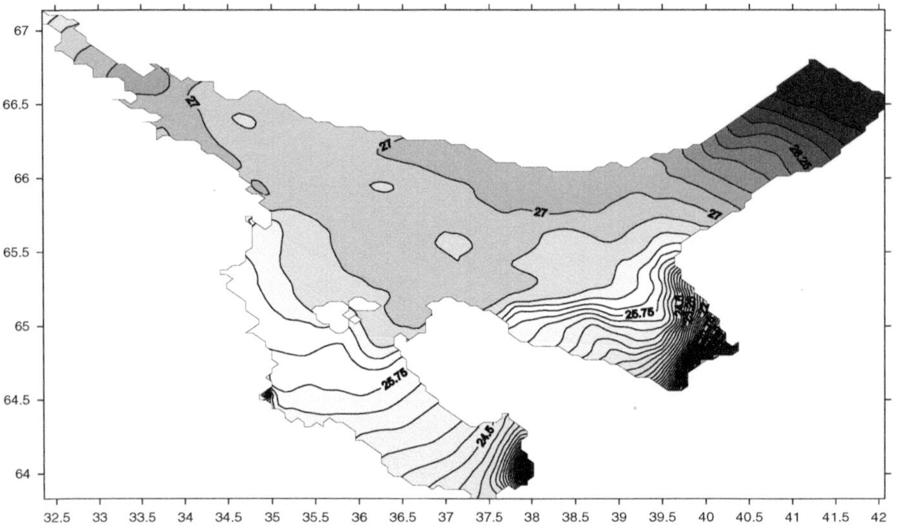

Fig. 6.15 The White Sea surface salinity (psu) in mid-June. The frontal zones in areas of discharge of rivers Dvina and Onega are seen (Semenov, 2004) (See also Plate 31 in the Color Plate Section on page 369)

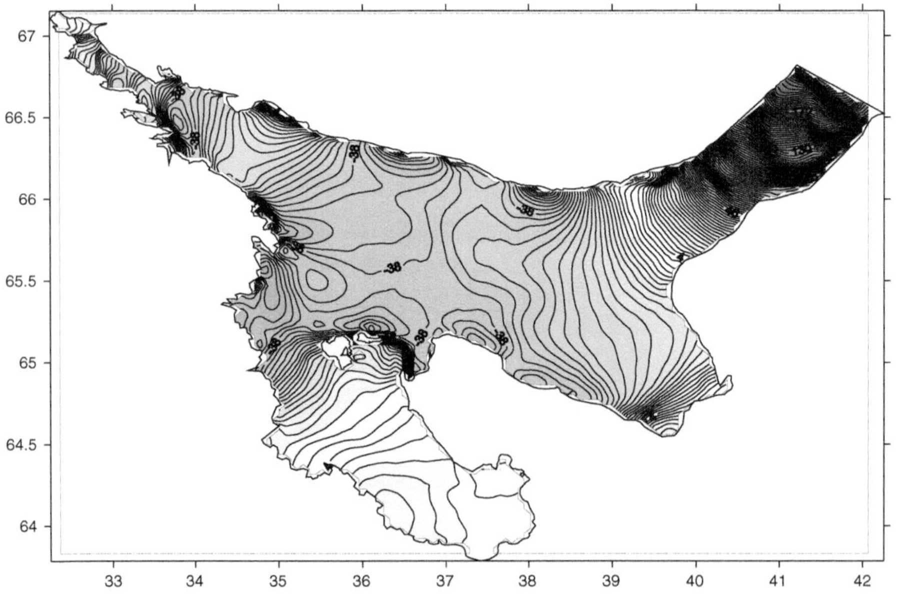

Fig. 6.16 White Sea surface topography (cm), obtained for the August in a moment of rising tide by using the nonlinear boundary condition (6.2.3) (Semenov, 2004) (See also Plate 32 in the Color Plate Section on page 370)

Another kind of frontal zones is formed due to river discharge. Such a front visible by salinity isolines, exists especially in the spring time (see Fig. 6.15).

Realistic frontal zones, strong vertical stratifications, cold and warm pooles and other stability characteristics well corresponding to observed data have been obtained in Semenov's calculations and are shown in these figures.

Figure 6.16 shows the White Sea surface topography in the moment of rising tide. This quite realistic picture is obtained due to the nonlinear kinematic free surface boundary condition.

In general, Semyonov's paper (2004) is the first serious step in modelling complicated nonlinear nonstationary physical processes of the White Sea. The numerical modelling of the White Sea has included many of the processes which regularly regenerate the geostrophic balance being distorted by tidal waves. These processes are the Kelvin waves, the internal waves of tidal period, etc. The totality of tidal movements generates the main stable hydrological structures of the sea: hydrological fronts, cold and warm pooles, a complicated type of stratification and the main peculiarities of the coastal water dynamics.

6.3 The Caspian Sea

6.3.1 Introduction

The Caspian Sea is the largest totally enclosed water body on Earth, constituting 44% of the global volume of lacustrine waters. Compared to other semi-enclosed and enclosed seas of the world, little is known of the Caspian Sea variability. The most urgent, yet unresolved questions relating to the Caspian Sea are: What is the three-dimensional general circulation of the sea and how does it affect transport of pollutants? How is this circulation generated? Through which climatic and dynamic mechanisms is the sea level variability controlled? The phenomenological evidence is too ambiguous or insufficient to give satisfactory answers to these questions.

The Caspian Sea has an elongated geometry (1,000 km in length and 200–300 km in width), where the Northern, Middle and Southern Caspian Basins (respectively the NCB, MCB and SCB) constitute the main geographic divisions, as illustrated by the model bottom topography in Fig. 6.17. The shallow NCB has a maximum depth of about 20 m, while the MCB and SCB have deep troughs with maximum depths of 788 m and 1,025 m respectively. Shelf areas with depths less than 100 m, mainly along the northern and eastern coasts, account for 62% of the total area. The underwater extension of the Apsheron peninsula forms a sill separating the MCB and the SCB, with a maximum depth of about 180 m. The SCB contains two thirds and the NCB makes up 1% of the total volume of water (Kosarev and Yablonskaya, 1994).

The sea surface temperature (SST) in the NCB ranges from below zero under frozen ice in winter to 25–26 °C in summer, while more moderate variability occurs in the SCB, changing from 7–10 °C in winter to 25–29 °C in summer. The seasonal

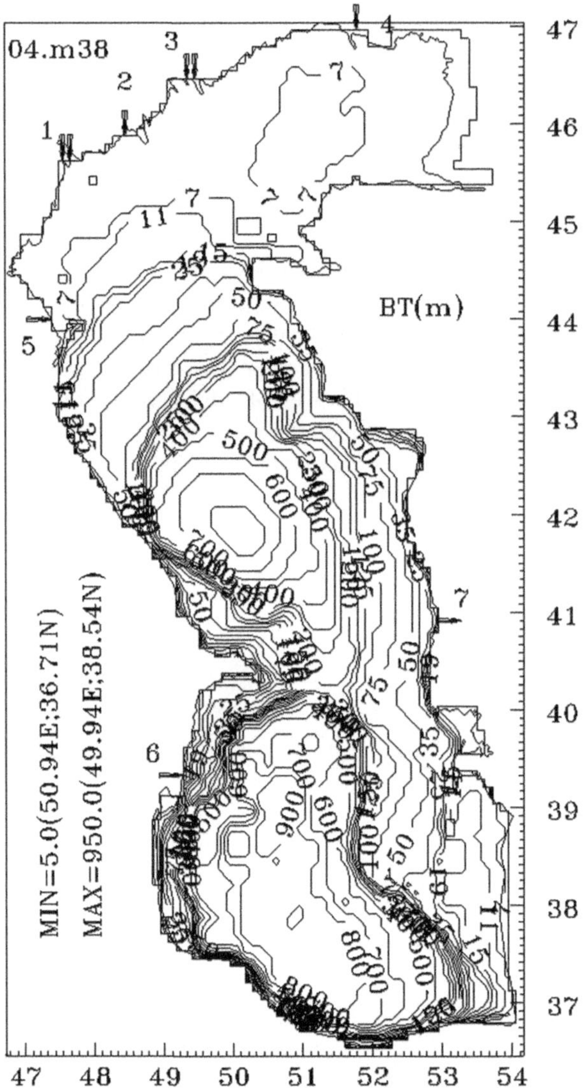

Fig. 6.17 Bottom topography of the Caspian Sea (depths in meters). *Arrows* indicate locations of open lateral boundaries, which include branches of the Volga river: 1 – Bakhtemir (Volga); 2 – Kamyzjak (Volga); 3 – Buzan (Volga), and 4 – Ural; 5 – Terek; 6 – Kura; 7 – outflow to Kara-Bogaz-Gol (Ibrayev et al., 1998)

thermocline occurs at a depth of 20–30 m during the warm season. Seasonal changes in thermal stratification typically reach a depth of 100 m in the SCB and to 200 m in the MCB, while convection is known to reach the bottom in parts of the MCB during severe winters.

The Caspian Sea has low salinity. In deep water areas, salinity varies little with depth (12.80–13.08 psu), and the density stratification largely depends on temperature changes (Terziev et al., 1992). Sharp gradients of salinity occur near the mouths of rivers such as the Volga, where it changes from 2 to 10 psu, typically at a distance of about 20–100 km from the coast.

The elongated geometry and specific topography of the basin, acted upon by variable wind forcing and baroclinic effects result in spatially and temporally variable currents in the Caspian Sea. The general circulation has been described to be cyclonic, based on the investigations carried out from the end of the 19th century till the 1950's, using indirect methods of current measurement (floats, bottles or the dynamic method), and mostly relying on hydrodynamical arguments. Especially standing out among these were the six instrumental surveys along the western coast of the Middle Caspian, carried out in 1935–1937 (Shtockman, 1938), showing predominantly southward currents along the western coast of the Middle Caspian, modified by wind-driven drift currents. A synthesis of these results led to the current scheme of Lednev (1943), which is presented in Fig. 6.18.

Since the 1950's, regular oceanographic observations and current measurements in coastal areas shallower than 100 m have confirmed some circulation features illustrated in Fig. 6.18. Accordingly, the southward currents have been well established along the western coast of MCB; while the northward currents indicated along the eastern coast contradict with summertime observations of surface southward currents in the same region. The cyclonic general circulation of Fig. 6.18 is partially supported by observed northward currents below a surface layer (7–8 m depth) of southward drift currents along the eastern coast. It appears that the southward surface currents along the eastern coast are driven by winds with a prevailing southward component in the eastern halves of the MCB and SCB from spring till autumn. The circulation indicated in the shallow NCB appears to be almost totally controlled by local winds (Terziev et al., 1992; Kosarev and Yablonskaya, 1994).

The Caspian Sea differs from other inland seas, e.g. the Mediterranean and the Black Seas, with respect to an extensive region of upwelling along the eastern coast of the MCB in summer, expressed by a well-defined pattern of cold water detected in satellite images (Sur et al., 1998) and also revealed by climatological temperature fields in the warm season. Although winter upwelling is also possible under favourable winds, detection by satellite remote sensing becomes more difficult in the cold season, as a result of smaller temperature contrasts with the surrounding waters.

The water budget of the landlocked Caspian Sea is extremely sensitive to climatic variability in the surrounding areas. With a large catchment area extending towards the Urals and Caucasia, river runoff dominates the water budget (with an annual average of $\sim 3 \times 10^{11}$ m^3 yr^{-1} and a range of 2.0–4.5 $\times 10^{11}$ m^3 yr^{-1} during the recorded period, (Terziev et al., 1992). Annual precipitation is about one third of runoff, while evaporation is of the same order as runoff. Runoff and evaporation each correspond to about 1 m/yr of sea level change. The water budget depends on climate, but anthropogenic effects such as water regulation schemes also had significant effects, leading to inter-annual, inter-decadal and longer term variations in sea

Fig. 6.18 Scheme of the Caspian Sea surface currents (Lednev, 1943)

level throughout the history of the Caspian Sea (Kosarev and Yablonskaya, 1994; Rodionov, 1994). During 1930–1977, the sea level decreased to −29 m relative to the mean sea level of the global ocean, from the earlier value of about −26 m lasting from the beginning of the century till the 1930's. From 1977 onwards, it increased once again to reach the pre-1930's levels. Rapid sea level change occurred in both of these periods, as indicated in Fig. 6.19 for 1977–1995. Superimposed on these inter-decadal changes, the sea level displays a clear seasonal cycle in Fig. 6.19, as a function of the net water budget (rainfall + river inflow – evaporation). The sea level reaches its lowest seasonal value in winter and increases in the May–July period, following the spring floods. The climatological mean seasonal range of sea level is about 30 cm.

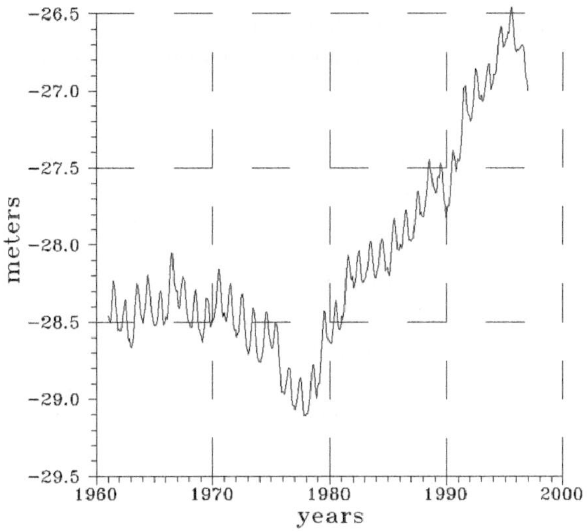

Fig. 6.19 Caspian mean sea level during 1961–1996 (Ibrayev, 2008)

Earlier models of sea level variability (e.g. Rodionov, 1994) generally considered the problem from a stochastic point of view. While stochastic models had a certain level of success, a more fundamental understanding can only be based on controls by hydrological as well as oceanic processes. The total water fluxes across the ocean surface and river mouths determine the mean sea level in an enclosed water body such as the Caspian Sea. Surface fluxes of momentum, water and heat are coupled together, and can strongly be modified by the surface temperature and circulation of the sea. On the other hand, these fluxes are the basic elements of the regional hydrological cycle coupled to the global climate. New findings suggest linkages of the Caspian Sea level to ENSO (El Niño – Southern Oscillation) via the Indian Monsoon, further supported by predictions of multi-decadal fluctuations and increased river discharges associated with global warming scenarios (Arpe and Roeckner, 1999).

In the past, modelling of the Caspian Sea circulation has been rather limited in scope, relying mostly on diagnostic models. A baroclinic diagnostic model (Sarkisyan et al., 1976) has shown the importance of wind stress and summer-time thermal stratification in establishing the circulation. The space-time variability of the summer circulation in response to prevailing north-west and south-east winds was studied by Badalov and Rzheplinski (1989), who combined results from non-stationary models of the NCB and of the upper ocean with a diagnostic model of the deep waters. Akhverdiev and Demin (1989), Kosarev and Yablonskaya (1994), and Terziev et al. (1992) presented a number of diagnostic studies of climatic and synoptic situations. The dynamically adjusted climatic seasonal circulation investigated by Trukhchev et al. (1995) and Tuzhilkin et al. (1997) showed persistent cyclonic and anticyclonic vortices respectively in the north-west and the south-east sectors of the MCB, and anticyclonic vortices in the north-west and the south-east of the SCB, to be the main elements of the circulation. The success of these diagnostic studies was limited by the spatial resolution and quality of the available hydrological data.

Considering the general lack of understanding of the Caspian Sea circulation and sea level variability, Ibrayev (2001) and Ibrayev et al. (2001) aimed (i) to develop a 3-D circulation model with variable water mass in the basin, including air-sea interaction and sea ice thermodynamics sub-models; (ii) to study the seasonal variability of the circulation and sea level, and in particular elaborate on the climatic, dynamical mechanisms controlling sea level.

A description of the model and its forcing is given in Sections 6.3.2 and 6.3.3. In Section 6.3.4 presents the analysis of the seasonal circulation, water budget and the resulting sea level variability in correspondence with heat and evaporation fluxes at the air-sea interface. In Section 6.3.5 the sensitivity of the model results to external forcing and model parameters are considered.

6.3.2 Model Description

6.3.2.1 General Remarks

The closed geometry and size of the Caspian Sea are convenient for numerical modelling. On the other hand, greater constraints are imposed in the formulation of boundary fluxes as compared to semi-enclosed or open seas, as improper accounting of mass or buoyancy fluxes could lead to unrealistic trends of total stored mass, heat and salt in the model.

Sea level change is a direct result of water balance, which depends on the quality of estimation of the river inflows and air-sea water fluxes. River inflows and precipitation, as remotely defined functions, are prescribed in the model, while evaporation depends on local air-sea interaction, i.e. atmospheric parameters (air temperature, humidity, wind speed) near the sea surface and SST. The air-sea interaction module used for computing fluxes is therefore an essential part of the model. Systematic errors in mass fluxes specified otherwise could rapidly contaminate the SST and lead to greatly differing estimates of sea level.

An essential for the model formulation is its capability to simulate the variability of total water mass in the basin. The authors used the kinematic equation at the sea surface allowing a time-varying thickness of the upper layer in correspondence with the continuity equation.

Estimates of the freshwater flow show that the hydrological turnover time for the Caspian Sea is of the order of 200 years and for the shallow NCB of the order of 1 year. As was discussed by Beron-Vera et al. (1999), use of ad hoc surface boundary conditions for salt balance, such as salt relaxation or "virtual" salt flux conditions are unphysical in nature because they create or destroy salt mass. The correct boundary conditions should include the fact that the vast majority of the salt particles remain in the sea during evaporation, and that the precipitated water is essentially pure freshwater. In formulation of boundary conditions for salt, heat and momentum fluxes we follow the approach of Beron-Vera et al. (1999), and of

Roulett and Madec (2000) and add to the usual formulation of air-sea fluxes the terms responsible for freshwater influence.

In the present study the free-surface, primitive equation, z-level numerical Model for Enclosed Sea Hydrodynamics (MESH) is used, described by Ibrayev (2001, 2008), Ibrayev et al. (2001) employing Boussinesq and hydrostatic approximations. The formulation of the free-surface condition in the model allows the propagation of surface gravity waves and mean sea surface elevation changes in response to a non-zero water balance.

6.3.2.2 Governing Equations

The basic equations of the model in spherical coordinates (λ – longitude, φ – latitude, z – depth) are the following:

$$\frac{\partial u}{\partial t} + (\mathbf{v}\cdot\nabla)u + w\frac{\partial u}{\partial z} - fv + a^{-1}tg\varphi u^2 = -(\rho_0 a\cos\varphi)^{-1}\frac{\partial p}{\partial\lambda} + \frac{\partial}{\partial z}\left(K_m\frac{\partial u}{\partial z}\right) + D_u \tag{6.3.1}$$

$$\frac{\partial v}{\partial t} + (\mathbf{v}\cdot\nabla)v + w\frac{\partial v}{\partial z} + fu + a^{-1}tg\varphi uv = -(\rho_0 a)^{-1}\frac{\partial p}{\partial\varphi} + \frac{\partial}{\partial z}\left(K_m\frac{\partial v}{\partial z}\right) + D_v \tag{6.3.2}$$

$$\frac{\partial p}{\partial z} = \rho g \tag{6.3.3}$$

$$\nabla\mathbf{v} + \frac{\partial w}{\partial z} = 0 \tag{6.3.4}$$

$$\frac{\partial T}{\partial t} + (\mathbf{v}\cdot\nabla)T + w\frac{\partial T}{\partial z} = \frac{\partial}{\partial z}\left(K_h\frac{\partial T}{\partial z}\right) + D_T + (\rho_o c_p)^{-1}I_z\cdot(1-A) \tag{6.3.5}$$

$$\frac{\partial S}{\partial t} + (\mathbf{v}\cdot\nabla)S + w\frac{\partial S}{\partial z} = \frac{\partial}{\partial z}\left(K_h\frac{\partial S}{\partial z}\right) + D_S \tag{6.3.6}$$

$$\rho = \rho(T, S) \tag{6.3.7}$$

where $\mathbf{v} = (u, v)$ is the horizontal velocity vector; w the vertical velocity; T, S, ρ the temperature, salinity and density of sea water; ρ_o – mean density; $f = 2\Omega\sin\varphi$ the Coriolis parameter, Ω representing the angular velocity of Earth's rotation; $\nabla\eta = (a\cos\varphi)^{-1}[\frac{\partial}{\partial\lambda}(u\eta) + \frac{\partial}{\partial\varphi}(v\eta\cos\varphi)]$ the two dimensional gradient operator; K_m, K_h the vertical turbulent viscosity and diffusion coefficients for momentum and scalars; D_u, D_v, D_T, D_S the horizontal turbulent viscosity and diffusion terms for momentum, heat and salinity; a the Earth's radius; c_p the specific heat of sea water; I the incoming solar irradiance; A – sea ice compactness. The UNESCO Tech. Pap. In Mar. Sci. was used for equation of state of sea water (see Fofonoff and Millard, 1983).

For stable stratification, we use the Richardson number dependent parameterization of the vertical mixing coefficients proposed by Munk and Anderson (1948):

$$K_m = a_{m0}(1 + \alpha Ri)^{-n} + a_{mb}$$

$$K_h = a_{h0}(1 + \alpha Ri)^{-n} + a_{hb}$$

where $a_{m0}, a_{mb}, a_{h0}, a_{hb}, \alpha, n$ are empirical constants, and Ri is the Richardson number defined as $Ri = g\frac{\partial\rho}{\partial z}\rho_0^{-1}[(\frac{\partial u}{\partial z})^2 + (\frac{\partial v}{\partial z})^2]^{-1}$. In the case of unstable stratification, water is mixed instantaneously with conservation of total heat and salt in mixed volumes of water.

Horizontal mixing terms (D_u, D_v, D_T, D_S) expressed in the form $D_\eta = (a\cos\varphi)^{-1}[\frac{\partial}{\partial\lambda}(A_\eta\eta_\lambda a^{-1}\cos^{-1}\varphi) + \frac{\partial}{\partial\varphi}(A_\eta\eta_\varphi a^{-1}\cos\varphi)]$, where η stands for either one of the velocity components u, v, temperature or salinity T, S, and A_η stands for the horizontal viscosity (A_m) or diffusion (A_h) coefficients.

6.3.2.3 Boundary Conditions

The sea surface evolution equation taking into account water fluxes is (Kamenkovich, 1973; Ibrayev, 2001):

$$w + \frac{\partial\zeta}{\partial t} = \rho_f^{-1}W \tag{6.3.8}$$

with $W = P + M - E$, where $\zeta(\lambda, \varphi, t)$ is the sea surface elevation; ρ_f – density of fresh water; W – water flux; P – precipitation; M – water flux due to ice melting / freezing; E – rate of evaporation.

Upper boundary conditions are specified at the sea surface $z = -\varsigma(\lambda, \varphi, t)$:

$$-K_m\left(\frac{\partial u}{\partial z}, \frac{\partial v}{\partial z}\right) + (u, v)\cdot\rho_f^{-1}W = \rho_o^{-1}(1 - A)(\tau^\lambda, \tau^\varphi) \tag{6.3.9}$$

$$p = p_a \tag{6.3.10}$$

$$-c_pK_h\frac{\partial T}{\partial z} + c_pT\rho_f^{-1}W = \rho_o^{-1}[Q_h^{aw}(1 - A) + Q_h^{iw}A] \tag{6.3.11}$$

$$-K_hS_z + S\rho_f^{-1}W = \rho_o^{-1}S^{iw}MA \tag{6.3.12}$$

where $(\tau^\lambda, \tau^\varphi)$ are the wind stress components; p_a – atmospheric pressure; Q_h^{aw}, Q_h^{iw} – air-water and ice-water heat fluxes; $S^{iw}M$ the rate of salt flux in the sea due to ice melting/freezing. The second terms of the left side of (6.3.9), (6.3.11)–(6.3.12) describe change of salt, heat and momentum content of the surface waters due to fresh water fluxes.

At the sea bottom, $z = H(\lambda, \varphi)$, the corresponding boundary conditions are:

$$w = u(a\cos\varphi)^{-1}\frac{\partial H}{\partial\lambda} + va^{-1}\frac{\partial H}{\partial\varphi} \tag{6.3.13}$$

$$\rho_o K_m \left(\frac{\partial u}{\partial z}, \frac{\partial v}{\partial z} \right) = (\tau_B^\lambda, \tau_B^\varphi) \tag{6.3.14}$$

$$K_h \left(\frac{\partial T}{\partial z}, \frac{\partial S}{\partial z} \right) = 0, \tag{6.3.15}$$

where $(\tau_B^\lambda, \tau_B^\varphi)$ are the bottom stress components.

At lateral walls, the free slip boundary condition and zero heat and salt fluxes are imposed:

$$\mathbf{v}_n = 0, \frac{\partial \mathbf{v}_\tau}{\partial n} = 0 \tag{6.3.16}$$

$$A_h \left(\frac{\partial T}{\partial n}, \frac{\partial S}{\partial n} \right) = 0, \tag{6.3.17}$$

where n and τ represent respectively in the normal and tangential directions to the surface. The model has also inflow and outflow open boundaries. At inflow boundaries horizontal velocity components as well as temperature and salinity are prescribed:

$$(u, v, T, S) = (u^{in}, v^{in}, T^{in}, S^{in}) \tag{6.3.18}$$

while at outflow boundaries only the horizontal velocity components are prescribed

$$(u, v) = (u^{out}, v^{out}) \tag{6.3.19}$$

and the scalars are allowed to advect out of the region with this velocity.

6.3.2.4 Air-Sea Interaction and Sea Ice Models

The heat fluxes at the sea and ice upper boundaries are the sum of long-wave backward radiation, sensible and latent heat fluxes at the sea surface. The momentum, sensible heat and evaporation fluxes are calculated through the air-sea interaction sub-model based on the Monin-Obukhov similarity theory. The bulk transfer coefficients depend on universal functions relevant to the given stability conditions of the atmospheric boundary layer. Inputs for the air-sea interaction sub-model are the air and dew point temperature at 2 m above the sea surface, wind speed at 10 m and the sea surface temperature. The method of iterative flux calculations is based on the approach of Launiainen and Vihma (1990).

Whenever thermal conditions are favourable to form ice, air-sea fluxes are modified to account for the effects of sea-ice, based on the thermodynamic sea-ice submodel of Schrum and Backhaus (1999).

6.3.2.5 Penetration of Solar Radiation

Although more than half of the incoming solar radiation that enters the ocean in the long wave spectral band is absorbed within the top half meter, the remaining short wave fraction, as it penetrates through the surface waters, modifies SST by absorption, which in turn affects the rate of evaporation, leading to an impact on the water balance of the sea. The subsurface profile for solar radiation is computed using the two-band approximation of Paulson and Simpson (1977):

$$I(z) = Q_s[R \cdot \exp(-z/\zeta_1) + (1 - R) \cdot \exp(-z/\zeta_2)] \tag{6.3.20}$$

where Q_s is the downward flux of incoming solar radiation; R is an empirical constant; ζ_1, ζ_2 are respectively the attenuation lengths for long wave and short wave spectral bands of solar radiation. For a one-dimensional model, Martin (1985) has found the simulations sensitive to the optical properties of the given type of seawater. For enclosed and semi-enclosed seas, Timofeev (1983) adopts a value of $R = 0.53$ for the empirical constant. Attenuation lengths for long wave radiation are typically small ($\zeta_1 = 0.033$ m is used, as proposed by Timofeev (1983)), so that total absorption occurs in the first model layer. The attenuation length for short wave band of solar radiation strongly depends on turbidity and differs between coastal and offshore regions. For the Caspian Sea its value is estimated to be about 10–15 m in the central parts of the MCB and SCB, and about 1–5 m in the NCB (Terziev et al., 1992). The short wave attenuation length is parameterized depending on local depth,

$$\zeta_2 = 15\,\text{m}, \text{ for H} > 100\,\text{m and } \zeta_2 = (15\,\text{m}/100\,\text{m})H, \text{ for H} < 100\,\text{m},$$

where H is the depth.

6.3.2.6 The Model Resolution

The grid resolution of the model is $(1/12)^\circ$ in latitude and $(1/9)^\circ$ in longitude, which gives a grid size of about 9.3 km. There are 22 vertical model levels defined at depths of 1, 3, 7, 11, 15, 19, 25, 35, 50, 75, 100, 125, 150, 200, 250, 300, 400, 500, 600, 700, 800, 900 m.

The maximum depth in the model is 950 m, and a minimum depth of 5 m occurs in the shelf region of the NCB. The bottom topography and coastline correspond to the conditions during 1940–1955, when the mean sea level was 28 m below the global ocean level. The model bottom topography in Fig. 6.17 realistically represents the flat NCB shelf, the steep topographic slopes of the SCB and of the western part of the MCB, as well as a number of islands.

6.3.2.7 Initial Conditions

The model is initialized from a state of rest corresponding to the climatologic state of the sea in November (Fig. 6.20) in order to overcome the lack of temperature and

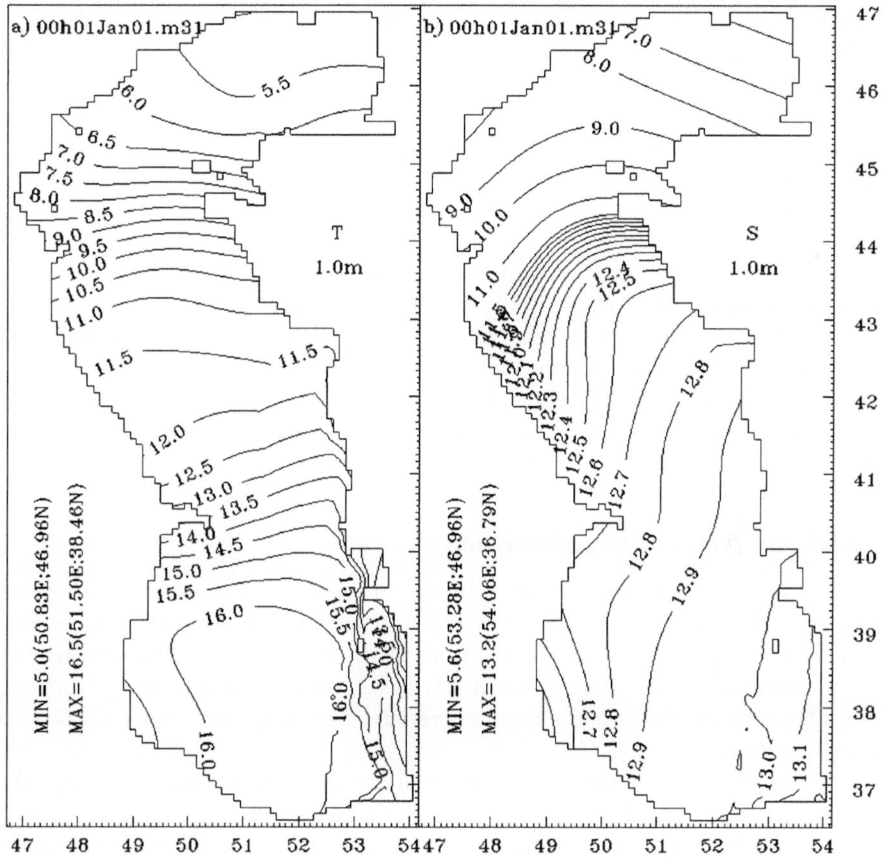

Fig. 6.20 Monthly mean sea surface (**a**) temperature (°C) and (**b**) salinity (psu) based on climatology for November (Ibrayev et al., 2001)

salinity data in winter in regions under ice cover. The data of sea surface temperature and salinity for January is shown in Fig. 6.20.

The values of vertical and lateral mixing coefficients were selected to have the following values: $(a_{m0}, a_{mb}, a_{h0}, a_{hb}) = (50., 1., 10., 0.02) \cdot 10^{-4}\, m^2 s^{-1}$, $\alpha = 1$, $n = 1$, $A_m = 150\,\mathrm{m^2 s^{-1}}$, $A_h = 0.1\,\mathrm{m^2 s^{-1}}$. The time step of integration was 30 min.

The model was run for four years with perpetual seasonal forcing, to ensure that the basin averaged kinetic energy, temperature and general circulation reach quasi-stationary periodical states.

6.3.3 External Forcing

The model forcing is computed from monthly mean atmospheric surface variables based on the ECMWF ERA15 reanalysis data (wind velocity at 10 m height, air and dew point temperatures at 2 m height, incoming solar radiation and thermal back radiation).

The seasonal dynamics was simulated by applying perpetual yearly forcing corresponding to a selected year. Since drastic sea level changes in the last two decades have resulted from imbalances in the external forcing, we select a year with the lowest net sea level change in the period of interest covered by the ECMWF data. Analyses of hydro-meteorological data from Makhachkala, Fort-Shevchenko, Krasnovodsk and Baku indicate 1982 to be a year with small change in mean sea level, the temperature difference between January and December is +6.75 °C. For testing the validity of ECMWF ERA15 data we compare them with climatologic data and statistics from hydrometeorological atlases of Samoilenko and Sachkova (1963) and the books of Kosarev and Yablonskaya (1994) and Terziev et al. (1992) (hereinafter briefly referred to as SS, KY and TKK).

Monthly mean river runoff data were obtained from routine hydrometeorological observations.

6.3.3.1 Air Temperature and Humidity

The characteristic air temperature patterns in winter and summer are shown in Fig. 6.21. In winter, the temperature has a meridional gradient, decreasing from about +8 °C in the SCB to −1 °C in the NCB, with a local minimum near the mountainous west. Air temperature in July has a zonal gradient resulting from contrasts between the desert and mountain regions, increasing from about 22 °C in the northwest to 27 °C in the east.

Throughout the year, vapour pressure is higher in the SCB compared to the other sub-basins and also in the interior of the sea compared to the coastal regions. Maximum vapour pressure occurs in July, reaching values of 27 and 23 mb respectively at the centers of the SCB and the NCB, and decreasing to 11–15 mb along the eastern coast. In February, the vapour pressure decreases from 8 mb at the center of SCB to 1–2 mb in the NCB and in the coastal regions.

Monthly mean air temperature and vapour pressure distributions for 1982 are close to the climatology provided by TKK, except for winter in the NCB, where air temperature from ECMWF is 3–5 °C higher than the values given by the climatology.

6.3.3.2 Wind

The wind speed is typically about 4 m/s during the summer and increases up to 5–6 m/s in winter. Wind speed in winter increases from south to north, exceeding 6.5 m/s in the north (Fig. 6.22a). In summer the maximum wind speed occurs to the east of the Apsheron peninsula. The annual cycle of the monthly mean wind can be divided into three periods: (a) December–January with convergence of winds in the MCB and SCB resulting from the high land-sea temperature contrast in winter, producing local cells of atmospheric circulation with upward motion of the relatively warmer air in the middle of the basin (Fig. 6.22a). (b) February–July when

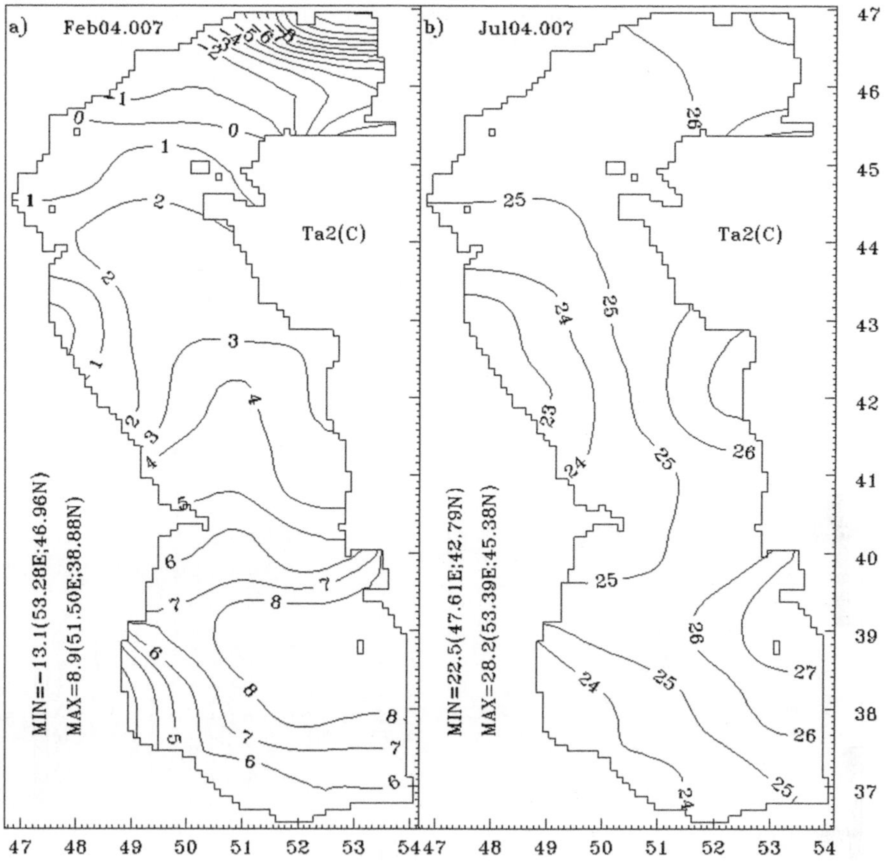

Fig. 6.21 Air temperature (°C) at 2 m height in (**a**) February and (**b**) July of 1982, based on the ECMWF ERA15 data set

large-scale anti-cyclonic winds prevail over the Caspian Sea (Fig. 6.22b), with south south-southwest-ward winds and divergence in the SCB. The local atmospheric circulation in summer in the SCB appears to be the opposite of the winter situation, as a result of the reversed land-sea temperature differences, when the land temperature in the surrounding deserts and steppes exceed 30–40 °C, while the sea is relatively cool. (c) August–November, when average wind direction gradually changes from south-, southwest-ward to westward (not shown here).

Substantial agreement is observed between monthly mean winds computed from the ECMWF reanalysis data for 1982 and the climatologic winds provided by SS on the basis of measurements made at ships and 72 coastal meteorological stations. The consistency between the climatologic means of SS from the 1950's and those derived from ECMWF ERA15 data for the 1980's suggest relatively small climatic change in the character of winds during the 30 years.

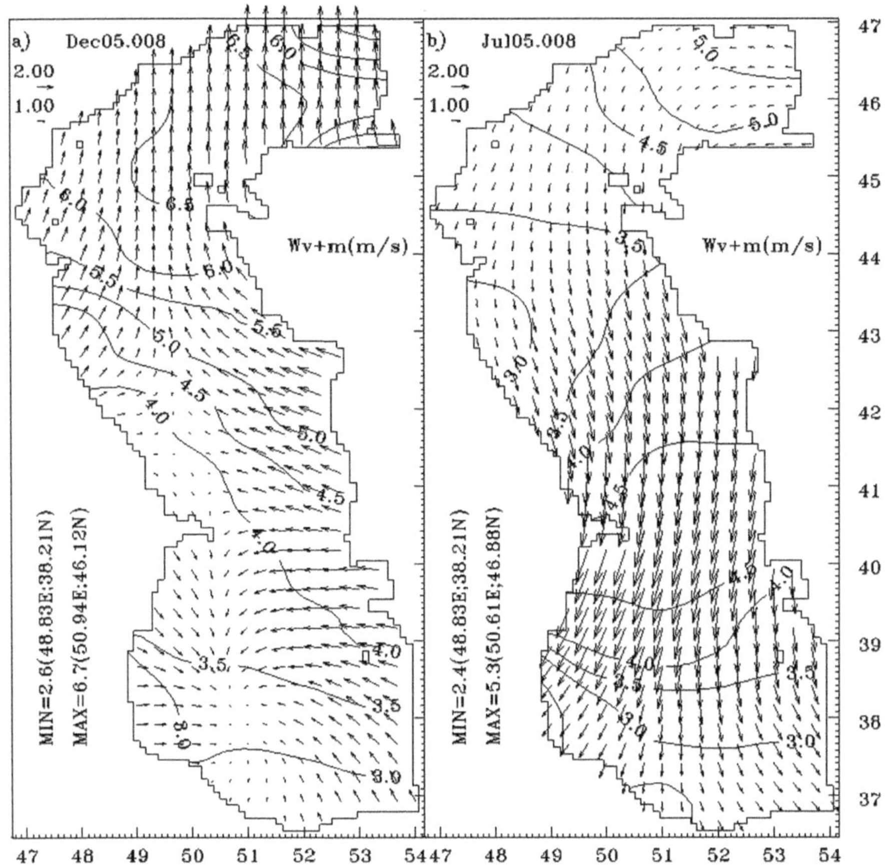

Fig. 6.22 Monthly mean wind velocity vectors and wind speed isolines in (**a**) December and (**b**) July 1982, based on the ECMWF ERA15 data set. The units are m/s

6.3.3.3 Precipitation

Precipitation over the sea is extremely non-uniform. The southwest receives up to 10 times more rainfall compared to the rest of the basin. The data for January is shown in Fig. 6.23a. A maximum value of 238 mm/month in November 1982 occurs in the ECMWF data, while the climatologic data of SS and TKK respectively indicate 300 mm/month and 255 mm/month, both being in October. Summer time rainfall in the western part of SCB is normally about 30–40 mm/month (Fig. 6.23b), but almost completely vanishes in some years. The annual mean precipitation of 340 mm/month in 1982 is close to the maximum value of 366 mm/month in 1993, and much greater than the minimum of 247 mm/month in 1986. Large-scale precipitation based on ECMWF reanalysis data shows good correlation with the 1900–1960 rainfall data of SS and with the climatologic data of TKK.

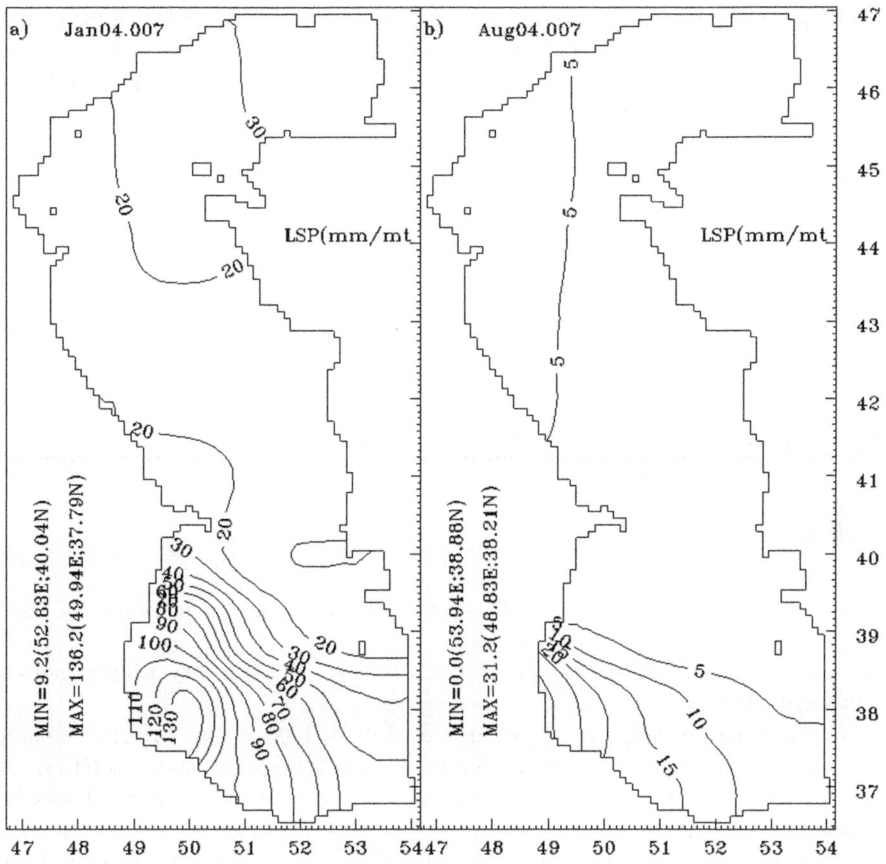

Fig. 6.23 Precipitation (mm/month) in (**a**) January and (**b**) August of 1982, based on the ECMWF ERA15 data set

6.3.3.4 Radiation Fluxes

Radiation flux for the Caspian Sea has a minimum in December and a maximum that typically occurs in June (Fig. 6.24). The most distinguishing feature of the net radiation flux pattern is the decrease from west to east in summer. The annual mean net radiation flux (defined by the sum of solar and thermal radiation fluxes) is 73 Wm^{-2} for the ECMWF data set, which is lower than the climatologic estimates of SS (79.7 Wm^{-2}) and of TKK, the latter one having 101–136 Wm^{-2} for different parts of the sea. Annual radiation flux in 1982 is the lowest in the analysed period, which has a maximum of 76.8 Wm^{-2} in 1985.

The radiative heat flux plays an important role in the heat and water budgets of the Caspian Sea. Because the net radiation flux of ECMWF reanalysis is lower than the climatological estimates, we have increased the solar radiation by 5% to yield a corrected annual mean net radiation flux of 80.6 Wm^{-2}, a value close to estimate of

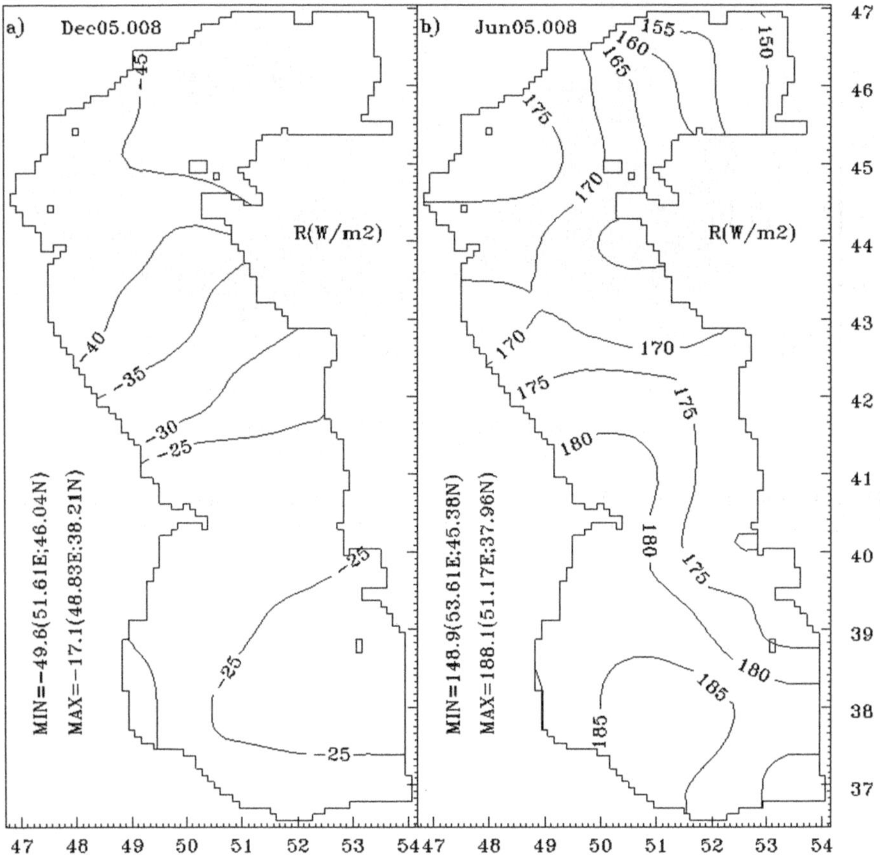

Fig. 6.24 Surface net radiation flux (Wm^{-2}) in (**a**) December and (**b**) June of 1982, based on the ECMWF ERA15 data set

SS. The sensitivity of the model to radiation flux is further discussed in the following sections.

The largest inflow of fresh water comes from the Volga River, accounting for about 80% of the climatological mean river discharge of 250 km^3 yr^{-1} (Kosarev and Yablonskaya, 1994). The mass, momentum and buoyancy inputs from the Volga river all play important roles in the dynamics of the sea. The three main branches of the Volga river delta are idealized in the model as shown in Fig. 6.17. All other sources of freshwater with net annual water flux greater than 10 km^3 yr^{-1}, namely the Ural, Terek and Kura rivers are represented by their corresponding discharges of fresh water. The Kara-Bogaz-Gol on the arid eastern coast acts as an important sink in the water balance, largely as a shallow evaporation basin connected to the sea. Altogether, lateral fluxes are specified at 7 input/output ports in the model. Monthly mean runoff for 1982 are specified for the Volga, Ural, Terek and Kura rivers. There was no outflow to Kara-Bogaz-Gol Bay in 1982. Water temperature at all rivers was taken to be the same as the Volga River, using averages for the 1960–1990 period at the Verkhnee Lebjazhie station, reported in the Water Cadastral Reference book of the USSR Rivers. Seasonal variability in Volga river runoff is extremely high. From July till April, the Volga River discharge is about 5,000–6,000 $m^3\ s^{-1}$. A sharp increase in run-off of up to 15,500 $m^3\ s^{-1}$ occurs in May–June during the spring flood. Different estimates show that up to 3–5% of Volga runoff measured at the Verkhnee Lebjazhie station, located upstream of the delta, is lost due to evaporation in the vast delta of the river (Terziev et al., 1992). To account for this loss, we corrected the Volga river runoff, assuming that the river discharge at the coast amounted to 96% of the inland measurements.

6.3.4 Seasonal Variability of the Caspian Sea Dynamics

6.3.4.1 Three Dimensional Currents

Monthly mean currents at 1 m depth in December, May and August in Fig. 6.25a–c exhibit the dominant seasonal patterns of surface circulation. In December and January, sub-basin-scale cyclonic gyres entirely cover the MCB and SCB, while a number of anti-cyclonic and cyclonic eddies are found in the NCB. This pattern of surface currents is expected, in view of the strong westward wind along the eastern coast and the convergence areas in Fig. 6.22a. The subsurface (25–100 m depth)

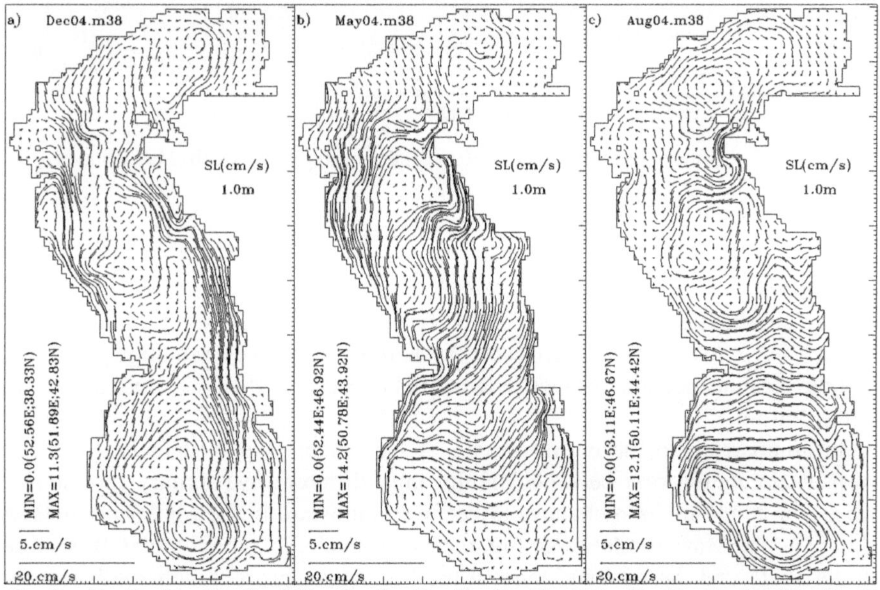

Fig. 6.25 Monthly mean sea surface currents (cm/s) for the months of (**a**) December, (**b**) May and (**c**) August (Ibrayev, 2008)

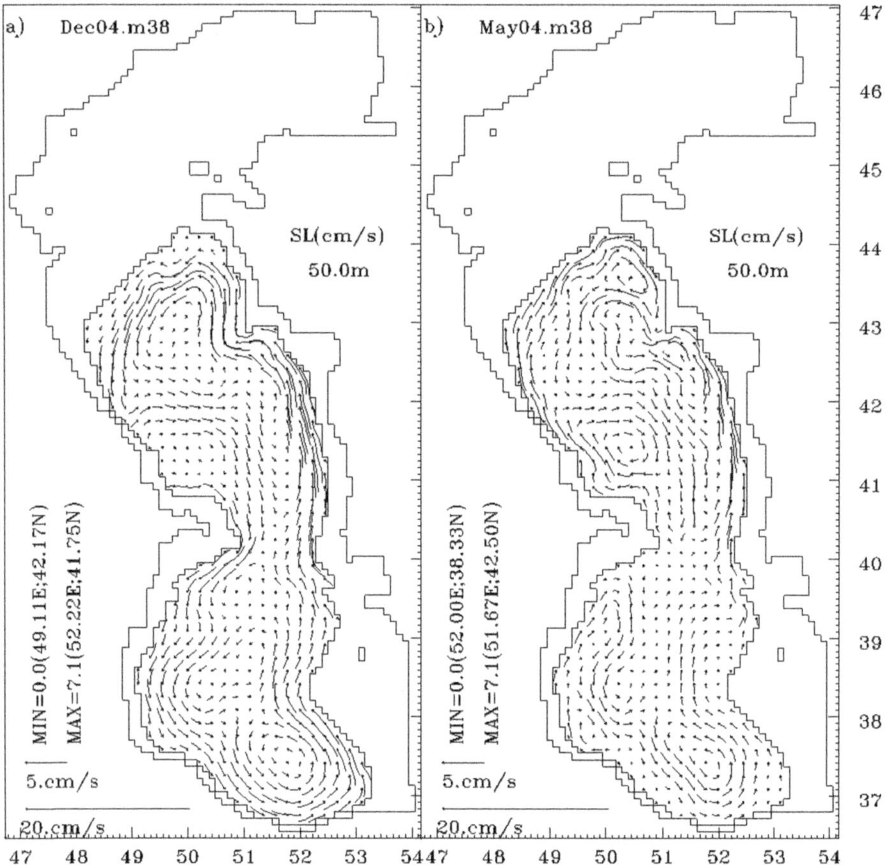

Fig. 6.26 Monthly mean currents (cm/s) at 50 m depth for (**a**) December, (**b**) May (Ibrayev, 2008)

circulation (Fig. 6.26a and b) in December and January correlates well with the surface circulation patterns of Fig. 6.25. Both the MCB and the SCB are occupied by cyclonic gyres connected across the Apsheron sill. Deeper circulation plots reveal that the currents become weaker with depth but preserve their structure all the way to the bottom.

In comparison to the earlier month of December displayed in Fig. 6.25, the surface circulation first becomes significantly different in February (not shown), when the wind direction changes to become southward in the MCB and SCB. As shown by the May circulation in Fig. 6.25b, the cyclonic gyre in the SCB is shifted to the south and the cyclonic gyre in the MCB disappears. South-southwestward Ekman drift currents dominate the deep-sea regions, superimposed on southward coastal currents along the eastern and western shelf regions. The main features of the circulation in May are representative of the period from February till July, which indicates additional small changes in currents along the eastern part of the NCB and cyclonic and anticyclonic eddies near the NCB-MCB boundary.

In February and March the subsurface circulation is gradually modified to become more like Fig. 6.26b, a pattern which is characteristic of the warm period from April till October. The main differences from the cold period are the appearance of anticyclonic eddies west of the MCB and northwest and south of the SCB with a corresponding decrease in the sizes of cyclonic eddies. Below the depth of Apsheron sill (not shown), the cyclonic circulation is preserved in the SCB (~2 cm/s at 300 m depth), but becomes much weaker in the MCB (Ibrayev et al., 1998).

Of particular interest is the presence of southward coastal currents along either the eastern and western coasts. Both current systems are dominant from February till July, when the wind-induced southwestward drift currents at the surface result in offshore transport near the east coast and onshore transport near the west coast, resulting in upwelling and downwelling respectively on these coasts to compensate the surface drift. Both current systems span the continental shelf/slope regions in the form of coastal jets. The coastal current along the west coast and the upwelling along the east are well-documented (Terziev et al., 1992; Kosarev and Yablonskaya, 1994), but often conflicting evidence is found on the eastern coastal current, as a result of its poorly understood horizontal and vertical structure.

The eastern coastal current near the surface occupies a coastal belt shoreward of the 50 m isobath. At a distance 50–100 km from the coast, the alongshore surface current turns offshore to join the surface drift, which is compensated by onshore motion in the subsurface layer, as shown in Fig. 6.27a. In the subsurface layer, slightly offshore of the core of the eastern southward coastal current, exists a northward countercurrent, attached to the slope between 50 and 100 m isobaths, as shown in Fig. 6.27b. A very similar, but narrower subsurface current flowing northward takes place under the western coastal current. The core of the counter-current coincides with the pycnocline, which is stronger and shallower in summer.

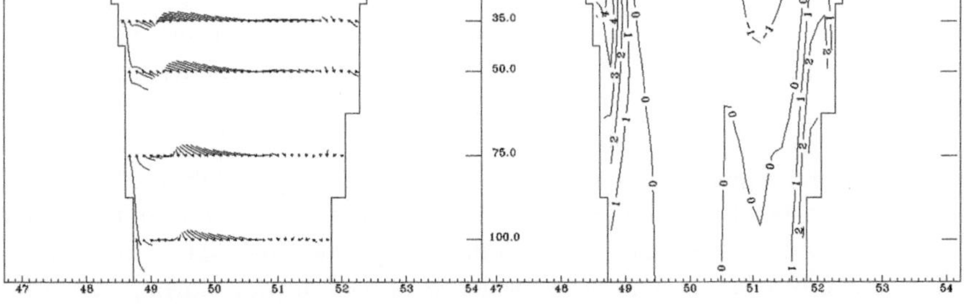

Fig. 6.27 Vertical cross section of currents along 42 °N in May: (**a**) current vectors in the plane of the cross section, (**b**) isotachs of velocity component perpendicular to the section (northward is positive, southward is negative) (Ibrayev, 2008)

After August, the circulation pattern is gradually modified towards the December pattern reviewed earlier. The changes in the circulation are correlated with the changes in wind direction from southward to westward in the MCB and becoming southwestward in the SCB. The earlier southwestward drift in the MCB becomes more west and northwest oriented, as indicated in Fig. 6.25c. In the western part of the MCB, a number of cyclonic eddies start to appear, gradually increase in size, and finally merge together to form the sub-basin scale cyclonic gyre of the MCB in December. In the same period, the eastern coastal current in the MCB becomes weaker and finally turns northward in November. The cyclonic eddy in the SCB once again fills the basin.

6.3.4.2 Water Mass Characteristics

In the shallow NCB, minimum SST occurs in February, when the sea is covered with ice (Fig. 6.28a). With a time lag, the minimum temperatures in the MCB and SCB occur in March, when the NCB starts warming. The winter time SST increases from lower temperatures near the coast to about 12 °C in the interior of SCB. In autumn and winter, the shelf areas are always 5–6 °C cooler than the interior regions, because of the smaller heat capacity of shallow waters.

From December till March, SST in the MCB is characterised by a north-south gradient, and tongues of warm water along the eastern coast and cold water along the western coast. In December and January, these coastal SST anomalies are apparently related to the cyclonic circulation shown in Figs. 6.25a and 6.26a.

A paradox with the tongue of warm water along the eastern coast is that it doesn't disappear in February and March, at a time when the surface currents (Fig. 6.25b) are directed southward. The transport of warm water along the coast is maintained by the northward subsurface current identified earlier (Fig. 6.26b). The warming of surface waters is a result of two mechanisms working in parallel: the upwelling of warm subsurface vein attached to the shelf (Fig. 6.27a) and by vertical mixing of this warm subsurface water with colder surface water.

The west to east increase of SST in the MCB in winter is a characteristic feature of the Caspian Sea climatology (Kosarev and Yablonskaya, 1994). A tongue of warm water extending from the SCB to the MCB along the eastern shelf has been one of the supportive arguments for the existence of the northward current and, hence, of the general cyclonic circulation pattern at the sea surface. While the surface current system in MCB and SCB in December and January supports the scheme of Fig. 6.18, then the existence of a southerly flowing coastal current along the east coast and offshore drift currents, at first appears totally to contradict that scheme. The above analysis shows that the warm water tongue of the eastern shelf of the MCB is produced by a more complex mechanism, i.e. the northward transport of warm water by the subsurface current, followed by upwelling and mixing between surface and subsurface waters. The existence of a northward flowing subsurface counter-current under the southward flowing surface current has been discussed by Kosarev and Yablonskaya (1994). The simulated structure of the currents along the east coast is also supported by observations made to the north of 43°N (Bondarenko, 1993).

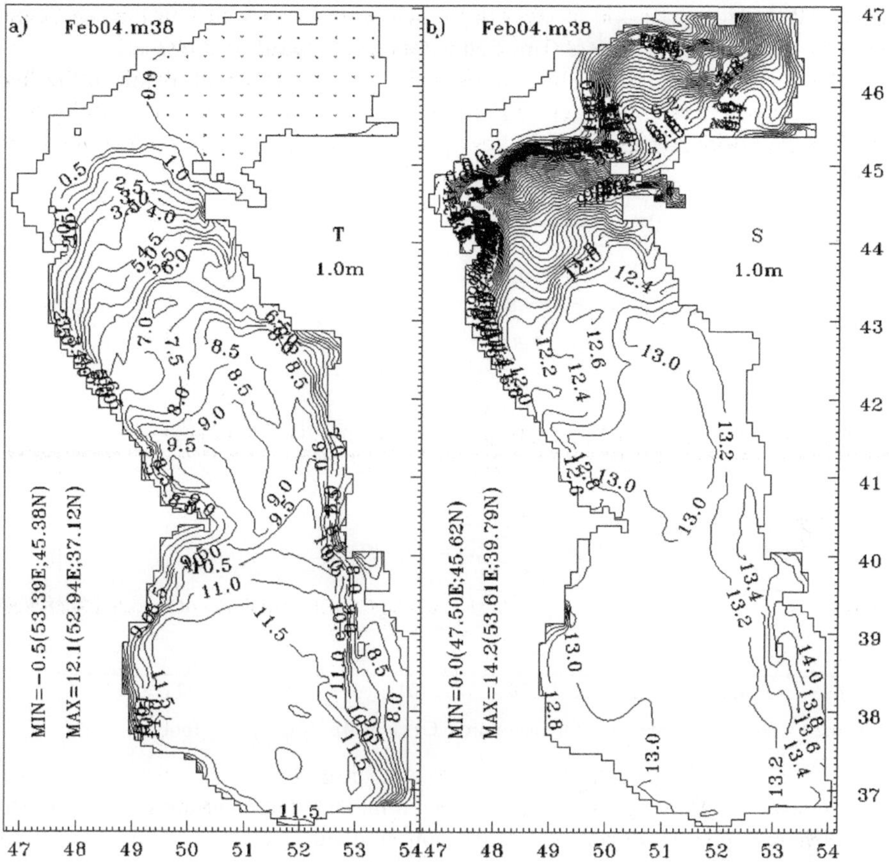

Fig. 6.28 Monthly mean (**a**) sea surface temperature (°C) and (**b**) salinity (psu) in February. Dotted regions in (**a**) indicate presence of ice (Ibrayev, 2008)

The effects of freshwater input from the Volga River, intensive evaporation along the eastern coast of the SCB, combined with the southward flowing coastal current along the western coast create three major salinity fronts (Fig. 6.28b) in the Caspian Sea. Merging of the saline waters from the MCB and the fresh waters discharged into the NCB by the Volga as well as other rivers creates a wide front between them, further enhanced by the depth difference between the two regions. The second, less sharp, salinity front is created on the eastern shelf of the SCB. Here the interior waters meet the more saline water of the shelf produced by excessive evaporation in the region. The third, meridionally stretched front is formed in the MCB, between low salinity waters of NCB transported south along the west coast and the higher salinity waters of the MCB interior.

Winter mixing in the MCB and SCB is strongest in March and reaches 50–75 m depth in the interior regions of the MCB and 100–200 m along the shelf slopes (Fig. 6.29). Newly formed cold water with temperature of about 7–8 °C occupies the upper part of the basin interior, while the temperature is close to zero along the

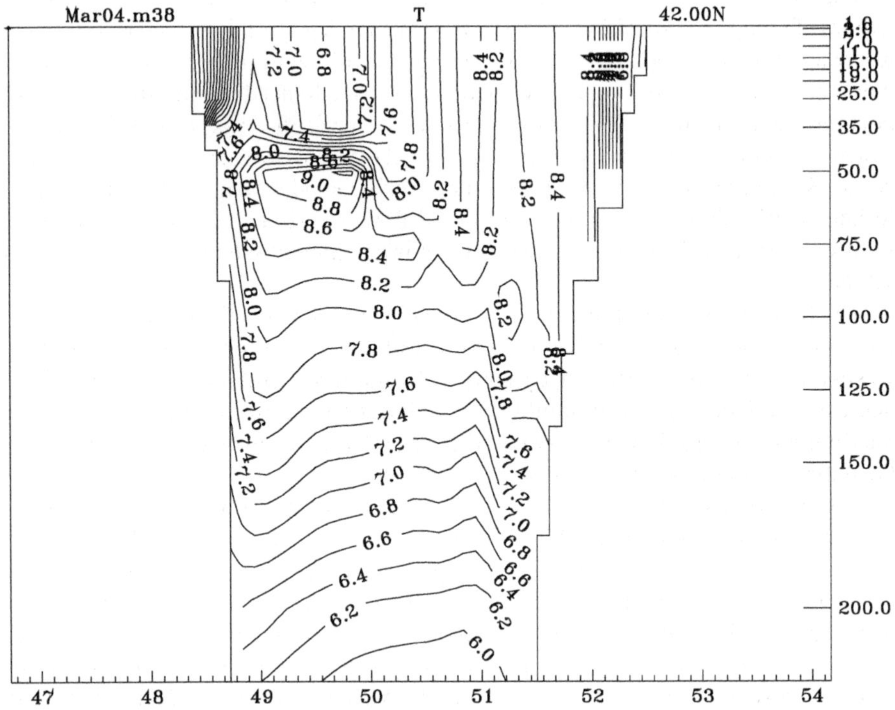

Fig. 6.29 Vertical cross section of temperature (°C) along 42 °N in March (Ibrayev, 2008)

west coast. Observations of dissolved oxygen indicate large variations in the depth of convective mixing, depending on the bottom depth, location and the severity of winter. During moderate and mild winters, convective mixing is shown to reach depths of 150–200 m, as opposed to severe winters when it reaches all the way to the bottom of the MCB.

Winter conditions in the NCB and the coastal regions of the MCB are favourable for freezing of seawater. Observations indicate that ice area and thickness strongly depend on the severity of winter. In the model, ice appears in the first half of December and reaches its maximum extent in the middle of February (Fig. 6.30), when the NCB is totally covered by ice. Maximum ice thickness of more than 70 cm is reached in the beginning of March. Model simulated ice growth, i.e. features such as the start of ice formation, periods of maximum ice cover and thickness, etc. are in agreement with the available observations. On the other hand, comparison with climatologic data shows that the ice edge is often about 50 km more to the south than indicated by the model, often extending south along the eastern coast of the MCB in moderate winters such as 1982. A smaller ice covered area in the model as compared to the observations could be a result of (i) higher air temperature in the NCB obtained from ECMWF data compared to the climatology, (ii) modified heat capacity of the extremely shallow areas of the NCB, artificially made deeper due to numerical stability considerations.

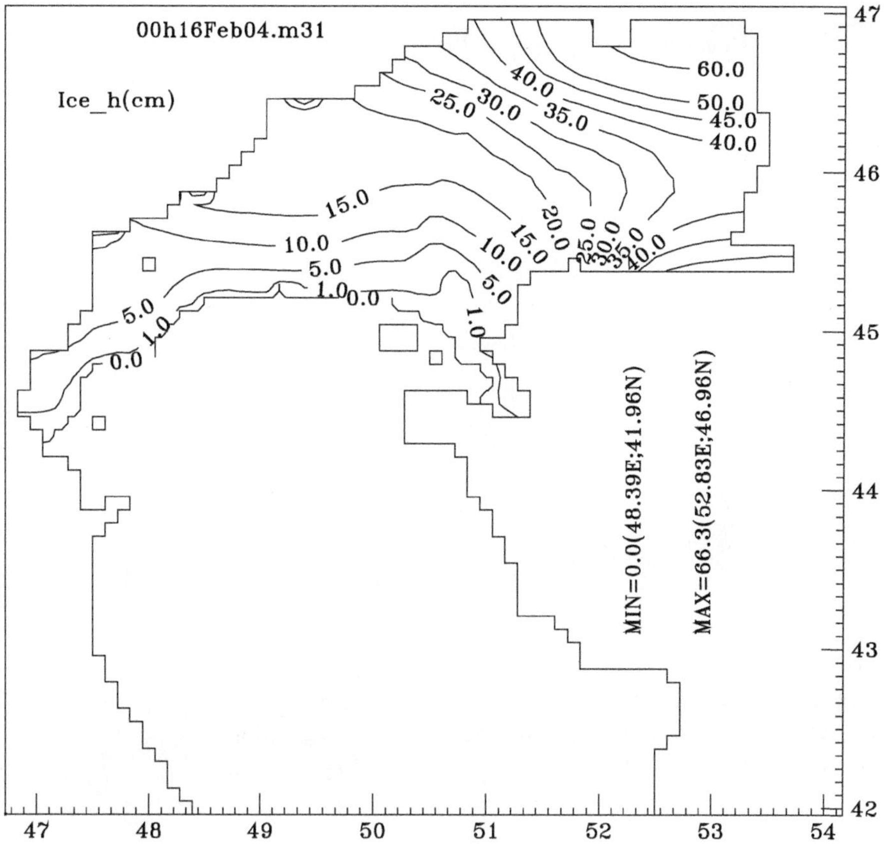

Fig. 6.30 Sea ice thickness (cm) in February (Ibrayev, 2008)

As surface waters start warming in March, the temperature difference between the shelf and the deep-sea regions start to decrease. In the beginning of April, after the melting of ice, the last cold water patch remains in the north-eastern part of NCB, undisturbed by the weak circulation in the region.

The average SST values in July are around 25–26°C, 22–23°C and 25°C respectively in the interior regions of the NCB, MCB and SCB (Fig. 6.31). The surface waters in the shallow coastal regions of the SCB are often much warmer, especially along the eastern shelf, where the temperature approaches 30°C and the salinity is increased by high rates of evaporation near the desert. Cold water with temperature of 14–16°C appears along the eastern shelf of the MCB, in the well-known upwelling region of the Caspian Sea, as a result of the surface drift directed away from the coast. Upwelling along the east coast is a quasi-permanent circulation feature of the warm season, supported by climatologic and satellite (Sur et al., 1998) observations.

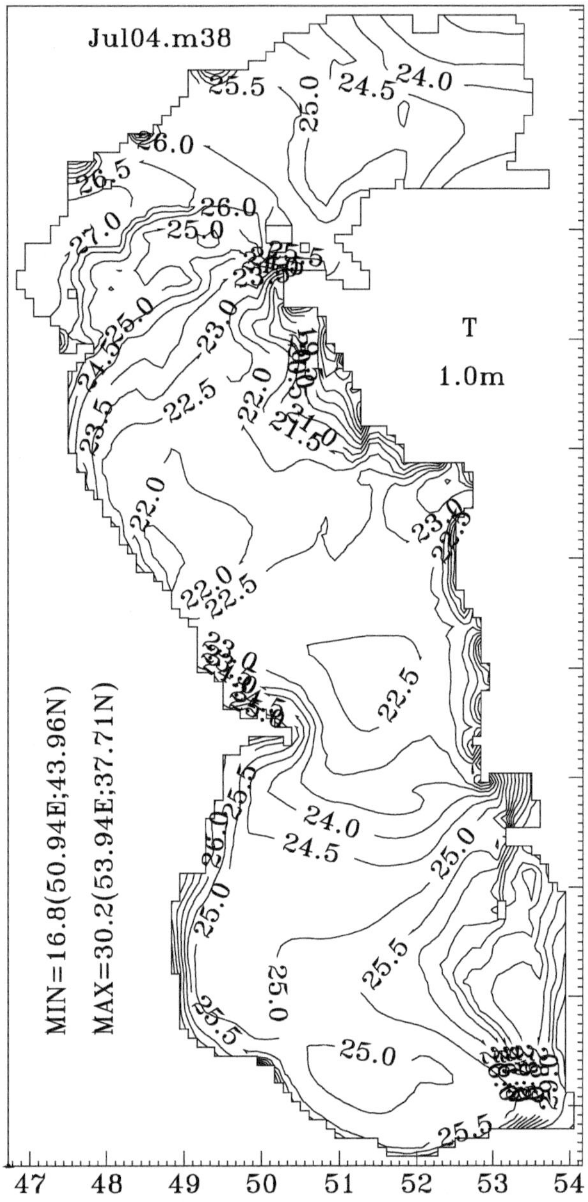

Fig. 6.31 Sea surface temperature (°C) in July (Ibrayev, 2008)

The sea surface salinity distribution in spring and summer is qualitatively similar to the winter (Fig. 6.28b), except for the following minor differences: (i) the spring floods cause the northern salinity front to be moved by up to 50–100 km away from the mouth of the Volga River; (ii) the salinity front of the eastern shelf of the SCB, locked to the slope topography in winter, becomes a much wider transition zone between the saline coastal waters (S = 15.6 psu) and the less saline interior waters

(S = 13.2 psu). Noticeable salinity changes occur along the western coast of MCB in August, when the coastal current is deflected from the west coast and therefore the southernmost penetration of low salinity waters is limited only to as far south as 42 °N.

The surface mixed layer thickness changes from 15 to 25 m on the western coast of MCB and SCB to about 10 m on the eastern coast of MCB, with essentially a two-layer density stratification developing in summer. A vertical temperature difference of 14°C develops across the seasonal thermocline, with the density increasing from 6.7 kg m^{-3} in the upper layer to 9.5 kg m^{-3} at the bottom (Fig. 6.32).

6.3.4.3 Air-Sea Fluxes of Heat and Mass

In this section we focus our attention on the seasonal air-sea fluxes of heat and moisture at the air-sea interface. Our model development for the Caspian Sea is unique in many respects: Flux estimates in the past often have been based on bulk formulae using extremely non-uniform and coarse data sets. We implement a more rigorous, interactive flux estimation scheme making use of oceanic and atmospheric surface properties respectively obtained from the Caspian Sea model and the ECMWF re-analysis atmospheric model relying on global data assimilation. A thermodynamic sea ice model is further used to modify the surface fluxes when ice is formed. The

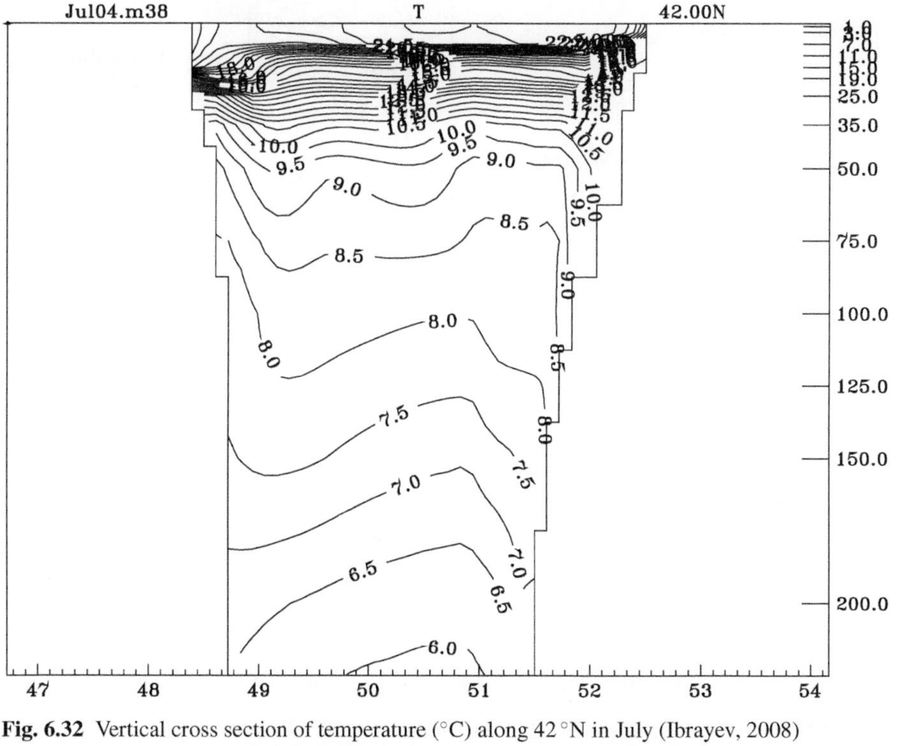

Fig. 6.32 Vertical cross section of temperature (°C) along 42 °N in July (Ibrayev, 2008)

requirement for mass conservation is relaxed in the model to account for a non-zero water budget of the sea with respect to the river and surface volume fluxes.

Estimation of surface heat flux from bulk formulae using monthly mean values of surface wind speed, humidity, air and sea surface temperature, has been shown to differ by less than 10% from computations using all samples by Esbensen and Reynolds (1981), and has been confirmed to have a ratio of 1.02–1.09 for different parts of the Caspian Sea by Panin (1987). We thus employ a correction factor of 1.09 for sensible and latent heat fluxes computed from monthly fields.

Evaporation

In winter, a region of high evaporation in the eastern part of MCB (Fig. 6.33a) results from the combined effects of (i) cold and dry air intrusions from the eastern coast and (ii) warm water from the SCB advected along the eastern coast. The

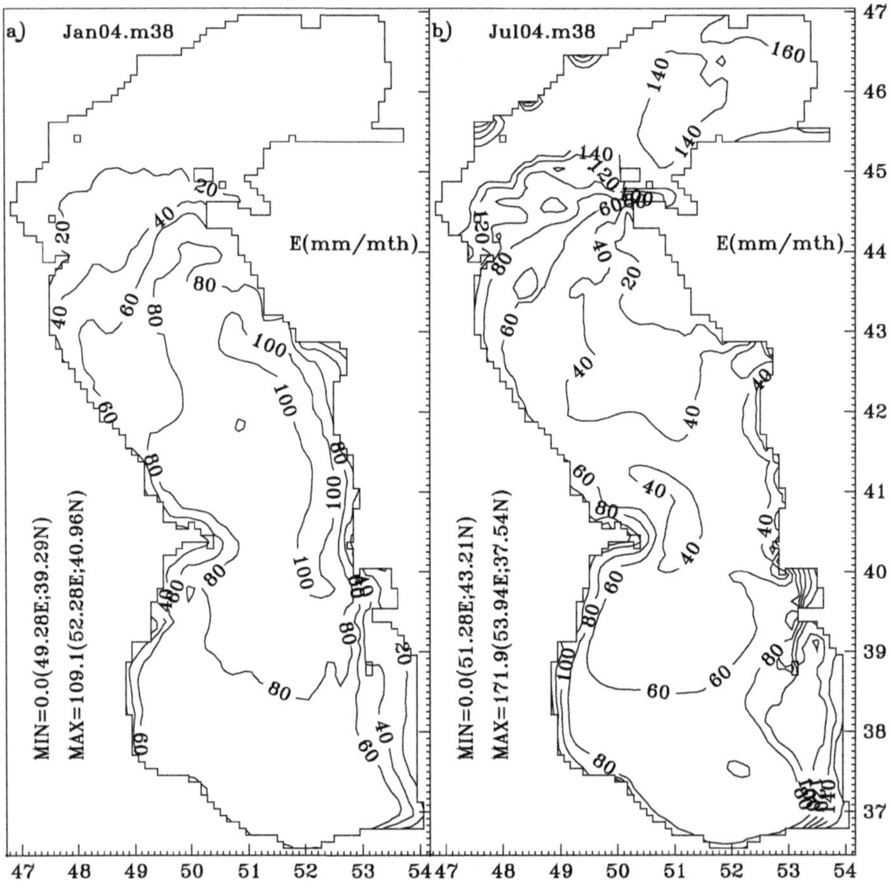

Fig. 6.33 Evaporation (mm/month) in (**a**) January and (**b**) July (Ibrayev, 2008)

summer evaporation pattern is the opposite (Fig. 6.33b): the cold water along the eastern shelf of MCB produces very little evaporation. Evaporation in summer has an increasing trend from north to south, except in the shallow NCB where evaporation is increased to almost twice the deep basin values. Analyses of monthly mean evaporation in the Caspian Sea made on the basis of 150,000 observations (Panin, 1987) are in good agreement with the simulated evaporation both in terms of distribution and magnitude. For example, for the eastern part of MCB, Panin gives $E = 100$–110 mm/month in January and $E < 40$ mm/month in July, which are consistent with our estimates. On the other hand, Panin's estimate of $E > 180$ mm/month in June–August in the southern part of the SCB is much higher than ours in the same region.

Sensible and Latent Heat Fluxes

Monthly and annual mean heat budget components are given Table 6.1. In the annual mean budget, heat influx by solar radiation, amounting to 160.1 W/m^2, is balanced by the outgoing thermal radiation (−79.8 W/m^2), latent heat (−69.9 W/m^2) and sensible heat (−10.4 W/m^2) flux components.

The seasonal cycle of latent heat flux follows that of evaporation. In summer the sensible heat flux almost vanishes as a result of the small difference between SST and air temperature. SST is higher than the air temperature in the NCB and SCB (Panin, 1987), while in the MCB, SST is usually lower than the air temperature as a consequence of upwelling. Sensible heat flux becomes relatively more significant in the heat budget in the autumn and winter seasons, when it is close to the net radiative heat flux.

In winter, the turbulent heat flux (sum of sensible and latent heat fluxes) at the sea surface (Fig. 6.34a) has a maximum along the east coast of the MCB, produced by the interaction of the warm water tongue with overlying cold air (Figs. 6.21a

Table 6.1 Monthly mean heat flux components (W/m^2) at the sea surface. Q_s = incoming solar radiation, Q_b = (long wave) thermal radiation, $Q_s + Q_b$ = total radiative heat flux, H = sensible heat flux, LE = latent heat flux, Q = total heat flux

Month	Q_s	Q_b	$Q_s + Q_b$	H	LE	Q
1	65.1	−79.2	−14.0	−22.6	−59.2	−95.8
2	94.4	−76.5	17.9	−23.2	−57.3	−62.7
3	146.3	−72.9	73.5	−4.9	−38.3	30.3
4	208.3	−68.0	140.3	4.3	−25.7	118.8
5	247.9	−73.5	174.4	−2.3	−43.5	128.6
6	269.5	−84.7	184.8	−0.3	−63.1	121.4
7	257.1	−84.4	172.7	0.9	−76.0	97.5
8	228.9	−91.7	137.2	−1.8	−109.9	25.5
9	175.7	−90.5	85.2	−7.4	−109.6	−31.9
10	108.7	−79.7	29.0	−21.6	−109.9	−102.4
11	65.9	−73.4	−7.6	−27.5	−90.2	−125.2
12	57.0	−82.6	−25.6	−18.6	−56.7	−100.8
Annual mean	160.4	−79.8	80.6	−10.4	−69.9	0.3
Standard deviation	77.5	7.0	84.5	10.9	27.9	94.4

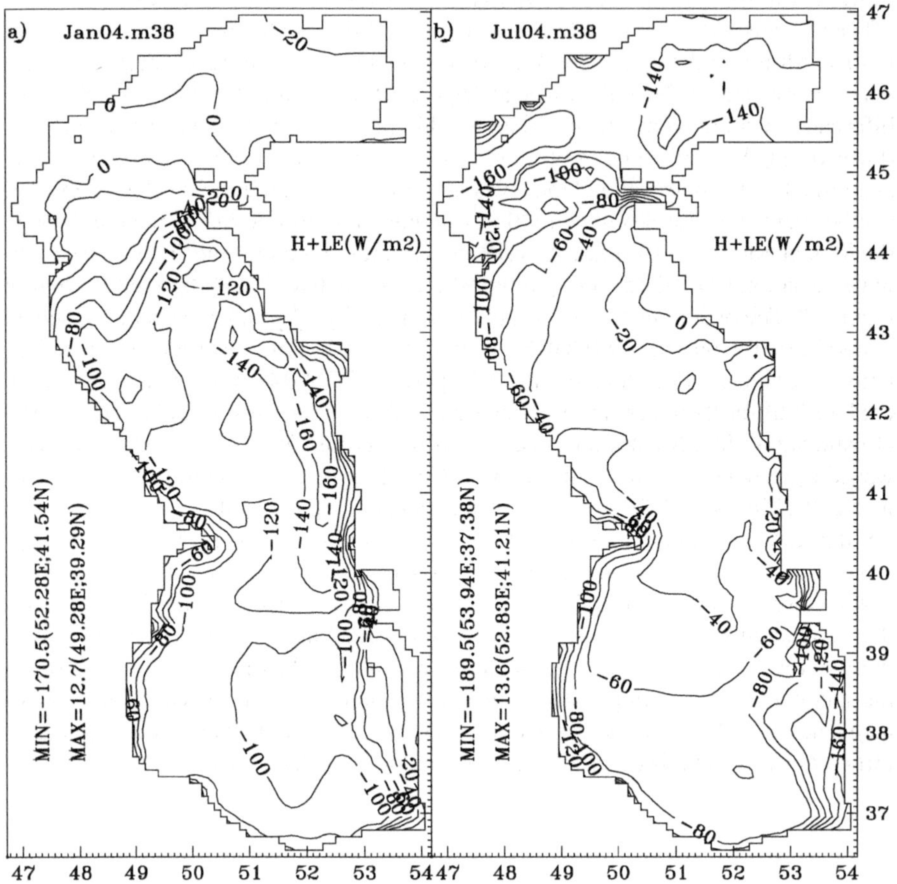

Fig. 6.34 The sum of sensible and latent heat fluxes in (W/m^2) in (**a**) January and (**b**) July (Ibrayev, 2008)

and 6.28a). In this area both terms of the turbulent heat flux are of the same order ($-250\,W/m^2$ for latent and $-150\,W/m^2$ for sensible heat flux), whereas in other parts of the sea, latent heat flux is often 2–5 times larger than the sensible heat flux. In July (Fig. 6.34b), sensible heat flux in the same area decreases to about (-10)-$(+20)\,W/m^2$ and latent heat flux dominates in total flux.

6.3.4.4 Sea Level Variability and Water Budget

Mean Sea Level Variability

Time series of model simulated and observed sea level anomaly at four Caspian Sea stations are shown in Fig. 6.35 for the year 1982. Station locations are shown in Fig. 6.36. Common features of the sea level time series at all four stations are the minimum in September–October, the rising trend from autumn to spring, followed

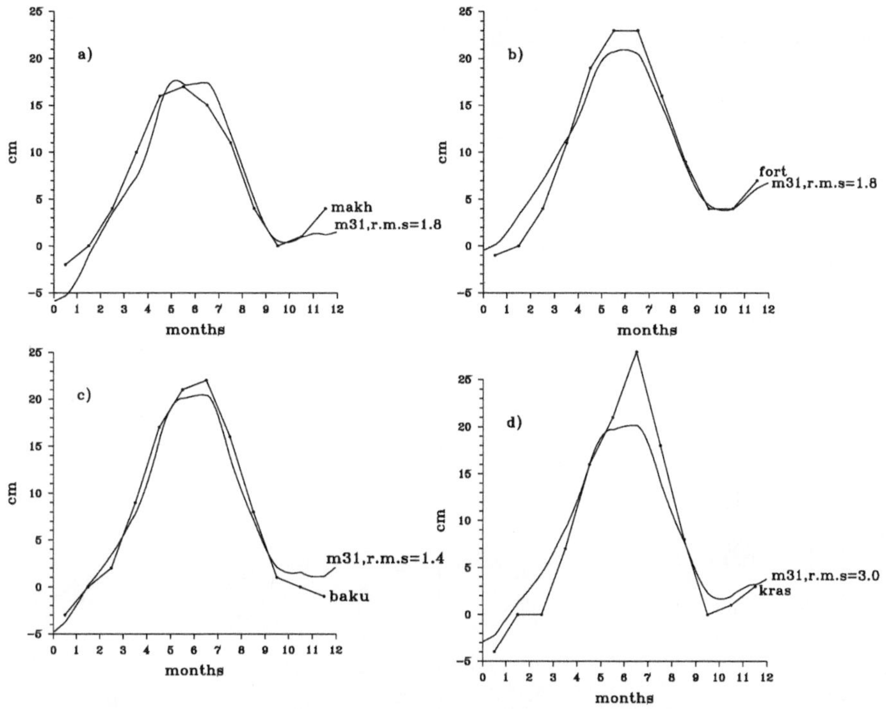

Fig. 6.35 Sea level anomaly (cm) in 1982 at different locations around the Caspian Sea: (**a**) Makhachkala; (**b**) Fort-Shevchenko; (**c**) Baku and (**d**) Krasnovodsk. *Smooth curves* show model results, the *broken lines* observations (Ibrayev, 2008)

by the fall in summer. There is a net rise in sea level at the end of the presented one year period because the sea level is on a rising trend in the longer term. Good correlation in terms of amplitude and of phase between simulated and observed sea level demonstrates the model capability to reproduce key hydro- and thermo-dynamical processes of the Caspian Sea, of which the sea level is an integral measure. Root mean square difference between the simulated and observed curves changes from 1.4 cm at Baku to 3.0 cm at Krasnovodsk stations, representing 6 and 13% of the annual sea level amplitude at the respective sites.

Spatial Variability of the Sea Surface Topography

The leading factors creating the observed spatial structure of sea surface topography in the model are baroclinicity and wind setup. Although the atmospheric forcing is variable in space and time, the following features emerge from a study of the seasonal cycle: East-west asymmetry in buoyancy resulting from the distribution of fresh water flux components (run-off, precipitation and evaporation). All major rivers enter the sea along the northwest coast. In the absence of other forcing, the fresh water introduced by the Volga river tends to flow along the western coast, as a result of deflection by the Coriolis force. Precipitation and evaporation over

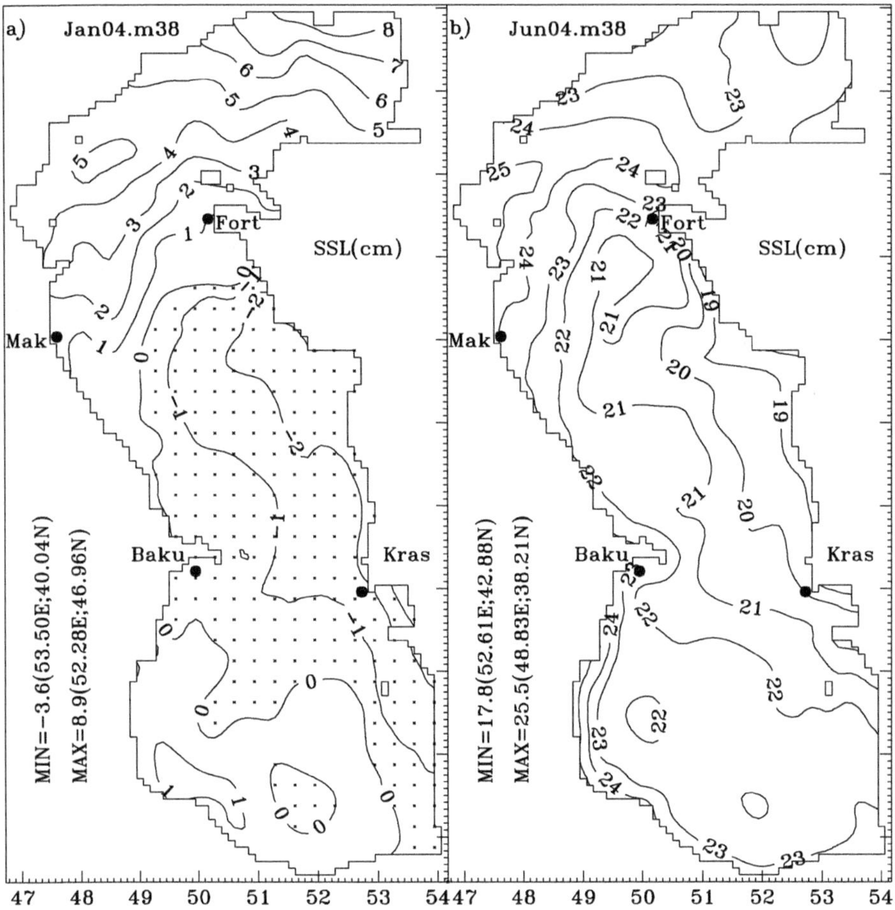

Fig. 6.36 Sea surface topography (cm) in (**a**) January and (**b**) June. The locations of hydrometeorological stations Makhackala (Mak), Fort-Shevchenko (Fort), Baku and Krasnovodsk (Kras) are also indicated (Ibrayev, 2008)

the sea are extremely non-uniform. The southwest part of the SCB receives rainfall that is up to 10 times greater than in the other parts of the sea (Fig. 6.23), while the evaporation has somewhat smoother variation over the basin. The predominant southward winds in the MCB and SCB are favourable for drifting surface water off the eastern coast, thus producing typical coastal upwelling of cold and saline sub-surface waters along this coast. All of the above factors support asymmetrical distribution of salinity, yielding high density waters on the eastern shelf and relatively low density waters on the north and southwest parts of the sea, supporting the model produced west-east slope of the sea surface topography in Fig. 6.36. The spatial range of sea surface topography is minimum in July (7.5 cm) and maximum in December (15 cm), which is 2–3 times smaller than the seasonal range of sea level variations.

Water Budget

Estimates of Caspian Sea water balance terms based on various sources (Table 6.2) do not differ significantly from each other. The model simulated evaporation rate, of 86.5 cm/yr, is close to the estimate of Bortnik and Nikonova (1992) for the period 1978–1990 and to the climatologic estimate of 96.3 cm/yr of Panin (1987).

The river runoff in 1982 was significantly lower than in the other years, amounting to only 73% of mean runoff value for the 1978–1990 period. It should be noted that inter-annual variability of river runoff is extremely high. For example, the difference between maximum and minimum yearly runoff in the last century is about 260 km^3/yr or more than 100cm/yr in terms of sea level rise (Bortnik and Nikonova, 1992).

Low river runoff in 1982 is compensated by high precipitation. The ECMWF reanalysis data for 1982 gives a value 28% higher than the estimates of Bortnik and Nikonova (1992) for the 1970–1977 and 1978–1990 periods. Precipitation in ECMWF reanalysis actually could have been underestimated, if one takes into account similar estimates elsewhere at around the same latitude, yielding 20% less precipitation compared to observations (Betts et al., 1999).

Underground water flux into the sea is the least known component of the water balance, with different authors' estimates in the range 0.3–49.3 km^3/yr (Bortnik and Nikonova, 1992). As most estimates are relatively small, on the order of 3–5 km^3/yr or about 1 cm/yr of mean sea level change, ground water inflow was not taken into consideration in the model.

In Table 6.3.2, also given the mean sea level increment estimated from observed river runoff data and precipitation/evaporation based on ECMWF reanalysis data. The difference of 9.2 cm/yr between ECMWF estimated and observed sea level change gives a measure of error of the modern estimations of the water balance.

6.3.5 Sensitivity Experiments

As already pointed out, the water budget of Caspian Sea is extremely sensitive to climatic variability. In fact, the present sea level variations are much smaller than the contributing terms of river runoff, precipitation and evaporation, which tend to balance each other. Therefore, small differences in water budget components can lead to large changes in sea level. It is a widespread opinion that inter-annual variability of sea level is controlled by river runoff while anomalies of precipitation and evaporation have less impact on the range of sea level fluctuations (Rodionov, 1994).

A set of experiments was performed by R. Ibrayev to ascertain the sensitivity of the circulation and sea level change to external forcing and model parameters. We refer to the above experiment with the parameters and forcing given earlier, as the control run (CR). In the following the experiments are presented with further variations in model parameters and of external forcing, using the beginning of the fourth year of the CR as initial conditions.

Table 6.2 Water budget of the Caspian Sea. Water fluxes are given in cm/yr, expressed in units of mean sea level change

Period and source	River runoff	Underground inflow	Precipitation	Evaporation	Outflow into Kara-Bogaz-Gol	Sea level increment
1970–1977 (Bortnik and Nikonova, 1992)	66.7	1.1	24.3	−103.9	2.0	−13.8
1978–1982 (Baidin and Kosarev, 1986)	85.8	–	25.7	−97.9	0.7	14.0
1978–1990 (Bortnik and Nikonova, 1992)	81.9	1.1	23.0	−91.8	0.5	13.7
1982 (model)	59.5	0	34.0	−86.5	0	7.0
1982 (ECMWF data and observations)	59.5 (observed)	0	34.0 (ECMWF)	−95.7 (ECMWF)	0	−2.2

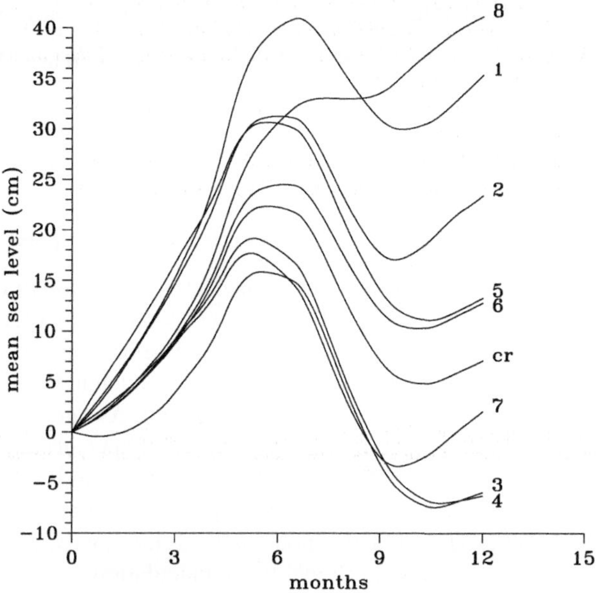

Fig. 6.37 Mean sea level seasonal cycle (cm) obtained from sensitivity experiments (Ibrayev, 2008)

In the first two experiments was examined the sea level change and circulation in response to variations in prescribed water budget components, i.e. river runoff and precipitation. Next are the experiments with variations of atmospheric parameters, which determine the evaporative flux, i.e. the air and dew point temperature, and wind speed. In further experiments the sensitivity of the model to solar radiation and to parameterisation of the solar radiation penetration into the sea was examined. In the last experiment, was analysed the sensitivity of the model to variations of vertical mixing. In most of these sensitivity experiments was implemented external forcing, which varied from the control run comparable with the observed inter-annual variability in them. Seasonal sea level changes derived from the sensitivity experiments are shown in Fig. 6.37.

6.3.5.1 Experiment 1: 50% Increase in River Runoff

Runoff of all rivers was increased by 50%. The reaction of the water budget and of sea level is quite expected and is almost linear. Mean sea level compared to the CR is increased by 28.3 cm/yr, corresponding to a value less than the 29.8 cm/yr that would be obtained by linear extrapolation. The small non-linear response of the water budget is related to increased stability of the water column followed by an increase of SST, which leads to excessive evaporation compared to the CR. The main consequence of increased runoff on circulation is the extension of western coastal current towards the south.

6.3.5.2 Experiment 2: 50% Increase in Precipitation

Based on linear extrapolation one expects an increase of 17 cm/yr in sea level when precipitation is increased by 50%, the model gave an realistic increase of 16.4 cm/yr. As in the first experiment, there is more stable water column especially in the SCB, due to excessive precipitation.

6.3.5.3 Experiment 3: 50% Increase in Wind Speed

Much stronger non-linear reaction occurs in the case of increased wind speed. Based on linear extrapolation, the sea level would be expected to drop by 43.2 cm/yr, amounting to 50% increase in annual evaporation. Because mixing is enhanced by increased wind stress, both the sensible and latent heat fluxes are affected, thus considerably modifying the expected reaction. As cooling is increased, SST decreases by 1–2 °C compared to the CR, decreasing the surface humidity yielding a smaller specific humidity difference at the sea surface and at 10 m height ($q_{surface} - q_{10m}$). The increased wind stress also changes the circulation. The overall decrease of mean sea level produced by the model is 13.0 cm/yr, which is about 3.3 times smaller than expected from a proportional linear calculation.

6.3.5.4 Experiment 4: 5°C Warmer Air Temperature

When air temperature is warmer and the dew point temperature is the same as that used in CR, the obvious reaction of the model is an increase of SST (up to 2°C in summer) and corresponding increase of specific humidity at the sea surface, resulting in more unstable atmospheric boundary layer with an increased Danton number C_E by a factor of 1.2–1.3. More intensive evaporation, as a result of larger difference $C_E(q_{surface} - q_{10m})$, results in a sea level drop of 13.4 cm/yr as compared to CR.

6.3.5.5 Experiment 5: Warmer and More Humid Air

An increase of the dew point temperature by 5°C compared to Experiment 4 affects the water balance in the opposite direction. Higher specific humidity of air prohibits the excessive evaporation observed in the previous experiment, leading to 6.2 cm/yr higher rise of sea level compared to the CR and of 19.6 cm/yr as compared to Experiment 4. As a result of the restricted latent heat flux, the surface waters are warmer by 2°C.

6.3.5.6 Experiment 6: Solar Radiation Heat Flux Without Correction

In the CR the used ECMWF solar radiation heat flux was corrected by a factor of 1.05 to ensure better correspondence to climatologic estimates. The primary influence of that correction was on the SST. This experiment was made without correction. The resulting SST was for 1°C lower as compared to climatology, though the seasonal cycle of currents and of upper mixed layer depth show very little change. In response to the lower SST there was lower evaporation, with an annual budget giving 5.7 cm/yr higher rise of sea level as compared to CR.

6.3.5.7 Experiment 7: Absorption of Solar Radiation at the Sea Surface

In this experiment the sensitivity of the seasonal cycle of sea level to parameterization of solar penetration into the sea was checked. In Fig. 6.36 the attenuation length for the short fraction of the solar radiation was put equal to 0.033 m, such that all the solar radiation is absorbed in the first model layer. Compared to the CR, SST starts to increase much faster in the spring to produce a sharper thermocline and a shallower upper mixed layer. As a result, higher values of SST and of evaporation were obtained in the period from January till September. On the other hand, the balance between sensible and latent heat release from the sea is modified such that relatively larger part of the heat flux was accounted by sensible heat flux. In May, the sensible and latent heat fluxes are equal to $(-10.8; -65.2)\,\mathrm{W/m^2}$ compared to $(-2.3;\ -43.5)\,\mathrm{W/m^2}$ in the CR, which means that the heat formerly released through evaporation or stored in the upper mixed layer is now released through sensible heat. In the period from September till December, the balance between sensible and latent heat fluxes is changed. In November there were $(-19.0; -73.6)\,\mathrm{W/m^2}$ for sensible and latent heat fluxes as compared to $(-27.5;\ -90.2)\,\mathrm{W/m^2}$ in the CR. The reduced penetration of heat into the sea results in stronger surface currents and more pronounced influence of baroclinicity on the circulation pattern.

6.3.5.8 Experiment 8: Constant Vertical Mixing Coefficients

This experiment is designed to explore how the specification of mixing coefficients modifies the circulation. Values of K_m and K_h were constant now and equal to the maximum values of $(50., 10.) \times 10^{-4}\,\mathrm{m^2s}^{-1}$ used in the CR. The reaction of the sea circulation was dramatic, sea currents become almost barotropic. The major effect of increased vertical mixing is a lowering of the SST by about 11°C in August and a corresponding decrease of evaporation from the sea surface as compared to the CR. Annual sea level rise increased by 34.1 cm/yr compared to the CR. The fall of sea level in August-October typical of the seasonal cycle in the CR is now absent, due to insufficient heat stored in the summer.

6.3.6 Summary and Conclusions

A coupled sea hydrodynamics – air/sea interaction – sea ice thermodynamics model has been developed by Ibrayev (2001) to simulate the intra-annual variability of the Caspian Sea circulation and sea level. Complex bottom topography including large shallow areas, wide shelf and slope regions, interconnected sub-basins, large freshwater inflows, sensitive response to atmospheric forcing, sea ice formation, and the observed level of climatic variability combined with man-made changes in hydrology make the Caspian Sea a challenging basin for testing and improving coupled models.

The model results and climatologic/observational data are analysed (Ibrayev et al., 2001) together to compare and discuss basic features of the simulated circulation and sea level change. The analyses reveal some fundamental features of the Caspian Sea circulation: summer upwelling along the eastern coast, west-east asymmetry in temperature which reverses between summer and winter in response to changes in upwelling and mixing dynamics, and an additional asymmetry in salinity resulting from fresh water influence. An important conclusion regarding circulation is the existence of cyclonic currents only in a depth averaged sense, considering the upper 50–100 m. The earlier arguments for a cyclonic circulation in the MCB were partly based on the transport of warm water along the eastern shelf in winter. These results show persistent northward transport by subsurface currents along the eastern shelf slope, while the surface currents are more often directed southward. The analysis of observation made in the 1960's (Kosarev and Yablonskaya, 1994) and more recent times (Bondarenko, 1993) support the simulated current structure of the eastern shelf.

An important part of the work is the simulation of sea surface topography, yielding verifiable results in terms of sea level. Analyses of heat and water budgets confirm climatologic estimates of heat and moisture fluxes at the sea surface. Experiments performed with variations in external forcing suggest a sensitive response of the circulation and the water budget to atmospheric and river forcing, comparable with the inter-annual variability observed in these fields.

It is also shown that the model thermodynamics and the formulation of boundary conditions are not close to being perfect, although it is difficult to ascertain the sources of the discrepancies at this stage. For example, the correction of ECMWF derived solar radiation by a factor of 1.05 is justified partly by the corresponding climatologic estimate, and partly by closing the difference between simulated and climatologic fields of SST and sea level. It is of course quite possible that some drawbacks in model thermodynamics is hidden from the eye by this operative adjustment of solar radiation. It seems advisable for the future to further improve model physics (better specifications of exchange coefficients, water import and export processes, and air-sea fluxes, especially in shallow regions), as well as the quality of external forcing (to include diurnal variation and better air-sea coupling), with particular emphasis on the simulation of deep water ventilation processes and upwelling dynamics, especially on the eastern coast.

6.4 The Black Sea

6.4.1 The Black Sea Descriptive Hydrophysics

We refer to the paper of Ozsoy and Unluata (1997), one of the best reviews on descriptive thermohydrophysics of the Black Sea. The surface area of the Black Sea

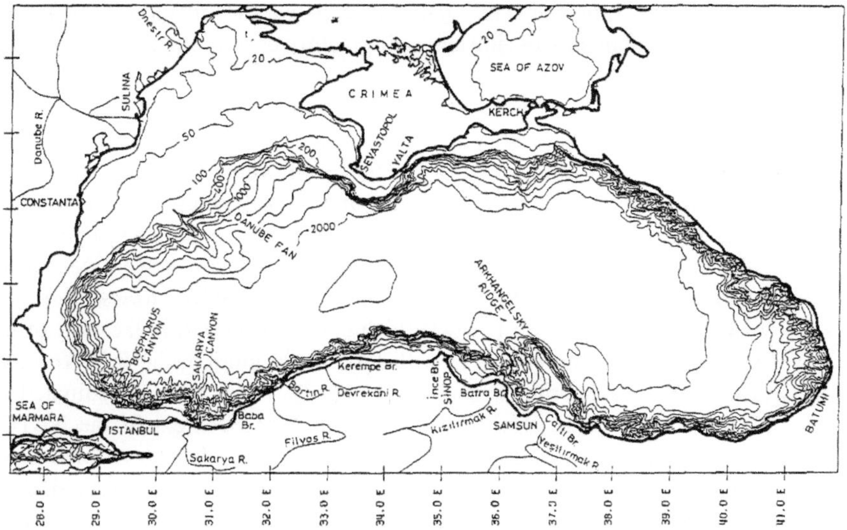

Fig. 6.38 The Black Sea bottom topography Depth contours are labeled in m (Ozsoy and Unluata, 1997)

is about 4.2×10^5 km^2 and the volume $\sim 5.3 \times 10^5$ km^3. The continental slope of the Black Sea is very complicated (see Fig. 6.38). Its waters have a restricted exchange with the world ocean via the Mediterranean Sea, Bosphorus, Dardaneles and the Sea of Marmara. As a result, the basin is almost completely anoxid, containing oxygen only in the upper 150 m (13% of its volume) and hydrogensulphide in the deep waters. A permanent halocline separates the oxid and anoxid waters.

In recent decades, the increasing anthropogenic inputs, and most significantly, mineralized nutrients from continental Europe, have driven a trend for eutrophication (Bologa, 1986; Chirea and Gomoiu, 1986; Mee, 1992), leading to alterations in the ecosystem, bottom hypoxia in the northwestern shelf region, changes in marine populations, invasion by opportunistic species (Tolmazin, 1985; Zaitsev, 1993), and changes in the nutrient structure (Tugrul et al., 1992; Saydam et al., 1993). It is most likely that collapses of the basin's fisheries (Kideys, 1994) is closely linked with the above processes, as well as with increased fishing.

Specific mechanisms determine the health of the marine environment in the Black Sea. The physical processes of circulation and mixing largely determine the redistribution and biochemical cycling of elements leading to biological productivity culminating in living resources.

The basin's oceanography is strongly influenced by freshwater inputs from rivers, active atmospheric forcing, thermohaline driving factors, fluxes through straits and sharp changes in topography. The investigation of the active Black Sea circulation, with rapidly changing jets and eddies, is crucial to determine its role in the transport of basic properties, the realization of primary production, and the growth, migration and entrainment of pelagic marine organisms. The study of mixing processes is essential in determining the stability of the existing stratification, the sources and redistribution mechanisms of nutrients, the factors contributing to new production and eutrophication processes.

A historical data base of hydrographic measurements exists in the Black Sea from the beginning of the 20th century till the end of it. compiled through the efforts of the former USSR and other countries. The recent years have evidenced systematic surveys with much increased coverage, and improved resolution and quality of data: some surveys were carried out by R/V ***Bilim*** along the Turkish coast during 1987–1989. Then, the first of the recent cooperative surveys were carried out in 1988–1989, with Turkish-USSR cooperation on board the R/V ***Kolesnikov*** and R/V ***Dmitriy Mendeleev***. and with USA-Turkish cooperation during the visit of the US R/V ***Knorr*** in 1988 (Murray, 1991) with R/V ***Bilim*** guiding the R/V ***Knorr*** tracer surveys near the Bosphorus (Ozsoy et al., 1993). After 1990 coordinated multi-institutional surveys were carried out, first within the context of the NATO TU-Fisheries program, continued later within the CoMSBlack international program, and currently within the NATO TU-Black Sea program, resulting in intercalibrated and pooled data sets (Aubrey et al., 1992; Oguz et al., 1993; Konovalov et al., 1994; Ivanov et al., 1997). During the recent cooperative studies, oceanographic stations were located with a nominal spacing of 20 km, and covered either the western part, or the entire basin in most cases after 1990.

The flat abyssal plain (depth >2,000 m) of the Black Sea (Fig. 6.38) occupies more than 60% of the total area. The maximum depth is about 2,300 m, and the average depth of the basin was calculated to be 1,240 m. The abyssal plain is separated from the margins by steep continental slopes, excluding the gentler slopes near the Danube and Kerch fans. Continental shelves (depth < 200 m) constitute about 25% of the total area. The wide northwestern continental shelf (mean depth – 50 m) occupies the region between the Crimean Peninsula and the west coast, and extends along the western and southwestern coasts of the Black Sea, with a depth of – 100 m at the shelf break. This continuous region of flat topography decreases in width towards the south and reaches an abrupt termination at Sakarya Canyon, where the depth suddenly increases from 100 m to about 1,500 m. The continental shelf in the remaining part of the Black Sea rarely exceeds a width of 20 km and occurs as narrow stretches along the coasts of Anatolia, Caucasus and Kerch, often separated by canyons or steep slopes adjoining the land. In addition to many canyons along the continental slope, prominent deep features, such as near the Arhangelsky ridge (depth > 400 m), present further complications of the peripheral topography, especially along the Anatolian and Caucasian coasts.

Wind conditions over the Black Sea are variable in winter. The dominant wind direction is northnortheast in the western part, whereas southerlies dominate the eastern part of the basin. Gales from the northwest are common in winter.

The summer months are warmer, with more uniform distribution of air temperature over the Black Sea. Air temperature decreases sharply in late October and November, and reaches a minimum in January and February. In winter, the air temperature has a strong north-south gradient. The daily average temperature can decrease to about 8 °C in the southem Black Sea, while negative temperatures are common in the northern parts, particularly in the region between the Danube basin and Crimea.

The Black Sea has a positive water balance, in which the inputs from freshwater sources exceed losses by evaporation. Although there is a large variation in the estimates reported, current estimates, based on a review of literature can be given as 300 km^3/yr for precipitation, 350 km^3/yr for runoff waters, -350 km^3/yr for evaporation from the sea surface (Unluata et al., 1989). The net flux through the Bosphorus accounts for the remaining component of the water budget.

The two-layer flows through the Bosphorus (the exchange flows to and from the Black Sea) have been estimated in various literature sources, based on the mass budgets, e.g. the Knudsen relations expressing the salt budget. A critical review and improved estimates are given by Unluata et al. (1989). Based on long-term averages of salinity at the strait entrances and using the steady-state mass and salt balances, the average fluxes at the Black Sea end of the Bosphorus (Fig. 6.39) have been computed to be – 600 km^3/yr ($-20,000\,m^3/s$), outflowing from the Black Sea and 300 km^3/yr ($-10,000\,m^3/s$), flowing into the Black Sea, respectively. The steady-state salt budget of the Black Sea requires that the ratio $Q_1/Q_2 = S_2/S_1, = 35.5/17.9 = 2$, where Q_1, S_1 and Q_2, S_2, are the upper (1) and lower (2) layer volume fluxes and salinities defined at the Black Sea entrance of the Bosphorus.

Although the average fluxes must satisfy the mass budgets, the exchange flows at any instant of time greatly differ from these estimates, as a result of the time-dependent meteorological and hydrological forcing originating from the adjacent basins. The transience of the Bosphorus transports on various time scales has been quantified by repeated measurements (Latif et al., 1991, 1992; Oguz et al., 1990; Unluata et al., 1989).

The Bosphorus operates in the full range of weak to strong barotropic forcing in either direction. Blocking of the flows in either layer occurs during extraordinary events, lasting for a few days each time (Fig. 6.40). The lower layer blocking

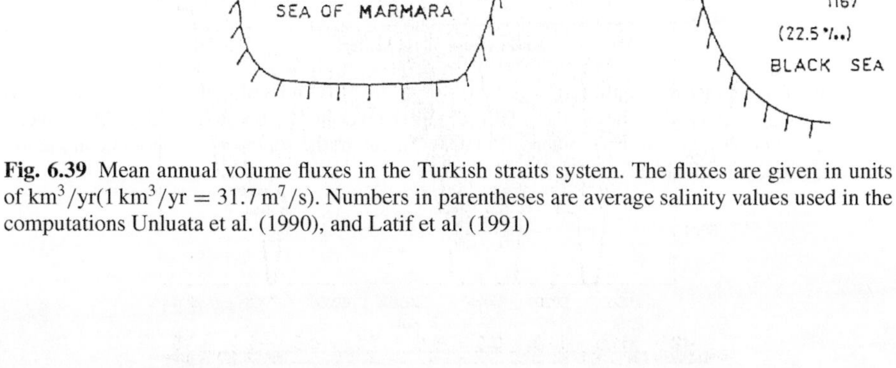

Fig. 6.39 Mean annual volume fluxes in the Turkish straits system. The fluxes are given in units of $km^3/yr(1\,km^3/yr = 31.7\,m^7/s)$. Numbers in parentheses are average salinity values used in the computations Unluata et al. (1990), and Latif et al. (1991)

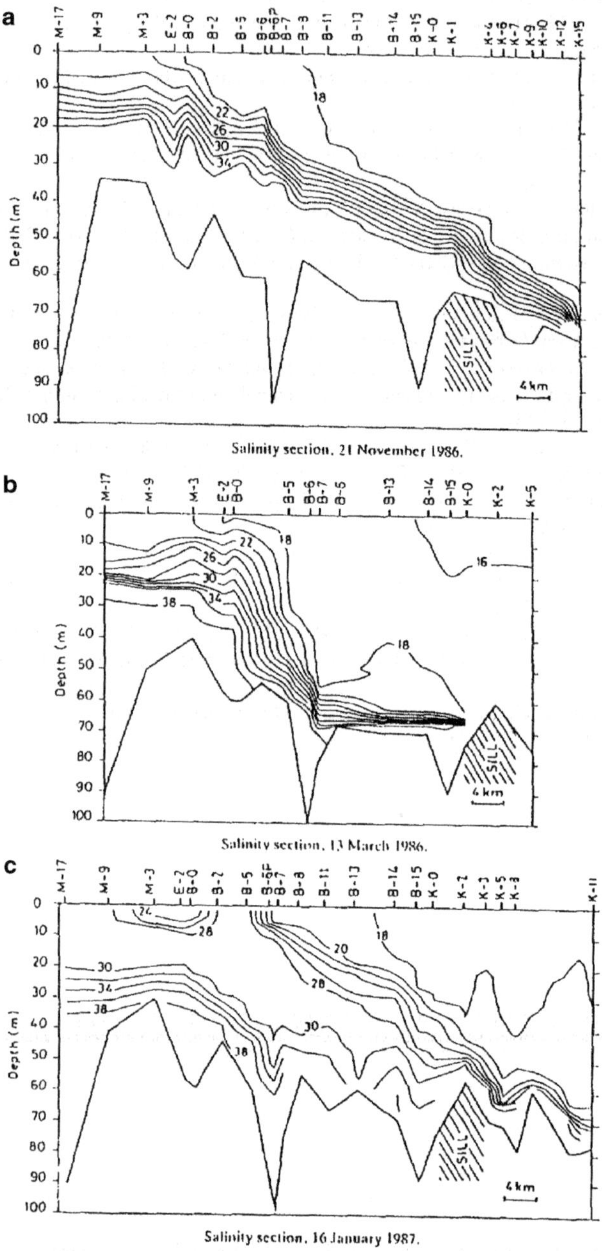

Fig. 6.40 The salinity distribution (psu) in the Bosphorus: (**a**) "normal" two-layer exchange, (**b**) lower layer flow blocked at the northern sill, (**c**) upper layer blocked, with resulting three-layers. Hydraulic controls apply at the northern sill (st. K-2), and at the southern Bosphorus contraction (st. B-7) (Ozsoy and Unluata, 1997)

typically occurs during the spring and summer months, when the net freshwater influx into the Black Sea increases. The upper layer blocking events (identified locally as Orkoz), occur in the autumn and winter months, when the surface flow reverses.

With two supercritical transitions at a contraction and a sill, the Bosphorus is the foremost example of a strait with maximal exchange (Unluata et al., 1989), verified by numerical computations (Oguz et al., 1994). Two important characteristics determine the exchange in the specific case of the Bosphorus: (1) suitable reservoir conditions in the adjacent basins and two hydraulic control sections lead to maximal exchange; and (2) the flow system is asymmetrical and sensitive to geometry when the sill is located nearer to the smaller density basin.

A peculiar vertical stratification is maintained in the Black Sea, with colder, fresher surface waters overlying warmer, more saline deep waters. The low salinity at the surface results from freshwater influence, while the higher salinity in deep waters is an imprint of the Mediterranean influence. The density in the subsurface waters is largely determined by salinity in the equation of state (except within the thin surface mixed layer of depth < 10–$30\,\text{m}$, where temperature effects predominate in summer). As a result of the limiting effects of the salinity stratification on convection, the halocline and the pycnocline coincide at a typical depth interval of 100–200 m, which further coincides with the lower boundary for the Cold Intermediate Water (CIW), characterised by the 8°C limiting isotherm. The oxycline and the chemocline also occur in the same depth intervals as the halocline, because similar mechanisms determine the vertical exchange of these scalar properties.

A thin ($-30\,\text{m}$) mixed layer of low salinity (-18 psu) responds strongly to seasonal heating and cooling at the surface. The Cold Intermediate Layer (CIL), characterized by the CIW, with minimum core temperatures of $-6\,^{\circ}\text{C}$ occurs between the permanent halocline and the seasonal thermocline. Because the CIW is capped by a warm surface layer in summer, it appears in the form of a subsurface temperature minimum. In winter, cooling and the ensuing convection establish an isothermal layer reaching depths of 70–80 m or deeper, with minimum temperatures of 6–7 °C in most areas of the Black Sea, e.g. along the Turkish coast of the westem Black Sea.

Because the Black Sea is an enclosed basin, it responds sensitively to interannual and longer-term climatic variability in atmospheric fluxes, which are well recorded in the structure of its stable pycnocline

Long-term data (more than 40 years in duration) suggest strong climatic changes in the features of the upper ocean influenced by convection in synchronism with the adjacent seas. For example, an extreme event of cooling evidently took place in 1987, when similar effects were noted in the surrounding seas, e.g. dense water intrusion into the Marmara Sea from the Aegean (Besiktepe et al., 1994), and deep water formation in the Rhodes Gyre region.

The changes in air temperature (T_a) and sea surface temperature (SST) during the recent years in 2-degree squares in the Black Sea, centred at 29°E 43°N and in the Levantine Basin of the eastern Mediterranean Sea, centred at 29°E 35°N, are shown in Fig. 6.41, based on the COADS data set. The extreme cold values of the surface

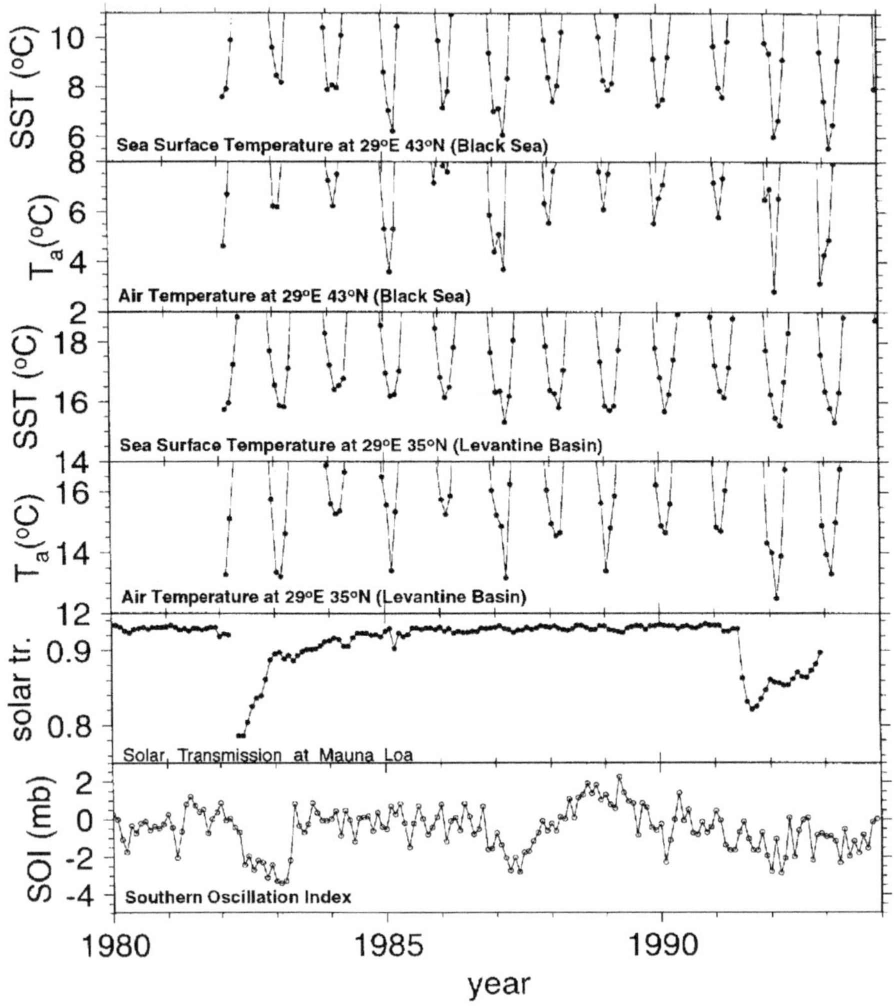

Fig. 6.41 Time series of surface temperature (SST) and air temperature (T_a) in 2-degree squares centred at 29 °E, 43 °N in the Black Sea (*uppermost panels*) and at 29 °E, 35 °N in the Levantine basin of the eastern Mediterranean Sea (*middle panels*, solar transmission at Mauna Loa, and the Southern Oscillation Index (*lower panels*) (Oszoy E. and U.Unluata, 1997)

air temperature, corresponding to severe winters, occurred simultaneously in both areas in 1982, 1985, 1987, 1992 and 1993, with corresponding minimum values reflected in the SST. As indicators of global effects, the solar transmission time series measured at Mauna Loa, and the Southern Oscillation Index (SOI), i.e. the mean sea-level pressure difference between Darwin and Tahiti.

In the case of winter 1991–1992, the cooling event appears to be linked with the persistent atmospheric anomaly pattern that occurred in the eastern Mediterranean/Black Sea region following the eruption of Mount Pinatubo in June 1991.

The Pinatubo eruption was an event of global significance with effects on decreased solar energy input, especially in the Northern Hemisphere, and anomalous atmospheric conditions in the following years. The surface atmospheric and sea temperature anomalies, showed similar patterns of cold anomalies in the "Middle East" region in the winters of both 1992 and 1993. In Turkey, the winter of 1992 was the coldest in the last 60 years (Turkes et al., 1995), and in Israel, it was the coldest in the last 46 years (Genin et al., 1995).

It is clear that, at least during the recent observation periods, there has been good correspondence between eastern Mediterranean and Black Sea cooling events. Some of the cold years in Fig. 6.41 correspond well with ENS0 events signified by the negative values of the Southern Oscillation (El Nino) Index (e.g. 1982–1983, 1986–1987, 1991–1992, and other anomalous cold years (e.g. 1992–1993) appear connected with the Pinatubo volcanic eruption, although positive physical linkages cannot be proven at present.

The cross-shelf spreading of the Mediterranean inflow in the Black Sea has been described earlier by Tolmazin (1985) and Yuce (1990), and a full account, based on carefully designed experiments has been given by Latif et al. (1992).

The layer of water with minimum temperature at depths of 50–75 m in Fig. 6.42 is the Cold Intermediate Water (CIW). By mixing with the overlying CIW, the temperature at the core of Mediterranean Water rapidly decreases from 14.5°C at the northern end of the Bosphorus to about 8°C at the shelf break. Similarly, the salinity in the core declines from about 37‰ at the Bosphorus exit to a maximum of 22.8‰ along the shelf break at the head of a canyon feature at around 41°35′N and 29 °E (Ozsoy et al., 1993).

Bosphorus water comes into direct contact with the CIW, and entrains it. Figure 6.43 shows the changing temperature- salinity characteristics at selected stations (along the same section as Fig. 6.42), where CTD data were obtained within a few meters off the bottom, to detect the Mediterranean inflow. Direct mixing of the Mediterranean Water with the CIW results in a linear evolution of the bottom water modelled by the dashed line. When the dense bottom water reaches the shelf edge (Station 51), it has become colder, yet more saline than the environment on the continental slope (Station 5), where both temperature and salinity increase with depth. The initially warm and salty Mediterranean Water is thus transformed into water differentiated from the ambient waters with a cold anomaly. Finally, when this water sinks along the continental slope to form intermediate depth intrusions, it continues to carry this signature of the Cold Intermediate Water impressed upon it on the shelf (Ozsoy et al., 1993).

Intrusions of anomalous waters in the vicinity of the Bosphorus have often been noted in the past. Bogdanova (1961) and Boguslavskiy et al. (1982) have claimed to have observed warm lenses of Mediterranean water within the interior of the basin. Recently, measurements with more accurate instruments have also shown fine structures in the region with cold water anomalies.

The cascading of the shelf-modified cold dense water along the shelf slope apparently results in a series of intrusions in the vicinity of the Bosphorus. The unstable intrusions are driven by salinity-temperature contrasts of the sinking modified

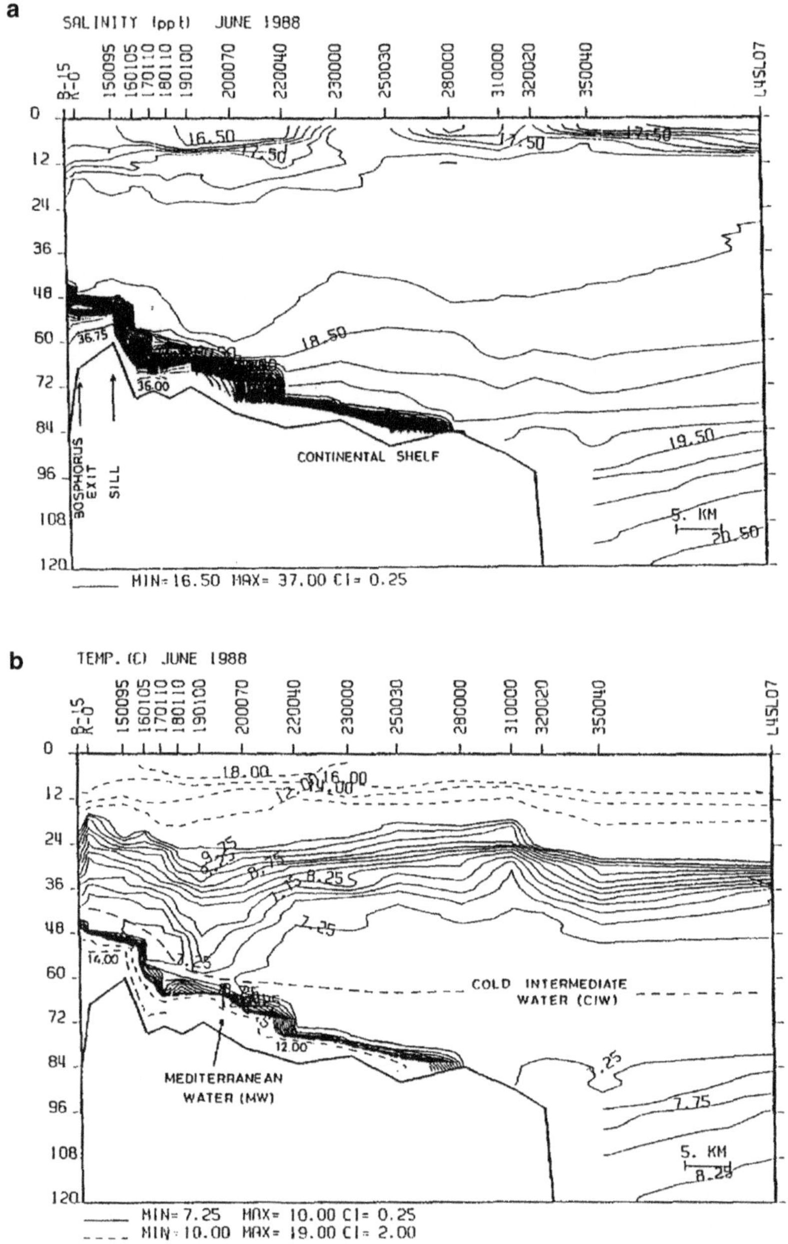

Fig. 6.42 (**a**) Salinity and (**b**) temperature cross-sections across a transect extending from the Bosphorus towards deep water, and following the bottom channel carrying the Mediterranean inflow, denoted by the *dashed line* (Ozsoy et al., 1993)

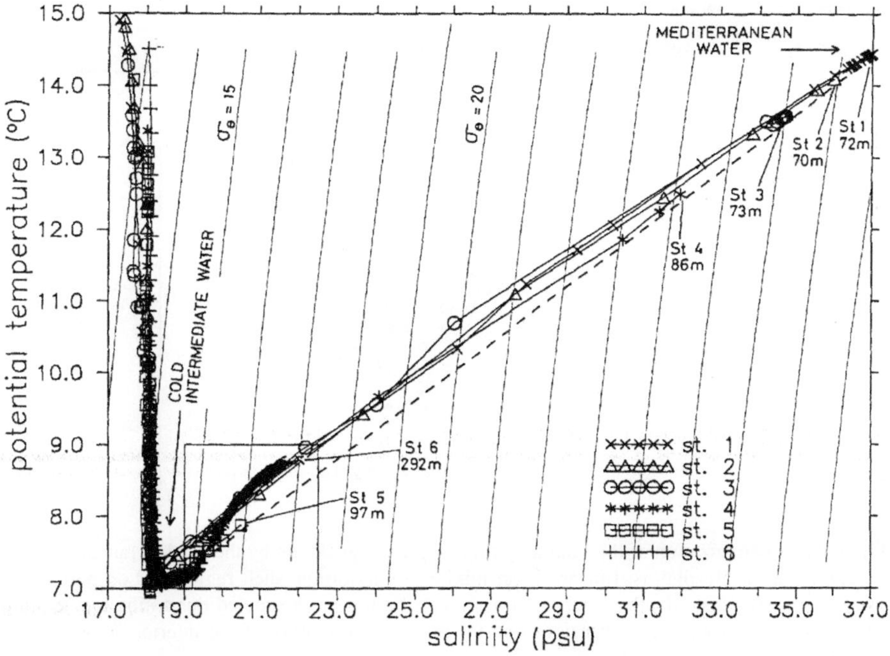

Fig. 6.43 Evolution of temperature – salinity across the shelf. Stations 1–5 extend from the Bosphorus to the shelf break. Station 6 is a deeper station immediately offshore. The dashed line models the changes in the "Mediterranean effluent" at the bottom. At the shelf break (station 5), the modified bottom waters are colder than the waters at comparable depths of the continental slope (station 6) (Ozsoy et al., 1993)

shelf waters, and aided by the double diffusive instabilities of the interior (ambient) stratification (Turner, 1978). As a result, a unique convection pattern is generated adjacent to the southwestern margins of the Black Sea. The boundary mixing processes resulting from the Bosphorus inflow, its shelf mixing and the double diffusive convection, in the form of laterally penetrating intrusions along the continental slope are schematized in Fig. 6.44 (Ozsoy et al., 1993).

The "Christmas-tree" pattern of double diffusive convection schematised in Fig. 6.44 is in many ways similar to the pattern triggered by two-dimensional effects (e.g. lateral boundaries) and by buoyancy sources in stratified environments with two diffusing properties, characterized by a series of alternating diffusive/fingering interfaces (Turner, 1973, 1978).

Additional factors influencing the intermittency and filamentation appear to be the interaction of the sinking flow with many canyon features and local currents in the region. The interaction of the currents with the abrupt topography of Sakarya Canyon is shown to have singular effects on the cross-shelf transports in the immediate vicinity.

In the temperature versus salinity diagrams (Fig. 6.45), the intrusions are identified first as a cold sheet of water on the continental slope (dashed lines), then

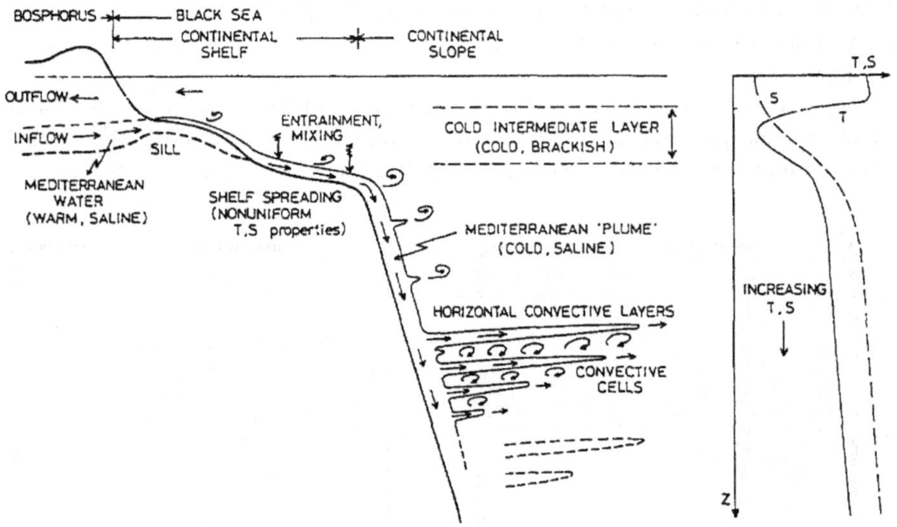

Fig. 6.44 Schematization of the boundary mixing processes driven by the Mediterranean effluent issuing from the Bosphorus. Linear, direct mixing occurs on the shelf region and on part of the slope. At intermediate depths, double diffusive instabilities are generated due to the temperature and salinity contrasts of the intrusions and the potential instability of the interior (Ozsoy et al., 1993)

as discrete layers of anomalous characteristics in the entire neighbourhood of the Bosphorus, extending hundreds of km east of the source.

Transport is motivated by the horizontal spreading of intrusions. The most direct evidence of transport originating from the shelf is given by light transmission measurements, and has been verified by independent measurements of Chernobyl radiotracers and shelf-derived particulates. The perfect coincidence of seawater, particulate and nutrient anomalies, such as shown in Fig. 6.46, indicate a common source of the materials that can be traced back to the southwest margin of the Black Sea. Much diluted imprints are also found further along the Anatolian coast (Kempe et al., 1991).

A basin-scale, coherent, cyclonic boundary current (referred to as the "Main Black Sea Current" in former-Soviet literature, and as the "Rim Current" in Oguz et al., 1992, 1993) is the main feature of the Black Sea general circulation (Fig. 6.47). This basic circulation occasionally encompasses partial (double or triple) cells occupying the cyclonic central part, a number of anticyclonic eddies along the periphery, and a quasi-permanent anticyclonic circulation (the Batumi eddy) in the easternmost corner of the basin. The cyclonic general circulation with two cells was first described by Knipovich (1932) and Neumann (1942), with further aspects of variability added later to the description of the circulation by Filippov (1968), Boguslavskiy et al. (1976), Blatov et al. (1984). A common deficiency of these climatologically based earlier studies is their non-synoptic nature with coarse sampling resolution, restricting the description of currents to relatively larger-scale features.

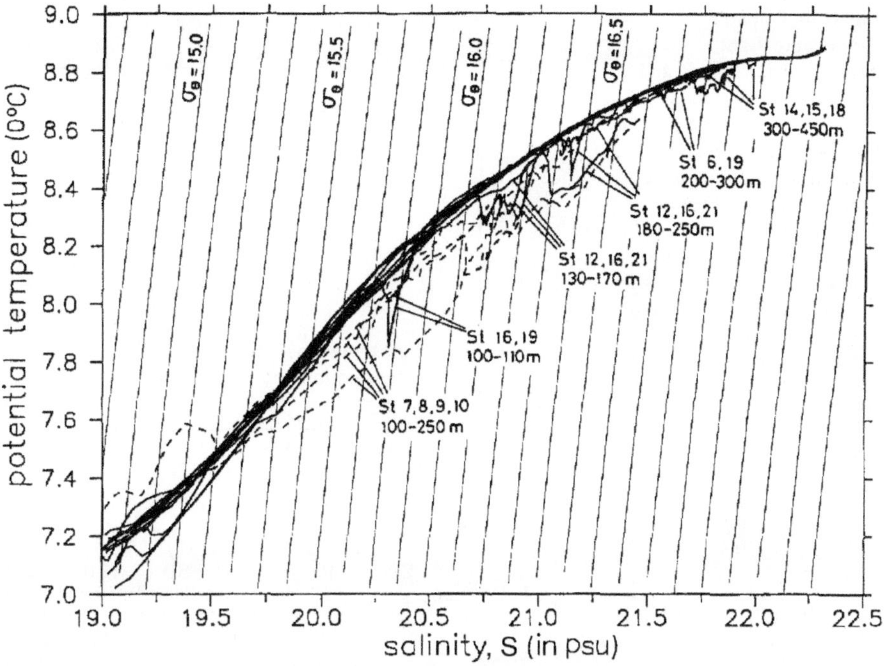

Fig. 6.45 The potential temperature-salinity relationship for stations near the southwestern shelf of the Black Sea. *Dashed lines* represent stations closest to the continental slope, i.e. within the boundary layer. The intrusive features at other stations offshore of the shelf region occur in the form of discrete layers spreading into the interior (Ozsoy et al., 1993)

Observations show that the location of the current and its corresponding density front generally coincide well with the continental slope region, and it is therefore natural to expect the slope currents to be controlled by topography. On the other hand, the pycnocline depth, typically at depths of 100–150 m in the peripheral region also coincides with the depth of the shelf break, especially along the wide western shelf. The impact of the joint effect of baroclinicity and bottom relief (the "JEBAR" effect) would therefore expected to be very important in the Black Sea circulation, as originally suggested by Gamsakhurdiya and Sarkisyan (1975).

6.4.2 *Modelling the Variability of Black Sea Physical Characteristics*

The first serious research of the Black Sea hydrography was done by Knipovich (1932) that is why in many soviet scientific papers on the Black Sea circulation two large cyclonic gyres of the Sea was called the "Knipovich glasses". Black Sea waters circulation, based on the reference level method calculations, was presented in a paper of Neumann (1942). A review of many papers on this subject was given

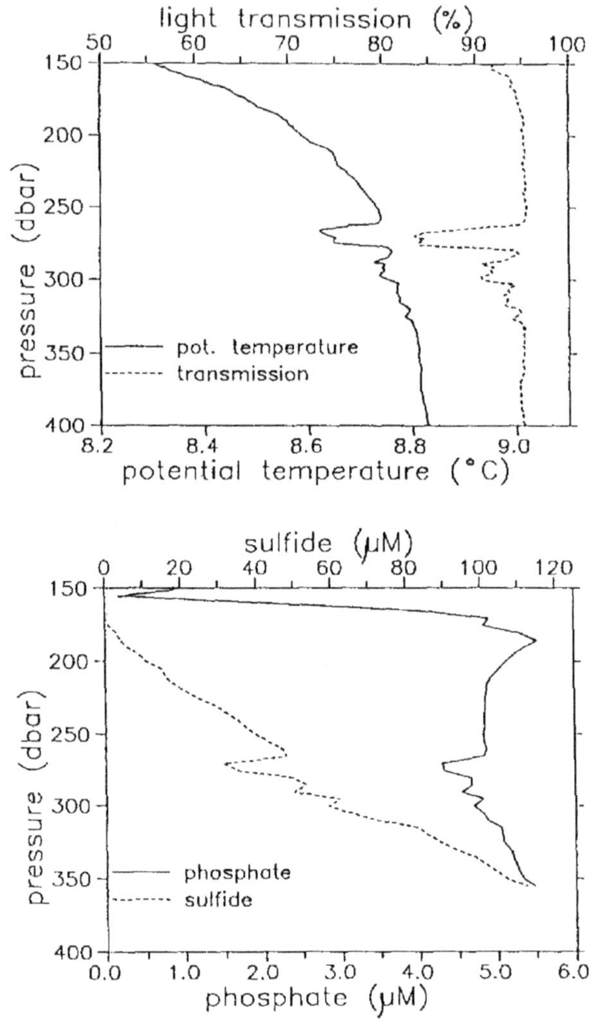

Fig. 6.46 Potential temperature, light transmission, sulphide and phosphate profiles in the southwestern Black Sea. The intrusions advect the water properties modified on the shelf and the continental slope, into the interior. Because the intrusions are below the pycnocline (or the oxycline). they contribute to the mixing across the halocline (Ozsoy et al., 1993)

in the monograph of Filippov (1968). First diagnostic calculations of the Black Sea three dimensional circulation were published by Sarkisyan and Dzhioev (1974) and Gamsakhurdiya and Sarkisyan (1975).

The first diagnostic calculations have shown their principal privileges as compared to the dynamic (reference level) method. Some of them were obvious a priori because the formulation of the problem itself is geophysically more adequate in diagnostic mode. Namely, in diagnostic calculations one takes into account bottom topography, boundary conditions, wind-stress etc. Besides the requirement of mass conservation in diagnostic models allows to calculate the flow velocity vertical component. The sea surface flow velocity in average is for some 10–20 cm/s higher in diagnostic calculations, besides, naturally, the reference level methods velocity

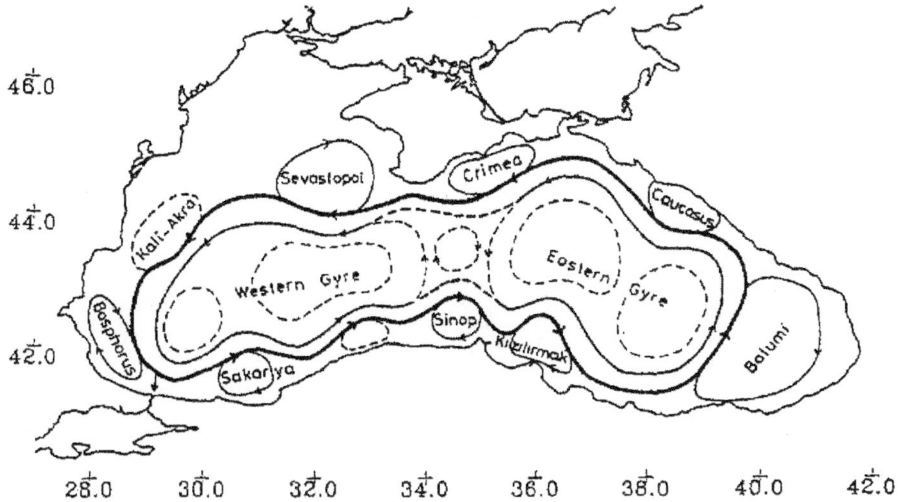

Fig. 6.47 Schematization of the main features of the upper layer general circulation based on a synthesis of past and recent studies. *Solid* (*dashed*) lines indicate quasi-permanent (recurrent) features of the general circulation (Oguz et al., 1993)

rapidly decays with depth. Figure 6.48 shows the basin-wide circulation as a result of diagnostic calculations. This picture already clearly displays three large-scale features, which now are well known: general cyclonic transport (the rim current), two inside cyclonic eddies (the "Knipovich glasses") and the eastern side anticyclone (the Batumi anticyclone). The sea surface flow velocity reaches 40–50 cm/s and equals 15–25 cm/s at 300 m depth in diagnostic calculations. The first attempt of baroclinic sea prognostic investigation, in which the density field is also being calculated, was done by Dzhioev and Sarkisyan (1976). The results in general correspond to diagnostic calculations, but because of the coarse resolution the density gradients, as well as the gradient currents, are for some 30% lower.

So the diagnostic method puts higher demand to hydrologic data while the prognostic one – to the grid mesh.

Several modelling studies of the Black Sea dynamics (Oguz and Malanotte-Rizzoli, 1996), showed that a realistic formulation of forcing functions is one of the main tasks, especially forcing at the sea surface: wind stress, heat and salt fluxes. Normally modelling of the Black Sea dynamics (as well as the oceans) aims at reproduction of general circulation and, in particular, of the sea surface temperature if the heat and water fluxes are prescribed at the sea surface. In that case, the solution depends on the quality of the numerical model and the quality of the buoyancy fluxes through the sea surface. The buoyancy fluxes in turn are an output of sea – atmosphere interaction models. This is the subject of coupled ocean – atmosphere model. However looking on ecological modelling the simulation of the general circulation is more important than the simulation of the sea surface temperature; therefore it seems reasonable to prescribe values of temperature (T) and salinity (*S*).

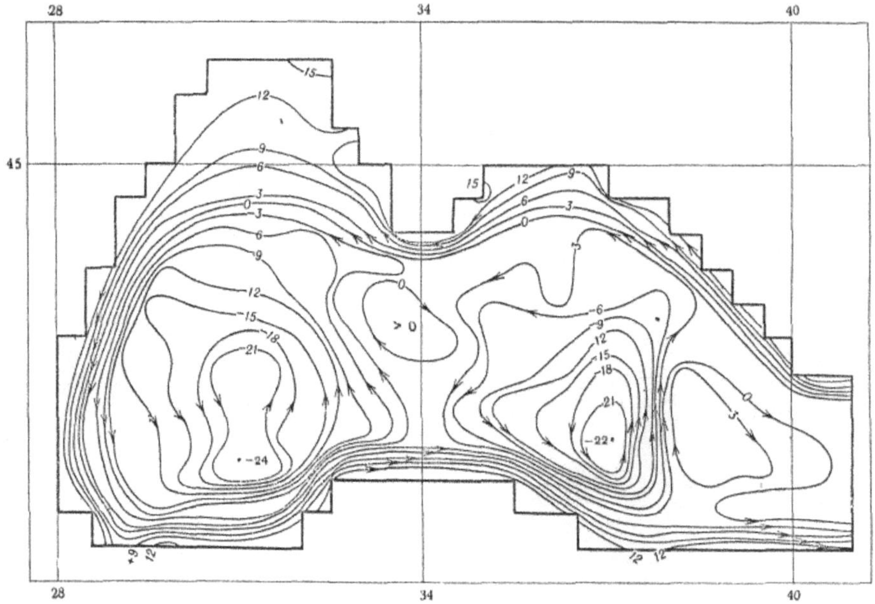

Fig. 6.48 The sea surface topography in cm. Mass flux via liquid boundary was taken into account (Gamsakhurdiya and Sarkisyan, 1975)

It is well known that the temperature and salinity of water are most reliable and directly measured characteristics of the sea. Using the Dirichlet boundary condition at the sea surface we exclude possible errors arising in the alternative buoyancy flux (Neumann) boundary condition.

Seasonal variability of the Black sea circulation was studied by Ibrayev and Trukhchev (1998) using a general circulation model. Attention is paid to the simulation of the upper layer of the sea, where the Cold Intermediate Layer (CIL) exists at the depths of 30–75 m throughout the year. It is a primitive equation ocean general circulation model with Richardson number dependent parameterization of vertical turbulent mixing and nonlinear horizontal viscosity. Horizontal grid spacing is 15′ longitude and 12′ latitude. There are 21 vertical levels. Forcing of the model includes climatological monthly temperature, salinity and wind stress at the sea surface. Climatological monthly values of velocity are set on the open lateral boundaries at river Danube and the Bosphorus Straits.

Detailed descriptions of the model and numerical procedure is given in Demin and Ibrayev (1986); Ibrayev (1993) (see Chapter 4), and is presented above partly in the Section 6.3.

We add some further information on turbulence and boundary conditions.

The horizontal subgrid-scale diffusion is defined as

$$F_\alpha = (a\cos\varphi)^{-2}\frac{\partial}{\partial\lambda}\left(A_\alpha\frac{\partial\alpha}{\partial\lambda}\right) + a^{-2}\cos^{-1}\frac{\partial}{\partial\varphi}\left(A_\alpha\cos\varphi\frac{\partial\alpha}{\partial\varphi}\right) \tag{6.4.1}$$

where α stands for either temperature or salinity, A_α horizontal diffusion coefficient is given by

$$A_\alpha = \gamma A_m$$

For numerical experiments the following parameters are used

$$(C, \gamma, A_m) = (0.4,\ 0.2,\ 10^5 \mathrm{cm}^2/\mathrm{s})$$

A Richardson number dependent parameterization of the vertical mixing was used.

The coefficients for stable stratification proposed by Munk and Anderson (1948) are calculated by the empirical formulae:

$$K_m = a_{m0}/(1 + \alpha Ri)^n + a_{mb} \tag{6.4.2}$$

$$K_h = a_{h0}/(1 + \alpha R_i)^n + a_{hb} \tag{6.4.3}$$

where $a_{m0},\ a_{mb},\ a_{ho},\ a_{hb},\ \alpha,\ n$ are the constants, Ri – Richardson number written as

$$R_i = g\rho_0^{-1}\frac{\partial \rho}{\partial z} \Bigg/ \left[\left(\frac{\partial u}{\partial z}\right)^2 + \left(\frac{\partial v}{\partial z}\right)^2\right]$$

The constants $a_{mb},\ a_{hb}$ determine minimal values of $K_m,\ K_h$. The constants $\alpha,\ n$ define the sensivity of $K_m,\ K_h$ to the value of Richardson number.

For unstable stratification, instantaneous convective adjustment is incorporated in the model.

The UNESCO equation of state for sea water is used in the model (Seventh Report of the Joint Panel on Oceanographic Tables and Standards, 1976).

Boundary conditions at the sea surface are

$$\begin{aligned} -\rho_0 K_m \left(\frac{\partial u}{\partial z}, \frac{\partial v}{\partial z}\right) &= (\tau^\lambda, \tau^\varphi), \\ (T, S) &= (T_{SS}, S_{SS}), \\ w &= 0, \end{aligned}$$

where $(\tau^\lambda, \tau^\varphi)$ are zonal and meridional components of the wind stress, $(T_{ss},\ S_{ss})$ climatological temperature and salinity at the sea surface.

At the bottom, boundary conditions are

$$-\rho_0 K_m \left(\frac{\partial u}{\partial z}, \frac{\partial v}{\partial z}\right) = 0.0013(u, v)\sqrt{u^2 + v^2}$$

$$K_h \left(\frac{\partial T}{\partial z}, \frac{\partial S}{\partial z}\right) = 0,$$

$$w - (a\cos\varphi)^{-1} u \frac{\partial H}{\partial \lambda} - a^{-1} v \frac{\partial H}{\partial \varphi} = 0$$

At the rigid lateral walls, a free-slip condition and zero heat and salt fluxes are imposed. At the open lateral boundaries associated with water inflow, the depth averaged normal velocity, temperature and salinity are prescribed. When there is water outflow then temperature and salinity are calculated from the solution.

One of the peculiarities is tuning and modifying the Munk-Anderson parameterization of vertical mixing as has been done by Friedrich and Stanev (1988). It was shown by them that the stability dependent parameterization of vertical diffusion is essential for a realistic modelling of the Cold Intermediate Layer (CIL) in the Black Sea. Tuning the parameters of the vertical mixing model results in a good qualitative agreement between climatic and simulated temperature and salinity profiles. Ibrayev and Trukhchev tuned their vertical mixing model to reproduce climatic space-temporal variability of thermohaline fields of the upper 100 m.

Wind stress at the sea surface was evaluated from monthly averaged atmosphere pressure fields. For the basin scale, the used wind stress patterns are close to those of Hellerman and Rosenstein (1983), but in addition several local structures exist. As compared to the Hellerman and Rosenstein the specified wind shows pronounced cyclonic character in all seasons. The monthly climatological values of total freshwater input in the sea (see Altman, 1991) is prescribed as the river Danube run-off, see Fig. 6.49. Run-off is equal to 340.58 km^3 over the year. The temperature of the Danube water is also set to be monthly dependent and is equal to the water temperature at Primorskii hydrometeorological station (Demidov, 1991). The salinity is set to a constant value 0.115 ppt. At Bosphorus Straits, the input through the lower current is constant and equal to 25 km^3/month. Total Bosphorus input

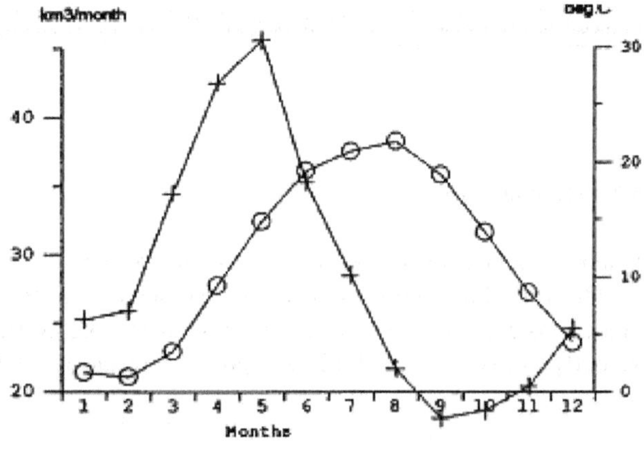

Fig. 6.49 Climatological monthly variations of total freshwater inflow of rivers (km^3/month, line with cross labels), and temperature of river Danube discharge (°C) (open circles) (Ibrayev and Trukhchev, 1998)

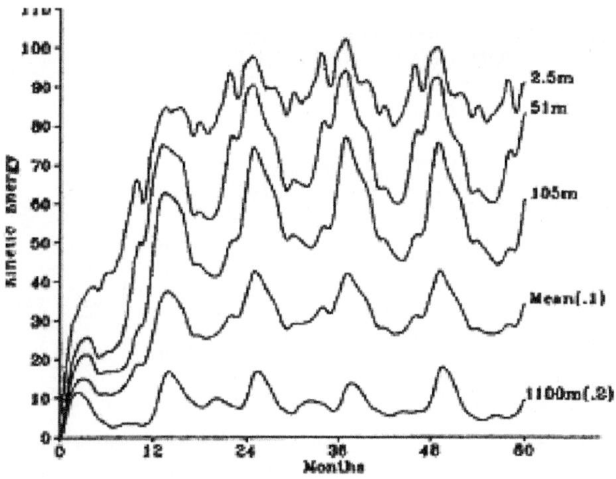

Fig. 6.50 Average kinetic energy ($u^2 + v^2$) at different depths in cm^2/s^2. Values of kinetic energy of the sea higher levels and at the depth of 1100 m must be multiplied by 0.1 and 0.2 respectively (Ibrayev and Trukhchev, 1998)

equal to 300 km^3/year is close to the estimate of Unluata et al. (1989). Boundary values of temperature and salinity are equal to 12 °C and 23 psu as in work of Oguz et al. (1990). The Bosphorus upper layer output equals the sum of Danube and lower Bosphorus current inflow at every moment. The net water flux connected with excess of evaporation over precipitation estimated to be equal to 53 km^3/year (Unluata et al., 1989) is small compared to other terms of the water budget, as well as an exchange through the Kerch Straits, therefore these fluxes were omitted. The model was initialised with the climatic profiles of $T(z)$, $S(z)$ and zero currents. The monthly varying forcing acted for 5 years. The time-variation of average kinetic energy is shown in Fig. 6.50. The experiments showed that the horizontal pattern of circulation is insensitive to the wide range of parameters of the vertical mixing model, which is not true for vertical thermohaline fields. An extended series of experiments with various values of parameters of the Munk-Anderson formula for the vertical mixing coefficient did not result in satisfactory correspondence with the climatic temperature and salinity data. The reason is, that winter convection, parameterized in the model by the mixing of unstable levels, penetrate down to the depth where water density is equal to sea surface density. The depth of penetration depends on the difference between salinity at 50 m and at the sea surface, because in winter a difference between temperatures at the depth of convection (−50 m) and at the sea surface is small. So one of the stipulation that winter convection would penetrate down to 50 m is that the salinity at this depth is close to the sea surface salinity. The deepening of isolines has to take place in summer, but strong thermocline in the model at 10–15 m separates the upper mixed layer from the deeper layer, suppressing the penetration of less saline water to the 30–40 m depths.

The monthly temperature and salinity data used for model forcing and validation, were taken from a climatological hydrological array (Eremeev et al., 1994). Original archive of data included 64,300 *T-S* profiles for the period from 1940 to 1990. The *T-S* data were compiled for 12 months on a grid with horizontal resolution 28′ longitude and 21′ latitude and on 25 vertical levels.

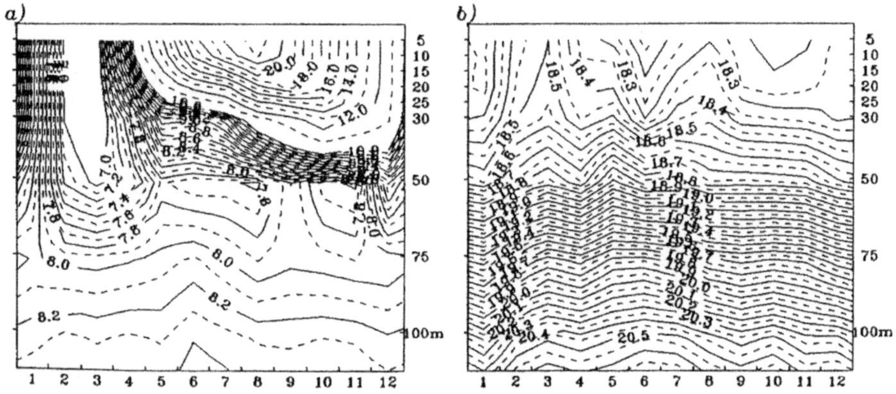

Fig. 6.51 Time-depth dependence of observed climatic (**a**) temperature (°C) and (**b**) salinity (psu) in the western deepwater region (Ibrayev and Trukhchev, 1998)

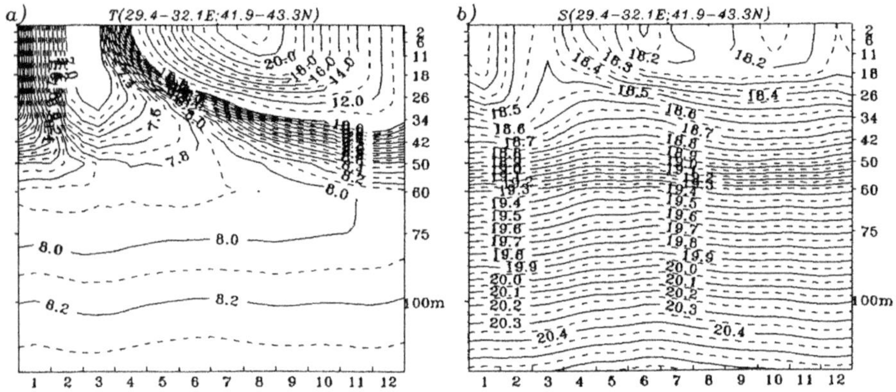

Fig. 6.52 The same as for Fig. 6.51 but for the model solution (Ibrayev and Trukhchev, 1998)

The sea surface data were preliminary smoothed. The monthly *T-S* data except some local features are similar to that of Altman et al. (1987).

General features of the *T-S* temporal evolution are the following. Heating of the sea surface terminates in July, accompanied by an increase of salinity in the surface 30 m layer in the deep sea regions. Starting from August, the cooling of surface water leads to deepening of the upper mixed layer (UML). In January-March, which is the period of maximum deepening, the UML penetrates down to 50–75 m depending on the region. Intensive mixing raises more salt waters to the surface. A depth of maximum variability of hydrological characteristics averaged over the sea is confined in the upper 75 m layer (Figs. 6.51 and 6.52). Detailed description of CIL formation is given in Blatov et al. (1984).

According to (6.4.2)–(6.4.3), the vertical mixing coefficient K_h tends to its minimum value when the Richardson number increases. In the Black Sea, the a_{hb} is about $0.01\,\mathrm{cm}^2/\mathrm{s}$, that is necessary to preserve the CIL throughout the year. This value of a_{hb} is too small for the diffusive penetration of the low saline water through the

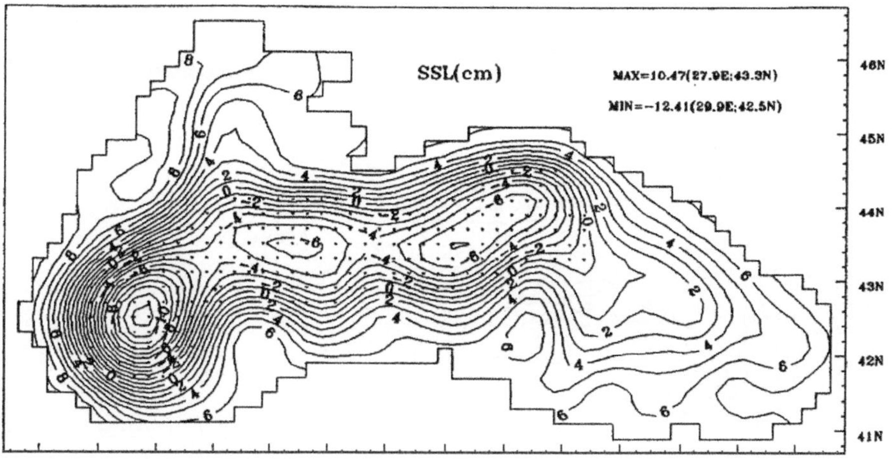

Fig. 6.53 Sea surface topography (cm) for the middle of October (Ibrayev and Trukhchev, 1998)

seasonal pycnocline. The climatic data show that seasonal pycnocline deepens from 10–15 m to 30 m from May to September. In November–December the pycnocline becomes less pronounced and deepens to the depths about 50 m being merged by the permanent pycnocline. In the model, the monthly averaged forcing at the sea surface excludes high frequency variability, which provides the episodic erosion of the seasonal thermocline and, as a consequence, the penetration of low salinity to the 30 m depth. In the Munk-Anderson formula the high frequency component of forcing is parameterized by a_{hb}. The depth dependent parameter a_{hb} varying from $1.0\,cm^2/s$ at the sea surface to $0.01\,cm^2/s$ at 50 m yields good space-temporal correlation of the CIL evolution between the model and climatic data. The time-depth dependence of averaged temperature over the central part of the western deep Black Sea reveals the main phases of CIL evolution (see Fig. 6.52).

The cyclonic cell over the deep part of the sea and the corresponding minimum of the sea surface elevation dominate the model circulation over the year. The cyclonic current system consists of a circular cell in the center of the western half of the sea and of a longitudinally elongated cell in the center of the sea, located between the Crimean peninsula and the Caucasus coast to the North and the Turkish coast to the South. These two cells constitute a permanently existing cyclonic current system confined within the topographic slope (Figs. 6.53 and 6.54).

Most of the seasonal variability is connected with cyclonic and anticyclonic eddies arising predominantly in spring, summer and autumn. The cyclonic cell covers totally the deep sea in winter constituting one large gyre. On the northwestern shelf the currents along northern and western coasts are directed cyclonically forming a northwestern stream that is parallel to the topographically confined Rim current. Winter season is the period of maximum mean kinetic energy in the basin (Fig. 6.50). These correspond with the diagnostic study of the seasonal circulation of the Black Sea (Ibrayev and Trukhchev, 1996).

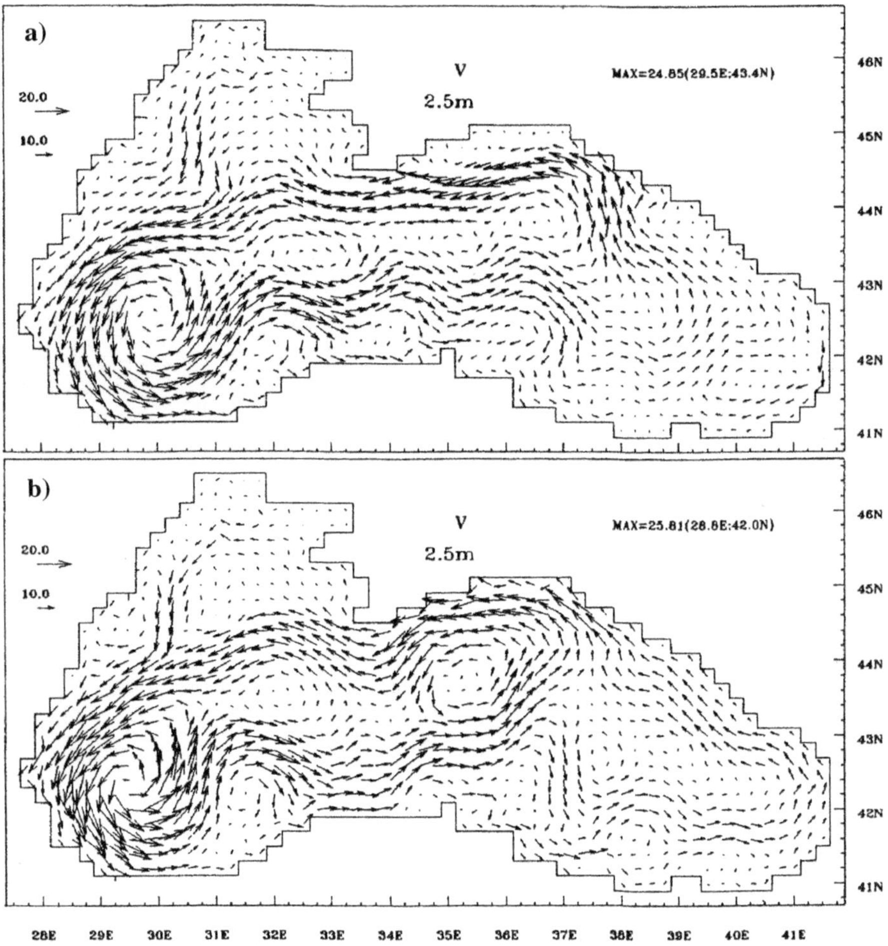

Fig. 6.54 Horizontal currents (cm/s) at depth 2.5 m for the middle of January (**a**), October (**b**) (Ibrayev and Trukhchev, 1998)

The April surface circulation (Fig. not shown here) exhibits the evolution of winter pattern showing appearance of an anticyclonic eddy near the Danube estuary. Spring and summer are the seasons when the maximal freshwater discharge in the northwestern shelf takes place (Fig. 6.49) accompanied by an increase of temperature of the river and consequently formation of the anticyclonic eddy on the northwestern shelf. The decrease of freshwater input in autumn is accompanied by a shifting of the Danube eddy to the South.

In spring-autumn anticyclonic Sakaria, Sinop and Kizilirmak eddies constitute a quasi permanent feature of the circulation near the Turkish coast sea (Oguz et al., 1993). The Batumi and Caucasus anticyclonic eddies are other characteristic features of spring and summer circulation. The existence of the above mentioned eddies is supported by Oguz et al. (1994).

The generation of anticyclonic eddies at the periphery of the Rim current during the warm season is accompanied by a splitting of the cyclonic deep water gyre into a number of cyclones, but nevertheless the general circulation in the sea maintains a pronounced cyclonic character.

As a byproduct of the study, we estimated the surface heat flux in the Black Sea. The general space and temporal characteristics of the heat fluxes at the sea surface (Fig. not shown here) are close to the estimates of Golubeva (1991), and Staneva et al. (1995). The anomalous high winter heat losses to the west and to the east of the Crimean peninsula correspond with the above mentioned estimates. The most striking difference is that the model gives a belt (width 40 km) of low heat gain in spring and summer along the coast of the sea, while in the deep sea the heat flux in the sea is 100–200 W/m^2 and close to the coast the flux is only 50–100 W/m^2. The model experiments reveal strong space and seasonal variability of heat fluxes. The wind stress patterns that are forcing the upwelling and downwelling obviously control local anomalous heat loss and gain regions.

6.4.2.1 Conclusions

Seasonal variability of the Black Sea circulation was studied by Ibraev and Trukhchev (1998) using the climatological wind stress, temperature and salinity distributions as forcing functions. They calculated the circulation for a medium space resolution with realistic boundary conditions including inflow/outflow through the open lateral walls. Besides, the sensitivity of vertical structure of thermohaline fields to the Richardson number dependent parameterization of vertical turbulent mixing was studied. With little modifications, the Munk-Anderson formulas are able to simulate the space-temporal variability of temperature and salinity of the upper 100 m of the Black Sea. Good agreement between the model results and climatological T, *S* time-depth fields in different regions of the sea gives confidence to a realistic representation of the general circulation.

The circulation pattern compared to other medium grid models- shows rather strong spatial variability. The model solution in case when wind stress, heat and salt fluxes are taken from the work of Staneva et al. (1995) give a more smoothed circulation pattern. This result supports the conclusion of Oguz and Malanotte-Rizzoli (1996) to rework sea surface forcing in the Black Sea.

An important output of the study is a new estimate of surface heat flux, which is in good qualitative agreement with estimates of Golubeva (1991), and Stanev et al. (1995).

6.4.3 Modelling the Seasonal Variability of Black Sea Climatic Characteristics by Means of Four-Dimensional Analysis

6.4.3.1 Simple Version of Modelling

A logical further step of the Black Sea diagnostic modelling is four-dimensional analysis based on data assimilation into dynamic models (Sarkisyan et al., 1986). Data assimilation techniques (Demyshev and Korotayev, 1992) ensure a mutual hydrodynamic adaptation of sea fields and the filtering of "noises" contained in measurements. Moreover, the continuous evolution of hydrophysical fields may be reconstructed by four-dimensional analysis as new observational data becomes available. It is consequent to use this method for the reproduction of the seasonal velocity fields, using climatic temperature and salinity arrays. This approach was implemented by Korotaev et al. (2000) and in several other papers.

The method is based on the nonlinear primitive equations. In a rectangular coordinate system (x,y,z) with eastwards-, northwards-, and vertically downwards-directed axes, respectively. The system of equations in the Boussinesq, hydrostatics, and sea water incompressibility approximations is written as follows (Knysh et al., 2001; see Chapter 4).

$$\frac{\partial u}{\partial t} + (\xi + f)v + w\frac{\partial v}{\partial z} = -\frac{1}{\rho_0}\frac{\partial (P+E)}{\partial x} + \nu_H \Delta^2 u + \nu_V \frac{\partial^2 u}{\partial z^2} \tag{6.4.4}$$

$$\frac{\partial v}{\partial t} + (\xi + f)u + w\frac{\partial v}{\partial z} = -\frac{1}{\rho_0}\frac{\partial (P+E)}{\partial y} + \nu_H \Delta^2 v + \nu_v \frac{\partial^2 v}{\partial z^2} \tag{6.4.5}$$

$$\frac{\partial u}{\partial x} + \frac{\partial v}{\partial y} + \frac{\partial w}{\partial z} = 0 \tag{6.4.6}$$

$$p = g\rho_0\zeta + g\int_0^z \rho dz \tag{6.4.7}$$

$$\frac{\partial T}{\partial t} + \frac{\partial (uT)}{\partial x} + \frac{\partial (vT)}{\partial y} + \frac{\partial (wT)}{\partial z} = K_H \Delta T + \partial (K_V \partial T/\partial z)\partial z + \frac{1}{REL}\left[1+\eta^2(z)\right](T_{KL} - T) \tag{6.4.8}$$

$$\frac{\partial S}{\partial t} + \frac{\partial (uS)}{\partial x} + \frac{\partial (vS)}{\partial y} + \frac{\partial (wS)}{\partial z} = K_H \Delta S + \partial (K_V \partial S/\partial z)\partial z + \frac{1}{REL}\left[1+\eta^2(z)\right](S_{KL} - S) \tag{6.4.9}$$

$$\rho = \varphi(T, S) \tag{6.4.10}$$

In these equations we designate: $\xi = \partial v/\partial x - \partial u/\partial y$ (vorticity), $E = \rho_0(u^2 + v^2)/2$ (kinetic energy); u, v, w are the vector components of the current velocity along the x-, y-, z-axes, respectively; T, S are the sea water temperature and salinity; p, ρ are pressure and density anomalies; f is the Coriolis parameter; g is the gravity acceleration; v_H and v_v are turbulent viscosities over the horizontal and the vertical; K_V and K_H are diffusion coefficients over the vertical and over the horizontal in heat and salt advection-diffusion equations; REL, $\eta^2(z)$ in an assimilation source of the Newton type in equations (6.4.8), (6.4.9) are relaxation time and the variance of measurement errors, which is normalized to the field dispersion; T_{KL}, S_{KL} are the climatic values of temperature and salinity, Δ, Δ^2 are the Laplace and biharmonic operators, respectively.

The conditional sea level ζ (see formula 2.2.4 of Chapter 2) is used in the model (6.4.4)–(6.4.10). The boundary conditions are: at the sea surface, z = 0

$$\nu_v \frac{\partial v}{\partial z} = -\tau^X, \ \nu_V \frac{\partial v}{\partial z} = -\tau^Y, \ w = -\frac{\partial \zeta}{\partial t} \tag{6.4.11}$$

$$K_V \frac{\partial T}{\partial z} = Q^T, \ K_V \frac{\partial S}{\partial z} = (PR - EV) \cdot S_0$$

at the bottom, z = H(x, y):

$$u = v = w = 0, \ \frac{\partial T}{\partial z} = 0, \ \frac{\partial S}{\partial z} = 0 \tag{6.4.12}$$

at the rigid lateral walls:
for the meridional parts of the boundary

$$u = 0, \ \frac{\partial v}{\partial y} = 0, \ \frac{\partial T}{\partial x} = 0, \ \frac{\partial S}{\partial x} = 0; \tag{6.4.13}$$

for the zonal parts of the boundary

$$v = 0, \ \frac{\partial u}{\partial x} = 0, \ \frac{\partial T}{\partial y} = 0, \ \frac{\partial S}{\partial y} = 0; \tag{6.4.14}$$

at the parts of the boundary with rivers and the Lower Bosporus Current inflow:
for the meridional parts

$$v = 0, \ \frac{\partial v}{\partial y} = 0, \ T = T^P, \ S = S^P \tag{6.4.15}$$

for the zonal parts

$$v = 0, \ \frac{\partial u}{\partial x} = 0, \ T = T^P, \ S = S^P \tag{6.4.16}$$

The conditions (6.4.13), (6.4.14) are applied for the Upper Bosporus Current.

In relations (6.4.11), (6.4.15) we designate: (τ^X, τ^Y) is the wind stress; $Q^T(x, y, t)$ is the heat flux; $PR(x, y, t)$ is precipitation; $EV(x, y, t)$ is the sea surface evaporation; T^P is the temperature at the mouths of rivers and in the Bosporus, S^P is the salinity in the Bosporus, both given by observations.

In this version of the hydrodynamic model Neumann conditions are used for the equations of heat and salt diffusion-transport at the sea surface.

A special problem is to specify the boundary conditions on lateral walls. Below the impact of rivers and the Bosporus is taken into account in a simple way.

The initial conditions are: the sea is at rest and the temperature and salinity are specified.

For solving the equations the methodology of Marchuk (1969, see Chapter 4) is used.

The Black Sea basin was divided into boxes with horizontal sizes 14.8 km × 14.8 km and 44 levels over the vertical.

Turbulent viscosity and diffusion coefficients over the vertical and over the horizontal were tuned in a set of numerical experiments. They turn out to be: $\nu_V = 5\,\mathrm{cm}^2/s$, $k_H = 5 \times 10^5\,\mathrm{cm}^2/\mathrm{s}$, the coefficient ν_H to a depth of 350 m is $5 \times 10^{18}\,\mathrm{cm}^4/\mathrm{s}$, at a greater depth it decreases linearly and reaches the value $6 \times 10^{19}\,\mathrm{cm}^4/\mathrm{s}$ at depths of 1,000–1,975 m. Then it increases linearly and reaches the value $4 \times 10^{19}\,\mathrm{cm}^4/\mathrm{s}$ at 2,000 m. The coefficient K_v to a depth of 50 m at the corresponding upper eight levels is: 2.5, 2.0, 1.5, 1.0, 0.9, 0.1, 0.05, 0.04, at a greater depth to the bottom $K_V = 0.03\,\mathrm{cm}^2/\mathrm{s}$. The time step in the model is 10 min.

The following rivers were taken into account: in the north-western part of the sea the Danube, the Dniester, the Dnieper, The Southern Bug; in the Caucasus, the Rioni; in Turkey the Eshil-Irmak, the Kizilirmak, the Sakarya. Salinity at the rivers mouths is set to be zero, the annual variation of water temperature is taken from "Hydrometeorology and Hydrochemistry of the USSR Seas", (1991).

Climatic temperature and salinity distributions for November are specified as initial data. Monthly average arrays from Staneva and Stanev (1998) were interpolated linearly into each time step between two neighbouring months for actual forcing.

The monthly climate values of temperature and salinity, which are available on the grids of 40′ by latitude and 60′ by longitude, were interpolated into the model grid. Then the obtained arrays of values were interpolated for each day of the year by an expansion in terms of time Fourier harmonics (12 harmonics).

The climatic values of temperature and salinity were used to recalculate the sources in equations (6.4.8), (6.4.9) at each time step, relaxation time is taken to be one day. At levels from 2.5 to 2,000 m the relative variance of the measurement errors varies from 0.05 to 1.0, respectively.

The convergence of the above assimilation procedure needs verification. For this purpose in the numerical experiment with a time step of 5 days the mean square differences σ_ζ between two sea level fields representing states of one year distance were calculated. The resulting sea level is shown in Fig. 6.55. As seen from Fig. 6.56, already in 320 days σ_ζ decreases substantially and its shape changes insignificantly from year to year. There is no pronounced trend in the behaviour of σ_ς either.

The calculation in the numerical experiment is performed for a period of 23 years upto sufficient stationarity. The computational results for the last 23rd year are used to analyse the intra-annual variability of hydrophysical parameters.

6.4.3.2 Seasonal Variability of the Climatic Black Sea Level

Unlike the previous studies (Ibrayev and Trukhchev, 1996; Trukhchev and Ibrayev, 1997) of climatic Black Sea circulation, in the following the continuous evolution of climatic fields was simulated at each time step (10 min). As key parameter for the analysis of the climatic seasonal circulation was considered the sea level because it characterizes the surface geostrophic dynamics.

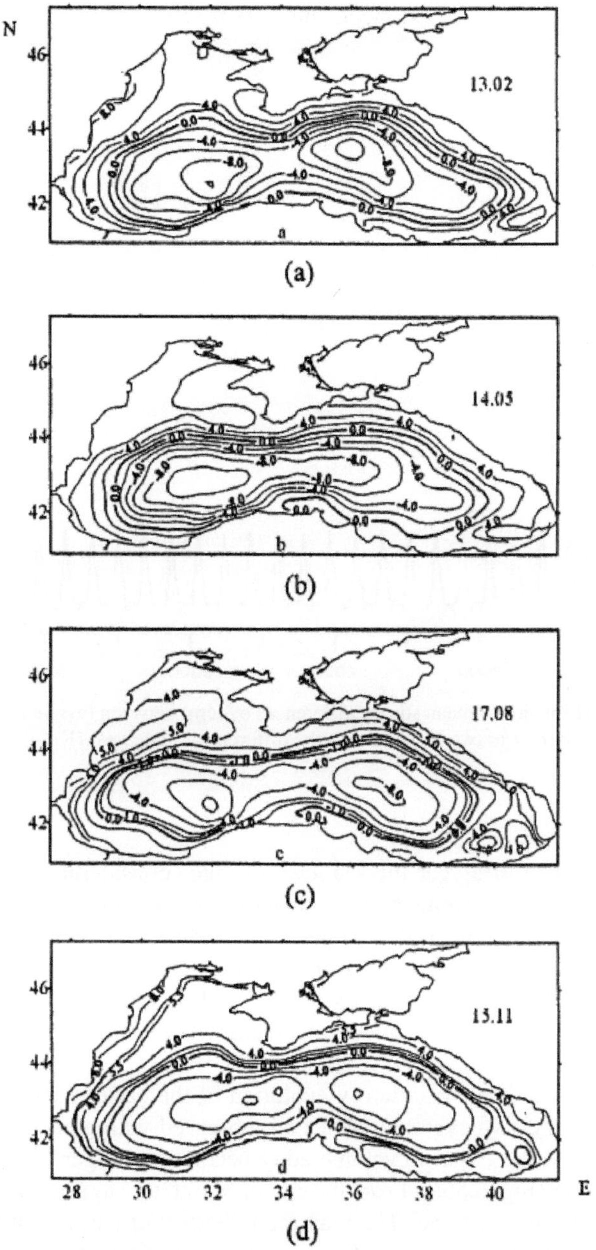

Fig. 6.55 Topography of the climatic Black Sea level (cm) in various seasons; (**a**) winter; (**b**) spring; (**c**) summer; (**d**) autumn (Knysh et al., 2001)

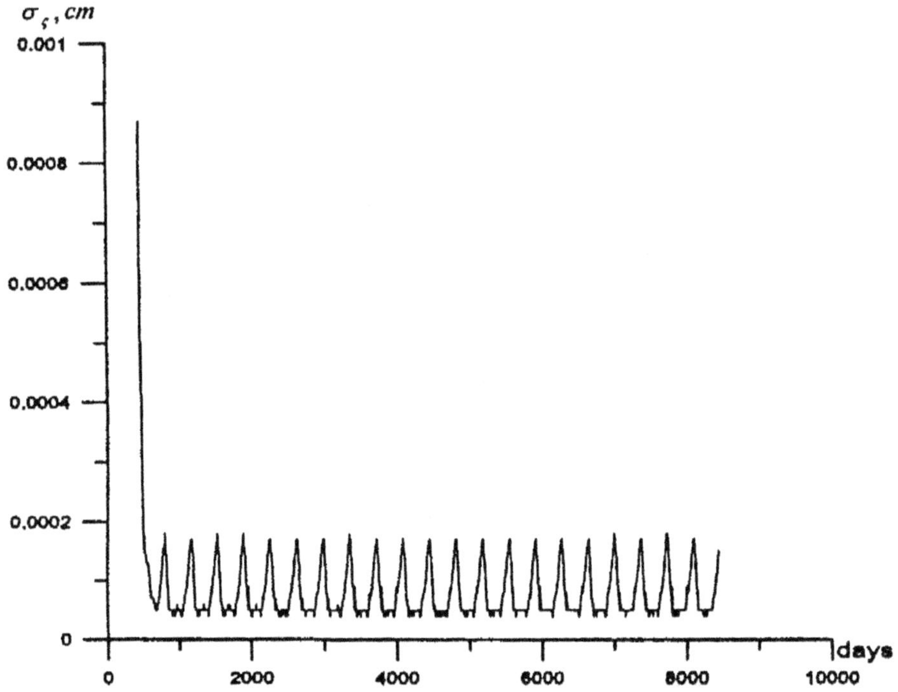

Fig. 6.56 Time behavior of mean square differences σ_ζ (cm) between two fields of the climatic sea level values which refer to two instants with the difference of one year (Knysh et al., 2001)

Over the course of a year the sea level changes considerably (Fig. 6.55). In all seasons one observes cyclonic rotation with two or one centers. The maximum of the sea level difference is observed in winter. Towards summer the sea level gradients on the shelf slope decrease and become minimal in August. In an autumn-winter period the sea level field intensity gradually increases. This character of the seasonal evolution of surface currents is in good agreement with satellite altimetric observations analysed by Korotaev et al. (1999, 2000, 2001).

There is a remarkable oscillatory evolution of the eastern and western cyclonic eddies. In January, two cyclonic eddies are observed with almost identical intensities. In February, the eastern cyclonic eddy becomes stronger and the western one becomes weaker (Fig. 6.55a). From the end of March to August, the western cyclone prevails over the eastern one. The maximum density of isolines in the western cyclone is in July. At the beginning of August two cyclonic centers are seen in the general cyclonic rotation. Their intensity gradually levels off. Then in September, the eastern cyclone becomes stronger again. In October, the western cyclone becomes stronger. At the beginning of December, there is a tendency for the eastern cyclone to become stronger and larger. This variability is evidently due to the propagation of Rossby waves generated near the eastern sea coast.

The climatic calculation reproduces the system of quasistationary anticyclonic eddies to the right of the jet of the Main Black Sea Current (MBSC) (Oguz et al., 1994). The south-east anticyclonic eddy is the largest of them. On sea-level maps it is observed from January to September.

The anticyclonic eddy formation south-west of the Crimea (the Sevastopol anticyclone) manifests itself over January–May and November–December, The anticyclonic eddy originated in the north-west shallow area of the sea at the beginning of May is combined with the Sevastopol anticyclone at the end of June. The eddy formed is observed up to the end of October. In November, the Sinop and Kizilirmark anticyclonic eddies are clearly reflected in the sea level field. Other anticyclonic eddies adjacent to the MBSC are discussed in the next section.

6.4.3.3 Seasonal Variability of Climatic Black Sea Currents

We consider now seasonal variation of the flow velocity fields, based on the calculation for the depth of 105 m. (The figures are not shown here). The currents at this level have a considerable seasonal variability. The maximum velocity is observed in winter and reaches 30 cm/s. In February, the western cyclonic rotation is stronger than the eastern one which consists of two eddies.

The velocities are low (3–5 cm/s) in the Sevastopol and Caucasian anticyclonic eddies as well as in the anticyclone north-east of the mouth of the Eshil-Irmak river. The south-east anticyclone with maximum current velocity at this level is clearly seen.

Towards spring the MBSC intensity at 105 m becomes somewhat weaker. In May, both the western and eastern cyclonic eddies are oriented in a latitudinal direction. The anticyclonic eddies are seen to the right of the MBSC south-west of the Crimea in the regions of Sinop, Samsun, and east of it, near the Caucasian coast. As in February, the anticyclonic eddy at south-east rotation with maximum velocity 30 cm/s is clearly present.

In summer months the main features of water circulation are retained. It is essential that circulation intensity becomes weaker. The number of cyclonic and anticyclonic eddies increases. Thus, in August, the Sevastopol, Crimean, Caucasian, Batumi, Kizilirmak, Sinop, and Sakarya anticyclones are observed on the MBSC periphery. The western and eastern cyclones consist of two cyclonic eddies; they are divided by a cyclonic one. The maximum velocity is observed in the north-western part of the MBSC jet (24 cm/s). The south-eastern anticyclone turns out to be extended in a north-east direction. A cyclone with low current velocities forms near it. Note that in this period of the year an anticyclonic eddy is also observed at depths of 15–20 m in the area of Constantsa (Kali-Akra).

In autumn, the MBSC is somewhat more pronounced. In November, the western circulation ring is extended in a zonal direction. Another characteristic feature is that in the area of the southern coast in the eastern part of the Black Sea from 35 °E to 40 °E. anticyclonic circulation is observed. The Kizilirmak eddy and an anticyclone in the north-east of the mouth of the Eshil-Irmak are clearly formed. The Sevastopol and Caucasian anticyclones are also observed. The maximum velocity in the jet of the main current is observed in the western part of the Anatolian coast and reaches 14 cm/s. The south-east anticyclone is shifted to the north. The cyclonic eddy formation, which is somewhat more intensive than that in August, is clearly seen south of it.

The main features of the general cyclonic rotation in all the seasons are retained in the layer to 500–600 m. At a greater depth the uniform current is not observed. The anticyclonic eddy in the south-east of the sea is observed at depths to 1,500 m throughout the year. This is in agreement with the results contained in "Hydrometeorology and Hydrochemistry of the USSR Seas" (1991). The anticyclone south-west of the Crimea at 105 m is more pronounced in spring, in summer, and in autumn.

Using the adopted computational procedure, one can characterize the climatic deep-water circulation at depths 1,500 m and 2,000 m. At these depths almost an identical number of eddies with cyclonic and anticyclonic rotations (as to current intensity and the occupied area) is observed. The current velocities here are not high (order of *0.1–1* cm/s).

The most significant feature is the existence of a deep jet which transports water from the western part of the basin to the eastern one. This jet forms east of the near-Bosporus area presumably due to salt water sinking in the Sea of Marmara and its subsequent propagation over the entire deep layer. Another feature of the currents is that the cyclonic rotation observed in the central part of the sea in winter becomes an anticyclonic one in summer. The foregoing is in qualitative agreement with the results of Korotaev (1997) and Whitehead et al. (1998).

6.4.3.4 Seasonal Variability of the Climatic Vertical Velocity

The vertical velocity fields at a greater depth than 1,100 m, which were obtained by Korotaev et al. (2000), show artifacts in the form of two-step waves because of the crude discretization along the vertical coordinate (20 levels). Calculations with high resolution (44 levels) and depth-dependant horizontal turbulent viscosity yield a more regular vertical velocity field in deep sea layers. This is an indicator of the calculation quality of deep-water circulation.

The space structure of vertical motions at 275 m in various seasons showed an appreciable correlation between the upwelling area and the intensity of cyclonic rotations. Thus, in February when the eastern rotation intensifies, the upwelling area in this eddy is larger than the total up-welling area corresponding to the western cyclone. In May, the reverse picture is observed: the upwelling area is larger in the region of the western cyclonic center. In August, the upwelling area in the eastern part of the sea is larger than that in the western part of the sea. In autumn, the upwelling areas in the eastern and western parts of the sea are practically identical. Of interest is the downwelling area corresponding to the south-east anticyclone.

An important characteristic of the vertical structure are the annual average profiles of w(z) obtained by the spatial integration in two regions of the Black Sea: the area adjacent to the slope and bounded by 100 and 1,000 m isobaths and the open part of the sea, where depths are larger than 1,000 m and smaller than 2,000 m.

Figure 6.57 shows the graphs of $\overline{w}(z)$. The papers of Korotaev (1997) and Whitehead et al. (1998) show that the salt water inflow from the Sea of Marmara into the

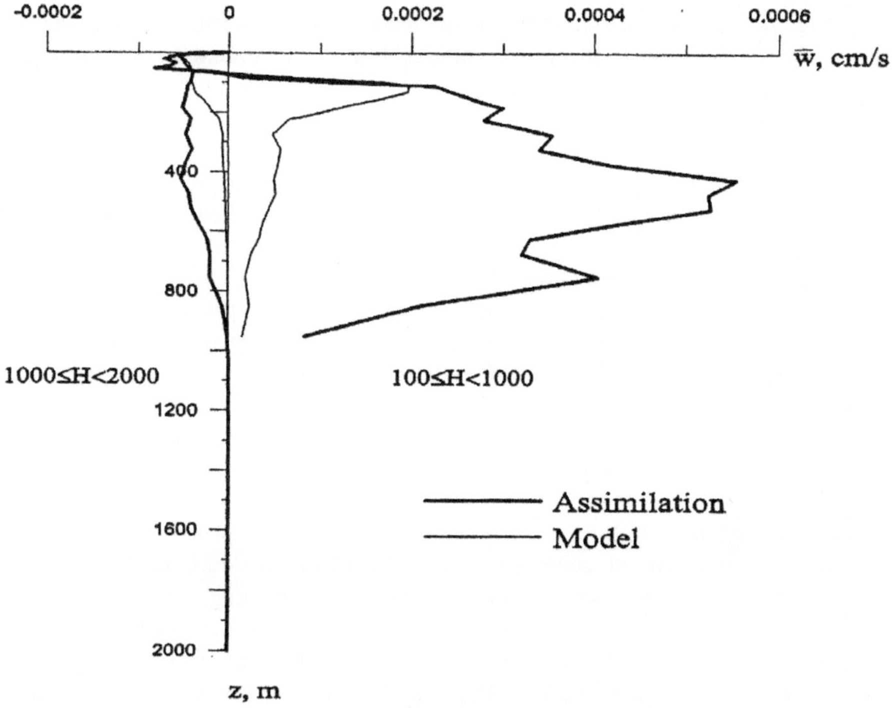

Fig. 6.57 Annual average vertical velocity profiles $\overline{w}(z)$ (Knysh et al., 2001)

Black Sea must result in intensive upwelling in the central part of the sea and downwelling near the coast. Indeed in the coastal area annual average downwelling at all depths is observed. To the contrary, in the center of the sea annual average upwelling dominates. This result is also confirmed by a purely hydrodynamic calculation when the observational data is not assimilated.

Of course, Fig. 6.57 does not allow us to definitely state that upwelling in the central part of the sea is due to a salt water inflow through the Bosporus because the annual average cyclonic wind vorticity causes upwelling at the lower boundary of the Ekman layer as well. However, the general theory implies that there is no exact correspondence between the sign of the vertical velocity at the upper and lower boundaries of a pycnocline because isopycnal surface displacements considerably damp the impact of the Ekman pumping.

6.4.3.5 Seasonal Variability of the Black Sea Cold Intermediate Layer (CIL)

The most striking feature of the thermohaline Black Sea water structure is the existence of an anomalously cold water with temperature lower than 8 °C in an active layer. Temperature maps with time increments of 5 days in a latitudinal section at 43.7 °N were made to analyse temperature variation processes in this layer of the sea. We briefly describe the characteristic CIL features and the time variability of temperature in the upper 200 m sea layer over the course of a year, starting from January. Analysis of the temperature evolution in a vertical section shows that cold water forming in the north-western part of the basin due to winter convection downwells to the CIL depth and is transported by the MBSC over the whole sea water area. Simultaneously with the onset of winter thermal convection at 43.7 °N the layer of temperature jump gradually breaks and the mixing process with time extends to the earlier existing CIL spots (Fig. 6.58). Hence, the CIL is also renewed by local convection (Ovchinnikov and Popov, 1984) along with advective transport of water forming in the north-west of the sea.

In the vertical temperature section at 43.7 °N, the above processes are observed from the beginning of December to the end of March. In this period the lower boundary of cold water reaches depths of 100 m and more. The surface temperature except for a small warm water spot east of 38 °E is everywhere lower than 8 °C.

In a period from mid-February to the end of March at the 43.7 °N the lower cold water boundary slowly deepens, in upwelling areas a domelike deviation of temperature isolines is observed.

As surface water with higher temperature than 8 °C due to heating at the end of March appears, the process of the new CIL formation starts in the east. This process involves gradual surface water heating, downwelling, and the horizontal leveling of the 8-degree isotherm. It lasts to the end of April. In this period the water temperature in the discussed section in the layer from the surface to a depth of 15 m is about 11 °C. The upper CIL boundary is at depths somewhat lower than 50 m, the lower boundary is at a depth of about 100 m,

In May–June, the surface temperature increases from 13 °C to 21 °C and the CIL slowly deepens. In summer the space-time evolution of the CIL occurs so that its thickness changes insignificantly. Since August, the CIL thickness gradually decreases mainly due to the increasing depth of the upper boundary. The process of decreasing the thickness of the cold water layer lasts to the second decade of November. In November, the CIL's disappears.

In December, cold water ($T < 8°$) manifests itself at depths of 50–100 m as separate disconnected spots with thickness of order of 20–25 m. At the end of December, water with $T < 8°$ in a zonal section forms in the north-west of the Black Sea, which is most likely due to convection. From January, the described process of the renewal and formation of cold water in an active sea layer is repeated.

6.4.3.6 Some More Detail on the Black Sea Cold Intermediate Layer (CIL)

The CIL formation is investigated in more detail by Demyshev et al. (2004b). The description of that primitive equation model and numerical scheme is presented above. Here we describe the results of calculation.

The horizontal grid spacing was 14.8 km × 14.8 km. In calculating the temperature, salinity, and horizontal velocity, 44 basic vertical levels were used. The vertical velocity was calculated between these levels. The turbulent exchange by the horizontal momentum was parameterized as a biharmonic operator with the coefficient $\nu_H(z)$. Its values (in cm^4/s) were taken to be 5×10^{18} from 2.5 to 350 m, 1.25×10^{19}

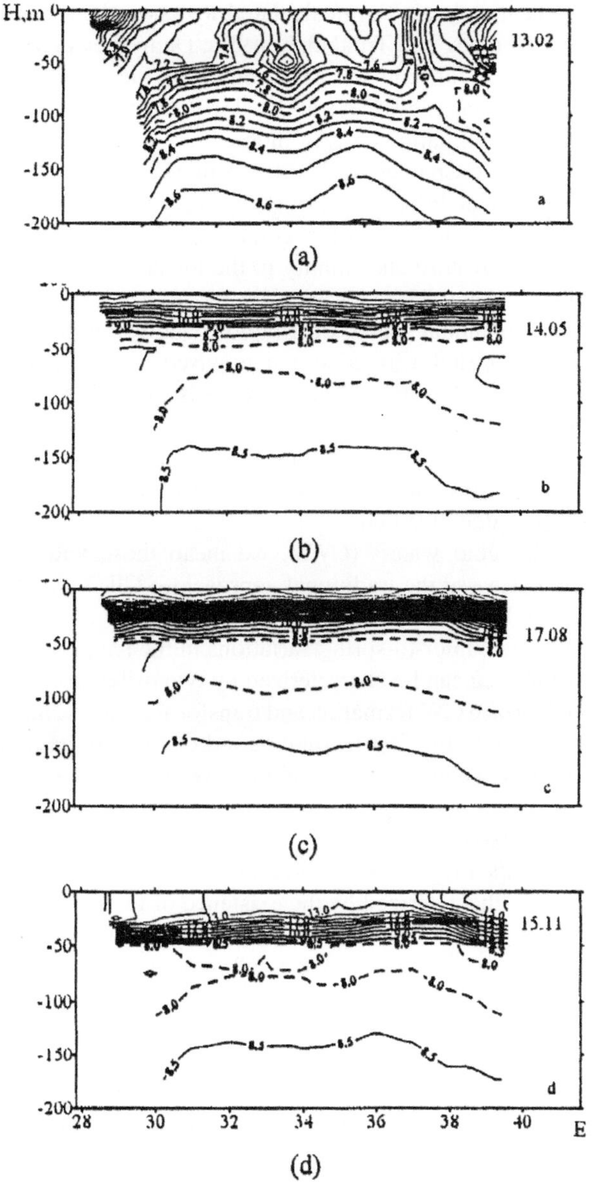

Fig. 6.58 Climatic temperature distribution (°C) in vertical section at 43.70N in various seasons: (**a**) winter; (**b**) spring; (**c**) summer; (**d**) autumn (Knysh et al., 2001)

from 450 to 600 m, 6×10^{19} at 650 m, and 6×10^{19} from 700 m to the bottom. The horizontal diffusion coefficient for heat and salt was equal to $10^6\,\text{cm}^2/\text{s}$. Two experiments were performed.

The first experiment was based on the numerical model (6.4.4)–(6.4.16), disregarding the assimilation of temperature and salinity (the last terms in Eqs. (6.4.8) and (6.4.9)). In addition, the vertical turbulent viscosity and the diffusion coefficient of heat and salt were calculated by using the formulae with the Richardson number (Ibrayev and Trukhchev, 1996; Demyshev et al., 2004b). The surface boundary condition for temperature accounts for the climatic solar radiation absorbed by the sea (Bryan, 1969). This allowed to perform a prognostic calculation with seasonally varying climatic fluxes of momentum, heat, precipitation, and evaporation (Ibraev, 1993, see Chapter 4). Unlike Demyshev et al. (2004b), this experiment took into account the influence of rivers by specifying the climatic temperature (Bryan, 1969) and no salinity (boundary conditions (6.4.15), (6.4.16)) at their mouths. The values of temperature and salinity in the lower Bosphorus current were taken to be equal to 16 °C and 35‰, respectively.

The annual spatial distributions and variations of the temperature, salinity, flow velocity, and vertical turbulent viscosity and diffusion values at sections along 43.7 °N, 28.5 °E and at certain levels are analyzed below basing on the seventh year of calculations. The numerical convergence was controlled through the temporal evolution of rms deviations of sea-surface fields of two subsequent years. In this time period, the rms deviation tends to the quasi-stationary oscillation mode with a peak value of 0.1 cm.

By cold waters (CWs), we mean those with a temperature lower than 8 °C. We also use the traditional approach to CIL separation as a hydrological construct bounded by the isothermal surface of 8 °C.

The winter-to-spring variations in the temperature field of the northwestern area of the sea can be characterized by two different time periods. The first is the period of surface CW formation and transformation, ending with the most active phase (late January to the first ten-day period of March), while the second is the period when surface waters start to heat and CWs gradually stop flowing from the northwestern area into the zone of die western continental slope (the second ten-day period of March to early May).

In the first period, the structure of the surface circulation in the western area of the sea is characterized by the existence of two jets (Figs. 6.59a and b). The current of the first jet is directed from the southwestern Crimea to a northwestern shallow area, then turns to the west and south, moving along the Romanian and Bulgarian coasts (the northern side of the Black Sea Rim Current (BSRC)). The second current moves from the southern Crimean coast westward. These two stream jets begin to interact between 43.5 °N and 44 °N and can be seen at all depths up to 50 m inclusive.

An analysis of the wintertime temperature evolution shows that the belt of cold surface waters with a minimum temperature of 5 °C in the northwestern area emerges as early as the first ten-day period of December (Fig. 6.58a) and, then, covers the whole western area of the sea (Fig. 6.58b). It can also be seen from Fig. 6.58a that, in the east-cyclone zone, the sea-temperature field includes CW areas (contoured by the 8 °C isotherm) generated by free convection.

In the numerical experiment, the free convection was simulated in a commonly accepted way (Bryan, 1969; Ibrayev, 1993, see Chapter 4). Checking the hydrostatic stability is followed, if necessary, by hydrostatic adaptation (correction of temperature and salinity aimed at potential energy conservation). Mixing at each

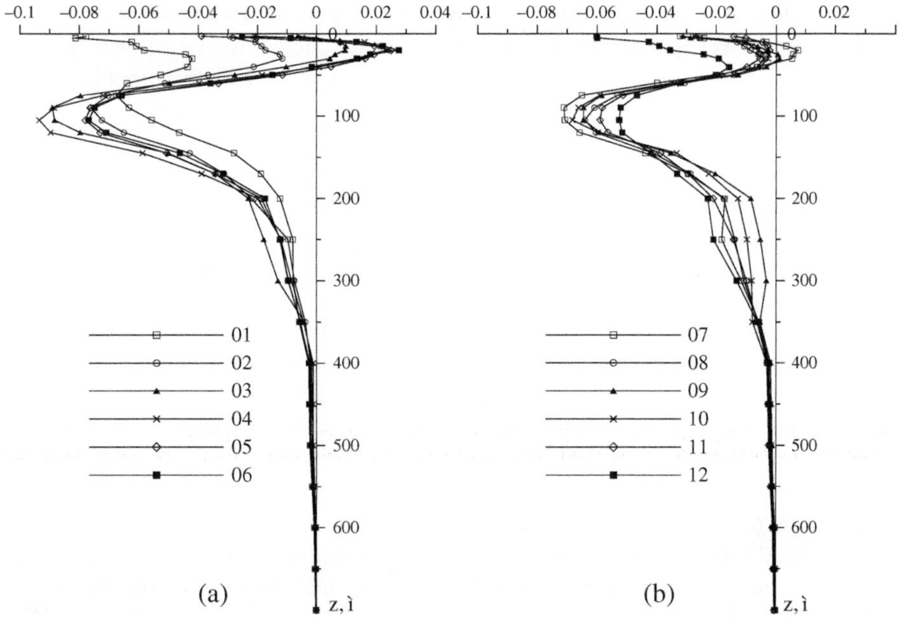

Fig. 6.59 Vertical profiles of monthly mean values of the coefficients $k_S(z, t_m)$ in the first (**a**) and second (**b**) half of year (Knysh et al., 2007)

water column persists until all inversions are eliminated. As a result, the density is smoothed vertically.

In late November, the sea-surface temperature was higher than 8 °C, and the upper and lower boundaries of the CIL in the eastern cyclone zone were substantially raised to the surface due to upwelling. Because of thermal convection, the temperature at the surface and below this cyclonic area became lower than 8 °C on December 10. After this, water convection and horizontal advection processes led to the enlargement of the CW plume.

6.4.4 Data Assimilation by the Method of Double Correction

6.4.4.1 Introduction

In a series of publications (Knysh et al., 2002, 2005), a new method for reconstructing the seasonal currents in the Black Sea on the basis of assimilation of climatic fields of temperature and salinity into models of sea-water circulation was proposed. Archive hydrologic data were assimilated either at each time step or after a definite number of steps. This assimilation delivered the continuous evolution of climatic sea fields in the annual cycle.

Another approach used the monthly mean fields of the dynamic level of the Black Sea obtained from assimilation of altimetry data into the sea model (Dorofeev et al.,

2004, 2006). The sea surface topography reflects specific features of the seasonal sea dynamics over approximately the past 12 years. Together with the monthly mean climatic hydrologic arrays of data on temperature and salinity, these fields can be used for reconstruction of the seasonal currents of the Black Sea. This is important for both studies of the basin climate and further improvement of the method of assimilation of sea-level data into the model. In this regard, as has been outlined by Dorofeev and Korotaev (2004), Mellor and Ezer (1991, see Chapter 4), Knysh et al. (1996, see Chapter 4), Korotaev et al. (1998, see Chapter 4), it is very important to use reasonable statistical information for extrapolating data on the climatic sea level to the deep layers. For example, Mellor and Ezer (1991, see Chapter 4) first transformed the data on ocean level anomalies (deviations from annual mean values) into equivalent anomalies of temperature and salinity at different depths. For this purpose, a model climate and correlations of the sea level anomaly with temperature and salinity anomalies are used. Knysh et al. (1996, see Chapter 4) took the estimates for cross correlations between the errors of sea level and density forecasts to extrapolate the measurements of the ocean level to the corresponding corrections of density and flow velocity at different layers levels. Korotaev et al. (1998, see Chapter 4) calculated the cross-correlation function of deviations of the sea level and seawater salinity from climatic data, and the level of the Black Sea was obtained via the adaptation method. Dorofeev and Korotaev (2004) assumed the statistical characteristics of errors of temperature and salinity forecasts to be proportional to the statistical characteristics of the observed fields.

As a new approach, we propose a method of combined assimilation of climatic hydrologic fields of temperature and salinity and the climatic dynamic level of the Black Sea (Dorofeev et al., 2006) into models of sea dynamics (Knysh et al., 2007). The statistical characteristics of errors of sea level, temperature, and salinity forecasts were assumed to be proportional to the statistical characteristics of the differences between monthly mean climatic fields of the sea level, temperature, and salinity that are calculated from the assimilation of altimetry observations of the sea level into model of dynamics and analogous climatic hydrologic fields. Knysh et al. (2007) could reproduce (in current fields) quasi-stationary synoptic anticyclonic eddies located along the periphery of the Black Sea Rim Current.

The following designations will be used for the description of the method.

$T^{Cl}(\vec{x}, t_M)$, $S^{Cl}(\vec{x}, t_M)$ are the new monthly mean arrays of climatic temperature and salinity of the Black Sea that are obtained from of archive observational data over the period from 1903 to 2002;

$T^{Cl}(\vec{x}, t)$, $S^{Cl}(\vec{x}, t)$ are the climatic temperature and salinity used for assimilation into the model of seawater circulation.

$\zeta^A(x, y, t_M)$, $T^A(\vec{x}, t_M)$, $S^A(\vec{x}, t_M)$ are the monthly mean arrays of climatic dynamic level of the sea, temperature, and salinity that are calculated from the results of assimilation of altimetry data into the model.

$\zeta^A(x, y, t)$ is the climatic altimetry level of the sea that is used for assimilation into the model of seawater circulation.

$\zeta^{Cl}(x, y, t_M)$ are the monthly mean arrays of the climatic dynamic level of the sea that are calculated from the results of assimilation of climatic temperature and salinity into the model of currents (Demyshev et al., 2005).

$T(\vec{x}, t)$, $S(\vec{x}, t)$, $\zeta(x, y, t)$ are the temperature, salinity, and sea level that are calculated in the numerical model of seawater circulation.

$K_T(z, t)$ and $K_S(z, t)$ are the coefficients appearing in the expressions for the sources $Q_{T\zeta}(\vec{x}, t)$ and $Q_{S\zeta}(\vec{x}, t)$ in equations (6.4.22) and (6.4.23) at each time step in the numerical model of seawater currents. The values of these coefficients were obtained by a linear interpolation of monthly mean values of the coefficients $K_T(z, t_M)$ and $K_S(z, t_M)$.

6.4.4.2 Description of the Model and Algorithm of Assimilation of Climatic Data

The fields of currents in the Black Sea were reconstructed from climatic temperature, salinity, and sea level by means of the three-dimensional numerical model of Demyshev et al. (2005).

For the incompressible seawater, the model equations in the Boussinesq and hydrostatic approximations are written as

$$\frac{\partial u}{\partial t} - (\xi + f)v + w\frac{\partial u}{\partial z} = -g\rho_0\frac{\partial \zeta}{\partial x} - \frac{1}{\rho_0}\frac{\partial}{\partial x}(P' + E) + \frac{\partial}{\partial z}\left(\nu_V\frac{\partial u}{\partial z}\right) + F^u, \tag{6.4.17}$$

$$\frac{\partial v}{\partial t} + (\xi + f)u + w\frac{\partial v}{\partial z} = -g\rho_0\varsigma_y - \frac{1}{\rho_0}\frac{\partial}{\partial y}(P' + E) + \frac{\partial}{\partial z}\left(\nu_V\frac{\partial v}{\partial z}\right)_z + F^v, \tag{6.4.18}$$

$$\frac{\partial u}{\partial x} + \frac{\partial v}{\partial y} + \frac{\partial w}{\partial z} = 0, \tag{6.4.19}$$

$$\frac{\partial \zeta}{\partial t} + \int_0^H \left(\frac{\partial u}{\partial x} + \frac{\partial v}{\partial y}\right) dz = \frac{\Pr - Ev}{\rho_1}, \tag{6.4.20}$$

$$P = g\rho_0\varsigma + g\int_0^z \rho\, d\mu = g\rho_0\varsigma + P', \tag{6.4.21}$$

$$\frac{\partial T}{\partial t} + \frac{\partial}{\partial x}(uT) + \frac{\partial}{\partial y}(vT) + \frac{\partial}{\partial z}(wT) = \frac{\partial}{\partial z}\left(\kappa^T\frac{\partial T}{\partial z}\right) + F^T + Q_T + Q_{T\zeta}, \tag{6.4.22}$$

$$\frac{\partial S}{\partial t} + \frac{\partial}{\partial x}(uS) + \frac{\partial}{\partial y}(vS) + \frac{\partial}{\partial z}(wS) = \frac{\partial}{\partial z}\left(\kappa^S\frac{\partial S}{\partial z}\right) + F^S + Q_S + Q_{S\zeta}, \tag{6.4.23}$$

$$\rho = \rho_0 + \alpha_1^T T + \alpha_1^S S + \alpha_2^T T^2 + \alpha^{ST} ST, \tag{6.4.24}$$

where

$$Q_T(\vec{x},t) = \frac{1}{REL[1+\eta^2(z)]}[T^{Cl}(\vec{x},t) - T(\vec{x},t)],\ E = \rho_0(u^2+u^2)/2,$$
$$\xi = \frac{\partial v}{\partial x} - \frac{\partial u}{\partial y} \tag{6.4.25}$$

$$Q_{T\zeta}(\vec{x},t) = \frac{1}{REL1}[K_T(z,t)\delta\zeta^*(x,y,t) - \frac{1}{\Omega^z}\int\limits_{\Omega^z} K_T(z,t)\delta\zeta^*(x,y,t)d\omega], \tag{6.4.26}$$

$$Q_S(\vec{x},t) = \frac{1}{REL[1+\eta^2(z)]}[S^{Cl}(\vec{x},t) - S(\vec{x},t)], \tag{6.4.27}$$

$$Q_{S\zeta}(\vec{x},t) = \frac{1}{REL1}[K_S(z,t)\delta\zeta^*(x,y,t) - \frac{1}{\Omega^z}\int\limits_{\Omega^z} K_S(z,t)\delta\zeta^*(x,y,t)d\omega], \tag{6.4.28}$$

$\eta^2(z)$ is the variance of measurement errors normalized by the field variance,
REL and REL1 are the relaxation parameters,

$$K_T(z,t) = P_{T\zeta}(z,t)/[\sigma_\zeta^2(t) + \sigma_{\zeta_m}^2], \tag{6.4.29}$$

$$K_S(z,t) = P_{S\zeta}(z,t)/[\sigma_\zeta^2(t) + \sigma_{\zeta_m}^2], \tag{6.4.30}$$

$$\delta\zeta^*(x,y,t) = \zeta^A(x,y,t) - \zeta(x,y,t) - \frac{1}{\Omega^0}\int\limits_{\Omega^0}[\zeta^A(x,y,t) - \zeta(x,y,t)]d\omega,$$

$P_{T\zeta}(\cdot)$, $P_{S\zeta}(\cdot)$ are cross-covariance functions, $\sigma_\zeta^2(\cdot)$ is the variance of level differences, $\sigma_{\zeta_m}^2$ is the variance of errors in level measurements, Ω^z is the area of the sea at the horizon, Ω^0 is the area of the sea at the surface, and Ev and Pr are the evaporation and precipitation at the sea surface. The second term in square brackets in formulas (6.4.29) and (6.4.30) are introduced to avoid fictitious heat and salinity fluxes into the deep sea (Dorofeev et al., 2004).

The coefficients of turbulent momentum exchange and vertical diffusion are calculated using the Pacanowski–Philander approximation (Pacanowski and Philander, 1981; Demyshev et al., 2004a):

$$\nu_V = \nu_0^V(R_0+Ri)^{-2} + \nu_1^V,\quad \kappa^S = [(\nu_0^S(R_0+Ri)^{-2} + \nu_1^S]/(R_0+Ri) + \kappa_1^S,$$
$$\kappa^T = [(\nu_0^T(R_0+Ri)^{-2} + \nu_1^T]/(R_0+Ri) + \kappa_1^T \tag{6.4.31}$$

where ν_0^V, ν_0^S, ν_0^T, ν_1^V, ν_1^S, ν_1^T, R_0 are the specified constants; κ_1^S, κ_1^T are the specified functions of time; and Ri is the Richardson number. If $\nu_0^V = \nu_0^S = \nu_0^T$, $\nu_1^V = \nu_1^S = \nu_1^T$, and κ_1^S, κ_1^T are constants, then, relations (6.4.31) are exactly consistent with the vertical-mixing model proposed in (6.4.30).

The horizontal viscosity and diffusion are described with the help of both a biharmonic representation of turbulence terms in the equations of motion and a harmonic operator in the equations of heat and salinity diffusion:

$$F^u = -\nu_H \nabla^4 u, \quad F^v = -\nu_H \nabla^4 v, \quad F^T = \kappa_H \nabla^2 T, \quad F^S = \kappa_H \nabla^2 S. \qquad (6.4.32)$$

Equation (6.4.20) is equivalent to a linearized kinematic condition at the sea surface ($w = -\varsigma_t + \frac{\text{Pr} - Ev}{\rho_1}$).

Boundary and Initial Conditions

At the surface $z = 0$:

$$\nu_V \frac{\partial u}{\partial z} = -\tau^x, \; \nu_V \frac{\partial v}{\partial z} = -\tau^y, \; k^y \frac{\partial T}{\partial z} = -Q^T, \; k^y \frac{\partial S}{\partial z} = \frac{E_v - \text{Pr}}{\rho_1} S_0 + \gamma(S^{Cl} - S_0) \qquad (6.4.33)$$

Here Q^T is the total kinematic heat flux minus climatic solar radiation absorbed by the sea and S_0 is the salinity at the surface.

$$\text{At the bottom } z = H(x, y): u = v = w = 0, \; \frac{\partial T}{\partial z} = 0, \; \frac{\partial S}{\partial z} = 0 \qquad (6.4.34)$$

At the rigid lateral walls, we have

$$u = \nabla^2 u = \frac{\partial v}{\partial x} = \nabla^2 \frac{\partial v}{\partial x} = 0, \quad \frac{\partial T}{\partial x} = 0, \; \frac{\partial S}{\partial x} = 0, \qquad (6.4.35)$$

for meridional segments of the boundary and
for zonal segments

$$v = \nabla^2 v = \frac{\partial v}{\partial y} = \nabla^2 \frac{\partial v}{\partial y} = 0, \quad \frac{\partial T}{\partial y} = 0, \; \frac{\partial S}{\partial y} = 0, \qquad (6.4.36)$$

In the boundary segments where water flows in, the following conditions are used:

$$u = u^p, \nabla^2 u = \frac{\partial v}{\partial x} = \nabla^2 \frac{\partial v}{\partial x} = 0, \quad T = T^p, \; S = S^p, \qquad (6.4.37)$$

for meridional segments of the boundary and
for zonal segments

$$v = v^p, \nabla^2 v = \frac{\partial u}{\partial y} = \nabla^2 \frac{\partial u}{\partial y} = 0, \quad T = T^p, \; S = S^p. \qquad (6.4.38)$$

For the upper Bosphorus Current and the Kerch Strait, when the current is directed from the Black Sea to the Azov Sea:

$$v = v^s, \nabla^2 v = \frac{\partial u}{\partial y} = \nabla^2, \frac{\partial u}{\partial y} = 0, \quad \frac{\partial T}{\partial x} = 0, \frac{\partial S}{\partial x} = 0. \tag{6.4.39}$$

u^p, v^p, u^s, v^s are the velocities in river estuaries and straits, respectively, and T^p, S^p are the temperature and salinity in river estuaries and straits, respectively. We assume that the salinity of river waters is nonzero and equals 7 psu.

The initial conditions have the form

$$u = v = 0, \quad \varsigma = \varsigma^0, \quad T = T^0, \quad S = S^0 \text{ at } t = t_0. \tag{6.4.40}$$

Model Parameters

For each day, the components of wind stress were determined from 6-hourly sea surface pressure fields for the period 1988–1998 (Dorofeev and Korotaev, 2004). Analysis of the vorticity of the wind revealed its substantially higher intensity in comparison with the seasonally varying wind used in Korotaev et al. (2000), Knysh et al. (2002, 2005) and prepared by Staneva and Stanev (1998).

The seasonally varying total climatic heat flux at the sea surface and the climatic solar radiation absorbed by the sea were taken into account in the same manner as in Demyshev et al. (2004a,b). The parameter γ in condition (6.4.33) was 1 day^{-1}.

The calculations were performed on a 14.8 km × 14.8 km grid, with 44 irregular vertical levels The time step was 5 min.

The coefficient $\nu_H(z)$ (dimension cm^4/s) varies with depth in the following way: $\nu_H = 5 \times 10^{18}$ at 2.5–350 m, $\nu_H = 1.25 \times 10^{19}$ at 400 m, $\nu_H = 2 \times 10^{19}$ at 450–600 m, $\nu_H = 6 \times 10^{19}$ at 650 m, and $\nu_H = 10^{20}$ at 700–2,000 m.

The parameters in formulas (6.4.31) were chosen on the basis of numerical experiments. The calculation results were compared to observation-derived seasonal features of the temperature regime of the active layer of the Black Sea. The optimal values of the parameters turned out to be as follows: $\nu_0^V = \nu_0^S = \nu_0^T = 30\,\text{cm}^2/\text{s}$, $\nu_1^V = 5\,\text{cm}^2/\text{s}$, and $\nu_1^S = \nu_1^T = 1\,\text{cm}^2/\text{s}$. The formation of the Cold Intermediate Layer (CIL), the position of its boundaries in the vertical, and its structure turned out to be sensitive to the seasonal variability of the parameter κ_1^T. The CIL characteristics proved to be close to the climatic ones for the following values of κ_1^T changing monthly from January to December, respectively: 4, 6, 3, 1, 0.05, 0.005, 0.003, 0.003, 0.005, 0.008, 0.01, and 1 cm^2/s.

The vertical position of the pycnocline was close to the climatic position for the values of κ_1^S decaying from 1.8 at the sea surface to 0.03 cm^2/s at depths to 112.5 m. For depths below 112.5 m, $\kappa_1^T = \kappa_1^S = 0.03\,\text{cm}^2/\text{s}$.

6.4.4.3 Climatic Arrays

The new climatic array on temperature and salinity was created on the basis of observational data stored in oceanographic databanks of the Marine Hydrophysical Institute of the National Academy of Sciences of Ukraine and the Marine Department of the Ukrainian Research Hydrometeorological Institute.

How were the climatic altimetry fields compiled? First, the hydrophysical parameters of the Black Sea were reproduced with the assimilation of altimetry data into the circulation model described above for the period from 1992 to 2003 (Dorofeev and Korotaev, 2004). The finite-difference approximation of model equations was performed in on a grid with a horizontal step of 5 km (the inner internal radius of Rossby deformation in the open sea is about 25 km). There were 35 irregular vertical levels from 2.5 to 2,100 m. The boundary conditions at the sea surface were specified using the fields of wind stress for each 6 h. The period 1992–1998 was based on the above mentioned Ukrainian data, the period 1998–2003 on NCEP reanalysis (1998–2003). The climatic solar radiation was disregarded. The monthly variations of the total heat flux, precipitation, and evaporation were taken from Staneva and Stanev (1998).

The coefficient of vertical turbulent diffusion was a function of the vertical coordinate alone and provided a CIL seasonal variability close to the climatic one. The salinity in river estuaries was calculated using simplified equations of salt diffusion with given components of the river flow velocity.

The initial altimetry data were anomalies of the sea level along standard tracks of the ERS1 and ERS 2, TOPEX/Poseidon, Jason, EnviSat, and GFO satellites. The cycle periodicity was 35 days for ERS1, ERS 2 and EnviSat, 10 days for TOPEX/Poseidon and Jason, and 17 days for GFO. The algorithms for reproducing the dynamical topography of the sea surface on the basis of both altimeter data and the results of assimilation of altimetry data can be found in Dorofeev and Korotaev (2004), Korotaev et al. (1998, see Chapter 4).

The monthly mean climatic altimetry arrays were obtained through averaging over the period from 1992 to 2003 of the salinity, temperature, and sea-level fields reproduced from a continuous assimilation of altimetry data into the model of sea circulation.

The monthly mean climatic arrays $T^{Cl}(\vec{x}, t_M)$, $S^{Cl}(\vec{x}, t_M)$ were interpolated on the model's grid and, then, the Fourier coefficients were computed and the corresponding fields were reproduced for each day of the year. The monthly mean climatic fields of the altimetric sea level $\zeta^A(x, y, t_M)$ were also interpolated on the model's grid. The diurnal climatic fields of temperature, salinity, wind stress, monthly mean altimetry sea level, and monthly mean climatic fields of precipitation, evaporation, total heat flux (Staneva and Stanev, 1998), solar radiation, and parameter κ_1^T were interpolated linearly at each time step.

6.4.4.4 Computation of the Coefficients $K_T(z, t_M)$, and $K_S(z, t_M)$

The form of cross-covariance functions existing in expressions (6.4.29) and (6.4.30) for the coefficients $K_T(\cdot)$, $K_S(\cdot)$ follows from the relations of the Kalman multielement correlation algorithm (Knysh et al., 1996, see Chapter 4). Therefore an estimation of cross covariances between the errors of salinity (temperature) prediction and the errors of sea-level prediction is required. These statistical characteristics should be computed through a solution of the corresponding differential equations in the four-dimensional space-time domain. However, this is a very difficult and yet unsolvable problem.

It is assumed instead that the influence of the values of the altimetry climatic level at the grid points surrounding the fixed node (except the node itself) can be disregarded. In addition, the covariance functions are assumed to depend only on the vertical coordinate and month of the year. Thus, the fields of errors in salinity, temperature, and level relative to statistical characteristics are assumed to be horizontally homogeneous and isotropic.

The located estimate for the cross covariance $P_{S\zeta}(z, t_M)$, where t_m is the month of the year, was determined as

$$P_{S\zeta}(z, t_M) = \frac{1}{N(z) - 1} \sum_{ij} \delta S'_{ij}(z, t_M) \cdot \delta \zeta'_{ij}(t_M). \tag{6.4.41}$$

In expression (6.3.41), we have

$$\delta S'_{ij}(z, t_M) = \delta S_{ij}(z, t_M) - \overline{\delta S}(z, t_M), \tag{6.4.42}$$

$$\delta S_{ij}(z, t_M) = S^A_{ij}(z, t_M) - S^{Cl}_{ij}(z, t_M), \tag{6.4.43}$$

$$\delta \zeta'_{ij}(t_M) = \delta \zeta_{ij}(t_M) - \overline{\delta \zeta}(t_M),, \tag{6.4.44}$$

$$\delta \zeta_{ij}(t_M) = \zeta^A_{ij}(t_M) - \zeta^{Cl}_{ij}(t_M),, \tag{6.4.45}$$

$N(z)$ is the number of grid points at a given horizon.

The unbiased estimate for the cross covariance $P_{T\zeta}(z, t_M)$ was computed by formulas similar to (6.3.41)–(6.3.45). The variance of level differences was estimated by the formula

$$\sigma^2_\zeta(t_M) = \frac{1}{N(0) - 1} \sum_{i,j} \left[\delta \zeta'_{ij}(t_M)\right]^2. \tag{6.4.46}$$

where $N(0)$ is the number of grid points at the surface (the first level). The variance of measurement errors of the level $\sigma^2_{\zeta m}$ was $2\,\text{cm}^2$.

Figure 6.59 presents the dependence of the coefficient $K_S(z, t_M)$ on depth for all months of the year. One can see a seasonal variation in the profiles of this coefficient, especially in the upper 400-m layer of the sea. Significant values of the coefficient are observed at depths to ~ 600 m. Below this level, its values are small.

The coefficient $K_T(z, t_M)$ has almost the same behavior with depth in the first half of the year. In the summer–fall months, its values for the upper 50-m layer of the sea are positive. Significant values of $K_T(z, t_M)$ are also observed at depths to ~600 m.

6.4.4.5 Analysis of the Climatic Sea Fields Obtained by Method of Double Correction

Earlier, it has been shown (Demyshev and Knysh, 2004) that the scheme of periodic assimilation of climatic arrays of temperature and salinity delivers adjusted hydrophysical fields of the Black Sea. An optimal assimilation period of 3 h was found. Unlike data assimilation at each time step (Demyshev et al., 2005), the main feature of the adapted fields is that the structure of vertical velocity is smooth, almost without small-scale noise. The maximum values of vertical velocity decreased by about one order of magnitude compared with the results of Demyshev et al. (2005).

The prognostic calculations with the combined 3-hourly assimilation of climatic data of temperature, salinity, and altimetry sea level were conducted for a period of four years. The value of the parameter REL was 24 h. The relaxation parameter REL1 was varied and taken to be 24, 10, and 6 h. The initial fields of temperature, salinity, flow velocity, and sea level were taken from the results of calculations with assimilation of climatic arrays of temperature and salinity (Demyshev et al., 2005). It should be pointed out that the results almost independent of the initial fields. Two main criteria were chosen to determine the optimal variant of calculations. These take into account qualitative and quantitative characteristics of the fields of vertical velocity. According to the first criterion, the structures of the vertical velocity must not contain small-scale disturbances of a wavelike character. The maximum value of vertical velocity should be lower than 0.5×10^{-2} cm/s, from the continuing equation. The variant of calculations with REL1 $=$ 6 h turned out to match the optimization criterion.

The curves in Fig. 6.60 characterize the process of passing to a periodic mode of calculations The settling time was 70 days. The calculation data for the last year were used in the subsequent analysis of the seasonal variability of the climatic circulation in the sea.

An important characteristic of surface geostrophic currents are the elevations of the free sea surface (Fig. 6.61). The maps indicate the differences between the climatic altimetry level (Figs. 6.61c and f) and the level reproduced by the method of double correction (Figs. 6.61b and e). The differences are reflected in the smoothing of synoptic anticyclonic eddies right of the Black Sea Rim Current. This smoothing is due to a fast geostrophic adaptation of the hydrophysical fields (Demyshev et al., 2006).

Along with this, the maps in Figs. 6.61b and e show a greater number of small-scale eddies than in Figs. 6.61a and d, thus being indicative of the influence of the climatic altimetry level in the course of assimilation.

It can be seen from Fig. 6.61e that, in summer, are found almost all, except Caucasian (Sochi and Tuapse), anticyclonic eddies that are seen in Fig. 6.61f. These are Yalta, Caucasian (Anapa), Batumi, Kizilirmak, and Sinop mesoscale eddy-formation processes. The location and configuration of the western and eastern cyclonic gyres are quite close to the gyres shown in Fig. 6.61f. Similar results were obtained for the remaining seasons.

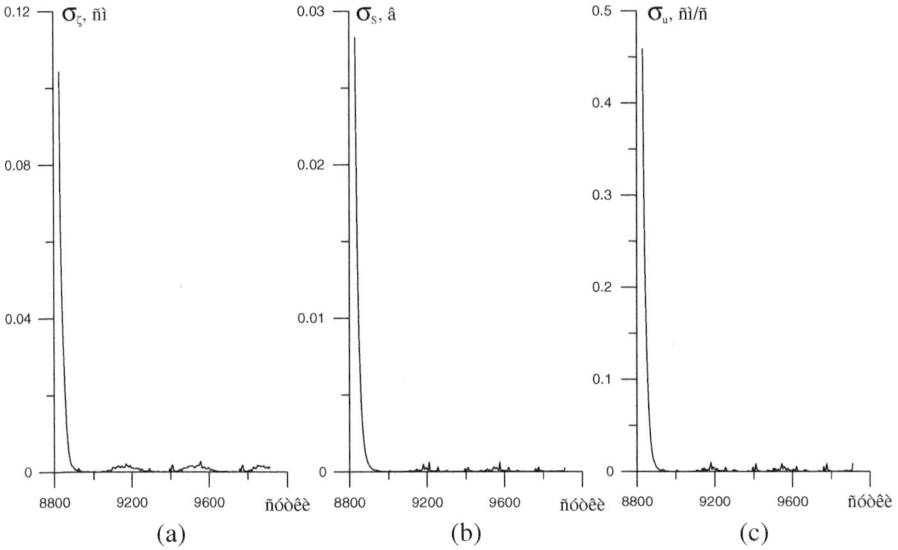

Fig. 6.60 Time-series of the square mean year to year differences of climatic values (**a**) of the sea surface height SSH (cm), (**b**) salinity (psu) and (**c**) the flow velocity u-component (cm/s) (Knysh et al., 2007)

The differences between the temperature fields can be seen in Fig. 6.62. It is known that, in the Black Sea, the vertical temperature distribution has a clearly defined CIL in all seasons of the annual cycle. Its upper and lower boundaries are determined by the isotherm 8°C (Ovchinnikov and Popov, 1984). In summer (when a new CIL is being formed), the cold reserve of this layer in the method of double correction is higher than for assimilating only temperature and salinity into the model (Figs. 6.62a and b). This is comparable with the cold reserve of the CIL in the distribution of climatic altimetry temperature (Fig. 6.62). The latter was not assimilated but was used to estimate the cross covariance. It can be concluded that the spring cold reserve of the CIL in the method of double correction is caused by the assimilation of the climatic altimetry level. This follows from analysis of summer temperature distributions at the section of 43.7°N (Figs. 6.62d–f). The growth of the cold reserve of the CIL in spring is possibly caused by the fact that, over the past ~20 years, the years with moderate and intensive formation and renewal of the CIL are prevailing.

In the section of interest, one can also see the thermocline features in spring (Figs. 6.62a–c) and summer (Figs. 6.62d–f). In the distribution of the climatic altimetry temperature, the summer thermocline is slightly smeared in depth (Fig. 6.62f). In the method of double correction, the thermocline is narrower and preciseowing to the assimilation of climatic temperature and salinity (Figs. 6.62d and e). In Figs. 6.62d and e, one can clearly see a thin homogeneous surface layer with a thickness between 10 and 12 m in the central and western part of the sea. In the eastern area, this layer is thinner (~7 m).

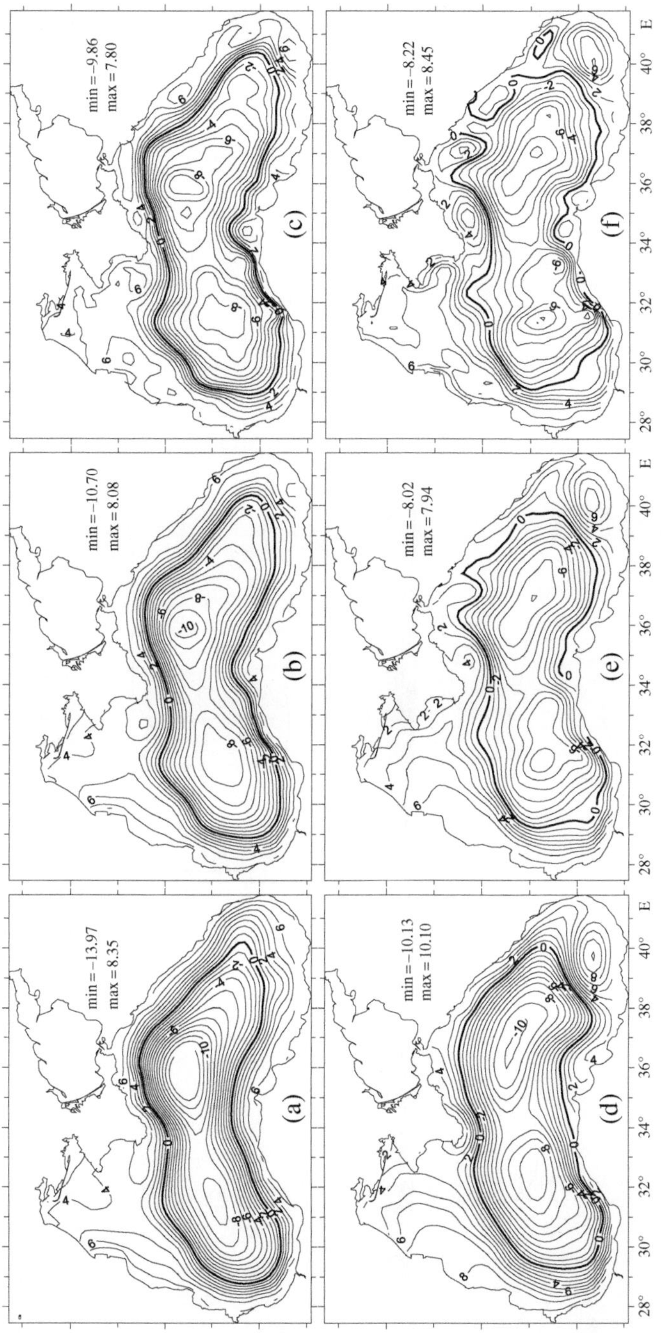

Fig. 6.61 The climatic SSH (cm) reconstructed by assimilating only the climatic values of the temperature and salinity for **(a)** February and **(b)** August; the same, but by assimilation of temperature, salinity and climatic altimetric SSH for **(c)**, February and **(d)** August; the climatic altimetric values of SSH for **(e)** February and **(f)** August (Knysh et al. 2007)

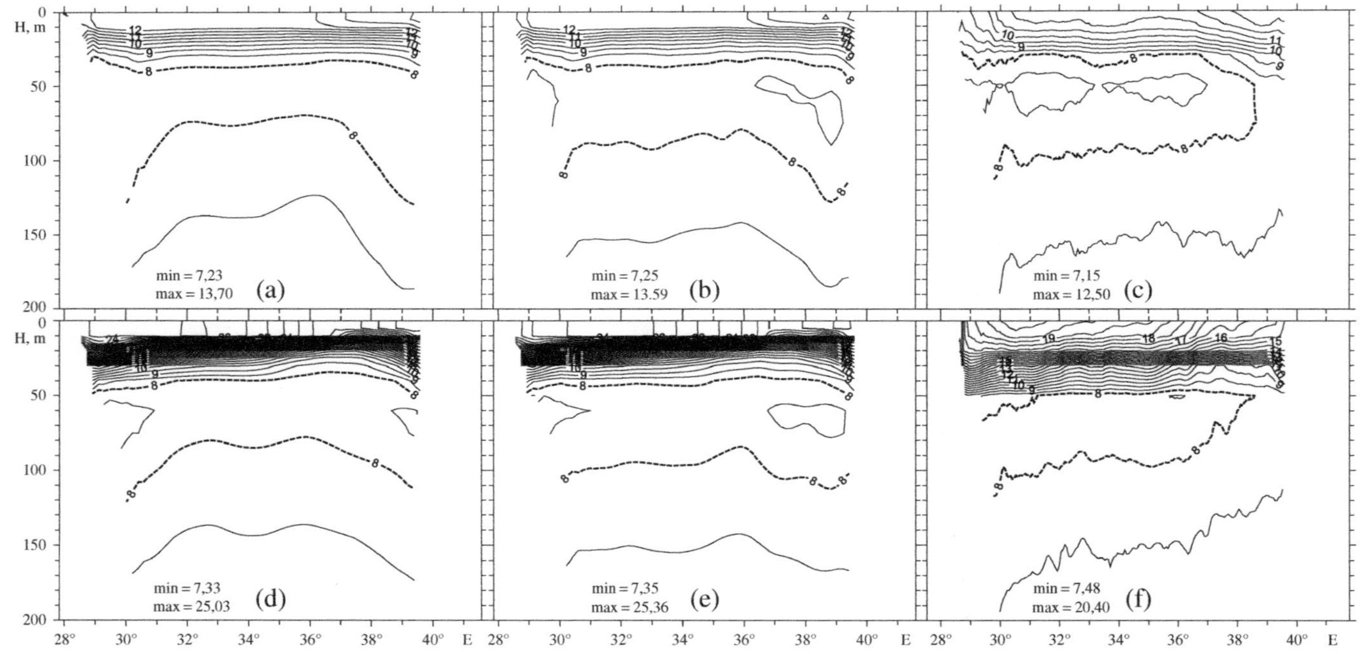

Fig. 6.62 Vertical cross-sections of temperature (°C) along the 43.7 °N in upper 200 m layer for the following cases: reconstructed by standard assimilation of climatic T and S for (**a**) May and (**d**) August; same by method of double correction for (**b**) May and (**e**) August; same by extrapolation of altimetric data for (**c**) May and (**f**) August (Knysh et al., 2007)

In winter, differences between the structures of temperature fields in the above versions of reconstruction of hydrophysical fields are smaller. In fall, as in summer, the CIL thickness in the method of double correction is slightly higher than for the assimilation of temperature and salinity alone.

The distribution of salinity in the layer between 0 and 200 m at the section at 43.7 °N is shown in Fig. 6.63. The position of the main halocline in the method of double correction (Figs. 6.63b and e) is close to its position obtained in the climatic altimetry salinity (Figs. 6.63c and f). In the assimilation of temperature and salinity alone, the lower boundary of the main halocline is elevated (Figs. 6.63a and d).

A minimum of the altimetry salinity at the section is obtained at the surface in the northeastern area (i.e., close to the Danube estuary). This result may be due to the simplified treatment of rivers in the numerical model. According to climatic hydrologic data, the salinity in this area is considerably higher (Figs. 6.63a and d). This fact is reflected in the salinity distribution in the given area obtained by the method of double correction (Figs. 6.63b and e).

The intraannual variability of intensity of the climatic velocity of currents in the Black Sea is shown in Fig. 6.64 as vertical-time plots of the kinetic energy (KE) averaged over the water area. The plots separate five layers: the surface Ekman layer (~10 m), a layer with small vertical KE gradients (~10–60 m), a layer with relatively high vertical KE gradients (~60–150 m), a layer with a gradual decrease of KE (~150–1, 100 m), and a deep-sea layer (~1, 100–2, 000 m) characterized by the barotropic velocity.

The upper Ekman layer of the sea reflects the formation of climatic currents as a result of seasonally varying fields of wind stress and the assimilation of data on temperature and salinity (Fig. 6.64a,b) and on temperature, salinity, and level (Fig. 6.64c,d). In this layer, the KE reaches maximum values of 93.7 (Fig. 6.64a) and 81.3 cm^2/s^2 (Fig. 6.64d). In Fig. 6.59, one can see a clearly defined correlation between the distributions of KE and wind vorticity in the seasonal cycle. At times with particularly high values of wind vorticity, its influence can be traced to depths of about 1,200 m.

The 10–60-m layer is generally characterized by a much slower variation of KE with depth than in layers below and above it. This is true especially from mid-May through August. The weak dependence of currents on depth in the 10–60-m sea layer can be due to an intensified free convection in the fall–winter–spring period. Let us note that, convective mixing, which is caused by unstable stratification, was parameterized through averaging of the temperature and salinity of unstable layers under the condition that the potential energy is conserved.

The currents in the layer of a constant halocline are characterized by their substantial attenuation with depth (Figs. 6.64b and d). The regions with extreme KE can be clearly identified with maximum on March 25 and minimum on September 11.

In the layer between approximately 150 and 1,100 m, the KE in the annual cycle has an oscillatory character and decreases gradually with depth. The wavelike character of the KE curves in the main halocline and below indicates the propagation of baroclinic Rossby waves.

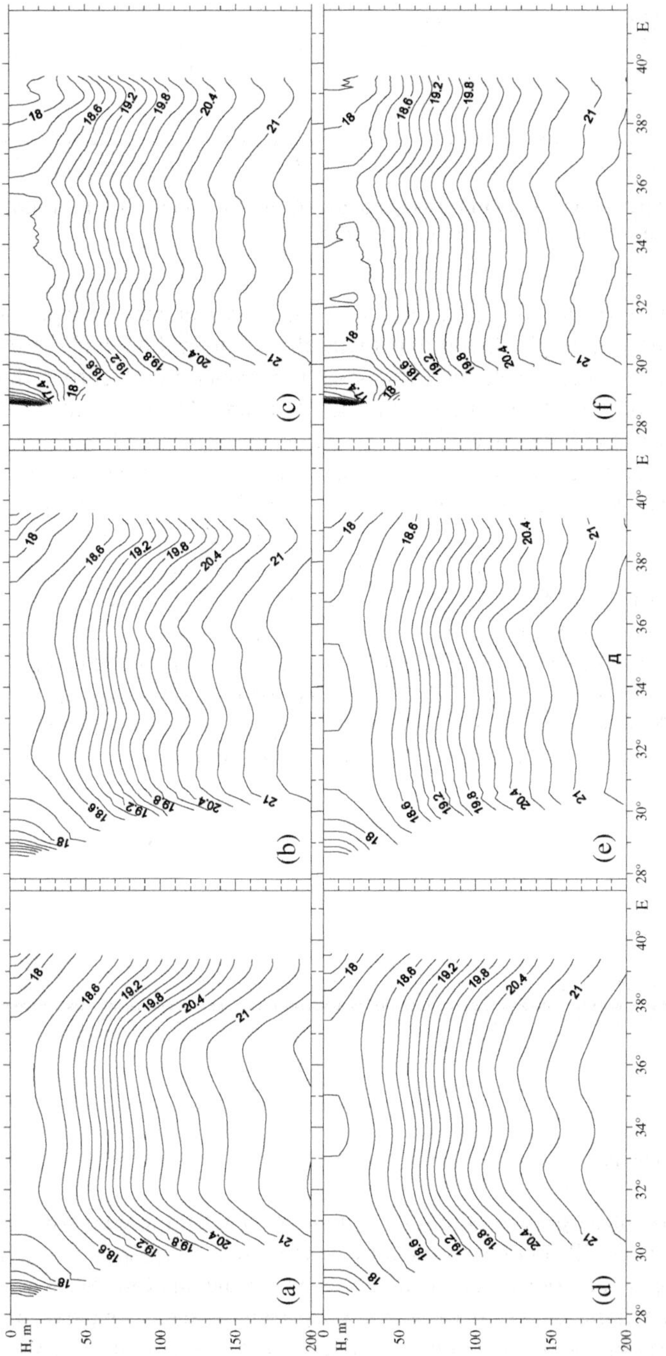

Fig. 6.63 Same as on Fig. 6.62, but for the salinity (psu) (Knysh et al., 2007)

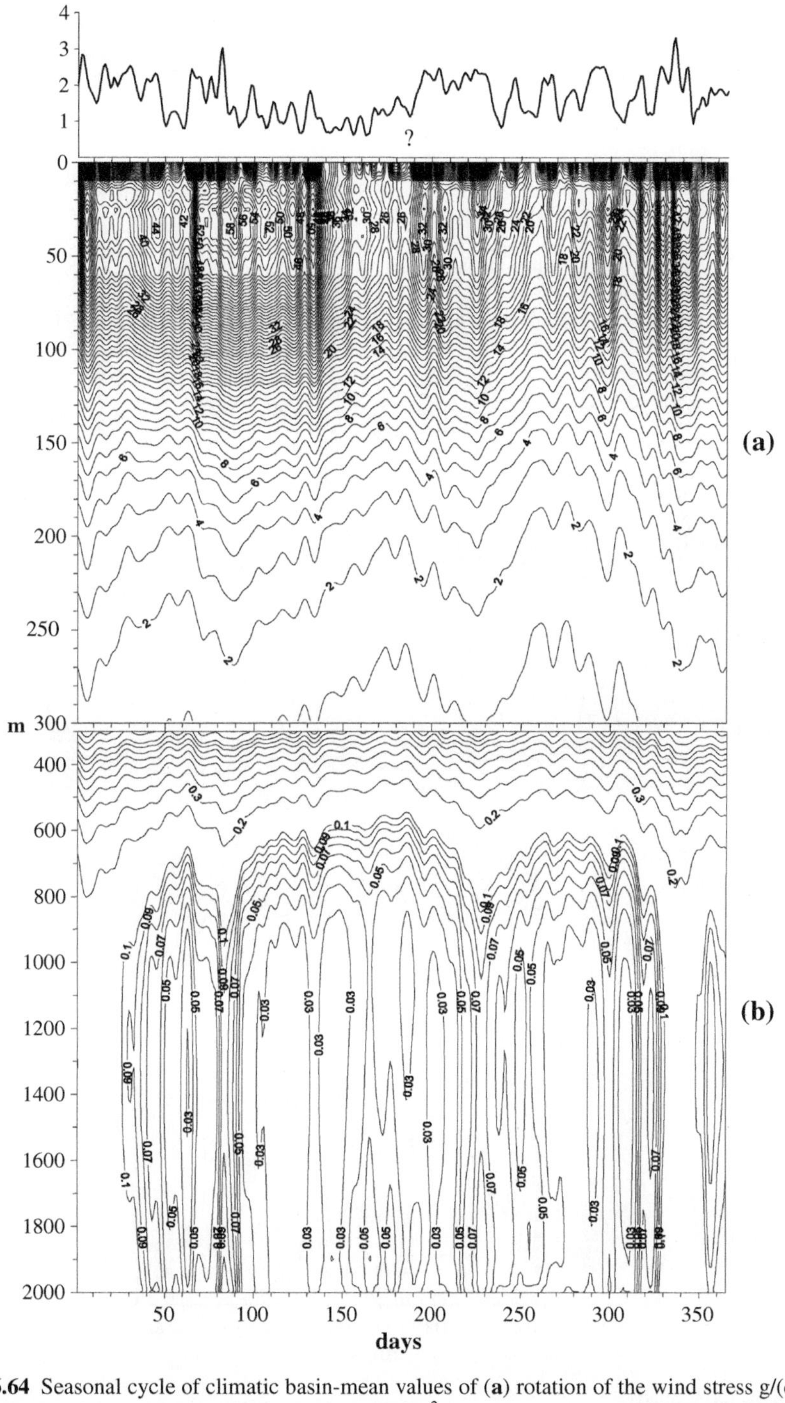

Fig. 6.64 Seasonal cycle of climatic basin-mean values of (**a**) rotation of the wind stress g/(cm/s) (**b**) vertical distribution of the kinetic energy (cm^2/s^2) constructed by assimilation climatic temperature and salinity; c same as in cases b, but for assimilation of data constructed by altimetric data extrapolation (Knysh et al. 2007)

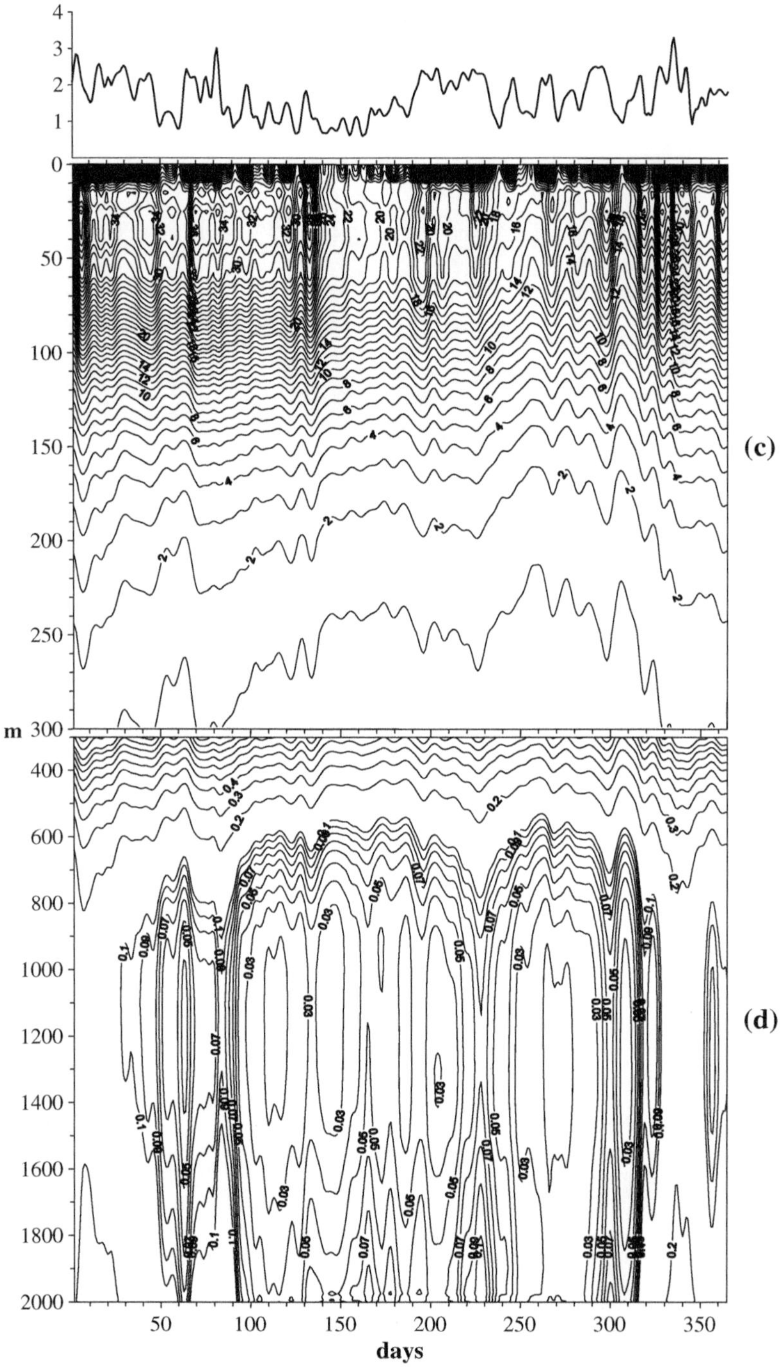

Fig. 6.64 (continued)

The KE of deep layer between approximately 1,100 and 2,000 m is most noticeable. Here, in the annual cycle, there is a clear alternation of vertical homogencity with maximum and minimum values of KE. In the last decade of March, the maximum KE is observed at the bottom.

The depth structure of currents in the layer between 1,100 and 2,000 m at a specific time remains unchanged. However, despite small wind-velocity magnitudes (with a maximum of 20 m/s), the seasonal variability of currents in this layer is substantial.

The difference between intraannual KE variabilities for the case of temperature and salinity assimilation and for the method of double correction lies in the following. The currents in the latter case are less intense in the layer between 0 and 500 m (Figs. 6.64b and d). In deeper layers, the differences are small. At the same time, as can be seen from Fig. 6.65 in all seasons except winter, the horizontal variability of currents in the method of double correction is considerably higher than that obtained from hydrologic climatic data. In the horizontal currents fields deduced from climatic altimetry the main halocline involves a larger number of cyclonic and anticyclonic eddies (especially in summer). The Black Sea Rim Current is narrower, while its intensity is highest near the west Anatolian coast (Figs. 6.61b, d, f and h). In summer, a chain of anticyclonic eddies can be traced along the Caucasian coast and the northern boundary of the basin.

The reproduction of appropriate fields of vertical velocity in the Black Sea becomes of special significance for applied problems, for example, the estimation of long-term spreading of oil spill hydrocarbons.

The spatial distribution of the climatic vertical velocity obtained by the method of double correction is characterized by the presence of alternating zones of upwelling and downwelling (Figs. 6.66b, d, f and h). This is due to the eddy structure of horizontal currents. In the fall–winter–spring period, upwelling zones are prevailing. In summer, one can separate two major areas: First, the downwelling zones in the northwestern half of the basin, in the area adjacent to eastern Anatolia, and in the Batumi anticyclone; Second, the upwelling zones in the eastern half of the sea and in an area adjacent to western Anatolia. The extreme values of vertical velocity at the depth of 113 m vary between 0.3 and 0.5×10^{-2} cm/s, which is close to integral estimates.

Nearly the same structure of vertical velocity and its extreme values were obtained in the case of assimilation of only climatic thermohaline fields into the model (Figs. 6.61 a, c, e and g). The difference is that the alternation of upwelling and downwelling zones virtually does not occur in this case. This may be due to a smaller number of mesoscale eddies in the fields of horizontal velocity, when double correction is not used.

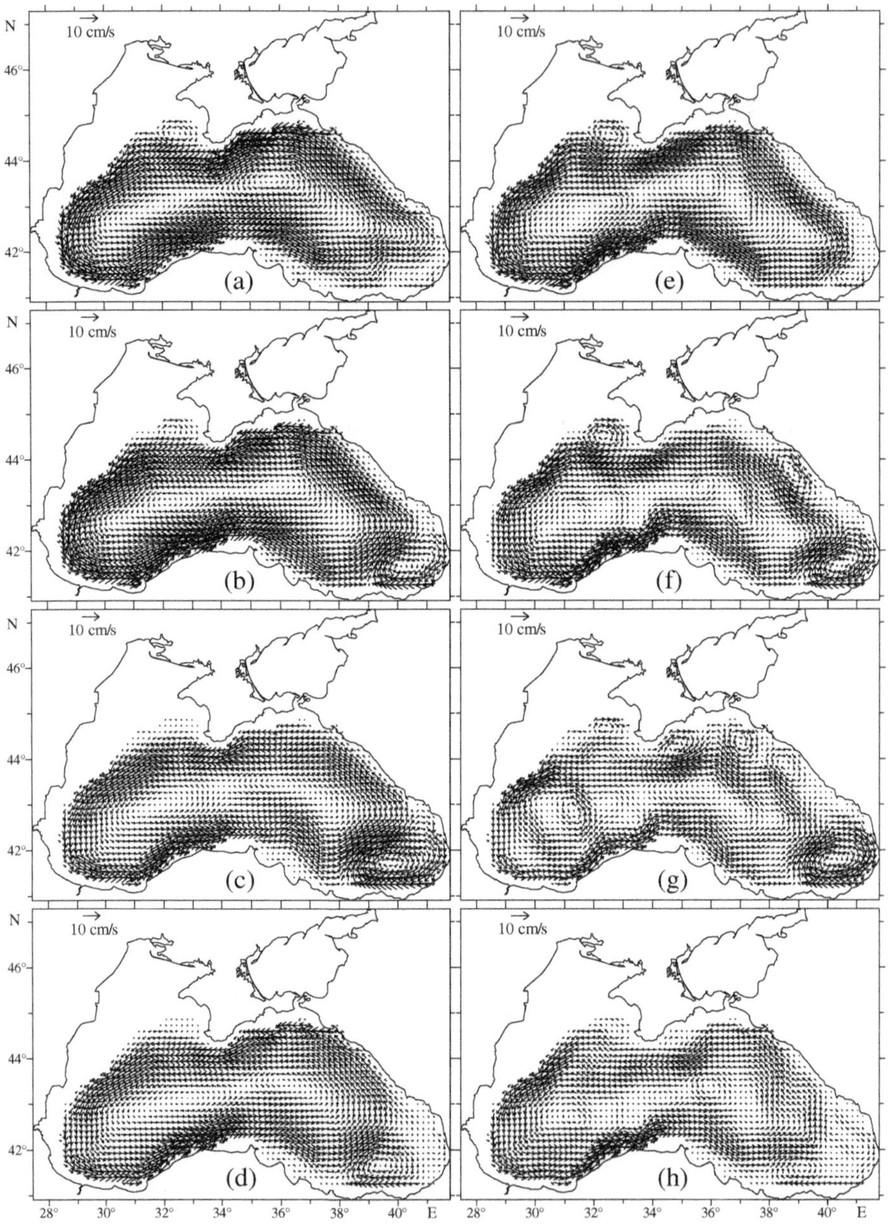

Fig. 6.65 Climatic currents at 105 m depth, reconstructed by assimilation only of the temperature and salinity (a, c, e, g) and by combined assimilation of temperature, salinity and altimetric sea surface high (b, d, f, h): **a,b** – for February, **c,d** – May, **e,f** – August, **g, h** – November (Knysh et al., 2007)

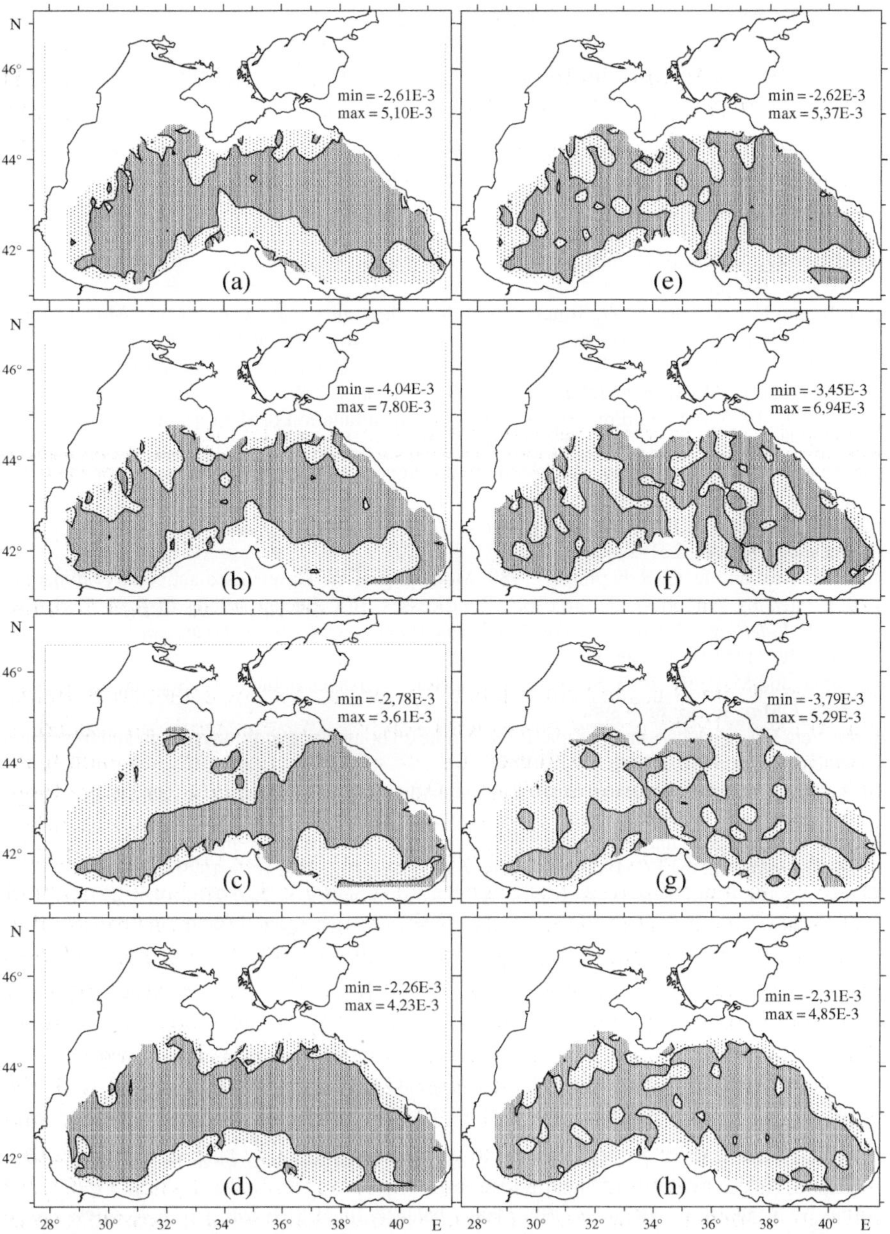

Fig. 6.66 Climatic vertical velocity (cm/s)at the depth 113 m (see Fig. 6.65 for definitions). Areas of negative values are shadowed (Knysh et al., 2007)

References

Akhverdiev I. O. and Y. L. Demin, 1989. On of the Caspian sea currents in the summer season as result of diagnostic calculation. In book of A.N. Kosarev, *The Caspian sea: The Water Structure and Dynamics*, Nauka, Moscow, 5–15 (in Russian).

Altman E. N., 1991. Water balance. In: A.I. Simonov and E.N Altman (eds.), *Hydrometeorology and Hydrochemistry of the USSR Seas. Vol.4: The Black Sea*, Hydrometioizdat, St.-Petersburg, 103–124 (in Russian).

Altman E. N., I. F. Gertman, and Z. A. Golubeva, 1987. Climatological fields of salinity and temperature in the Black Sea, *Report, State Oceanogr. Inst., Sevastopol Branch*, Sevastopol, Ukraine, 109 pp.

Arpe K. and E. Roeckner, 1999. Simulation of the hydrological cycle over Europe: Model validation and impacts of increasing greenhouse gases. *Adv. Water Resour.* 23: 105–119.

Aubrey D. G., Z. Belberov, A. Bologa, V. Eremeev, and U. Unluata, 1992. A coalition to diagnose the patient: CoMSBlack and the Black Sea. *Mar. Technol.* 2: 5–8.

Backhaus J., 1985. A three-dimensional model for the simulation of shelf sea dynamics. *Dt. Hydrogr. Z.* 38: 165–187.

Backhaus J. and D. Hainbucher, 1987. A finite difference general circulation model for shelf seas and its application to low frequency variability on the North European shelf. In: Nihoul J. and B. Jamart (eds.), *Three-Dimensional Models of Marine and Estuarine Dynamics*. Elsevier Oceanogr. Ser. 45, 221–244.

Badalov A. B. and D. G. Rzheplinski, 1989. Modelling of the Caspian sea active layer dynamics generated by synoptic-scale atmospheric processes. In collected articles of A. S. Sarkisyan (Ed.) *Modelling of Hydrophysical Fields and Processes in Closed Basins and Seas*. Nauka, Moscow, 31–51 (in Russian).

Becker G. and M. Pauly, 1996. Sea surface temperature changes in the North Sea and their causes. *ICES J. Mar. Sc.* 53: 887–898.

Beron-Vera F. J., J. Ochao, and P. Ripa, 1999. A note on boundary conditions for salt and freshwater balances. *Ocean Modelling* 1: 111–118.

Besiktepe S., H. I. Sur, E. Ozsoy, M. A. Latif, T. Oguz, and U. Unluata, 1994. The circulation and hydrography of the Marmara sea. *Prog. Oceanogr.* 34: 285–334.

Betts A. K., J. H. Ball, and P. Viterbo, 1999. Basin-scale surface water and energy budgets for the Mississippi from the ECMWF reanalysis. *J. Geophys. Res.* 104, D16: 19293–19306.

Blatov A. S., N. P. Bulgakov, V. A. Ivanov, A. N. Kosarev, and V. S. Tuzhilkin, 1984. *Variability of Hydrophysical Fields of the Black Sea*, Hydrometizdat, Leningrad, 240 pp. (in Russian).

Bogdanova A. K., 1961. The distribution of Mediterranean waters in the Black Sea. *Okeanologia* 1: 983–992. [English Translation (1963): *Deep-Sea Res.* 10: 665–672].

Boguslavskiy S. G., A. S. Sarkisyan, T. Z. Dzhioyev, and L. A. Koveshnikov, 1976. Analysis of Black Sea current calculations. *Izv. Atmos. Oceanic Phys.* 12: 205–207 (Engl. translation).

Boguslavskiy S.G., Y. A. Agafonov, and L. S. Isayeva, 1982. Exploration of the Black Sea during the 23rd Cruise of the R/V *Akudemik Vernadskiy. Oceanology* 22: 385–386.

Bologa A. S., 1986. Planktonic primary productivity of the Black Sea: A review. *Thalassia Jugosl.* 22: 1–22.

Bondarenko A. L., 1993. *The Caspian Sea Currents and the North Caspian Waters Salinity Formation*, Nauka, Moscow, 122 pp. (in Russian).

Bortnik V. N. and R. E. Nikonova, 1992. Water balance. In: F. S. Terziev, A. N. Kosarev, and A. A. Kerimov, *Hydrometeorology and Hydrochemistry of Seas. Vol.6: The.Caspian Sea. Iss.1: Hydrometeorological Conditions*, Hydrometeoizdat, St. Petersburg, 211–221.

Bryan K. A., 1969. Numerical method for the study of the circulation of the world ocean. *J. Comput. Phys.* 4: 347–376.

Cane M., V. Kamenkovich, and A. Krupitsky (1998). On the utility and disutility of JBAR. *J. Phys. Oc.* 28: 519–526.

Charnock H., K. Dyer, J. Huthnance, P. Liss, J. Simpson and P. Tett, eds., 1994. *Understanding the North Sea System*, Chapman & Hall, London, 222 pp.

Chirea R. and T. Gomoiu, 1986. Some preliminary data on the nutrient influx into western Black Sea. *Cercet. Mar., IRCM Constanta* 19: 171–189.

Demidov A. N., 1991, Water temperature and salinity. In: A. I. Simonov and E. N. Altman (eds.), *Hydrometeorology and Hydrochemistry of the USSR Seas. VoL 4: The Black Sea*, Hydrometeoizdat, St.-Petersburg, 373–397 (in Russian).

Demin Y. L. and R. A. Ibrayev, 1986. A numerical method of calculation of currents and sea surface topography in multiply-connected domains of the oceans. *Sov. J. Numer. Anal. Math. Modelling* 4; 211–225.

Demyshev S. G. and G. K. Korotayev, 1992. C-grid numerical energy-balanced model of baroclinic currents in the ocean with a rough bottom. In: Numerical Models and Results of Calibration Calculations of Currents in the Atlantic Ocean (IVM RAN, Moscow, 1992), 163–231 (in Russian).

Demyshev S. G., G. K. Korotaev, and V. V. Knysh, 2004a. Modelling the seasonal variability of the temperature regime of the Black Sea active layer. *Izv. Akad. Nauk, Fiz. Atmos. Okeana* 40: 259–270 [*Izv., Atmos. Ocean. Phys*. 40: 227–237 (2004)].

Demyshev S. G. and V. V. Knysh, 2004. Reconstruction of an adapted vertical velocity of the Black Sea on the basis of the synthesis of a circulation model and climatic data on temperature and salinity. In: *Ecological Safety of Coastal and Shelf Areas and a Complex Use of Shelf Resources* (EKOSI-Gidrofizika, Sevastopol, 2004), No. 11, 93–104 (in Russian).

Demyshev S. G., V. V. Knysh, and A. S. Sarkisyan, 2004b. Some characteristic features of the climatic waters circulation and the formation of the cold intermediate layer of the Black Sea. *Izv. Atmosph. and Oceanic Phys*. 40(5): 569-582 (English trans).

Demyshev S. G., V. V. Knysh, and G. K. Korotaev, 2006. Calculation of adapted Black Sea fields on the basis of assimilation of climatic temperature and salinity data into the model. *Izv. Akad. Nauk, Fiz. Atmos. Okeana* 42: 604–617 [*Izv. Atmos. Ocean. Phys*. 42: 555–567 (2006)].

Demyshev S. G., V. V. Knysh, and N. V. Inyushina, 2005. Seasonal variability and depth transformation of climatic horizontal currents of the Black Sea from results of assimilating new climatic data on temperature and salinity into models. *Marine Hydrophys. Jour*. 6: 28–45.

Dorofeev V. L. and G. K. Korotaev, 2004. Assimilation of the data of satellite altimetry in an eddy-resolving model of the Black Sea circulation. *Marine Hydrophys. Jour*. 1: 52–68.

Dorofeev V. L., G. K. Korotaev, M. V. Martynov, and Y. B. Ratner, 2004. System of monitoring hydrophysical fields of the Black Sea in a quasi-operative mode. In: *Ecological Safety of Coastal and Shelf Areas and a Complex Use of Shelf Resources* (EKOSI-Gidrofizika, Sevastopol, 2004), N 11, 9–23 (in Russian).

Dorofeev V. L., V. V. Knysh, and G. K. Korotaev, 2006. Estimation of the long-term variability of hydrophysical characteristics of the Black Sea on the basis of assimilation of climatic hydrologic and altimetric fields. *Marine Hydrophys. Jour*. 4: 3–17.

Dzhioev T. K. and A. S. Sarkisyan, 1976. Numerical computations of the Black Sea currents. *Izv.Acad. Sci. USSR Atmos. Oceanic Phys*. 6: 217–223 (Engl Transl).

Edwards M. and P. Reid, 2001. Implications of wider Atlantic influences on regional seas with particular reference to phytoplankton populations and eutrophication. OSPAR Convention for the Protection of the Marine Environment of the North Atlantic, Meeting of the Eutrophication Task Group (ETG), London, 10 pp.

Eremeev V. N., Ivanov, V. A., Kosarev, A. N., and Tuzhilkin, V. S., 1994. Annual and semi-annual harmonics in the climatic salinity fields of the Black sea. In: Eremeev, V. N. et al. (eds.), *Diagnosis of the State of Marine Environment of the Azov-Black Sea Basin*, NAS of the Ukraine, MHI, Sevastopol, 89–101.

Esbensen S. K. and R. W. Reynolds, 1981. Estimating monthly averaged air-sea transfers of heat and momentum using bulk aerodynamic method. *J. Phys. Ocean*. 11: 457–465.

Filippov D. M., 1968. *Circulation and Structure of the Waters in the Black Sea*, Nauka, Moscow 136 pp. (in Russian).

Fofonoff N. P. and R. C. Millard, 1983, *Algorithms for Computation of Fundamental Properties of Seawater*. UNESCO, Tech. Pap. In Mar. Sci., 44, 53 pp.

Friedrich H. J. and E. V. Stanev, 1988. Parameterization of vertical diffusion in a numerical model of the Black sea. In: J. C. J. Nihoul and B. M. Jamart (eds.), *Small-Scale Turbulence and Mixing in the Ocean, Proc. 19th Liege Colloquium on Ocean Hydrodynamics. Elsevier Oceanography Series 46*, Elsevier, Amsterdam, 151–167.

Gamsakhurdiya G. R. and A. S. Sarkisyan, 1975. Diagnostic calculations of current velocities in the Black Sea. *Oceanology* 15: 164–167 (Engl. Transl.).

Genin A., B. Lazar, and S. Brenner, 1995. Vertical mixing and coral death in the Red Sea following the eruption of Mount Pinatubo. *Nature* 377: 507–510.

Golubeva Z. A., 1991. Heat balance of the sea surface layer. In: A. I. Simonov and E. N. Altman (eds.), *Hydrometeorology and Hydrochemistry of the USSR Seas. Vol.4: The Black Sea*, Hydrometeoizdat, St.-Petersburg, 129–146 (in Russian).

Haidvogel D. and A. Beckmann, 1998. Numerical models of the coastal ocean. In: Brink, K. and A. Robinson (eds.), *The Sea*, Wiley & Sons, New York, 457–482.

Hainbucher D., J. Backhaus, and T. Pohlmann, 1987. Transport of conservative passive tracers in the North Sea: First results of a circulation and transport model. *Cont. Shelf Res.* 7(10); 1161–1179.

Hellerman S. and M. Rosenstein, 1983. Normal monthly wind stress over the World ocean with error estimates. *J. Phys. Oceanogr.* 13: 1093–1104.

Huthnance J., 1984. Slope currents and JBAR. *J. Phys. Oc.* 14: 795–810.

Hydrometeorology and Hydrochemistry of the USSR Seas, 1991. Hydrometeorology and Hydrochemistry of the USSR Seas. Vol. IV. Black Sea. Iss.l. Hydrometeorological Conditions, Hydrometeoizdat, St. Petersburg, 103–262 (in Russian).

Ibrayev R. A., 1993. Reconstruction of the climatic characteristics of the gulf stream. *Izv. Akad. Nauk, Fiz. Atmos. Okeana* 29(6): 803–814.

Ibrayev R. A., 2001. Model of enclosed and semi-enclosed sea hydrodynamics. *Russ. J. Numer. Anal. Math. Modelling* 16(4): 291–304.

Ibrayev R. A., 2008. Mathematical modelling of thermodynamical processes of the Caspian Sea. GEOS, Moscow Publ., 127 pp. (in Russian).

Ibrayev R. A. and D. I. Trukhchev, 1996 A diagnosis of the climatic seasonal circulation and variability of the cold intermediate layer in the Black Sea. *Izv. Akad. Nauk, Fiz. Atmos. Okeana* 32(5): 655–671.

Ibrayev R. A. and D. I. Trukhchev, 1998. Model study of the seasonal variability of the Black Sea circulation. In: L. Ivanov and T. Oguz (eds.), *Nato TU-Black Sea Project Ecosystem Modelling as a Management Tool for the Black Sea, Symposium on Scientific Results*, Kluwer Academic Publishers, Dordrecht, V 2, 179–196.

Ibrayev R. A., A. S. Sarkisyan, and D. I. Trukhchev, 2001. Seasonal variability of the circulation of the Caspian Sea reconstructed from mean multi-year hydrological data. *Izv. Atmos. Oceanic Phy.* 37(1): 103–111.

Ibrayev R. A., E. Ozsoy, A. S. Sarkisyan, and H. I. Sur, 1998. Seasonal variability of the Caspian Sea dynamics: Barotropic motions driven by climatic wind stress and river discharge. In: Oceanic Fronts and Related Phenomena. International Oceanographic Commission. Workshop Report N 159, 212–217.

Ivanov L., S. Besiktepe, and E. Ozsoy, 1997. The Black Sea cold intermediate layer. In: Ozsoy, E. and A. Mikaelyan. (eds.), *Sensitivity to Change: Black Sea, Baltic Sea and North Sea*, NATO AS1 Series 2, Environment 27, Kluwer, Dordrecht, 253–264.

Kamenkovich V. M., 1973. *Ocean Dynamics Fundamentals*, Gidrometeoizdat, Leningrad, 270 pp. (in Russian).

Kempe S., A. R. Diercks, G. Liebezeit, and A. Prange, 1991. Geochemical and structural aspects of the pycnocline in the Black Sea (R/V Knorr 134, 8 Leg 1, 1988). In: Izdar, E. and J. M. Murray (Eds.). *The Black Sea Oceanography*. NATO/ASI Series, Kluwer, Dordrecht, 89–110.

Kideys A. E., 1994. Recent changes in the Black Sea ecosystem: The reason for the sharp decline in Turkish fisheries. *J. Mar. Syst.* 5: 171–181.

Knipovich N. M., 1932. The hydrological investigations in the Black Sea. *Proc. of Azovo-Chernomorsk. Scientific-Industrial Expedition*, 1932, 10, 274 pp. (in Russian).

Knysh V. V., G. K. Korotaev, S. G. Demyshev, and V. N. Belokopytov, 2005. Long-term variations in thermohaline and dynamic characteristics of the Black Sea as inferred from climatic data on temperature and salinity and their assimilation into models. *Marine Hydrophys. J.*, 3: 11–30.

Knysh V.V., O. A. Saenko, and A. S. Sarkisyan, 1996. Method of assimilation of altimeter data and its test in the tropical north atlantic. *Russ. J. Numer. Anal. Math. Model* 11(5): 333–409.

Knysh V. V., S. G. Demyshev, and G. K. Korotaev, 2002. Method of reconstructing the climatic seasonal circulation of the Black Sea on the basis of assimilation of hydrologic data into models. *Marine Hydrophys. J.* 2: 36–52.

Knysh V. V., S. G. Demyshev, G. K. Korotaev and A. S. Sarkisyan, 2001. Four-dimensional climate of seasonal Black Sea circulation. *Russ. J. Numerical Anal. Math. Modelling* 16(5): 409–426.

Knysh V. V., S. G. Demyshev, G. K. Korotaev, and A. S. Sarkisyan, 2007. Method and results of assimiliation of climatic data on temperature, salinity, and level into a numerical model of the Black Sea. *Izv. Russ. Acad. Sci. Atmosp. Ocean Phys.* 43(3): 398–412.

Konovalov S., A. Romanov, I. Salihoglu, O. Bagtiirk, S. Tugrul, and S. Gokmen, 1994. Intercalibration of CoMSBlack-93a chemical data, unification of methods for dissolved oxygen and hydrogen sulfide analyses and sampling strategies of CoMSBlack-94a Cruise. Inst. Mar. Sci., Middle East Tech. Univ., Erdemli, Icel. 26 pp.

Korotaev G. K., 1997. Circulation in semi-enclosed seas induced by buoyncy flux through a strait. In: *Proc. of the NATO Advanced Research Workshop on Sensitivity of North Seat Baltic Sea and Black Sea to Anthropogenic and Climatic Changes*. Kluwer Academic Publisher, Dordrecht, 395–401.

Korotaev G. K., O. A. Saenko, C. D. Koblinski, S. G. Demyshev, and V. V. Knysh, 1998, Accuracy estimation, methods, and some results of TOPEX/POSEIDON data assimilation in a model of the Black Sea general circulation. *Issled. Zemli Kosmosa* 3: 3–17.

Korotaev G. K., O. A. Saenko, C. J. Koblinsky, and V. V. Knysh, 1999. Satellite altimetry observations of the Black Sea. In: S. Besik-tepe et al. (ed.), *Enviromental Degradation of the Black Sea: Challenges and Remedies*, Kluwer Academic Publisher, Dordrecht, 225–244.

Korotaev G. K., S. G. Demyshev, and V. V. Knysh, 2000. Three-dimensional climate of the Black Sea ecosystem processes and forcasting. In: *Operational Workshop Meeting METU, IMS, ERDEMLI*, 1–10.

Kosarev A. N. and E. A. Yablonskaya, 1994. *The Caspian Sea*. SPB Academic Publishing, Amsterdam, 259 pp.

Kravets A. G., 1987. The tides and their modelling in the White Sea. In: *Arkhangelsk: The Problems of the White Sea*, Arkhangelsk, 36–38.

Laane, R., W. van Leussen, J. Berlamont, J. Sündermann, W. van Raaphorst, and F. Colijn, 1996. North-west European shelf programme (NOWESP): An overview. *Dt. Hydrogr. Z.* 48: 217–230.

Latif M. A., E. Ozsoy, I. I. Salihoglu, A. F. Gaines, O. Bagtiirk, A. Yilmaz, and S. Tugrul, 1992. Monitoring via direct measurements of the modes of mixing and transport of wastewater discharges into the Bosphorus underflow. METU – Inst. Mar. Sci., Erdemli, & el, Tech. Rep. 92–2, 98 pp.

Latif M. A., E. Ozsoy, T. Oguz, and U. Unluata, 1991. Observations of the Mediterranean inflow into the Black Sea. *Deep-Sea Res.* 38(Suppl. 2): S711–S723.

Launder B. E., D. B. Spalding, 1972. *Mathematical Models of Turbulence*. Academic Press, London, 162 pp.

Launiainen J. and T. Vihma, 1990. Derivation of turbulent surface fluxes – an iterative flux –profile method allowing arbitrary observing heights. *Environ. Softw.* 5(3): 113–124.

Lednev V. A., 1943. *The Caspian Northern and Medium Parts Flow Fields*, Morskoy Transport, Moscow, 97pp. (in Russian).

Marchuk G. I., 1969. Numerical solution of the poincare problem for the ocean circulation. *Dokl. Akad. Nauk SSSR* 165(5): 1041–1044.

Martin P. J., 1985. Simulation of the mixed layer at OWS November and Papa with several models. *J. Geophys. Res.* 90(C0): 903–916.

Mee L. D., 1992. The Black Sea in crisis: the need for concerted international action. *Ambio* 24: 278–286.

Mellor G. L. and T. Ezer, 1991. A gulf stream model and an altimetry assimilation acheme. *J. Geophys. Res.* 96: 8779–8795.

Munk W. H., E. R. Anderson, 1948. Note on the theory of the thermocline. *J. Mar. Res.* 7: 276–295.

Murray J. W. (Ed.), 1991. Black Sea Oceanography: Results from the 1988 Black Sea Expedition. *Deep-Sea Res.* 38(Suppl. 21): 1266 pp.

Neumann G., 1942. Die absolute Topographie des physikalischen Meeresniveaus und die Oberflächen-Strömungen des Schwarzen Meeres. *Ann. Hydrogr. Berlin* 70: 265–282.

Oguz T., D. G.Aubrey, V. S. Latun, E. Demirov, A. G. Kolesnikov, H. I. Sur, V. Diaconu, S. Besiktepe, M. Duman, R. Limeburner, and V. Eremeev, 1994. Mesoscale circulation and thermohaline structure of the Black Sea observed during HydroBlack'91, *Deep Sea Res.* 41: 603–628.

Oguz, T. E., E. Ozsoy, M. A. Latif, and U. Unluiata, 1990. Modelling of hydraulically controlled exchange flow in the Bosphorus Strait. *J. Phys. Oceanogr.* 20: 945–965.

Oguz T. and P. Malanotte-Rizzoli, P., 1996. Seasonal variability of the wind and thennohalme-driven circulation in the Black Sea: Modelling studies, *J. Geophys. Res.* 101(C7): 16551–16569.

Oguz T., P. E. LaViolette, and U. Unluata, 1992. The upper layer circulation of the Black Sea: its variability as inferred from hydrographic and satellite observations. *J. Geophys. Res.* 97: 12569–12584.

Oguz T., V. S. Latun, M. A. Latif, V. V. Vladimirov, H. I. Sur, A. A. Markov, E. Ozsoy, B. B. Kotovshchikov, V. N. Eremeev, and U. Unluata, 1993. Circulation in the surface and intermediate layers of the Black Sea. *Deep Sea Res.* 40: 1597–1612.

Orlanski I., 1976. A simple boundary conditions for unbounded hyperbolic flows. *J. Comput. Phys.* 12(3): 251–269.

Ovchinnikov I. M. and Y. I. Popov, 1984. On the problem of forming the cold intermediate layer in the Black Sea. *Dokl. Akad. Nauk SSSR* 29: 986–989.

Ozsoy E.and U. Unluata, 1997. Oceanography of the Black Sea a review of some resent results. *ELSEVIER Earth Sci. Rev.* 42: 231–272.

Ozsoy E., U. Unluata, and Z. Top, 1993. The Mediterranean water evolution, material transport by double diffusive intrusions, and interior mixing in the Black Sea. *Prog. Oceanogr.* 31: 275–320.

Pacanowski R. C. and S. G. H. Philander, 1981. Parameterization of vertical mixing in numerical models of tropical oceans. *J. Phys. Oceanogr.* 11: 1443–1451.

Panin G. N., 1987. *The Caspian Sea Evaporation and Heat Exchange*, Nauka, Moscow, 88 pp. (in Russian).

Paulson E. A. and J. J. Simpson, 1977. Irradiance measurements in the upper ocean. *J. Phys. Oceanogr.* 7: 952–956.

Phillips O. M., 1985. Spectral and statistical properties of the equilibrium range in wind-generated gravity waves. *J. Fluid Mech.* 156: 505–531.

Pohlmann T., 1996a. Predicting the thermocline in a circulation model of the North Sea: Model description, calibration and verification. *Cont. Shelf Res.* 16(2): 131–146.

Pohlmann T., 1996b. Calculating the annual cycle of the vertical eddy viscosity in the North Sea with a three-dimensional circulation model. *Cont. Shelf Res.* 16(2): 147–161.

Pohlmann T., 1996c. Calculating the development of the thermal vertical stratification in the North Sea with a three-dimensional circulation model. *Cont. Shelf Res.* 16(2): 163–194.

Pohlmann T., 1996d, Simulating the heat storage in the North Sea with a three-dimensional circulation model. *Cont. Shelf Res.* 16(2): 195–213.

Pohlmann T., 2003. Eine Bewertung der hydro-thermodynamischen Nordseemodellierung. Berichte aus dem Zentrum für Meeres- und Klimaforschung der Universität Hamburg, Reihe B, Heft 46, 141 pp.

Prandle D., ed., 2000. Pre-operational modelling of the shelf seas of Europe (PROMISE). *Coast. Eng.* 41, 359 pp.

Proctor R., 1997. NOMADS – North Sea Model Advection-Dispersion Study. Final Report, POL Internal Document 10, 855 pp.

Rodionov S. N., 1994. *Global and Regional Climate Interaction: The Caspian Sea Experience.* Water Technology Library V, Kluver Academic Publisher, Dordrecht, 241 pp.

Roulett G. and G. Madec, 2000. Salt conservation, free surface, and varying levels: a new formulation for ocean general circulation models. *J. Geophys. Res.* 105(C10): 23927–23942.

Ruddick K. (ed., 1997). Processes in regions of freshwater influence (PROFILE). *J. Mar. Syst.* 12, 323 pp.

Samoilenko V. S. and A. I. Sachkova (eds., 1963), *Hydrometerological Atlasses of Caspian and Aral Seas,* Leningrad: Gidrometeoizdat, 1963, 179 pp.

Sarkisyan A., 1977. The diagnostic calculation of large-scale oceanic circulation. In: Goldberg, E. (ed.). *The Sea, Vol. VI: Marine Modelling*, Wiley & Sons, New York, 363–458.

Sarkisyan A. S., B. R. Zaripov, A. N. Kosarev, and D. G. Rzheplinski, 1976. The Caspian Sea currents diagnostic calculation. *Izv. AN, Fizika Atmos. i Okeana* 12(10): 1106–1110 (in Russian).

Sarkisyan A. S., and T. Z. Dzioev, 1974. A numerical model and calculation of currents in the Black Sea. *Meteorologia I Gidrologia* 3: 70–76 (in Russian).

Sarkisyan A. S., Y. L. Demin, A. L. Brekhovskikh, and T. V. Shakhanova, 1986. *Methods and the Computational Results on World Ocean Circulation.* Hydrometeoizdat, Leningrad, 160 pp. (in Russian).

Saydam C., S. Tugrul, O. Basturk, and T. Oguz, 1993. Identification of the oxic/anoxic interface by isopycnal surfaces in the Black Sea. *Deep-Sea Res.* 40: 1405–1412.

Schrum C. and J. O. Backhaus, 1999. Sensitivity of atmosphere-ocean heat exchange and heat content in North Sea and Baltic Sea. A comparitive Assessment. *Tellus* 51A: 526–249.

Semenov E. B., 2004. Numerical modelling of the White Sea dynamics and the problem of monitoring. *Izv. RAN* 40(1): 128–141.

Semenov E. V., 1989. Ein numerisches Schema der vierdimensionalen Analyse von thermohalinen Feldmessungen im Ozean. Berlin: Beitr. Meereskd 60: 41–52.

Semenov E. V. and M. V. Luneva, 1999, On joint effect of tide, stratification and vertical turbulent mixing on hydrophysical fields formation in the White Sea, *Izv. AN, Atmos. and Ocean Phys.* 35(5): 660–678.

Seventh Report of the Joint Panel on Oceanographic Tables and Standards, 1976. *UNESCO Techn. Papers Mar. Sci.* 24, Appendix I, 39–54.

Stanev E. V., V. M. Roussenov, N. H. Rachev, and J. V. Staneva, 1995. Sea response to atmospheric variability. Model study for the Black Sea. *J. Mar. Sys.* 6: 241–267.

Staneva J. V. and E. V. Stanev, 1998. Oceanic response to atmospheric forcing derived from different climatic data sets. Intercomparison study for the black sea. *Oceanol. Acta* 21: 383–417.

Staneva J. V., E. V. Stanev, and N. H. Rachev, 1995. Heat balance estimates using the atmospheric 'I analysis data: a case study for the Black sea. *J. Geophys. Res.* 100(C9): 18581–18596.

Shtockman W., 1938. An investigation of kinematic of the medium parts of Caspian Sea western coastal currents. Transactions of the Azerbaidjan Scientific-Investigative Fishery station, Baku, V 1, 76 pp. (in Russian).

Sündermann J. (ed., 1994). *Circulation and Contaminant Fluxes in the North Sea*, Springer, Heidelberg, 654 pp.

Sündermann J., G. Becker, P. Damm, D. van den Eynde, R. Laane, W. van Leussen, T. Pohlmann, W. van Raaphorst, G. Radach, H. Schultz, and M. Visser, 1996. Decadal variability on the North-West European shelf. *Dt. Hydrogr. Z.* 48: 365–400.

Sündermann J., S. Beddig, I. Kröncke, G. Radach, and H. Schlünzen, 2001. The changing North Sea – knowledge, speculation and new challenges. Berichte aus dem Zentrum für Meeres- und Klimaforschung der Universität Hamburg, Reihe Z, Heft 3, 358 pp.

Sündermann J. and W. Lenz (eds., 1983). *North Sea Dynamics*, Springer, Heidelberg, 693 pp.

Sur H. I., E. Ozsoy, and R. Ibrayev, 1998, Satellite – derived flow characteristcs of the Caspian Sea. In: D. Halpern. *Satellites, Oceanography and Society* Elsevier Science B.V., Amsterdam, 289–297.

Terziev F. S., A. N. Kosarev, and A.A. Kerimov, 1992. *Hydrometeorology and Hydrochemistry of the USSR Seas, V6, The Caspian Sea, Issue 1 Hydrometeorological Conditions*, Hydrometeoizdat, St. Petersburg, 359 pp. (in Russian).

Timofeev N.A., 1983. *The Oceans Radiational Regime*, Naukova Dumka, Kiev (in Russian).

Tolmazin D. L., 1985. Changing coastal oceanography of the Black Sea, 11. Mediterranean effluent. *Prog. Oceanogr.* 15: 277–316.

Trukhchev D., A. Kosarev, D. Ivanova, and V. Tuzhilkin, 1995. Numerical analysis of the general circulation in the Caspian Sea. *Comptes Rendus de l'Academie Bulgare des Sciences*, Sofia 48(10): 35–38.

Trukhchev D. and R. Ibrayev, 1997. Seasonal variability of the Black Sea climatic circulation. Sensitivity to change: Black Sea, Baltic Sea and North Sea. NATO ASI Series, Kluwer Academic Publishers, Dordrecht, N 27, 365–374.

Tugrul S., O. Basturk, C. Saydam, and A. Yilmaz, 1992. Changes in the hydrochemistry of the Black Sea inferred from water density profiles. *Nature* 359: 137–139.

Turkes M., U. Siimer. and G. Kilig, 1995. Variations and trends in annual mean air temperatures in Turkey with respect to climatic variability. *Int. J. Climatol.* 15: 557–569.

Turner J. S., 1973. *Buoyancy Effects in Fluids*. Cambridge University Press, Cambridge, 367 pp.

Turner J. S., 1978. Double-diffusive intrusions into a density gradient. *J. Geophys. Res.* 83: 2887–2901.

Tuzhilkin V. S., A. N. Kosarev, D. I. Trukhchev, and D. P. Ivanova, 1997. Seasonal peculiarities of the Caspian Sea deeper parts general circulation. *Meteorologiya I Gidrologiya* 1: 91–99 (in Russian).

Unluata U., T. Oguz, M. A. Latif, and E. Ozsoy, 1989. On the physical oceanography of the Turkish Straits, In: J. Pratt (ed.). *The Physical Ocenography of Sea Smirs, NATO AS1 Scr, Ser. c*, Kluwer Academic Publishers, Nonvell, MA, pp. 25–60.

Whitehead J. A., G. K. Korotaev, and S. N. Bulgakov, 1998. Convective circulation in mesoscale abyssal basins. *Geophys. Astrophys. Fluid Dynamics* 89: 169–203.

Yuce H., 1990. Investigation of the Mediterranean water in the Strait of Istanbul (Bosphorus) and the Black Sea. *Oceanol. Acta* 13: 177–186.

Zaitsev, Y. P., 1993. Impacts of eutrophication on the Black Sea fauna, studies and reviews. *Gen. Fish. Counc. Medit.* 64: 59–86.

Color Plate Section

(a) (b)

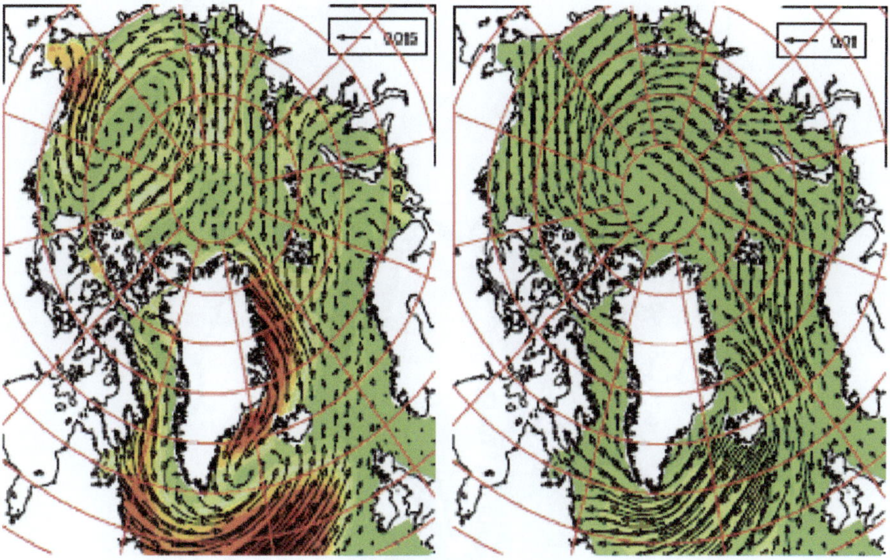

Plate 1 (**a**) Wind friction pattern averaged over the whole simulation period of 1948–2002, (**b**) wind friction restored by the 3d-mode eigenfunction in EOF analysis of a wind series. The samples of arrows in the right upper corner correspond to wind friction values 0.015 and 0.01 N/m^2. The background darkening corresponds to the friction vector modulus, darker areas indicate larger values of the modulus (Kuzin et al., 2006) (See also Figure 5.1 on page 158)

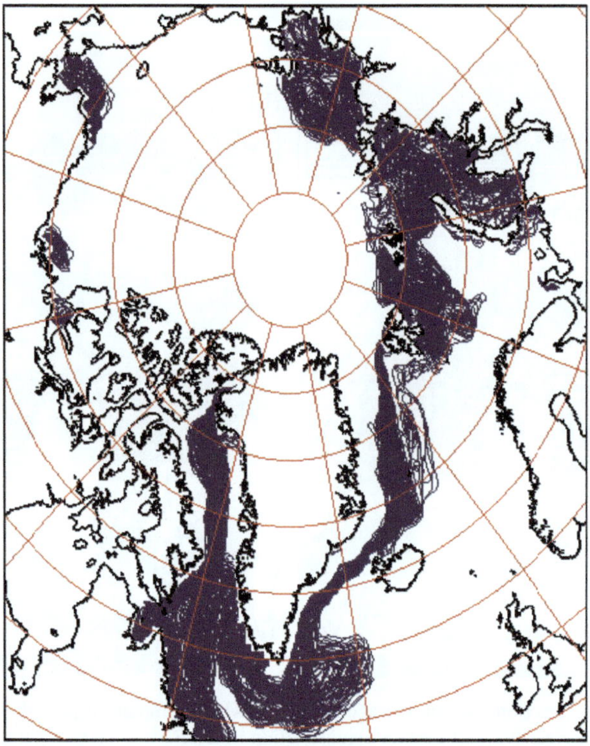

Plate 2 Monthly positions of the ice edge boundary in 1948–1960 (Kuzin et al., 2006) (See also Figure 5.2 on page 159)

(a) (b)

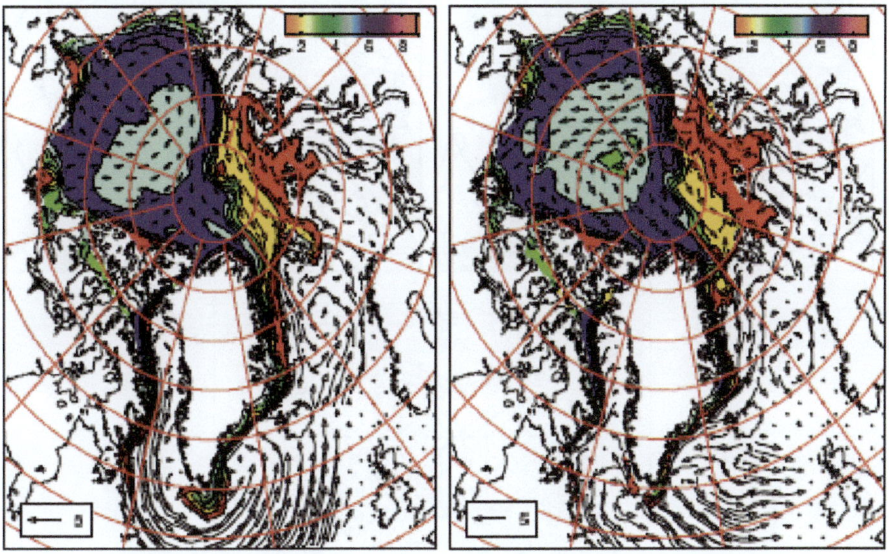

Plate 3 Ice thickness and drift patterns for a typical anticyclonic circulation (**a**) and for the circulation in the period of the anticyclonic component failure (**b**). The samples of arrows in the left lower corner correspond to the velocity of 5 cm/s. Ice thickness (m) is graduated by darkness in the figures (Kuzin et al., 2006) (See also Figure 5.3 on page 160)

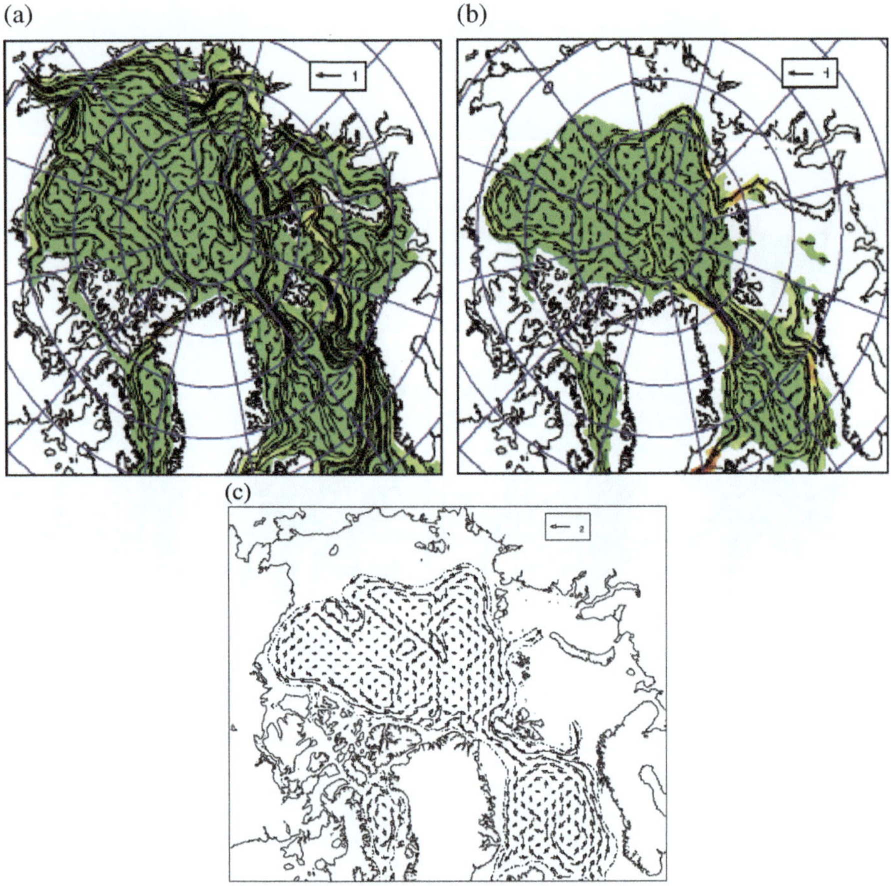

Plate 4 (**a**) Surface currents field. The samples of arrows in the right upper corner correspond to the velocity of 1 cm/s. The background dark color corresponds to velocity value; (**b**) currents at a depth of 250 m. The darker shades indicate higher velocity. (**c**) Circulation at 400 m depth obtained in experiment with Neptune parameterization, shows a steady, cyclonic, topography-steered current for the entire period of investigation (Kuzin et al., 2006) (See also Figure 5.4 on page 162)

(a) (b)

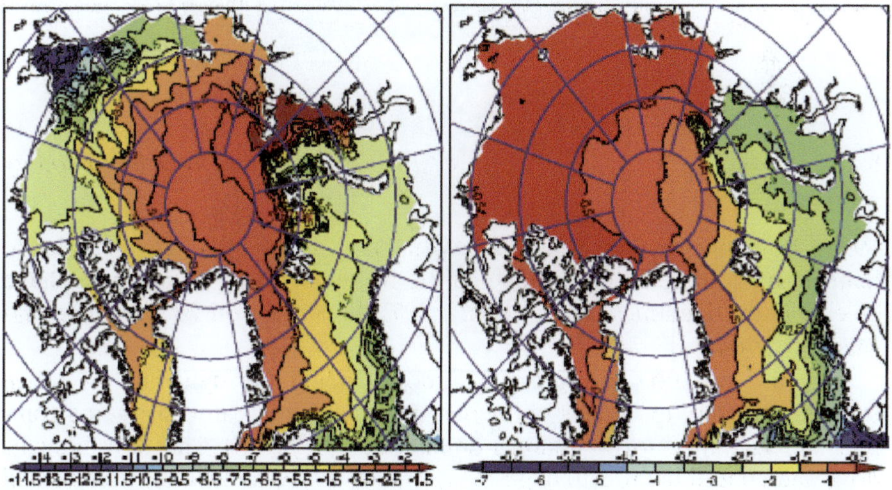

Plate 5 Passive tracer distribution 37 years after the onset of emission: (**a**) the Ob, (**b**) Pacific water of the Bering Straits. The isolines are the decimal logarithms of the corresponding tracer concentration (Kuzin et al., 2006) (See also Figure 5.7 on page 167)

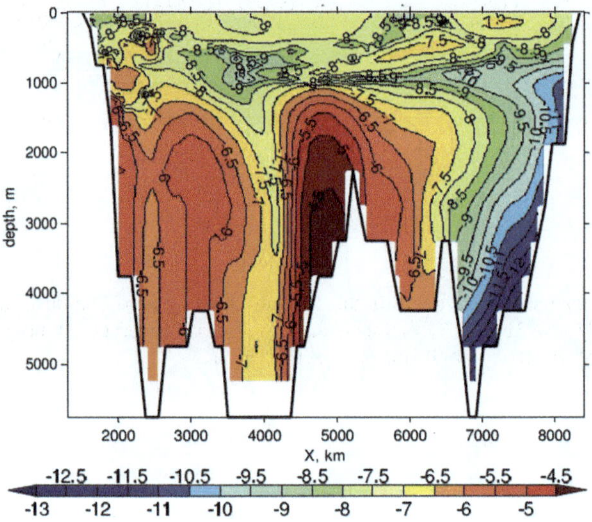

Plate 6 The vertical cross-section of the tracer concentration field of the Ob River in the North Atlantic along the parallel 30°N 37 years after the onset of emission. Left–west, right–east. The isolines are the decimal logarithms of concentration (Kuzin et al., 2006) (See also Figure 5.8 on page 167)

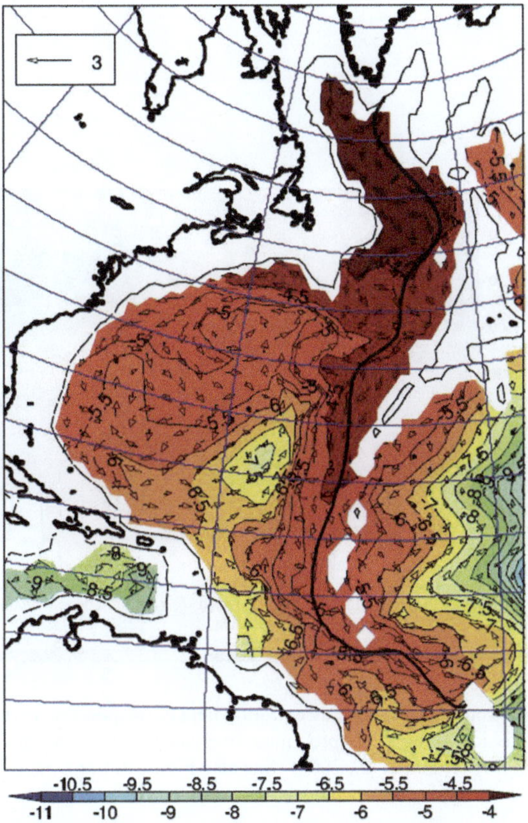

Plate 7 The tracer propagation of the Ob run-off in the North Atlantic at a depth of 3 km 37 years after the onset of emission. The arrows indicate the direction and velocity of the current (the *arrow* corresponding to the velocity 3 cm/s is indicated in the *left upper corner*) (Kuzin et al., 2006) (See also Figure 5.9 on page 168)

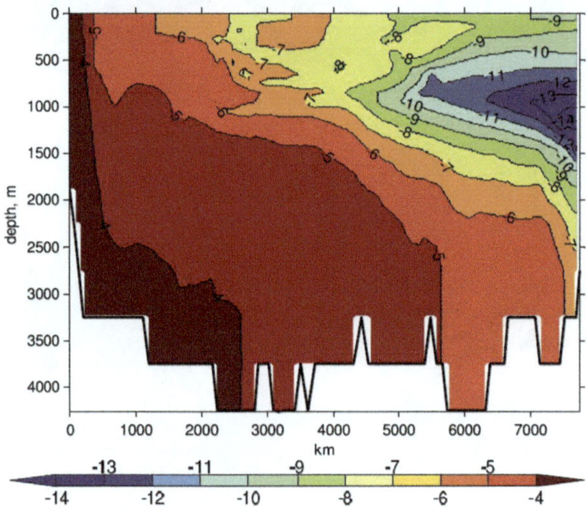

Plate 8 The cross-section of the tracer concentration field corresponding to the Ob run-off 37 years after the onset of emission along the approximate trajectory of its propagation, which is indicated in Fig. 5.9 by a *solid line* (Kuzin et al., 2006) (See also Figure 5.10 on page 169)

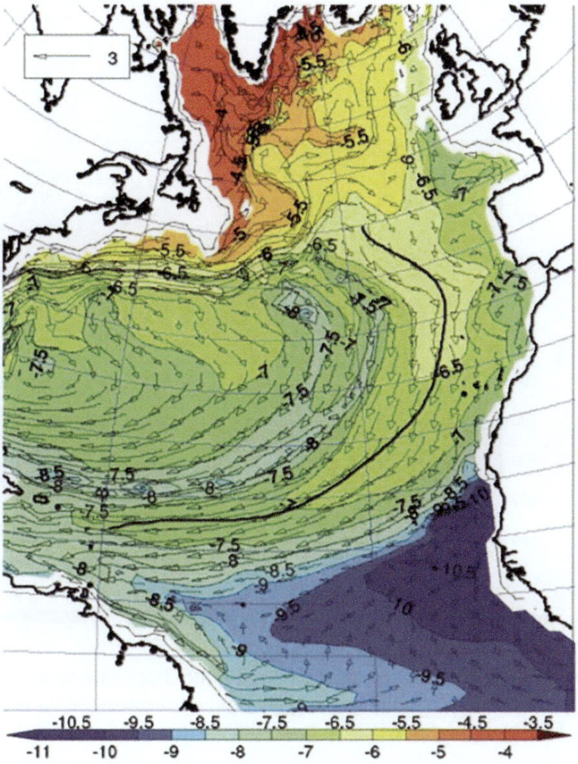

Plate 9 The tracer propagation in the Ob run-offin the North Atlantic at a depth of 300 m 37 years after the onset of emission. The arrows indicate the direction and velocity of the current (the arrow corresponding to the velocity 3 cm/s is indicated in the left upper corner) (Kuzin et al., 2006) (See also Figure 5.11 on page 170)

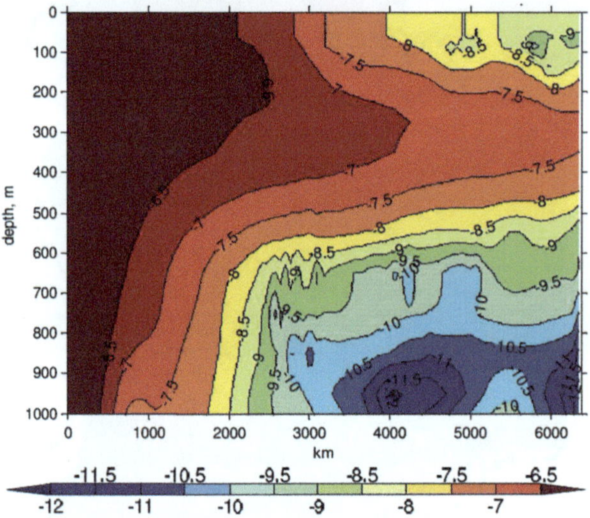

Plate 10 The cross-section of the tracer concentration field corresponding to the Ob run-off 37 years after the onset of emission along the approximate trajectory of its propagation, which is indicated in Fig. 5.11 by a *solid line* (Kuzin et al., 2006) (See also Figure 5.12 on page 171)

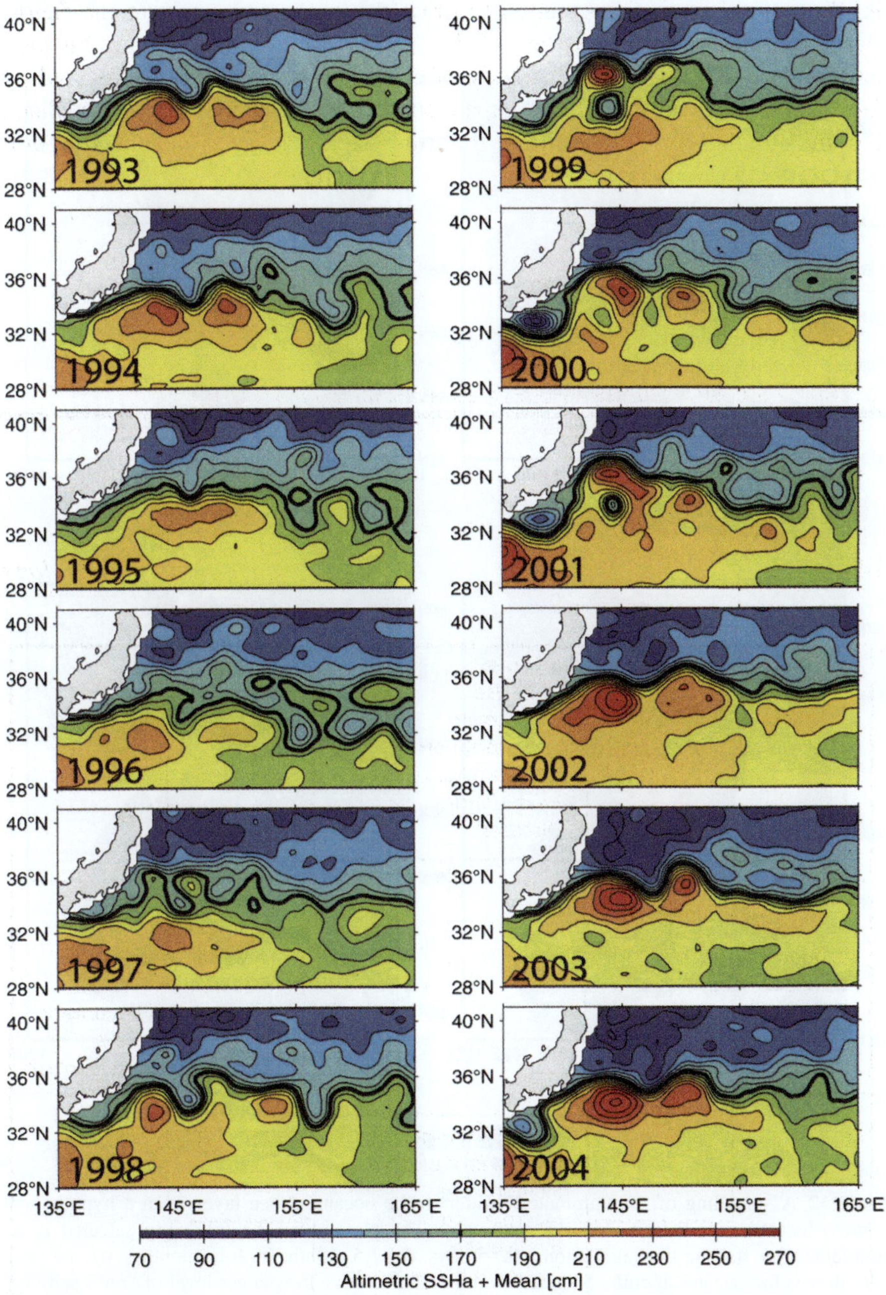

Plate 11 Maps of yearly averaged sea surface height field. Contour intervals are 10cm with the thick lines denoting the 170 cm contours (Qiu and Chen, 2005) (See also Figure 5.27 on page 191)

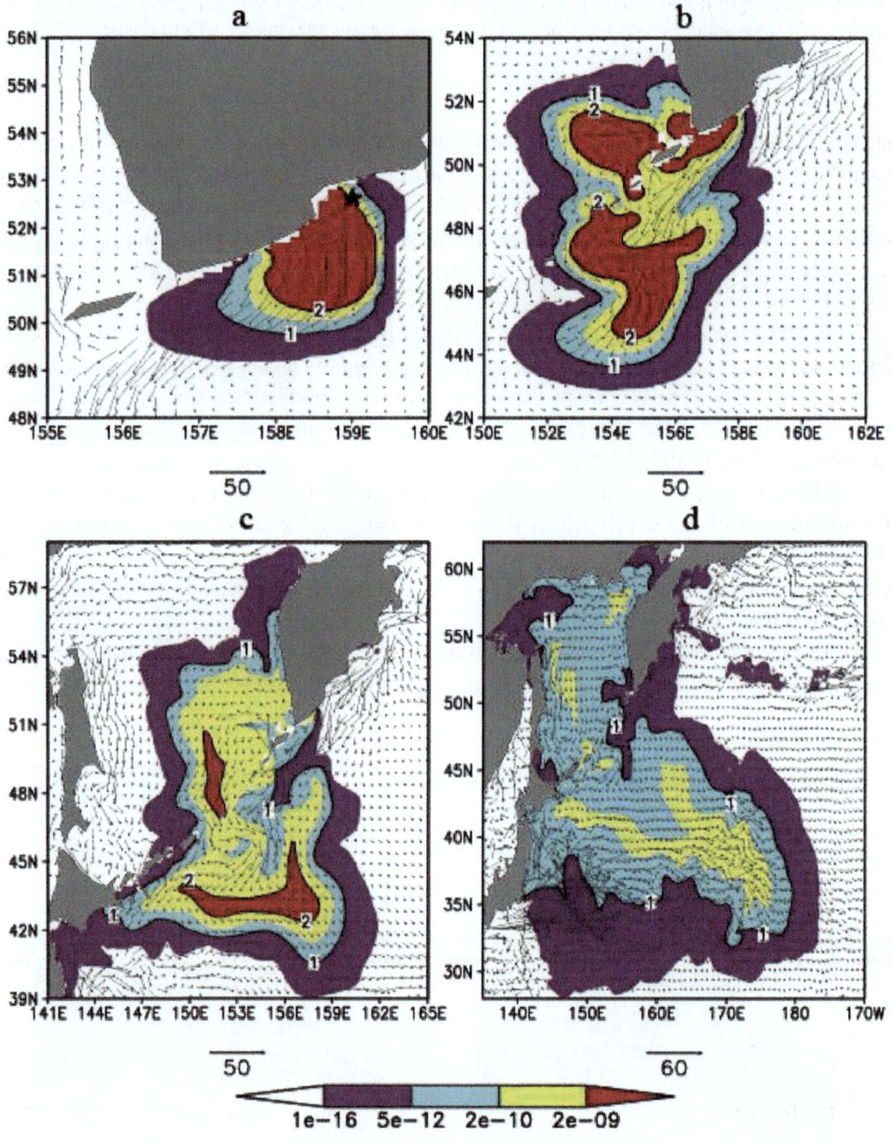

Plate 12 A spreading of "contaminated" water in the ocean surface layer from a hypothetical source, located near Petropavlovsk-Kamchatsky. The plots show the following selected time-moments after starting the calculations: (**a**) 5 days, (**b**) 1,5 month, (**c**) four months, (**d**) one year. The flow velocities are in cm/s. Numbers 1 and 2 characterize the danger level of "radioactivity" (Antipov et al., 2006) (See also Figure 5.31 on page 197)

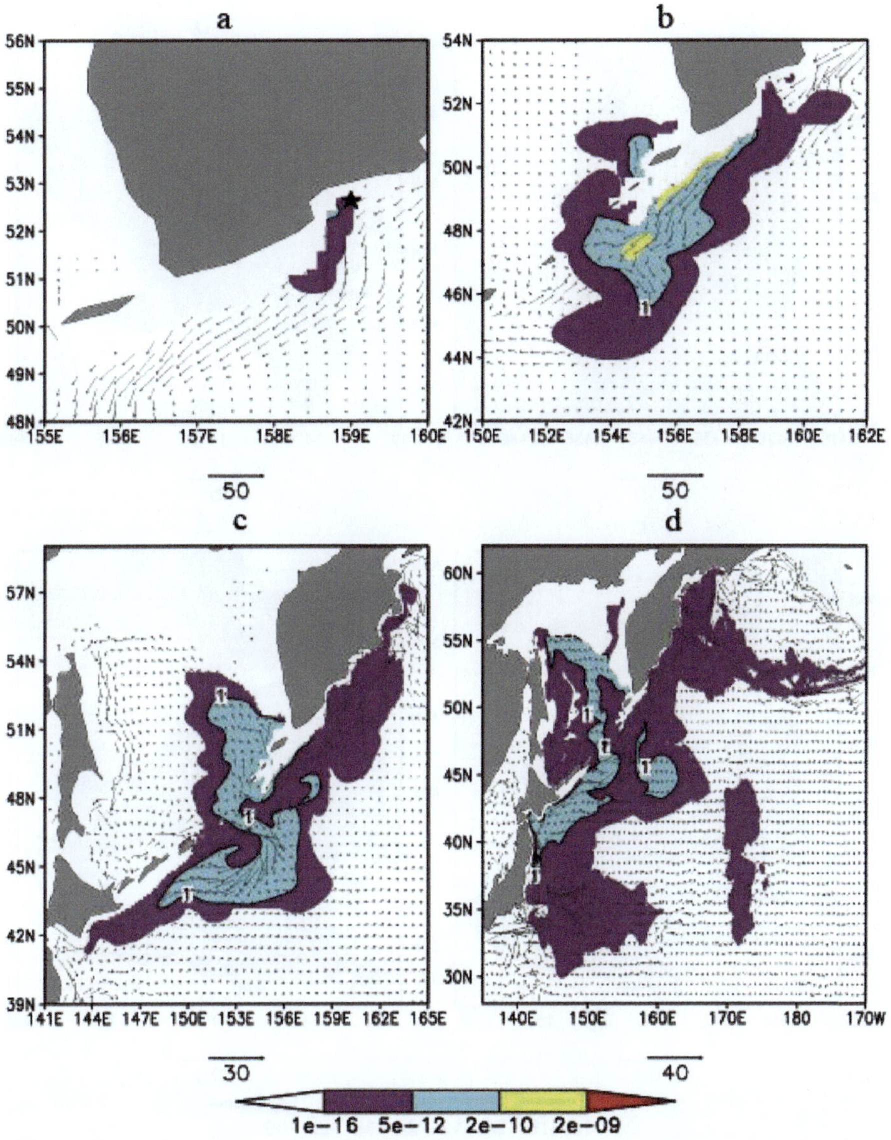

Plate 13 The same as in Fig. 5.31 but for the depth 500 m. The *asterisk* shows the location of hypothetic source of radioactivity (Antipov et al., 2006) (See also Figure 5.32 on page 198)

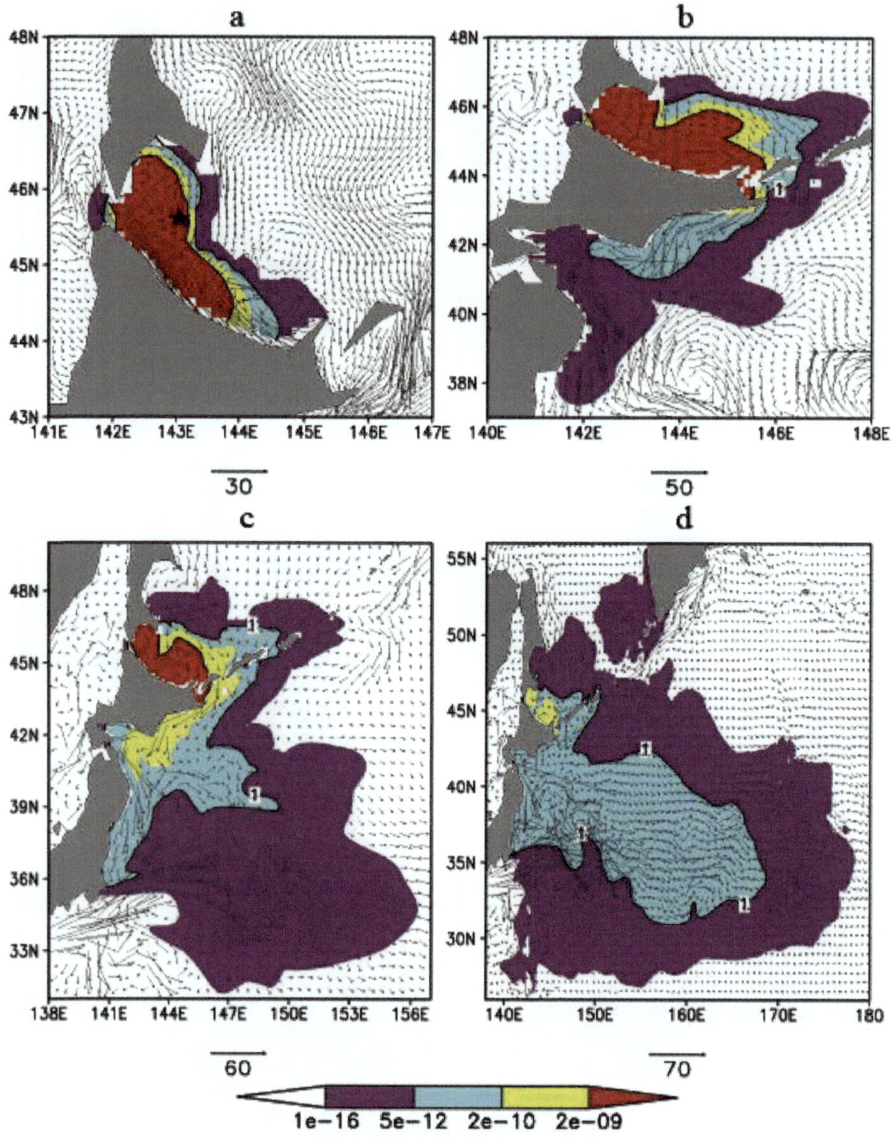

Plate 14 The same as in Fig. 5.31 but for the case when the source of radioactivity is near the city Vladivostok (Antipov et al., 2006) (See also Figure 5.33 on page 199)

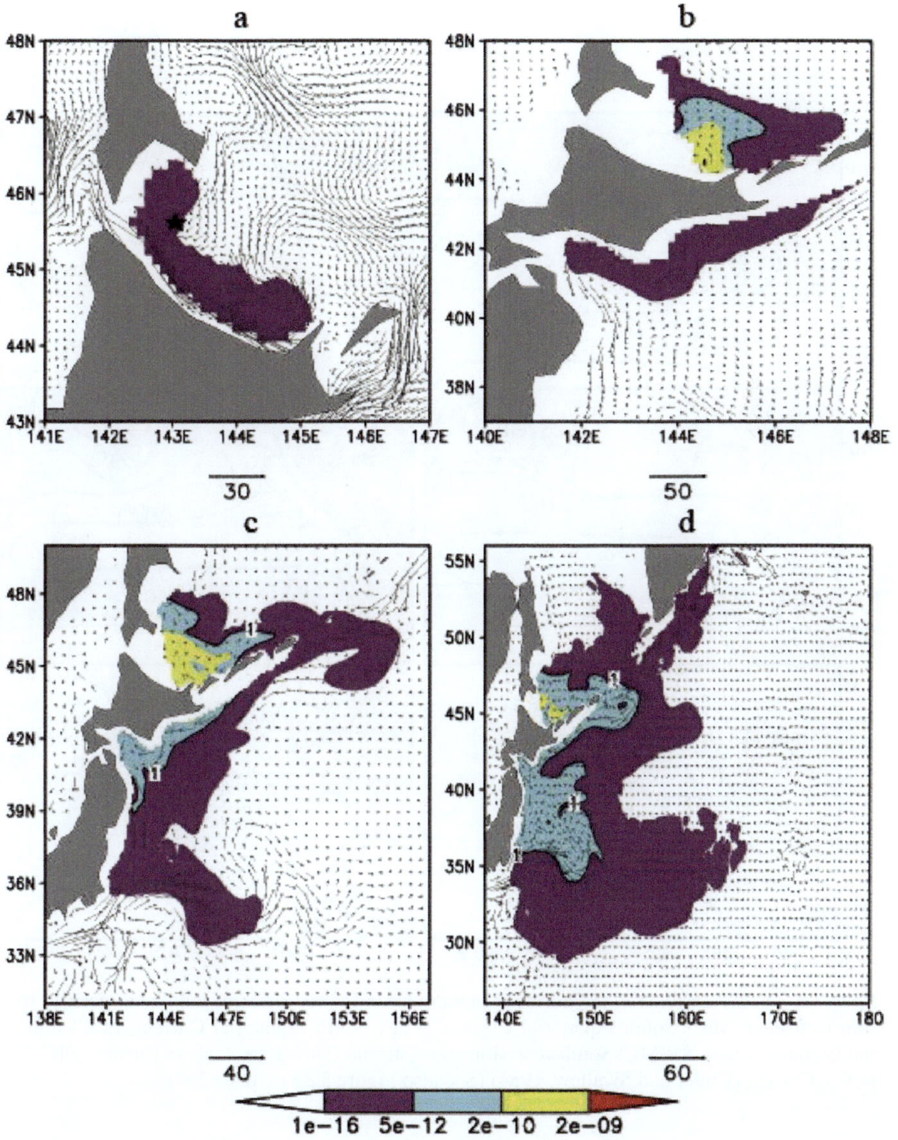

Plate 15 The same as in Fig. 5.33 but for the depth 500 m (Antipov et al., 2006) (See also Figure 5.34 on page 200)

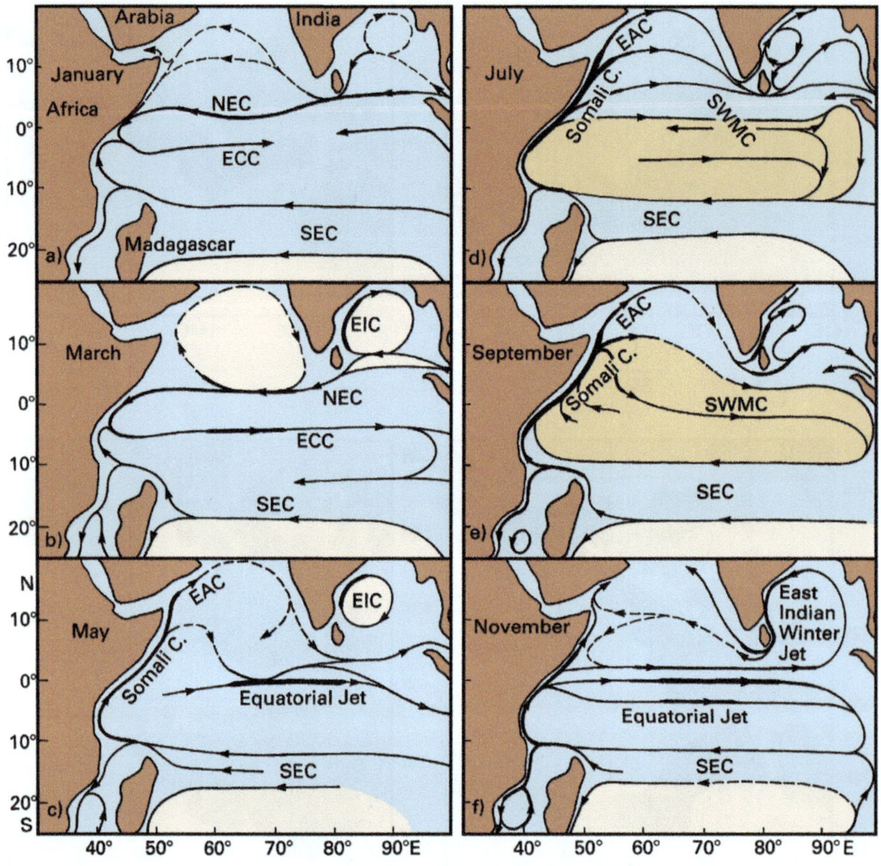

Plate 16 Surface currents in the northern Indian Ocean for six different months as derived from ship drift data. SEC: South Equatorial Current, NEC: North Equatorial Current, ECC: Equatorial Countercurrent, SWMC: Southwest Monsoon Current, EAC: East Arabian Current, EIC: East Indian Current (Cutler and Swallow, 1984) (See also Figure 5.35 on page 204)

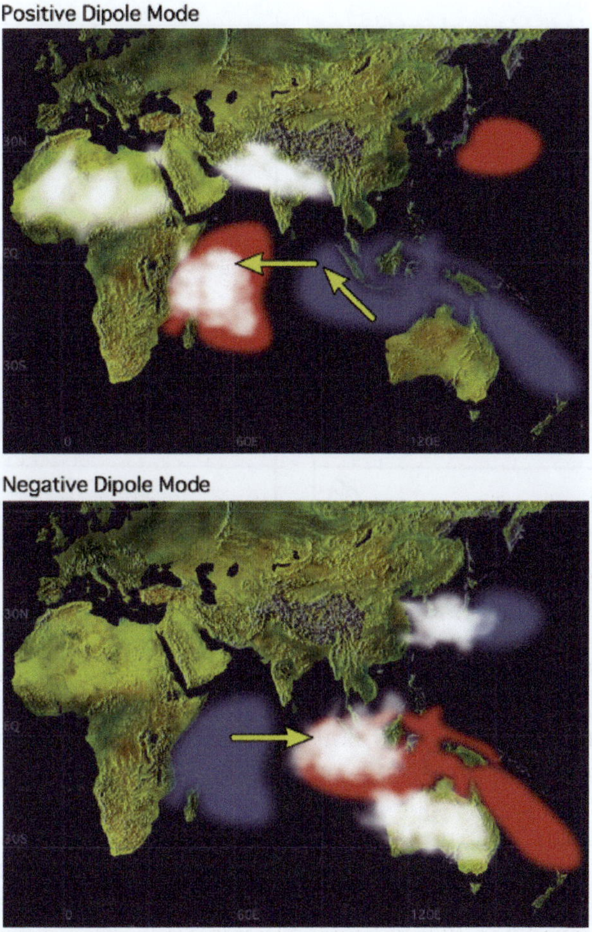

Plate 17 Schematic diagram of sea surface temperature anomalies (*red shading* denoting warming; *blue* cooling) during a positive Indian Ocean Dipole (IOD) event (*top*). White patches indicate increased convective activity with arrows indicating wind direction. The negative IOD (*bottom*) which is, in effect, the reversal of the positive IOD – complete with increased convective activity over Australia, Indonesia and Japan (Saji et al., 1999) (See also Figure 5.36 on page 209)

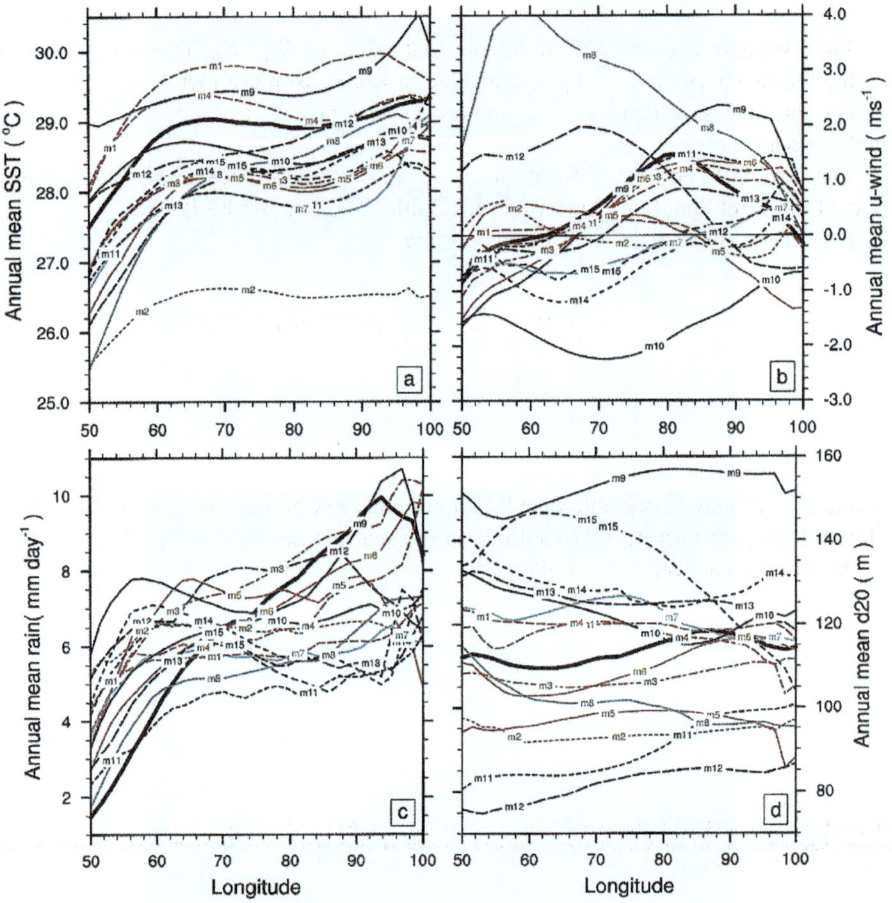

Plate 18 Annual-mean equatorial climatologies for (**a**) SST, (**b**) zonal wind, (**c**) rain, and (**d**) D_{20} in climate simulations and observations (*black curve*). For the observational curves, SODA is used for SST and D_{20}, CMAP for rain, and ERA40 for zonal winds. Model curves are colored in rainbow colors from red to violet in descending order of NinÕ-3 index variance (Saji et al., 2006) (See also Figure 5.37 on page 213)

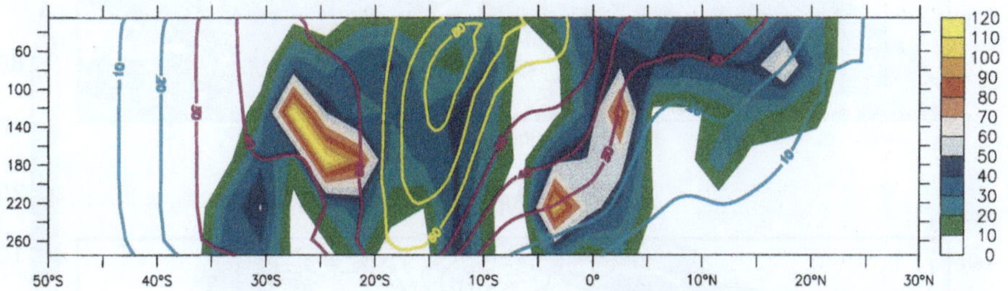

Plate 19 The meridional distribution of particles and tracers along 60 °E. Shaded values are the number of particles counted in $2^\circ \times 2^\circ \times 50$ m bins. Contour lines represent the tracer concentrations. The equatorward slope in the existence of ITF in the northern Indian Ocean is consistently evident from both the particles and tracers (Valsäla and Ikeda, 2007) (See also Figure 5.38 on page 216)

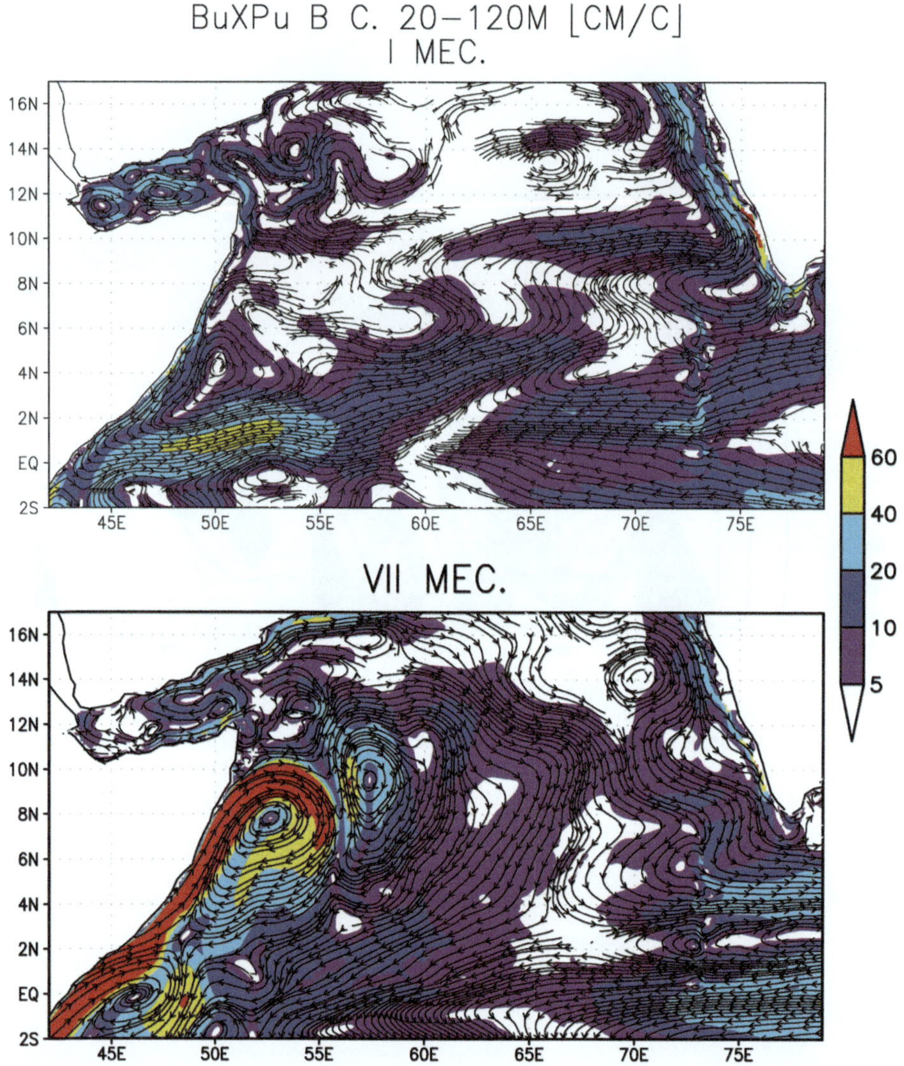

Plate 20 Mean currents for (*top*) January and (*bottom*) July in the surface 10 m layer of the Indian Ocean. The streamlines of the currents are shown. The velocities (cm/s) are shown with gradations of *gray color*. The *color scale* is shown on the right (Diansky et al., 2006) (See also Figure 5.39 on page 218)

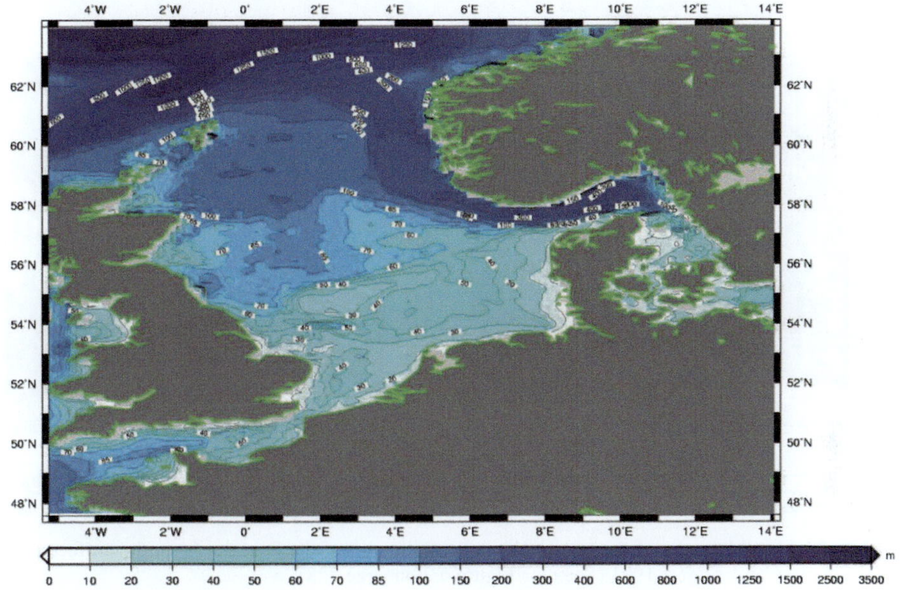

Plate 21 Topography of the North Sea (m) (See also Figure 6.1 on page 226)

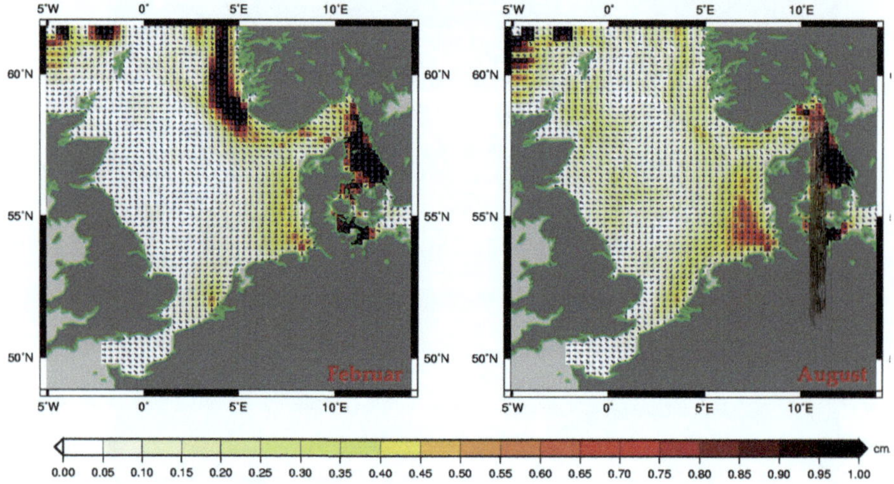

Plate 22 Monthly mean of the baroclinic-geostrophic part of the flow (cm/s) in the surface layer for February and August 1991 (See also Figure 6.2 on page 228)

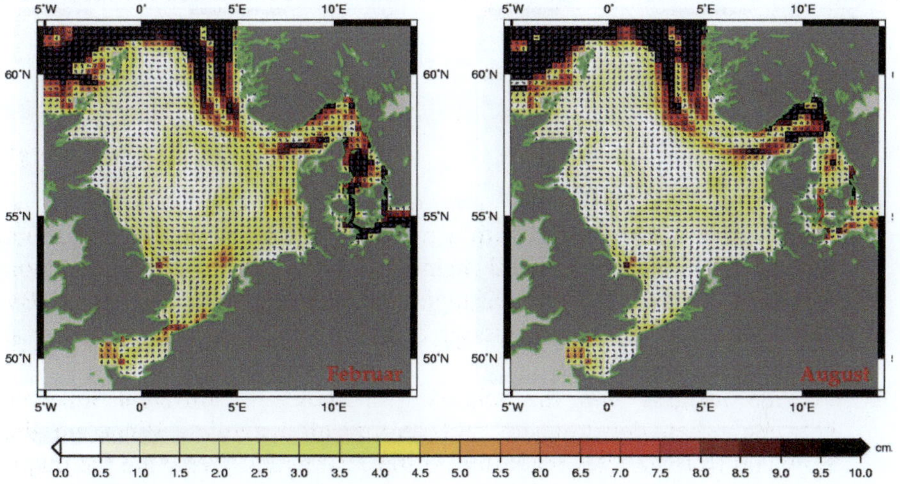

Plate 23 Difference between the monthly mean of the total flow and the baroclinic-geostrophic part (cm/s) in the surface layer for February and August 1991 (See also Figure 6.3 on page 228)

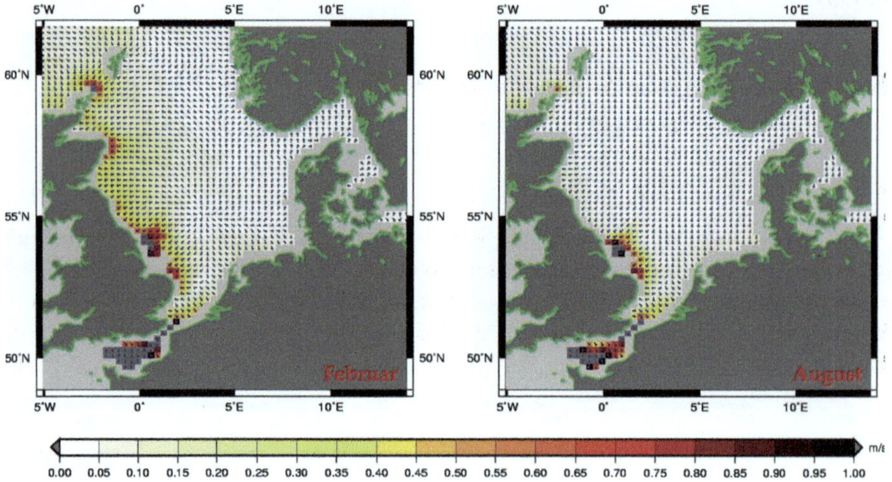

Plate 24 Monthly mean of the Ekman flow in 5 m depth (m/s) for February and August 1991 (See also Figure 6.4 on page 229)

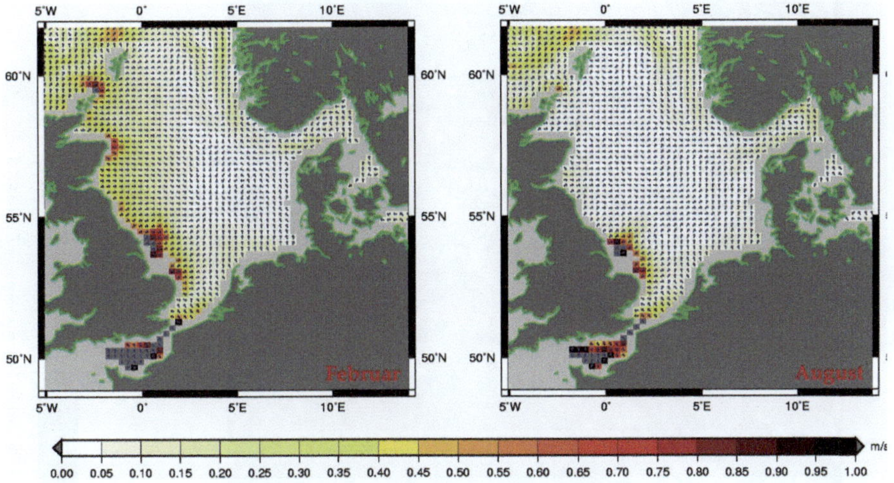

Plate 25 Difference of the monthly mean of the total flow and the Ekman part (m/s) in 5 m depth for February and August 1991 (See also Figure 6.5 on page 230)

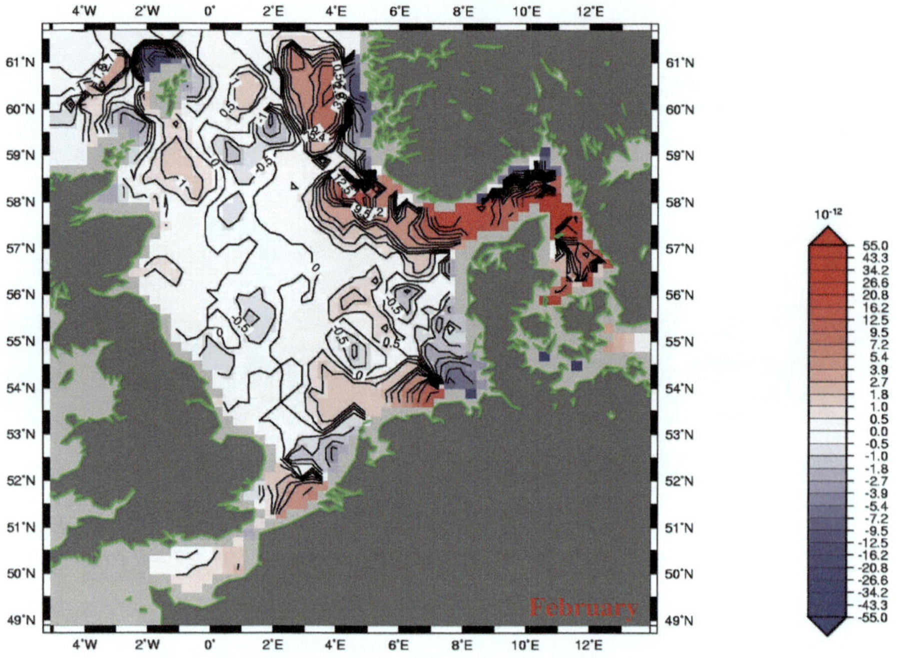

Plate 26 Monthly mean of the JEBAR term ($10^{-12}s^{-1}$) for February 1991 (See also Figure 6.6 on page 231)

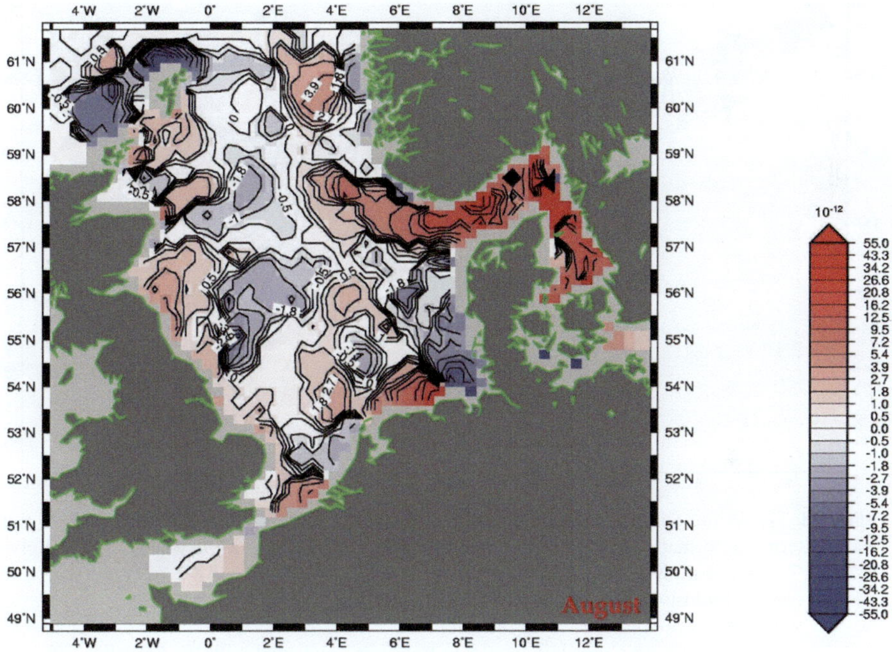

Plate 27 Monthly mean of the JEBAR term ($10^{-12}s^{-1}$) for August 1991 (See also Figure 6.7 on page 231)

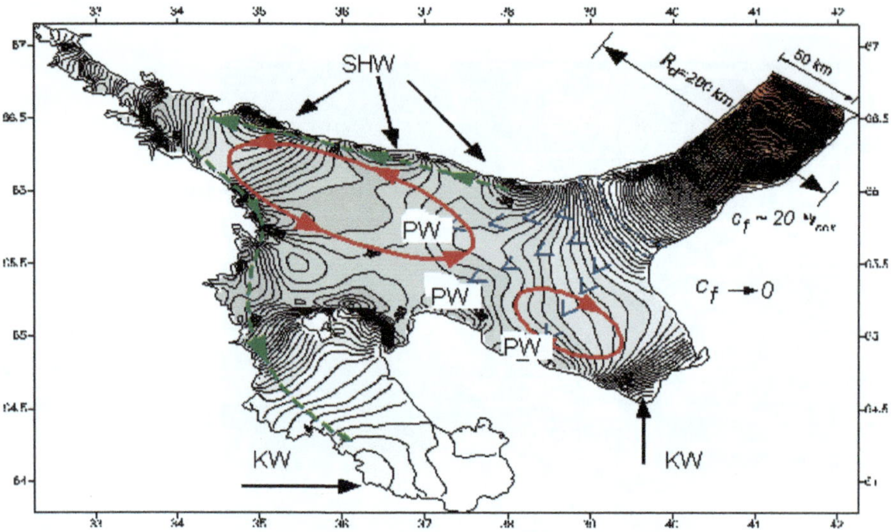

Plate 28 The White Sea bottom topography and its dynamical processes schematics: R_d – tidal wave external deformation radius in the White Sea Throat; SHW – shelf waves; KW – Kelvin waves; PW – Poincare waves; C – group velocity

———————- hydrological fronts at the exit of the throat;

>>>>>>>>>>>> Poincare waves radiation;

→→→→→→ Supposed alongshore cyclonic transport connected with the baroclinic effect of Poincare waves and shelf waves;

—-→—-→—-→ Large-scale quasigeostrophic flow (Semenov, 2004) (See also Figure 6.11 on page 235)

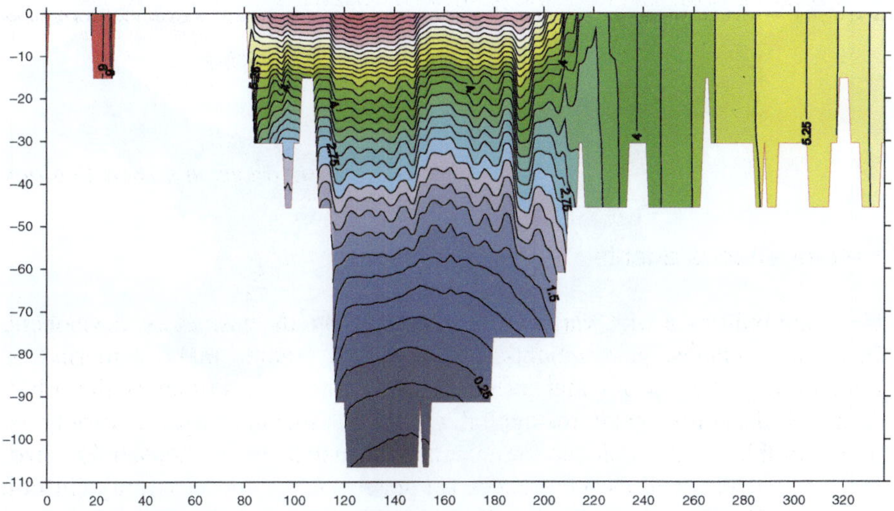

Plate 29 Vertical distribution of summer temperature (°C) across the section O-TH (Onega bay to the Throat) (Semenov, 2004) (See also Figure 6.12 on page 242)

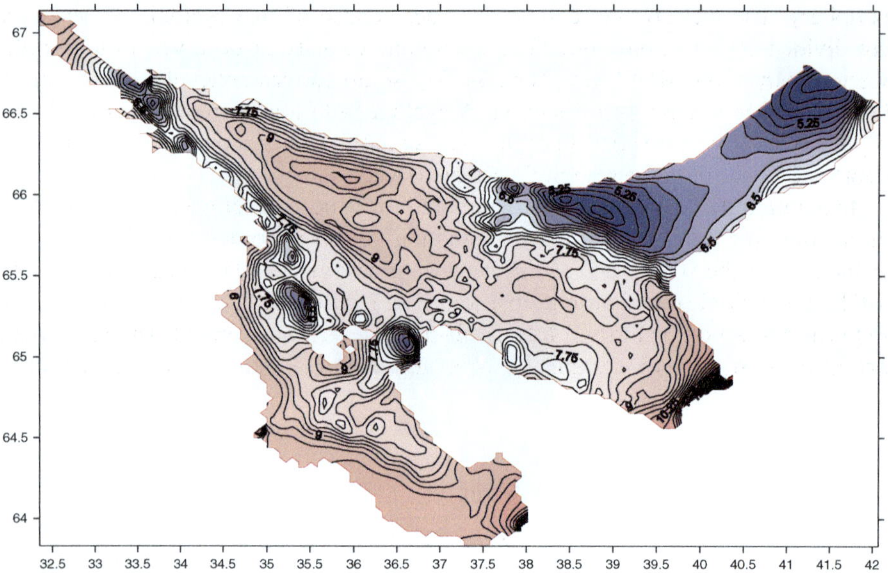

Plate 30 The White Sea surface temperature (°C) in August. Both eddies and frontal zones are depicted (Semenov, 2004) (See also Figure 6.14 on page 243)

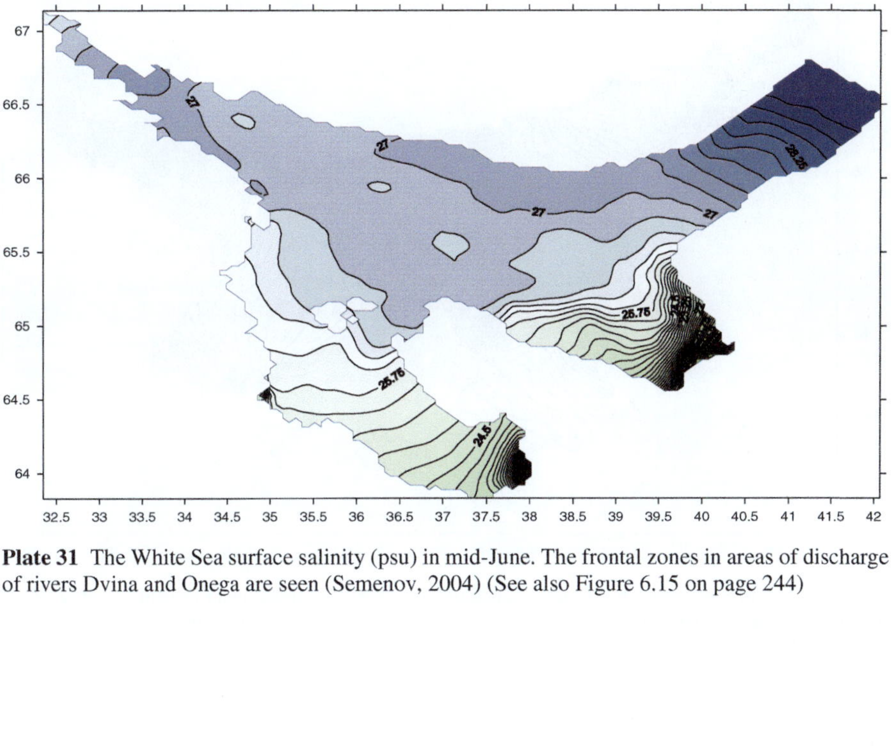

Plate 31 The White Sea surface salinity (psu) in mid-June. The frontal zones in areas of discharge of rivers Dvina and Onega are seen (Semenov, 2004) (See also Figure 6.15 on page 244)

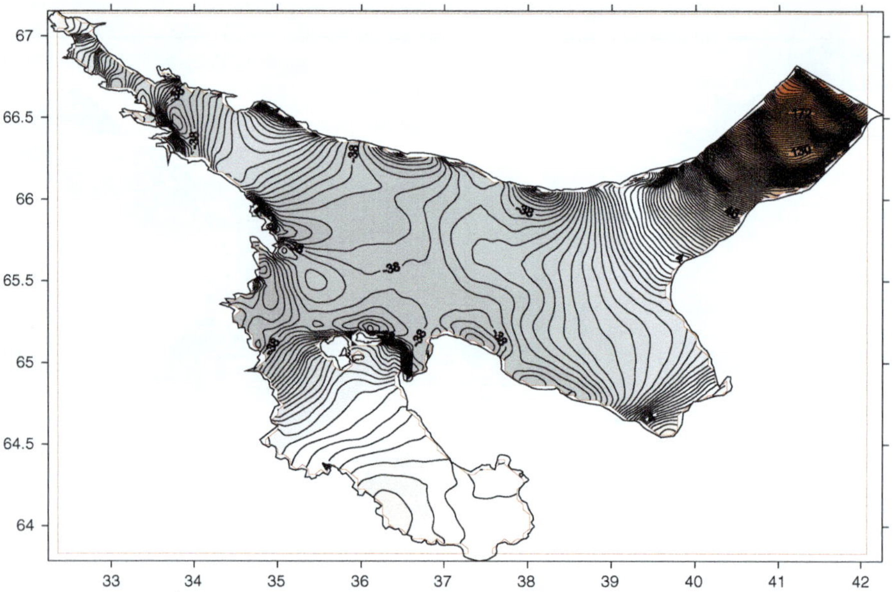

Plate 32 White Sea surface topography (cm), obtained for the August in a moment of rising tide by using the nonlinear boundary condition (6.2.3) (Semenov, 2004) (See also Figure 6.16 on page 244)

Index

GPSR Compliance

The European Union's (EU) General Product Safety Regulation (GPSR) is a set of rules that requires consumer products to be safe and our obligations to ensure this.

If you have any concerns about our products, you can contact us on ProductSafety@springernature.com

In case Publisher is established outside the EU, the EU authorized representative is:

Springer Nature Customer Service Center GmbH
Europaplatz 3
69115 Heidelberg, Germany

Batch number: 10584686

Printed by Printforce, the Netherlands